# The Einstein Dossiers

Siegfried Grundmann

# The Einstein Dossiers

## Science and Politics – Einstein's Berlin Period with an Appendix on Einstein's FBI File

Translated by Ann M. Hentschel
with 65 Figures

Professor Dr. sc. phil. Siegfried Grundmann
Weichselstraße 1, 10247 Berlin, Germany



Title of the German edition: S. Grundmann *Einsteins Akte.* Second Edition
© Springer-Verlag Berlin Heidelberg 2004
ISBN 3-540-20699-X Springer Berlin Heidelberg New York

Library of Congress Control Number: 2005930169

ISBN-10 3-540-25661-X Springer Berlin Heidelberg New York
ISBN-13 978-3-540-25661-8 Springer Berlin Heidelberg New York



Springer is a part of Springer Science+Business Media
springeronline.com

Printed in Germany



Cover design: Erich Kirchner, Heidelberg
Production: LE-TEX Jelonek, Schmidt & Vöckler GbR, Leipzig
Typesetting: Da-TEX Gerd Blumenstein, Leipzig

Printed on acid-free paper 55/3141/YL - 5 4 3 2 1 0

*For my grandchildren*
*Norma*
*and*
*Emil*

# Preface

In wise premonition of what was to come, the Prussian Ministry of Science, Arts and Culture opened a dossier on "Einstein's Theory of Relativity" in November 1919.[1]

The first indication I found of the existence of this dossier in 1960 was in Bruno Thüring's diatribe published in 1942 on "Albert Einstein's attempt to overthrow physics [...]."[2] As Thüring saw it, the mere fact that a ministry of the Weimar Republic[3] could compile such a file was evidence enough of the extent to which propaganda tools of the State were being exploited for the benefit of the Jew Einstein. Thüring wanted to use the dossier "to draw the activities of the Jewry [...] out of the dark or twilight in which they had intentionally been held in obscurity up to the National Socialist breakthrough" and to expose them to "the bright light of critical Aryan science."

After much effort and almost feverish searching, I discovered this file in 1962 in Merseburg, a small town near Leipzig (in the Merseburg branch of the former German Central Archive). The files of the Prussian Ministry of Culture had been removed there for storage during the war and only some of the files were accessible. The Einstein dossier became the basis of my doctoral thesis, which I defended in the Karl Sudhoff Institute for the History of Medicine and the Sciences at the University of Leipzig in 1964.[4]

In 1993 the file was returned to its original location, the former Secret Prussian State Archive (since then referred to as: Geheimes Staatsarchiv – Preußischer Kulturbesitz) in the Berlin district of Dahlem. It was long thought to be the only integral file explicitly dedicated to "Einstein" and "the theory of relativity."

Since then, it has become evident that *numerous* "Einstein" dossiers exist, including procedural documentation. By sheer volume these also deserve to be considered files in their own right:

1. The previously mentioned dossier: "Einstein's Theory of Relativity" in the Secret Prussian State Archive (Berlin).

2. and 3. Two files on "Lectures by Professor Einstein Abroad" in the Political Archive of the Foreign Office (Berlin).[5]

4. The "Einstein" section of the file "Expatriation"[6] at the Foreign Office.

5. to 11. Seven files in the official archives of the Community of Caputh.[7]

12. The file "Property Purchase for Professor Dr. Albert Einstein" in the Main Archive of the *Land* of Brandenburg (Potsdam).[8]
13. The file "Einstein, Albert… Citizenship Matters" in the Secret Prussian State Archive (Berlin).[9]

Other matters of business concerning Einstein – such as the expropriation of his sailboat[10] and the confiscation of his bank accounts[11] – are recorded in very many other files as well. The police file "Einstein" cited by Herneck[12] (hence still physically accessible after 1945), however, could not be located despite a most intensive search.

Other files on Einstein have to be considered lost: The Gestapo's Einstein dossier is one. A cataloguing card discovered in the Federal Archive in Berlin indicates that the Gestapo did, in fact, maintain a dossier on Einstein. The obvious assumption that this file could now be in the "NS Archive" of the former Ministry of State Security of the German Democratic Republic (now Archive of the Federal Commissioner for Files of the State Security Service of the former German Democratic Republic – abbreviated here as BStU) could not be verified. A search in the Special Archive in Moscow (where holdings from German archives were transferred to Moscow after 1945) was likewise unsuccessful. In any event, the response to an inquiry in this regard was negative; not even President Putin could assist me any further.

At least *one* complete Einstein dossier exists elsewhere: the FBI's Einstein file at the National Archives in Washington.[13] It was not available to me prior to publication of the first German edition of this book. The complete file is currently available on the Internet. I am grateful to Fred Jerome for first suggesting to me to peruse this source. (Jerome has since authored the book "The Einstein File.")[14]

The first of these files already broaches the topic of "science and politics."

It comprises a formidable 523 sheets and has the capacious breadth of about 10 cm. The temporal distribution of the individual issues in the file already documents the fluctuation in activity during the period between the November Revolution and the establishment of National Socialism in Germany.

If we consider recurrent identical enclosures as a *single* issue and furthermore treat enclosed brochures each as a *single* issue, the graph depicted above emerges: The jump in the number of events in the period after November 1919, when the results of the British solar eclipse expedition were made known, is clearly apparent. This was the politically unsettled "postwar period." The battles over Einstein's theory of relativity then subside and there is a lull in the mid-1920s – "Superficially, at least," as Max Born wrote in his memoirs.[15] It is no coincidence that this was simultaneously the more peaceful, apparently "golden" years of the Weimar Republic. The events accrue dramatically toward the end of the 1920s. Einstein again becomes a preferred right-wing target. Animosity toward him, Born continues, had smoldered on "until it flared up again openly in 1933."[16] We have arrived at the period of the global economic crisis and Hitler's seizure of power in 1933. The situation develops into an open conflict between Einstein and the German Reich; Einstein is driven into exile.

**Figure 1: Cover of the dossier "Einstein's Theory of Relativity" from the Prussian Ministry of Science, Arts and Culture – opened November 1919, closed October 1934. (A file cover dating to before the war was used: The ministry – commonly referred to in short as the "Prussian Ministry of Culture" – was formerly called "Prussian Ministry of Intellectual, Educational and Medicinal Affairs." Pursuant to a cabinet decree of 13 Mar. 1911, the entire administration of medicinal affairs – stripped of scientific and medical education – was transferred to the Ministry of the Interior. The ministry thus became a "Prussian Ministry of Intellectual and Educational Affairs." On 1 Nov. 1919 it was renamed again to "Prussian Ministry of Science, Arts and Culture." On 1 May 1934 the ministry was incorporated within the "Reich Ministry of Science, Education and Culture.")**

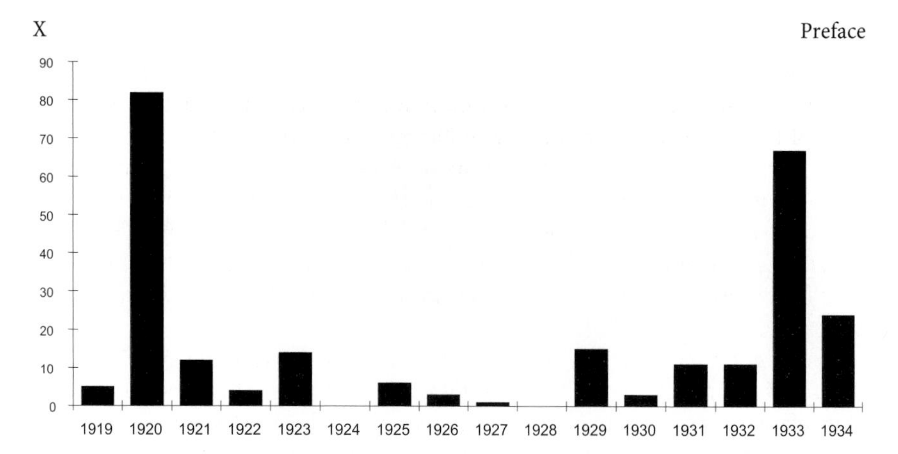


**Figure 2: Distribution of the number of issues covered in "Einstein's dossier" between November 1919 and October 1934.**

The two delimiting pillars of the dossier are downright symbolic.

Right at the front of the file (sheet 2) we find *Albert Einstein*'s letter to Prussian Minister of Science, Arts and Culture Haenisch from 6 December 1919. In it Einstein expresses his appreciation that the State Budgetary Committee of the Constituent Prussian Assembly had promised to make available 150,000 marks for empirical research in the area of the general theory of relativity. At the same time he mentions his reservations about such a resolution being made during a time of severe economic need with the potential of triggering bitter feelings among the public at large. He points out conditions under which an experimental test of the theory is possible without such high expenditures. The file closes with a letter by his worst enemy, the Nobel laureate Philipp Lenard, to Reich Minister of Public Information and Propaganda Goebbels from 8 October 1934, triumphantly demanding a complete settling of accounts with Einstein and his theory. Lenard calls for personal consequences for anyone still following Einstein, even if only on the scientific terrain, and the expulsion of all his sympathizers from academic chairs. He also uses political arguments in making his case against his archenemy. Nor should Einstein have any influence in science because it is "so politically harmful."[17]

*Based on the content* the dossier can be divided into three parts. The relevant topics would be:

1. Einstein is promoted.
2. Einstein is exploited.
3. Einstein is attacked and chased away.

These topics are interrelated. He was promoted *because* he was being exploited. And while he was being attacked politically by right-wing elements, and for that reason frequently traveled abroad, Einstein was still useful as a representative of German interests outside the country. Initially, his *promotion* predominates, then his *exploitation*, and finally open *enmity*. Despite the temporal overlapping of these three elements, personal promotion is generally attributable to the period of the Kaiserreich, political exploitation to the Weimar Republic (roughly until Germany's entrance in the League of Nations), and political animosity to the end of the Weimar Republic and Hitler's "Third Reich." Methodologically it is relatively easy to separate the Kaiserreich from the Weimar Republic. It is a much more complicated problem to divide Weimar and the Third Reich, primarily because the Weimar period was not such a sound world as is sometimes supposed.

For the German original editions of this book, I chose the title "Einstein Akte" with reference to this Einstein dossier in particular. It is now evident that this reference is too limited, for other files also played at least as important a role in the present work. But I would not like to relinquish completely any correlation to this foundation stone of my research. That is why I opted for the plural form, hence: "Einstein's *Dossiers*."[18]

The topic of inquiry I had chosen four decades ago still influences my approach to the personality Albert Einstein. It was not, and still is not my intention to write yet another biography of Einstein. My object here is rather: "science and politics, mind and might, intellectual and politician – as exemplified by Albert Einstein," within the context of Einstein's "Berlin period." I am fully aware that these themes are abstractions and polarities as do not and cannot really exist in so apodictic a form. There are such things as apolitical theories (Einstein's theory of relativity among them), but no such thing as apolitical science. The "intelligentsia" (in recent times particularly the scientific intelligentsia) has always helped itself to politics – whether to our benefit or detriment remains an open question.

So this is not a new biography. The existing ones will suffice.[19] Einstein's love life is of no interest here either, nor his manner of dress or eating habits. Einstein's character, style of life and opinions are taken into consideration only insofar as they are of *political* significance.

As the namesake of this work and a pivotal point of the action, Einstein constitutes *only half* of the title and only *one* pivotal point of the plot. In a much more general sense, the person Einstein is not even at the focus of this study.

This is not altered by the fact that the appendix (on "The FBI's Einstein file – reports on Albert Einstein's Berlin period") led me to delve quite deeply into the private life of Einstein. The *subject* of the reports, alleging Einstein's complicity in underground communist activities, compelled me to do so. I *had* to investigate his living conditions and office space, his family relations and relationships with other persons in order to assess claims that Einstein was politically active during his Berlin period and in order to be able to verify or falsify them accordingly.

The fact that Einstein was born on 14 March 1879, at 11 o'clock and 30 minutes in the morning in Ulm at Bahnhofstraße no. 135, may be of import elsewhere.[20] Here, however, it suffices to place his birth year roughly toward the end of the seventies of the nineteenth century, the period of Bismarck's socialist law, an unprecedented strengthening of the German Empire founded in 1871, the flowering of the capital Berlin, the "Gründerzeit," a time of promoterism[21] and of strife. A more *precise* dating of his birth only gains relevance when compared against other scientists who later became close collaborators of his – if not declared enemies. Both Otto Hahn and Max von Laue were born in the same year as Einstein. Laue later became one of Einstein's closest friends.[22] The physicists Lise Meitner (born 1878) and Max Born (born 1882) as well as the later Prussian Ministers of Culture Haenisch (born 1876) and Becker (born 1876) were about the same age as Einstein. The persons at least ten years his senior were: Fritz Haber (born 1868), Adolf von Harnack (1851), Leopold Koppel (1854), Hendrik Antoon Lorentz (1853), Walther Nernst (1864), Max Planck (1858), Friedrich Schmidt-Ott (1860), Wilhelm Siemens (1855), and Emil Warburg (1846). His bitterest opponent, Philipp Lenard (born 1862), was substantially older than him. Another opponent of his, Johannes Stark, was only five years his senior (born 1874). We might dispense altogether with mentioning the actual *date* of Einstein's birth, if it did not happen to coincide with that of Konrad Haenisch – Prussian minister of science, art and culture from 1918 to 1921.

The fact that Einstein was the father of the theory of relativity and an important scientist in other – related – areas of physics, does not have to take center stage in this work either. The quality and validity of the theory is, nowadays, no longer contentious. It suffices to know that Einstein counts as one of the greatest scholars in the history of mankind. *Today* virtually no one questions Fritz Haber's premonition at the 50th anniversary of Albert Einstein's birth: "In a few hundred years, the man on the street will know our times as the period of the World War, but the cultivated person will connect your name with the first quarter of this century, just as today the former remembers the wars of Louis XIV at the close of the 17th century as the latter remembers Isaac Newton."[23] The fact that Einstein happened to be the creator of the theory of relativity, as opposed to some other (hypothetically "equivalent") theory, is of secondary importance here as well. If reference is made to the substance of the theory, then only because *his contemporaries* drew such connections between the changing moral values of the period after World War I and the content of the theory.

My *initial assumption* is that author and reader are *in agreement on one thing only:* Albert Einstein was evidently and indisputably one of the greatest scholars in the history of mankind. In other respects, Einstein is simply treated like the famous "black box." We compare the input against the output. The difference between the two would then be interpreted as the regular consequence of Einstein's characteristics or as a result of chance. Einstein's characteristics, opinions and achievements are drawn into the discussion only where they are necessary for explaining the historical phenomena. Those characteristics, opinions and achievements of no *social* relevance will remain beyond the scope of consideration.

Such an axiomatic approach will not always be feasible, of course. For, it is the details and anecdotes that give the story its zest. Any such inconsistencies in treating this material will nevertheless do nothing to change the fact that the biographical element does *not* lie at the focus of this book.

What will be examined is *Albert Einstein's place in German politics* – the relationship between science and politics as is reflected in the case of Albert Einstein. Consequently, before treating Einstein himself, the analysis will be on the general *framing conditions* of his life and work. We must look at personal constellations of which Einstein formed a part, which he personally could not define, at best only could influence. The fact that he came to Germany was the work of *others*. If at all, it was only indirectly his doing, by virtue of his outstanding scientific accomplishments and talent.

The logic intrinsic in this narrative is that as the chapters goes by, Einstein becomes less and less manageable as an *object*. The further progress is made through time, the richer his biography becomes and the more *his* personality determines the course of things, whether or not he liked it. Anyone wanting something from Einstein had to know not just that he was a great scientist, but also what he thought and did, what stance he took in German politics and in the world. Thus it is necessary, after all, that we concern ourselves appropriately with *Einstein's biography*. The chapter on "Weimar" will do so much more than the chapter on the "Kaiserreich," especially in describing the transition to the "Third Reich." Without it we could not understand how and why his position in German politics changed.

I made an effort to be brief in my commentary of the events described and to use restraint in making assessments. The reader will make up his own mind when he learns of Einstein's occasional brushes with Bolshevism. The *fact* is, Einstein was a political leftist, sometimes even consorting with communists (one would have to distinguish between a Thälmann and a Paul Levi ...). It is also a *fact* that *at the end of the Weimar Republic* Einstein took on a pacifist stance that is summed up in the controversial quote by the writer Tucholsky: "soldiers are murderers" (from which he distanced himself, however, after the Nazis seized power). From this it does not necessarily follow that one must likewise espouse pacifism and a leftist inclination. Einstein was no god, nor did he want to be one. He was not infrequently mistaken and did not always realize what was happening to him. Not everything he did was praiseworthy, but one ought to take this also into account in forming one's own judgment on political issues, our view of the world, and morality. It will become evident that even though the events described here took place over sixty years ago, many constellations and types of behavior are still alive and well today.

Although the *goal* of my research was not to write another Einstein biography, one *result* of it (a by-product, but a very welcome one, at that) was that hitherto unknown or generally unfamiliar *Einstein documents* came to light. These include all of Einstein's letters to Wilhelm Solf and to the managing director of the Prussian State Library, Hugo Andres Krüss, as well as a large part of the correspondence at the Genevan Archive of the League of Nations. The floor plans of the Berlin apartment as well as of Albert Einstein's office space, published here for the first time in chapter 2 as well as in the appendix, are of interest as well.

I intentionally have many documents quoted in the text. First of all, because I do not see why other words have to be used to reflect what has already been aptly formulated elsewhere; secondly, because an original document always has a special significance of its own.

Because some documents were needed in a variety of contexts, it was not always possible to avoid repetitions. I count on the reader's acquiescence that I frequently present a relevant document *after* a given quote. It is then already partly familiar but provides a lot of new elements in addition.

In this way, personal interweavings in the government apparatus of essential relevance to the Einstein topic become visible. It is at least just as valuable as an indicator of the continuity from Kaiserreich to the Weimar Republic to the Third Reich. Many who at one time had cried out "Hosanna!" later were among those who were present or remained silent when the call of the moment was: "Crucify him!" A stance completely opposing Einstein's was mostly considered a duty of one's office and completely reconcilable with one's conscience. Schmidt-Ott deserves particular mention here. From 1920 to 1934 he was president of the Emergency Committee of German Science (Notgemeinschaft) and from 1920 to 1949 also chairman of the supervisory board of the dye factories F. Bayer & Co. Hugo Andres Krüss also, an official at the Prussian Ministry of Culture and subsequently (from 1925 to 1945) managing director of the Prussian State Library. Schmidt-Ott and Krüss were already on the scene when Einstein's appointment to Berlin was being negotiated in 1913. They were still enjoying the full honors and high rank of office when the Third Reich collapsed. Schmidt-Ott survived; in the clamor of the final battle – a hundred meters away from the front line – Hugo Andres Krüss took his own life on 28 April 1945.

I have every reason to thank numerous persons and institutions for their assistance. A grant by the Deutsche Forschungsgemeinschaft that was issued to me after the publication of the first edition of this book, facilitated and made possible perusal of the holdings of the Genevan Archive of the League of Nations as well as costly research on Einstein's FBI file and other archival work.

I would like to thank the following archives for permission to use their holdings and for the special assistance of the persons named:

- Archives de la Société des Nations à Genève (Mme. Ruser, Anna Svenson, M. Diara),
- Archiv beim Bundesbeauftragten für die Unterlagen des Staatssicherheitsdienstes der ehemaligen DDR (Frau Gehrke),
- Amtsarchiv Caputh (Carmen Hohlfeld),
- Archiv der Berlin-Brandenburgischen Akademie der Wissenschaften (Wolfgang Knobloch and Wibke Witzel),
- Archiv zur Geschichte der Max-Planck-Gesellschaft Berlin-Dahlem (Eckart Henning, Marion Kazemi and Dirk Ullmann),
- Brandenburgisches Landeshauptarchiv (Frau Nakath),
- Bundesarchiv – Abteilungen Potsdam (now Berlin) (Herr Ritter, Frau Pfullmann),
- Bundesarchiv – former Berlin Document Center (Herr Fehlauer),

- Bundesarchiv – Filmarchiv (Herr Vogt and Frau Buchholtz),
- Bundesarchiv – Stiftung Archiv der Parteien und Massenorganisationen der DDR (Frau Ulrich),
- Bundesarchiv Koblenz (Herr Pickro and Frau Franz),
- Firmenarchiv der Carl Zeiss Jena GmbH (Edith Hellmuth),
- Friedrich-Herneck-Archiv Dresden (Uwe Lobeck),
- Geheimes Staatsarchiv Preußischer Kulturbesitz – Bildarchiv (Herr Ludwig),
- Geheimes Staatsarchiv Preußischer Kulturbesitz (Frau Elsner, Herr Marcus),
- Landesarchiv Berlin (Herr Dettmer, Frau Welzing, Frau Erler, Frau Schroll),
- Politisches Archiv des Auswärtigen Amts Berlin (prev. Bonn) (Maria Keipert),
- Rijksarchief Utrecht – archief ex-keizer Wilhelm II (Frau Pennings),
- Staatliche Museen zu Berlin – Kunstbibliothek (Bernd Evers and Herr Mayer),
- Staatsbibliothek Berlin – Handschriftenabteilung (Eva Ziesche),

Whoever has worked in an archive knows that it is the archivist who holds the key to success. I feel the need to acknowledge their efforts, which too often remain anonymous.

I am indebted to Hubert Laitko for valuable comments on the manuscript. If the manuscripts submitted to press were much better than the foregoing drafts, it is due to his considerable and kind assistance, comprehensive expertise, and admirable meticulousness. I also received helpful suggestions regarding the content and structure of the manuscript from Wolf Beiglböck and Bernhard vom Brocke. Winfried Hansch (Berlin) and Marianne Polenz (Berlin) helped me with the translations of the Spanish and French texts. I am also grateful to the following persons for their various assistance: Fritz Kettmann (Caputh), Dietmar Strauch (Caputh), Elfriede Munzel (Caputh), Herrn Voß (Caputh), Bernd Noack (Berlin), Werner Schochow (Berlin), Ines and Steven Schmidt (Berlin), Herrn Thot (Berlin), Inge and Günter Tittel (Berlin), Frau Schönwald, Günter Wendel (Berlin), Johannes Grabinski (Regensburg), Ottokar Luban. Particular mention must be made to the employees of Springer Verlag, above all Wolf Beiglböck and Sabine Lehr.

Christa Kirsten, the long-time director of the Archives of the Berlin Academy and coeditor of the documentary edition: *Einstein in Berlin*,[24] gave me valuable tips in my search for sources. She already helped me find material forty years ago. But I also remember with gratitude Gerhard Harig, the former director of the Karl Sudhoff Institute for History of Medicine and Science at the University of Leipzig. Although one does not have to inquire at every opportunity about the origins of one's own existence: he actually inspired and encouraged me to work on the topic of the present book.

I thank my wife Rosmarie Grundmann. She was constantly involved in the project with love, patience and other assistance. I am convinced that it was of favorable benefit to the book as well. My grandchildren must also be thanked, whom I gave much of the most precious and scarcest commodities I have: time, and who gave me what a person, even one behind a desk, needs most – particularly in this internally shattered world: joy, perseverence and confidence.

Berlin, 25 December 2004 Siegfried Grundmann

**Postscript**

I regret that my English is, so I believe, not much better than the English of Albert Einstein. But I have read the text, checked many things, and reached the conclusion that the translator deserves the greatest respect. The questions that she asked and her own careful checking demonstrated that a highly committed and very knowledgeable person of an alert understanding was at work. For her achievement I extend to Ann Hentschel my heartfelt thanks.

Siegfried Grundmann

# Contents

## To the Reader

The formatting is intended to serve as an orientational aid

- `Monospaced` font is used to indicate documents rendered in full or nearly in full.
- Documents rendered in monospaced type *aligned* with both margins indicates a likewise justified original printed document.
- monospaced font *plus left-alignment* reflects originals with left justification (as a rule, typewritten text).
- monospaced font plus left-alignment *plus italics* indicates autograph original documents (in the majority of cases documents published for the first time by the author in the German edition of this work).

Hasty readers should know that they need not read the sans serif print and will not be short changed by doing so. They will advance more quickly, albeit foregoing the charm and profit that only an original document and details can afford. Numerous and occasionally lengthy biographical inserts on the actors also appear sans serif. They interject the line of narrative to inform about how the persons involved were otherwise made, where they came from and where they subsequently went.

# Chapter 1

# During the Kaiserreich

## 1.1 Military power and science – "sturdy pillars of Germany's might"

Albert Einstein left Germany at the age of fifteen on 29 December 1894, because he could not stand the spirit of militarism, drill and subservience anymore. He moved away from the Bavarian capital Munich to join his parents in Milan.

He hated his preparatory school in Munich especially much. Einstein was absolutely disgusted that his classmates could hardly wait to perform their service as "one-year olds" in the army right after finishing high school. Einstein had particularly serious personal differences with his class teacher. It is presumed that this teacher's attitude was nationalistic and anti-Semitic. One argument they had before Christmas 1894 ended with his suggesting Einstein leave school. Einstein did not need to be asked twice but was careful not to react too impulsively. He procured a doctor's official recommendation that he interrupt his schooling and certification by his mathematics teacher Ducrue that he had a comprehensive command of this subject and was generally a very exceptional pupil. His departure from Germany right then (two and a half months before schedule) was also especially convenient for him, because as a German citizen he had the right to leave the country freely only until age sixteen – thus, in his case, until 13 March 1895.[25]

In accordance with Einstein's wish, his father applied for his release from German citizenship; it was granted him on 28 January 1896.

Two decades after his emigration from Germany, on 29 March 1914, Einstein not only returned to Germany but even to the *Reich capital, Berlin*. He remained there until 1933. Why? Had the German empire changed in the interim into a democratic state of honest citizens? Had Berlin since stopped being the stronghold of militarism, drill and subservience?

Certainly not. "*Military power*" remained one of the "sturdy pillars of Germany's greatness."[26] Militarization had made good progress in the past nineteen years. Germany was one of the best armed countries of the world. Young as it was, the German Reich was intent on becoming a world leader and on expanding its influence.

The unification of Germany – the smaller nation-building solution excluding Austria – had been established only since 1871, centuries after the unions of its strongest rivals England, France and Russia. Along with unification came the (new) German empire, the Kaiserreich, under the leadership of Prussia.[27] Unity was not a fruit of peace but of war – the war of 1870/71 against the "hereditary foe" France. Prussified Germany had finally regained what standing in the world it had lost with the death of Frederic the Great in the Napoleonic wars. The prosperity on which the unprecedented economic boom of the 1870s was based was not purely a product of the country's own efforts but also of foreign plunder – covert though it was. The clash of weapons had brought victory. War was evidently a useful instrument for implementing political goals. It was, at last, once again a "promoter of politics by other means," as the former Prussian general and military theoretician Clausewitz phrased it – and a *successful* promoter of politics, at that. Weapons were henceforth to determine the fate of Germany.

A country's greatness and might were, at that time, synonymous with – territorial and colonial property and a strong army. When the German union took place under Prussian leadership, the world had already been divided up. Germany was too late in coming. Little was left over for the taking and even that could only be wrested away by threat and blackmail (the rights of "uncivilized nations" were not a matter of legal conscience for the "cultivated nations" anyway). German colonial holdings were considerable, but small compared against the colonies of France, Britain and Russia. Germany wanted more and the others did not see why they should say nothing while a new rival was gathering strength. Under such circumstances, the growing numbers of disarmament conferences held by the Great Powers Britain, France, Russia and Germany, around the turn of the century, were only an instrument for deceiving the public around the world and for weakening their opponents. Feverish rearmament of all the parties, especially, Germany, reveals what the real intentions were and what was in the offing. The incident that triggered the war, the escalating succession of events and the constellations of power that had formed might come as a surprise, but not the genocide.

Kaiser Wilhelm II had made enough sabre-rattling proclamations for his attitude to be known. When he saw the German expedition corps off at Bremerhaven on 27 June 1900, to put down the Boxer Rebellion in China, he held his notorious Hun speech, laying down the maxim: "There shall be no pardon, no prisoners shall be taken! Just as a thousand years ago the Huns under King Attila made such a name for themselves that they still seem mighty in the legends and stories come down to us today, thus you shall defend the name of Germans in the same way for a thousand years hence in China, so that never again does a chinaman even dare to look askance at a German."[28]

The enthusiasm throughout the nation at the beginning of the war indicates that most people were of one mind with their kaiser. His sermon on violence bore abundant fruit. Military power had become more than a political and economic agenda. It had captured and corrupted the nation's *spirit*. War interests affected every aspect of society and invaded deep inside the private sphere. Heinrich Mann's "loyal subject" Diederich Hessling was an exaggeration, but nevertheless typical of the time.[29] Much had changed during Einstein's absence from Germany: everything had gotten much worse.

Looked at this way, there was absolutely no reason for Albert Einstein to return to Germany, and least of all to move to the *Reich capital*. Politically, Switzerland was a much better home for him. Why did Einstein take this step anyway? Why did he ultimately put aside all his doubts?[30]

We find the answer in the rest of the above quote by the managing director of the Royal Prussian Library and confidant of the kaiser, Adolf von Harnack: "Military power and science are the two sturdy pillars of Germany's greatness and, following its glorious traditions, the Prussian state must assure that they both be maintained."

Military power *and science!* On one hand, science as the intellectual basis for growing productivity and therefore for more effective exploitation of resources in the *existing* territory (within the borders of the Reich), on the other hand, the intellectual basis of military power and *conquest of new territory*. Science and culture were supposed to prepare the rest of the world for subordination, consolidation and cover-up.

As always, science and culture were not being fostered and valued for their own sake. They were political instruments. *That* was why Einstein was lured to Berlin. Although he had other motivations for coming, by returning to Germany and doing his best to further science there, he became a supporting pillar of the system. Science was his profession and his life, the basis of his existence. The fact that science – even "pure" science – did not merely serve to advance knowledge and entertain its disciples was a secondary question for Einstein. The fact that Einstein was not conscious of the nationalistic function of his research changes nothing. *His* task was to do good work. Enthusiasm and idealism could only be helpful. Those who brought him to Germany took care of the rest. *They* had no idealistic illusions; *they* consciously wanted "*a nationalistic stamp impressed on every research result in science.*"

At the dawn of the new century it was not necessary to convince influential state officials and economic leaders that the Prussian State had a duty toward science. Its usefulness had been demonstrated long ago. After the founding of the Reich, there was a unique and intense collaboration between science and industry. History has apparently reserved the term "revolution" for other epochs in describing the development of production, technology and science. But the end of the nineteenth century was a veritable *revolution* in industry and technology. Inventions leading to major jumps in industrial productivity were mostly not the work of hobby tinkerers. They were the result of systematic research. From the academic point of view, the rapidly expanding young industry of electrotechnology was a product of science. And *Berlin* was its geographic center. Similar things could be said about the development of the chemical industry – with a different geographic focus.

Within this context fundamental research was not simply an ornament for a given country. Men of influence in politics and science knew that it prepared and secured the future of industry and export. And they also knew that ideas emanating from a country served as its light cavalry, preparing the ground for future lucrative exportation and profit.

It is not a coincidence that many other nobler voices are audible during a period of intense preparations for the next war.

Just before the outbreak of World War I, the German Reich Chancellor Bethmann-Hollweg demanded that Germany foster *politico-cultural endeavors* in a letter to the historian Karl Lamprecht dated 21 June 1913 (it was published later that year, on 12 December, in the 'Vossischen Zeitung'). He wrote: "I share your belief in the importance, indeed in the necessity of a foreign cultural policy. I do not deny the benefits that France's politics and economy reap from this cultural propaganda, nor the role that British cultural policy plays in holding the global British Empire together. Germany, too, must take this path, if it wishes to conduct politics [...]. I think that in this respect too few of us here recognize the importance of this task. We are a young nation, and perhaps still have much too naïve a reliance on force, underestimate the subtler means and do not yet know that what is gained by force can never be retained by force alone."[31]

Despite Germany's leap forward economically and militarily, a long and hard battle still needed to be fought for world leadership. Others also recognized the importance of science in this contest. autorindexHarnack[Adolf]'s memorandum to Wilhelm II lists the fields in which Britain, America, France and other countries were rapidly gaining ground on Germany and where they had already gained world dominance. His purpose was not scaremongering but a tactical emphasis on the impending threat. "We are falling gradually behind every year, although we have the human resources in sufficient numbers to master the largest and most comprehensive researches, if only the institutional and financial means were available." But Harnack did not stop at complaints and admonitions. He devised a concrete program for furthering science. His memorandum became the birth certificate of the "Kaiser Wilhelm Institute of Scientific Research" and the "Kaiser Wilhelm Society for the Advancement of the Sciences."[32] Harnack drew on earlier work by Althoff,[33] Rathenau, Krüss and others in writing his memorandum. A senior civil servant at the ministry, Schmidt (Schmidt-Ott), provided the final impetus for its submission.[34]

In his memorandum Harnack pandered to the kaiser's *ambition and thirst for recognition* by cleverly embedding the demands of science in praise of the "most gracious interest" the kaiser had always devoted "to the progress of the sciences and the need for research institutions." That was precisely what the German kaiser wanted to be seen as: a patron of science and a luminary spirit. He went down in history as the quintessence of martial affectation, with a strong dose of megalomania. But it remains an indisputable fact that he had *three* passions: the military, ballet and science.

The military has already been addressed, ballet may be left aside here, but his enthusiastic advocacy of science deserves some emphasis. Of course Wilhelm II was also thinking of his *fame*, but if we were to measure patronage of science and the research itself against the personal motives of its patrons and researchers, we would have to rewrite the history of science completely. For the sake of this passion, the kaiser even transgressed some social barriers (in *such* a case, it was possible!): Going against his own better judgment and anti-Semitic opposition among his advisors, he appointed Jewish scientists to high-ranking positions. He later regretfully confessed having not only "had Jews at his own table, [but] supported Jewish professors and helped them out" as well.[35] Even so, the kaiser did *personally* promote a Jewish convert to Christianity, Fritz Haber, to the rank of captain, against stiff resistance by the military. His very personal efforts contributed decisively toward the flowering and brilliance of Berlin science during the first third of the twentieth century. All academic disciplines, ranging from ecclesiastical history to classical antiquity and theoretical physics, benefited copiously from his patronage. It was fortunate for the history of science that the kaiser had Althoff, Harnack, Schmidt-Ott, Planck, Nernst and other outstanding managers of scientific research at his side, who were adept at encouraging his inclinations and exploiting them to the utmost.

Wilhelm II not only supported the founding of the new scientific society. In a way, he made the decisive move toward its coming to life. The initial idea to found a "German Oxford" on the State domain in Dahlem originated with Privy Councillor Althoff – director of the division for science and arts at the Prussian Ministry of Culture. Althoff died in 1908. It may be to the personal merit of the *kaiser* that the idea did not die with him. We gather this from a letter by Privy Councillor von Valentini, head of the cabinet for civil affairs, from 24 March 1909 to the Ministry of State (or even if Valentini had ably suggested the idea to him, it was the fruit of imperial *interest* in the advancement of science).

The deceased "Secret Minister of Culture's" executor was no nobody. It was the *kaiser* himself. What other head of state has ever done the like? (What other head of state – we might ask in most cases – had reason to do the like?)

It was extremely flattering that the planned research society should bear *Kaiser Wilhelm*'s name, after other ideas had been abandoned. (Initially it was to be called the "Royal Prussian Society for the Advancement of the Sciences," later "Imperial Society for the Advancement of the Sciences.")[36]

On 10 December 1909 Valentini was able to report to the author of the memorandum: "I hasten to inform you privately that His Majesty permitted me last Wednesday to read out your memorandum, verbatim, about the justification for a Kaiser Wilhelm Institute for Scientific Research and thereupon commissioned me to convey to you His Supreme warmest thanks for the magnificent work. The individual arguments as well as the suggestions you formulated were most enthusiastically welcomed without qualification by His Majesty."[37]

On 18 December 1909 the head of the cabinet for civil affairs informed Reich Chancellor Bethmann-Hollweg that the kaiser expected the "Royal Prussian Society for the Advancement of the Sciences" to materialize as quickly as possible; he requested a report.[38] An apathetic monarch would probably not have done so. It was thanks to this support that the Kaiser Wilhelm Society could be formed in the face of opposition by the Prussian Ministry of Finance. The "involvement of the person of His Majesty" was not, as the representative of the Ministry of Finance (R. Dulheuer) thought, "questionable,"[39] but highly fortuitous.

It is not astonishing that other people also wanted to be patrons. His Honorable Imperial and Royal Majesty had acted as an example to his subjects. Supporting science soon became a mark of good breeding. It did no harm to science that many major industrialists and bankers (and their heiresses) thought less about the fame and influence it brought than about the associated tax advantages and profit. Donations were made and the occasional high contributions bought honorable membership in a society for which the kaiser had assumed the official function of *promoter*.

The "Kaiser-Juden" – "the Jews at the very top, financially, in Germany"[40] – played prominently among these patrons. They included Eduard Arnhold, James Simon, Leopold Koppel, Franz von Mendelssohn and Walther Rathenau. A flowering of science in Wilhelmian Germany was to a large degree *their* making. The trees they planted bore abundant fruit even *after* the Great War, perhaps even especially so then, after rampant inflation had ruined many a former patron – for instance, James Simon.

From this it does not follow that the political class was ready and willing to cede the stage to *scientists*. Scientists were sought after. But they were not to be anything more than an instrument and a feather in one's cap. The regular meeting of the Academy of Sciences on 24 January 1912 was symptomatic of the place the academic elite occupied in the system of power. It took place on "Frederic's Day," the anniversary of the birth of Frederic II – this was a particularly special occasion, because it was the *bicentennial* of the Prussian king's birthday. On 6 November 1911 Wilhelm II instructed ("His Majesty the Kaiser and King [...] wishes"),[41] that the academy's festive meeting take place at the palace in Berlin. Academy Secretary Roethe sent his cordial thanks, expressing his concern that the group of academicians should attend the event "just as guests" and his wish that the "academic character of the celebration not fall too far into the background" and that the academy appear at the hall as a single body. His wishes were granted. The academy appeared as *one body* in the White Hall at the palace. It was presided over by the kaiser. Next to him were the insignia of his power, representatives of the Ministry of State and the princes. (Seating was found for the empress and royal princesses – the only women present – on the musicians' podium.) Across from His Imperial Highness sat the members of the academy. A "table decked with scientific antiquities" separated kaiser from academy. For the sake of decency, the military (together with the court officials and secretaries of state) was not placed center stage but flanking the membership of the Academy of Sciences on either side. Escape was out of the question. Neither would nor could they have done so.

It goes without saying that no room was made for the working public. After all, this "Frederic's Day" was an occasion in celebration of the founding of a *scientific* association. The more remarkable is the presence of the military. What business did it have to appear in such a large contingent at an academic event? The very presence of the military underscores the place society assigned to science.

Like the militaristic seating arrangement, the ceremony itself followed a strict order of procedure.[42] Not a single little detail was left to chance. Whoever was deemed worthy enough to participate knew in advance at what time and in which room he was to appear and in what attire (in full decoration, please!). He knew exactly when he was to enter the White Hall and through which door. The protocol was that of a state ceremony, as rigid as military formations on parade.

The *speeches* set forth in unmistakable terms the place science was supposed to occupy.[43] A military touch even permeates down to the details of the speeches. It is not easy to detect from the speeches that this was an *academic* event.

The academy's permanent secretary, Mr. Waldeyer, opened the meeting. (The title of this office: "*Sekretar*" now has a definitely antiquated ring to it.) His greetings and thanks to his kaiser and king were duly solemn and lengthy. This year, the emphasis at the annual "Frederic meeting" was unusually unacademic. The

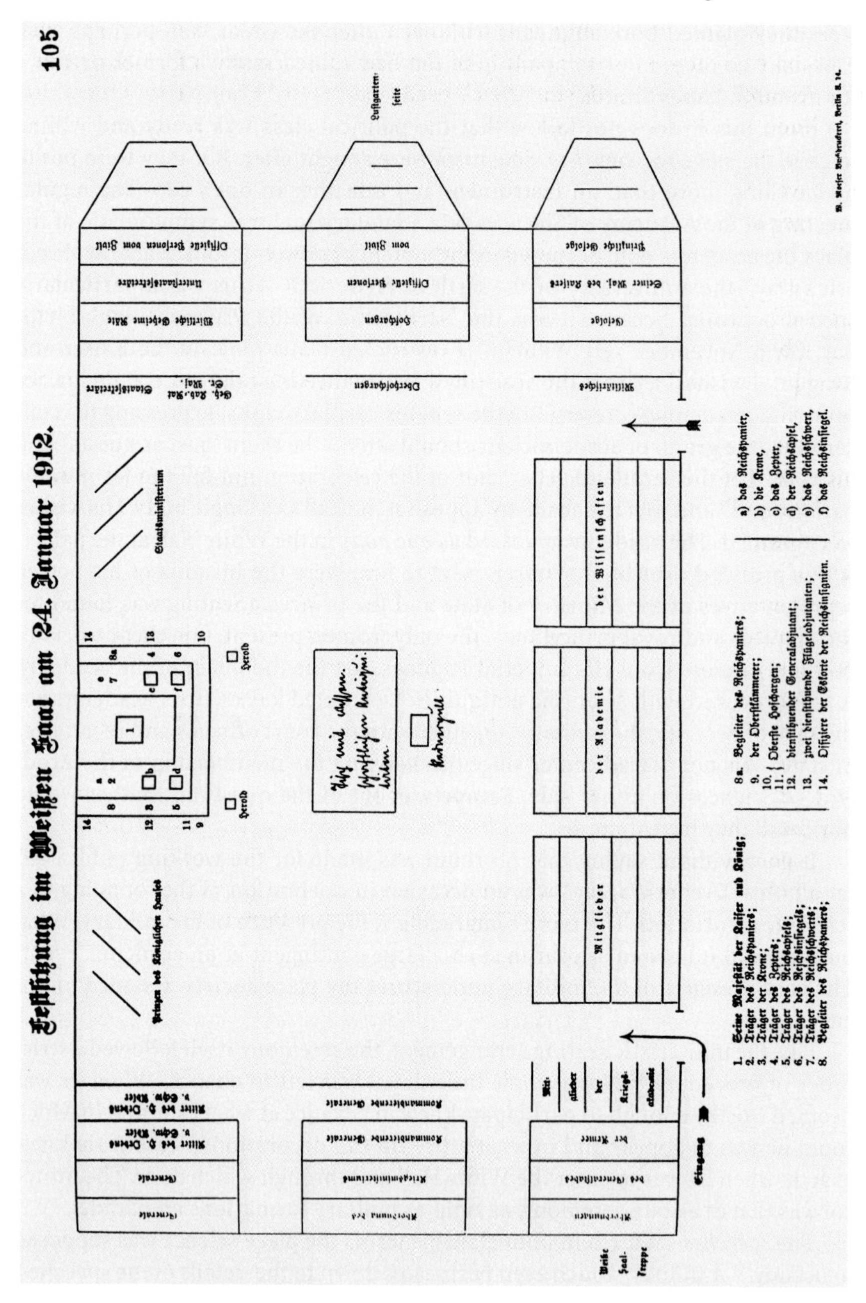

Figure 3: Seating arrangement for the festive meeting of the Royal Academy of Sciences in the White Hall of the Royal Palace on 24 January 1912 on the occasion of the "Bicentennial celebration of His Majesty of the days of yore, King Frederic the Great" – order of 6 December 1911.

speaker was particularly appreciative "that it pleased Your Majesty to call this occasion explicitly a festive meeting of the Academy of Sciences." Actual *scientific* research conducted during the elapsed year received but passing mention. Gratitude to the present monarch was the key note, in "happy awareness of how much of a share here too is due to our Most Gracious Kaiser." The kaiser could not have been otherwise than pleased. At any rate, the lauditory welcome did not damage his goodwill toward the academy.

In his reply Wilhelm II left no doubt about how he ranked the military and science: "The morning of this day was devoted to a celebration of the Army and above all of that division of the troops whose founding harks back to the 'King-Constable'; thus I welcome here the Academy of Sciences as an elite troop of the esprit, which Frederic the Great had mustered and assigned its honorable station." It was also good news for the academy to receive the following assurance from the kaiser: "And thus shall I too continue to take the Academy of Sciences under My special sovereign protection and be her helper in achieving its goals."

After praising Frederic the Great's military prowess, the official speaker Reinhold Koser[44] devoted his speech primarily to the "testamentary execution": Wilhelm II's continuation of Frederic II's legacy. Science was not at issue here. The subject was rather the spirit under which science was to be conducted in the new German Reich. Koser pointed to Germany's position as a great power and its amassed implements of power, "Germany's armored defense at sea" and the Prussians as a recently ascending race. Serving the fatherland was the highest honor and duty there is; neither "the ploughing field nor five yoke oxen, nor even the wife and child" may stand in its way. "Forever on the lookout!" is our lot, as in former times. "Today, as before, we must still be on guard against our neighbors and be ready to defend ourselves overnight against the deadly assaults of an enemy."

Albert Einstein was not yet among them. But it would not be long before he would take part in the "Frederic Day" festivities as a member of the academy: initially as obediently disciplined as a schoolboy and properly attired.

## 1.2 Einstein's path to Berlin

Where else could the young Einstein, that energetic genius, have found such a good working environment? In the German Kaiserreich he was best understood (also in part because his foreign language skills were not particularly good). Berlin, the capital city of science, an intellectually live metropolis in unsettled times.

The ground had already been prepared for Einstein's activities in the Reich capital before there was even any question of his being appointed:

Two institutions were inaugurated upon, or shortly after Einstein's appointment. Their creation had nothing to do with his nomination and eventual acceptance:

1. The position dedicated to scientific research at the Royal Prussian Academy of Sciences.
2. The Koppel Foundation.

The Prussian Academy of Sciences had two full-time positions with no specific duties attached to them besides the pursuit of research. One was established for research in the social sciences and the other for research in the natural sciences. Thus, evidently, it has long been known that there might be scientists whose highest goal in life was *not* research conducted under the constraints of a heavy teaching load. The persons occupying these positions automatically became members of the faculty of the University of Berlin. They had the right but no *obligation* to offer lecture courses. The positions were for life. When the chemist J. H. van 't Hoff died in 1911, one of these positions became vacant.

The academy secretary of the physico-mathematical class naturally had a decisive influence on the refilling of the post. *Max Planck* was serving in this capacity since 1912. In 1913/1914 he was, at the same time, president of the University of Berlin. His scientific reputation and standing, and probably also his selfless advocacy of the interests of science served as a guarantee that his petitions could be approved as well.

**Planck, Max** (23 Apr. 1858–4 Oct. 1947) Physicist, regular member of the Academy of Sciences 1894, full professor of theoretical physics at the University of Berlin 1892–1926, secretary of the physico-mathematical class of the Academy of Sciences 1912–1938, president of the University of Berlin 1913–1914, president of the Kaiser Wilhelm Society 1930–1937, chancellor of the medal *pour le mérite*. Nobel prize in physics 1918 (conferred 1919) "in acknowledgment of the advance made in the development of physics by his quantum theory."[45]

Planck's closest colleague at that time was probably Walther Nernst. His name appears together with Planck's on many proposals to the academy, and as we shall see, on almost all the ones with the aim of Einstein's appointment in Berlin.

**Nernst, Walther** (25 Jun. 1864–18 Nov. 1941). Chemist, regular member of the Academy of Sciences 1905, full professor of physical chemistry at the University of Berlin 1921–1922, president of the Bureau of Standards (PTR) 1922–1924, honorary professor of physics at the University of Berlin 1923–1924, full professor of physics at the University of Berlin 1924–1933. Nobel prize in chemistry 1920 (conferred 1921) "in acknowledgment of his thermochemical researches."[46]

We also come across the name Heinrich Rubens, of course. Rubens (30 Mar. 1865–17 Jul. 1922) was full professor of physics at the University of Berlin between 1902 and 1922. He was not supposed to be passed over. Neither was it permitted nor, indeed, possible to do so. – Proper Max Planck would not allow occasion to arise for him to be accused of breaking any such rules. Nernst was subordinate to Rubens, but in many respects he was much more important. First of all, he numbered among the most important physicists of the period. Nernst also had very good connections to the worlds of major industry and finance. Add to that a good helping of egotism and you have, not a fault, but a true blessing for science. Einstein later remembered that he owed the good conditions attached to his Berlin appointment more to Nernst than to restrained Planck – and occasionally even envied Nernst for this quality.[47]

Walther Nernst was a good pupil of Wilhelm Ostwald (1853–1932) not only in the area of chemistry but also in a much more comprehensive sense, who was not just a great chemist but also a genius organizer of science.[48] Ostwald realized that in the 20th century discovery and invention could no longer be left to chance. What he published in 1927 he undoubtedly already believed earlier: "In principle, under present-day conditions I must consider an organizer more important than a discoverer [...]; discovery in the sense of cultural advancement [is] only half the task. [...]." The implementation of new ideas is what is much more decisive. "Consequently I must consider an organizer's field much more important, because it is more difficult, than a researcher's."[49] "An organizer of science can only be one, however, if he has also been a discoverer, because otherwise he would have no yardstick for what he wants to organize."[50] Walther Nernst's scientific abilities, combined with his organizational talent, were importantly augmented for science policy, and especially for Einstein's promotion, by his good contacts in industry. He also knew people of political influence, to which "his good relationship with Kaiser Wilhelm II" is included.[51]

We shall see that Nernst also entertained very close relations with *Leopold Koppel,* a banker and major industrialist who was one of the leading patrons of the day.

**Koppel, Leopold** (born 20 Oct. 1854 in Dresden, deceased on 29 Aug. 1933). Banker, proprietor of the banking establishment Koppel & Co. and major stockholder of the Auer Company or Osram Works Co. Ltd., sponsor of the KWI of Physical Chemistry, Gold Leibniz Medal of the Academy of Sciences 1917. Koppel came from "modest circumstances." After becoming a very rich man, he saw it as his calling to be a benefactor of science.

Althoff explained in a letter to the kaiser that Koppel "did not want to enrich himself and his sufficiently secure family further; he was planning instead to invest the incoming capital proceeds from time to time in more substantial foundations for beneficial causes. The hard period of profit taking should now be followed by a finer one of giving."[52]

A petition regarding one of Koppel's foundations explained:

Koppel "worked his way up [...] from modest circumstances by his own energy and industry to his current wealth, which is estimated at about 20 million marks. After attending preparatory school through to the final year he learned the banking trade as an apprentice at Philipp Elimeyer's banking establishment in Dresden and in the middle of the sixties of the past century opened his own business there. In 1891 he moved residence to Berlin and has been operating here since then [...] as sole proprietor of his private banking business. [...] In 1900 he became Royal Councillor of Commerce for Saxony. In 1903 Royal Privy Councillor of Commerce for Saxony. No complaints concerning his character are on file [...]. [...] he has nowhere made himself conspicuous from a political point of view. He is of the Protestant faith." Koppel has, "following his own personal inclination and preference for science, come increasingly to regard the nurture and furtherance of science as the finest of goals for sponsorship."[53]

One reason behind the relationship between Koppel and Haber not mentioned in the petition but of particular significance to Einstein and others was that Leopold Koppel was a *Jew* (baptized as a Christian since his marriage).

All in all, very little is known about Koppel's biography. He preferred not to be mentioned in public and rather to let his money work for him. The kaiser's high regard and the esteem of recipients of his generosity together with his influence among the elite upper crust, were enough gratification for him. His patronage was driven by a true *passion* for science, away from the public eye. In those days there were still a few sponsors who put the interests of science ahead of their own renown.

A pretty pair indeed: self-effacing Koppel and egotistical *Nernst*. Another pretty pair is: self-effacing Koppel and equally egotistical *Haber*.

In 1905 Leopold Koppel set up a foundation of one million marks consisting of Berlin municipal bonds providing a yield of 3.5 percent. According to the articles of this "Koppel Foundation for the Promotion of Intellectual Relations between Germany and Foreign Countries" dated 25 November 1905, its purpose was: "to promote intellectual relations between Germany and other civilized states of the world by sending German scholars and teachers abroad and inviting foreign scholars and teachers to Germany, as well as by other suitable means, thus to contribute toward better mutual understanding and a lasting peaceful union of the civilized nations of the world."[54] The occasion prompting this new foundation was – what a clever move! – "the silver wedding of the imperial couple." And another clever move: the kaiser was invited "to decide on the allocation of the foundation funds."[55] Wilhelm II gave his blessing on 18 December 1905.[56]

Equipped with the requisite capital and contacts, Koppel had a perceptive view of the needs and demands of his class toward science. His say had much influence on Wilhelm II. His connections with leaders of the scientific community in Berlin were also exceptional, notably his relations with Walther Nernst (later also with Haber and Einstein). Koppel's role in the founding of the Kaiser Wilhelm Society for the Advancement of the Sciences was particularly prominent. He and the steel magnates Krupp, von Bohlen and Halbach donated the highest sums to the founding capital of the new society (700,000 and 400,000 marks). In return they were made senators of the society in the company of such industrialists and bankers as the Prince of Donnersmarck, von Siemens, Ludwig Delbrück, Franz von Mendelssohn and others, to decide on the interests of science and to evaluate the results of research. Koppel appeared on many other occasions as an influential voice, such as at the foundation of the Kaiser Wilhelm Institute of Physical Chemistry and Electrochemistry, the Kaiser Wilhelm Institute of Theoretical Physics and the Kaiser Wilhelm Foundation for War Technology.

Thus were the important social conditions and constellations of power *before* plans to appoint Albert Einstein to Berlin were devised.

**Figure 4: 'Paying homage to the Prince of Peace. The Three Kings from the Orient bearing their gifts.' From the – entirely serious – *Lustige Blätter*. The banker *Koppel*, the industrialist *Arnold* and the businessman *Simon* as the Three Kings from the East and – in representation of Christ – *Wilhelm II*. Frankincense and myrrh are not what the "Three Kings" are offering but – what is no less important – three buildings for science, including a "Kaiser Wilhelm Institute of Physics" (whose founding director Albert Einstein was to become, even though his "office" was located in his private apartment).**

The appointment procedure documents the close and fruitful interplay of influential representatives of science (esp. Nernst and Planck) and business – condoned and supported by politicians and by the kaiser personally. There was a third force, of course: Einstein himself, his achievements and his interests.

After leaving Germany at the end of 1894 Einstein suffered some disappointments and occasionally wandered astray. His path to success was not an easy one, but he continued to follow it with dogged persistence. After finishing off his schooling, he acquired a diploma (as "teacher of mathematical subjects") at the Polytechnic in Zurich and submitted his doctorate at the University of Zurich. His subsequent accomplishments secured him an eminent place in the history of science and of mankind beside Copernicus and Newton.

1905 was Einstein's "*annus mirabilis*" in his scientific career. By that point in time he had already authored a few research articles. But that year, three papers appeared in volume 17 of the journal *Annalen der Physik* that each became world famous:

- 'On the Movement of Small Particles Suspended in Stationary Liquids Required by the Molecular-Kinetic Theory of Heat' (a theory for the Brownian motion of molecules).
- 'On a Heuristic Point of View Concerning the Production and Transformation of Light' (this paper discussed the hypothetical existence of light quanta and became the basis of the award of his Nobel prize in 1921).
- 'On the Electrodynamics of Moving Bodies' – the first exposition of the special theory of relativity.

Einstein had been working since 23 June 1902 as a patent examiner at the Swiss Patent Office in Berne. He stayed there until 1909. His employment there had come about "more out of chance than out of any basic interest in patent law."[57] It was a way to earn a living, not a calling. After having soon "honored every physicist from the Northern Sea to the southern tip of Italy with an offer of [his] services,"[58] he had finally found a secure position at the Swiss Patent Office. The work there was a life saver but hardly the fulfillment of his dearest wishes.

Nonetheless, his publications from 1905 soon aroused interest among the community of scientists. The first name to come to mind is *Max Planck*. Even without considering his own outstanding achievements, Planck was the man Einstein was most indebted to for the furtherance of his career. We may disregard the fact that Planck was occasionally mistaken. What matters is that he recognized Einstein's talent early on and selflessly lobbied to engage him, despite his "youthful age." An ability to recognize genius among so many unknowns also demonstrates some talent, some genius, even.

"Planck" – a member of the editorial board of the *Annalen der Physik*, and since July 1906 its chief editor – "probably did not see Einstein's papers of 1905 prior to their publication, but he was certainly the first to study them thoroughly." In March or April 1906 Einstein received a letter from Planck regarding his paper 'On the Electrodynamics of Moving Bodies.' "After such a long period of waiting this was the first sign that his paper had been read at all. The young scholar's joy was the greater, since this acknowledgment of his work originated from one of the most important physicists of modern times ... at that moment, Planck's interest meant infinitely much to the young physicist as a boost to his morale."[59] "My papers are receiving much acknowledgment and are stimulating further investigations," he reported on 3 May 1906 to his friend Solovine. "Prof. Planck (Berlin) wrote me recently about it."[60]

Thus the stages of Einstein's career (University of Zurich 1909, University of Prague 1911, Zurich Federal Polytechnic 1912) are of less significance for Einstein's later appointment to Berlin than his *connections with leading physicists of his day*, primarily with *Max Planck* – who was not only the most important of them all, but also a very influential person. Planck had been the *first* to recognize Einstein's calibre. The others followed later. It is comprehensible that Planck's interest in Einstein would soon assume more concrete form.

Important stages along the way to professional recognition of Einstein as a theoretician:

- September 1906: Conference of the Society of German Scientists and Medical Doctors in Stuttgart. Planck advocated the theory of relativity. Sommerfeld was still unconvinced.
- 1907: Convention of the Society of German Scientists and Medical Doctors in Dresden: Sommerfeld delivered a talk – in support of – the theory of relativity.
- July 1907: In a long letter Planck reported to Einstein that supporters of the principle of relativity numbered but a "modest few."[61]
- 21 September 1908: Talk by the mathematician from Göttingen, Hermann Minkowski on the theory of relativity at the convention of German Scientists and Medical Doctors.
- April/May 1909: Max Planck's lectures at Columbia University in New York on the current state of physics. As Planck himself admitted in a letter, he was making "propaganda for the principle of relativity": "The revolution unleashed by this principle on our physical view of the world is probably only comparable in its scope and profundity to the one defined by the introduction of the Copernican system of the universe."[62]
- 20–24 September 1909: the first conference Einstein attended – the convention of German Scientists and Medical Doctors in Salzburg. On 21 September Albert Einstein delivered a talk before more than a hundred listeners. They included: Lise Meitner, Max Born, Max Planck, Arnold Sommerfeld and Johannes Stark.[63]
- March 1910: Nernst traveled to Zurich to inform Einstein about results in conformance with Einstein's theory.
- 25 September 1911: At the Scientists Convention in Karlsruhe Einstein made the acquaintance of Fritz Haber.
- 30 October to 4 November 1911: Einstein participated in the First Solvay Conference at Brussels. This "crisis conference and summit" was organized at

> the initiative of Walther Nernst. Nernst convinced the Belgian industrialist Ernest Solvay to sponsor the event. In reply to Nernst, Einstein wrote: "I am pleased to accept the invitation to Brussels and will gladly write the survey envisaged for me [...]. The whole enterprise appeals to me exceedingly and I scarcely doubt that you are the spirit behind it."[64] Other participants included: Hendrik Antoon Lorentz, Ernest Rutherford, James Jeans, Jean Perrin, Paul Langevin, Henri Poincaré, Marie Curie, *Walther Nernst, Max Planck, Heinrich Rubens.* Thus Einstein got to know the eminent researchers of physics – and they got to know him. This august "jury" held court, as it were, on their new discovery. The verdict was positive. Einstein was deemed worthy to bear the crown of science.

At the Solvay Conference, the dice on Einstein's future were thrown. "The consequence of the meeting with *Nernst and Planck* was that they, upon their return to Berlin, assumed the difficult and delicate task of luring Einstein away to Berlin."[65] *Emil Warburg,* director of the Bureau of Standards (PTR) and later co-signer of the proposal to elect Einstein into the Academy of Sciences, needed no convincing. He had already approached Einstein unsuccessfully himself.[66] And *Heinrich Rubens,* the fourth man in league had, like Planck and Nernst, also attended the Solvay Conference. *Fritz Haber,* director of the Kaiser Wilhelm Institute of Physical Chemistry and Electrochemistry since 1911, was likewise convinced.

Time was of essence. Solicitations from other places showed that. Gone was the time when Einstein had to beg his way through half of Europe for employment. Now it was Europe's turn to approach Einstein. The *briefness* of his periods of stay at Zurich and later at Prague demonstrated, though, that he had not yet reached his personal goal. He was still looking.

Planck and Nernst were not interested in any compromise solution. They wanted to do a proper job of it. It was not enough to lure Einstein to Berlin. He had to want to stay there over the long term (if not permanently) as well. So an attractive offer had to be made that catered to his interests.

Einstein was no enemy of teaching but he preferred not to have formal obligations. Nor was he an impresario of science – in this respect he was hardly comparable to Nernst and Haber. Einstein wished his day were not constantly burdened with regular obligations so that he could devote himself entirely to his research, freed from any material worries. Fritz Haber had "found out that, being completely emersed in his own research, he would gladly dispense with the major lecture course that he is obligated to teach [and Haber had ... ] furthermore arrived at the conviction that he had no basic objections to Berlin."[67]

At first, efforts were made to create a position for Einstein at the Kaiser Wilhelm Institute of Physical Chemistry (under Fritz Haber's directorship). Haber not only approved of this idea (assuming it had not originated with him); he even took the initiative to obtain the approval of the government and Mr. *Koppel*'s acquiescence.

It is understandable why Fritz Haber, as the director of the Kaiser Wilhelm Institute of Physical Chemistry, had high hopes about the appointment plans. He had in view the usefulness of Einstein's presence in Berlin for the development of physical chemistry (and thus he was pursuing goals slightly at variance with the ones Planck and Nernst had in mind).

The course of events took an unexpected turn in Planck's and Nernst's favor when, on 23 March 1912, Max Planck was elected secretary of the physico-mathematical class of the Academy of Sciences. An additional augmentation of his sphere of influence and his authority as a decision-maker came in 1913 with his appointment as president of the University of Berlin.

It was latest at this point that Planck favored the idea of appointing Einstein *to the Academy of Sciences* in Berlin. The formal opportunity to do so arose – as has already been mentioned – when *one* of the two positions funded by the academy became vacant. The occupant of that position, the Nobel laureate van 't Hoff had died on 3 March 1911.[68] After offering the position to Röntgen (purely as an unavoidable matter of form) and receiving his refusal on the grounds of advanced age, the way was clear for Einstein.

At the meeting of the physico-mathematical class of the academy on 12 June 1913, *Planck, Nernst, Rubens* and *Warburg* submitted a nominating proposal.

It did not go through completely uncontested. There was a weak undercurrent of opposition, a dim flicker of sheet lightning. On 20 June 1913 the ambassador of Switzerland confidentially wrote to Department Head *Schmidt-Ott* at the Ministry of Culture at his private address to report that Einstein "was probably of Semitic origin and in Zurich nothing had ever been heard about his parents."[69] The only possible meaning of this is: please reconsider this matter very carefully. Schmidt-Ott was no ordinary low-ranking official. He was a school friend of the kaiser, and a confidant both of Minister of Culture von Trott zu Solz and of the head of the Imperial Cabinet for Civil Affairs von Valentini.[70] The State leadership thus *knew* that Einstein did not quite fit into their world-view. They had certainly been warned. The fact that the decision did not come out otherwise shows that the upper echelons of power viewed Einstein's Jewishness (and as we shall see, also his Swiss citizenship) as a pardonable, minor hitch. A smart Jew was at least much more valuable than a stupid German.

Schmidt-Ott's attention and assistance was needed for another matter, too. A week later, on 27 June 1913, Max Planck wrote him to request that Einstein be exempted from military service – if need be, as an "act of grace."[71] Max Planck did not want the precious jewel he wanted to carry home to Germany to suffer any harm and possibly go to waste on some battlefield. Not even Max Planck's patriotism went so far as to let him serve his fatherland unconditionally. The letter was evidently written under the – legally correct – assumption that upon assumption of his official duties at the academy, Einstein would automatically become a subject of the German Reich. The files do not reveal what exactly Schmidt-Ott undertook. Whatever he did, the mere fact that there was an *attempt* to make special arrangements for Albert Einstein is interesting. A few months later, when Einstein had arrived in Berlin and the war broke out shortly afterwards, a more elegant solution to the problem was found. Einstein was treated as a foreigner. He was allowed to remain Swiss, thus obviating the issue of military service.

But first, before it came to that, the procedure for inviting Einstein to Berlin had to be worked out.

The signers of the election proposal to the academy had made good preparations. They knew that realization of this petition depended on something else, too: money.

Nernst evidently served as the mediator between the academy and Koppel. Who else but Nernst (or Haber) could have convinced Koppel to make his "offer"? Koppel certainly could not have judged for himself how valuable or promising Einstein was.

Nernst touched Koppel about supplementing Einstein's salary and met with success. Nernst later called this his greatest organizational coup.[72] In a private letter dated 3 June 1913, *Koppel* announced to Nernst,[73] and later also in a private letter to Planck,[74] his willingness to add the sum of 6,000 marks to the salary approved by the academy for Einstein, as of 1 April 1914, for a period of twelve years. Koppel specified behind closed doors his preference that this donation be treated as internal business.

Leopold Koppel's letter to Walther Nernst, 3 June 1913:[75]

Esteemed Privy Councellor,

Today we discussed the following:

It appears to be desirable for science to attract Mr. Einstein to Berlin. You thought it practicable to appoint him as a salaried member of the Academy of Sciences.

I am prepared to contribute to the salary that the Academy would be obligated to pay Mr. Einstein, insofar as this half would not amount to more that 6,000 marks and as long as Mr. Einstein retains his residence in Berlin.

In utmost respect,

Yours very truly,

Leopold Koppel

The minutes of the meeting of the physico-mathematical class of the Academy of Sciences on 12 June 1913 spell out the details.

Excerpt from the minutes of the meeting of the physico-mathematical class of the Academy of Sciences on 12 June 1913:[76]

The undersigned reads the proposal submitted jointly by Messrs. Nernst, Warburg and Rubens [...] on the election of the current full professor at the Polytechnic in Zurich, Dr. Albert Einstein, as regular member of the Academy, with a special personal salary of 6,000 marks.

Mr. Nernst provides confidential commentary on the details, the substance of which is that the plan of obtaining Einstein for the Academy has gained material form with the offer by Councillor of Commerce Koppel to supplement an annual amount of 6,000 marks over a period of 12 years so as to be able to offer the candidate an adequate

total salary. After Mr. Schwarzschild has also declared his support of the petition, Messrs. Fischer and Waldeyer suggest whether it would not be more conducive to the appointment conditions and at the same time more dignified for the Academy were the Class to place the full total of 12,000 marks at the candidate's disposal and were Councillor of Commerce Koppel to be asked to transfer his contribution of 6,000 marks to the Academy in the form of a gift. This proposal receives basic approval by the Class [...]. Mr. Nernst assumes the task of contacting Councillor of Commerce Koppel in the sense suggested. [...].
Planck

The deliberations by the physico-mathematical class proceeded a few days later, on 3 July 1913. The proposals were presented before the academy's Appropriations Committee for a vote. It was approved by a majority of the class.

Excerpt from the minutes of the meeting of the physico-mathematical class of the Academy of Sciences on 3 July 1913:[77]

The Class deliberates on the election of Professor of Theoretical Physics at the Federal Polytechnic in Zurich, Dr. A. Einstein as regular member of the Academy. [...]. Since a special personal salary is attached to the election proposal, it has been submitted beforehand to the Appropriations Committee of the Class, and this Committee suggests that the salary be set at 12,000 marks [...]. According to a confidential report by Mr. Nernst, Councillor of Commerce Koppel will regularly contribute half of the special salary disbursed by the Academy over a period of 12 years in the form of a gift to the Academy. [...].
Planck

The matter was again on the agenda before a plenum of the academy just a few days later, on 10 July 1913. This shows how important and urgent it was. It had to be settled before the beginning of vacation (from August to September). The plenum had conferred and decided on the basic issues on 12 June. Now it was a matter of specifying and confirming the details. Members concerned about "the involvement of a private person in the nomination of a new member of the Academy" were pacified with the assurance that the salary payments to Einstein should be treated completely independently of any involved private person (i.e., Leopold Koppel).

Excerpt from the minutes of the plenary meeting of the Academy of Sciences on 10 July 1913:[78]

[...]. Pursuant to the proposal by Mr. Brunner, the resolution about the salary shall take the following form: The special personal salary shall be set at 12,000 marks, on the condition, however, that in the event that Mr. Einstein should later receive another official income, a new agreement be negotiated. [...]. Receipt of a normal honorary salary of 900 marks is a matter of course and shall not be affected by these resolutions.

Messrs. W. Schulze, Erman, Ed. Meyer express reservations, in principle, about the involvement of a private person in the nomination of a new member to the Academy. In opposition to this, Messrs. Nernst and the undersigned point out the independence between the motivations behind the election proposal and the funding efforts for the salary. [...].

Planck

Once the necessary preparations were complete, Planck and Nernst (accompanied by their wives) set out to deliver the academy's offer to Einstein. They departed on 11 July – *the very next day after the academy meeting!* After a brief period of consideration, Einstein agreed.

On 28 July 1913 the academy informed the minister of intellectual and educational affairs that at their plenary session on 24 July 1913 they had elected Albert Einstein as a regular member of the physico-mathematical class and had approved in addition to the normal salary of 900 marks a special personal salary in the amount of 12,000 marks. The academy petitioned that the minister "most graciously bring the effectuated election to the attention of His Majesty the Kaiser and King and kindly arrange for Supreme confirmation."[79] The minister was informed that Einstein had "stated his willingness to accept the election on condition that the costs incurred by the move from Zurich and Berlin be reimbursed to him and that adequate provision be made for his beneficiaries in the event of his death."[80] (There is nowhere any mention, however, that Einstein *intended* to retain his Swiss citizenship. He evidently did not demand any written confirmation of this wish.) That summer the secretaries of the academy could leave on vacation with the gratifying feeling of a job well done.

The Ministry of Culture took its sweet time anyway. On 6 November 1913 the Ministry of Culture pleased "in deepest respect, to beg of His Imperial and Royal Majesty gracious confirmation of the election of Professor *Einstein* by benevolent execution of the enclosed draft as a Supreme decree."[81] By 12 November 1913 already, the Kaiser and King had confirmed Einstein's election as a member of the Academy of Sciences.[82]

The presiding secretary of the academy, Roethe, informed Albert Einstein of this on 22 November 1913.

Letter by the presiding secretary of the Academy of Sciences, G. Roethe, to A. Einstein, Zurich (excerpt):[83]

Berlin, 22 November 1913

Highly esteemed Sir,

On behalf of the Roy. Academy of Sciences, I have the honor of announcing to you that the same has nominated you as regular member of her phys.-math. class and that this nomination has been confirmed by His Majesty the Kaiser and King by Supreme decree on 12 November of this year; I now ask you please to state whether you accept this nomination.

In the event of your removal to Berlin, the Minister of Intellectual and Educational Affairs has stated his willingness to approve a reimbursal for moving costs in the amount of your real disbursements up to the legally defined amount for moving costs pursuant to the provisions for civil servants classed as fourth rank. [...].

The Academy has approved for you in addition to the usual salary of 900 marks a special personal salary of 12,000 marks annually; both salaries would be paid to you as of the 1st of the month in which you conduct your move to Berlin. [...].

Einstein sent his thanks on 7 December 1913.

Letter by A. Einstein to the Prussian Academy of Sciences:[84]

Zurich, 7 December 1913

To the Roy. Prussian Academy of Sciences.

I thank you cordially for having elected me a regular member of your organization, and declare herewith that I accept this election. I am no less grateful to you for offering me a post in your midst at which I may devote myself to scientific work free from any professional obligations. When I consider that each working day demonstrates to me the feebleness of my mental capacity, I can only accept the high distinction intended for me with a certain trepidation. But what encouraged me to accept the election was the thought that all that can be expected of a person is that he devote himself entirely and with all his might to a good cause; and I do feel capable of that.

You kindly left it up to me to choose the date of my moving to Berlin. With respect to that, I advise you herewith that I wish to take up my new duties during the first days of April 1914.

Respectfully,

*A. Einstein, Zurich*

On 10 December 1913 the secretaries of the academy, Roethe, Diels, Waldeyer and Planck, informed the minister that Einstein had accepted his nomination and would take up his new position in early April.[85]

Einstein then eventually arrived in the Reich capital on 29 March 1914. His office was located at Fritz Haber's institute: Faradayweg no. 4 in Dahlem on the outskirts of Berlin. His apartment – obtained with Haber's assistance – was very close by: Ehrenbergstrasse no. 33.

The persistent yet flexible efforts of Planck and Nernst, as well as of Warburg, Rubens, Haber and others, were crowned with success. Einstein's move to Berlin was a fortunate synthesis of scientific, political and personal interests. Thus not just Einstein's interests were served, equally so those of the Germano-Prussian Kaiserreich.

Formally, the moving technicalities were completed only by the end of that year. Although Einstein had set reimbursement of his moving expenses as a condition for his acceptance, he took his time submitting the appropriate paperwork. On 27 December 1914 he informed the Ministry of Intellectual and Educational Affairs in a handwritten letter that his move had cost 665 marks, thereof 515 for the shipment of the furniture and 150 for train tickets for himself and his family. He enclosed one letter of appointment and one bill by the mover. The indicated sender's address was: Wittelsbacherstr. no. 13. The ministry double-checked it and arrived at the result: "Directory: Dahlem, Ehrenbergstr. 33."[86] Albert Einstein had moved in the meantime. He had separated from his wife and

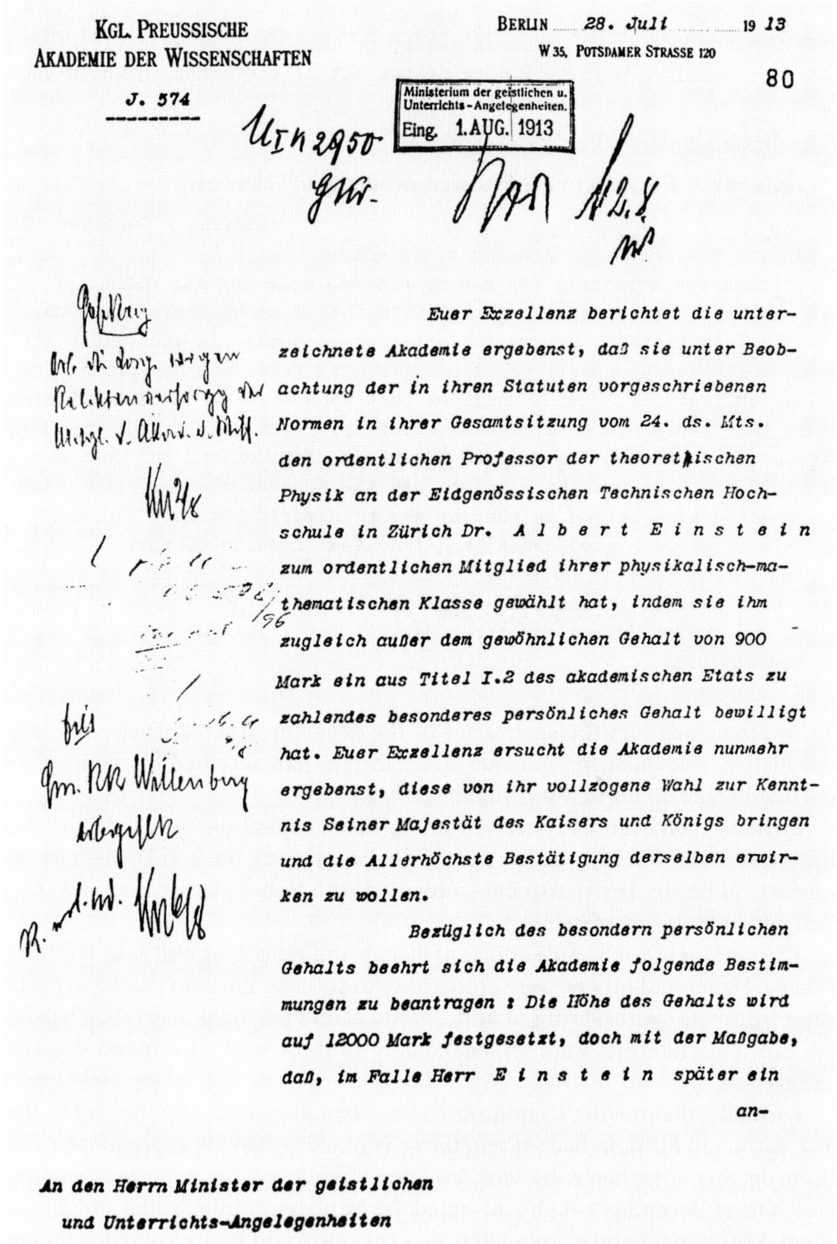

KGL. PREUSSISCHE
AKADEMIE DER WISSENSCHAFTEN

J. 574

BERLIN 28. Juli 19 13
W 35, POTSDAMER STRASSE 120

80

Ministerium der geistlichen u. Unterrichts-Angelegenheiten.
Eing. 1. AUG. 1913

Euer Exzellenz berichtet die unterzeichnete Akademie ergebenst, daß sie unter Beobachtung der in ihren Statuten vorgeschriebenen Normen in ihrer Gesamtsitzung vom 24. ds. Mts. den ordentlichen Professor der theoretischen Physik an der Eidgenössischen Technischen Hochschule in Zürich Dr. A l b e r t E i n s t e i n zum ordentlichen Mitglied ihrer physikalisch-mathematischen Klasse gewählt hat, indem sie ihm zugleich außer dem gewöhnlichen Gehalt von 900 Mark ein aus Titel I.2 des akademischen Etats zu zahlendes besonderes persönliches Gehalt bewilligt hat. Euer Exzellenz ersucht die Akademie nunmehr ergebenst, diese von ihr vollzogene Wahl zur Kenntnis Seiner Majestät des Kaisers und Königs bringen und die Allerhöchste Bestätigung derselben erwirken zu wollen.

Bezüglich des besondern persönlichen Gehalts beehrt sich die Akademie folgende Bestimmungen zu beantragen : Die Höhe des Gehalts wird auf 12000 Mark festgesetzt, doch mit der Maßgabe, daß, im Falle Herr E i n s t e i n später ein

an-

An den Herrn Minister der geistlichen
und Unterrichts-Angelegenheiten

**Figure 5: Einstein is elected to the Prussian Academy of Sciences. The Royal Prussian Academy of Sciences to the Minister of Intellectual and Educational Affairs, dated 28 July 1913, requesting the kaiser's stamp of approval to his election. Signed by Max Planck ... handwritten marginalia: administrative notes by Hugo Andres Krüss (Kr)**

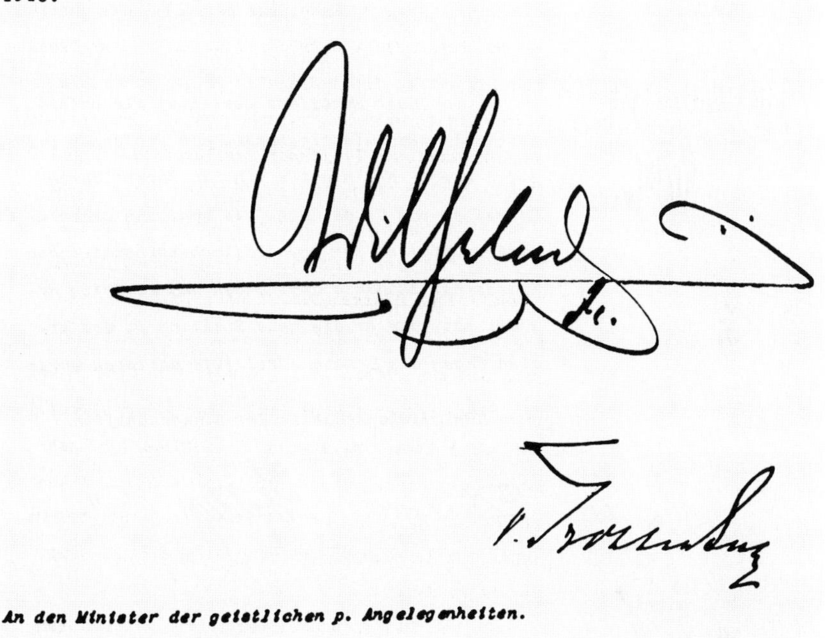

92

Auf Jhren Bericht vom 6. November d. Js. will Jch die von der Akademie der Wissenschaften zu Berlin vollzogene Wahl des ordentlichen Professors der theoretischen Physik an der Eidgenössischen Technischen Hochschule in Zürich Dr. Albert Einstein zum ordentlichen Mitgliede ihrer Physikalisch- mathematischen Klasse hiermit bestätigen. -- Berlin, den 12. November 1913.

An den Minister der geistlichen p. Angelegenheiten.

UIK 4054-13

**Figure 6: Confirmation by Wilhelm II of Einstein's election, 12 November 1913. Countersigned by Minister of Culture von Trott zu Solz**

children and was living as a bachelor in the suburb of Wilmersdorf (Wittelsbacher Str. 13). His subsequent address – from 1917 until the end of his "Berlin period" – was: Schöneberg, Haberlandstrasse no. 5, in the same building and thereafter even in the same apartment as his cousin – his lover and later wife Elsa Einstein.

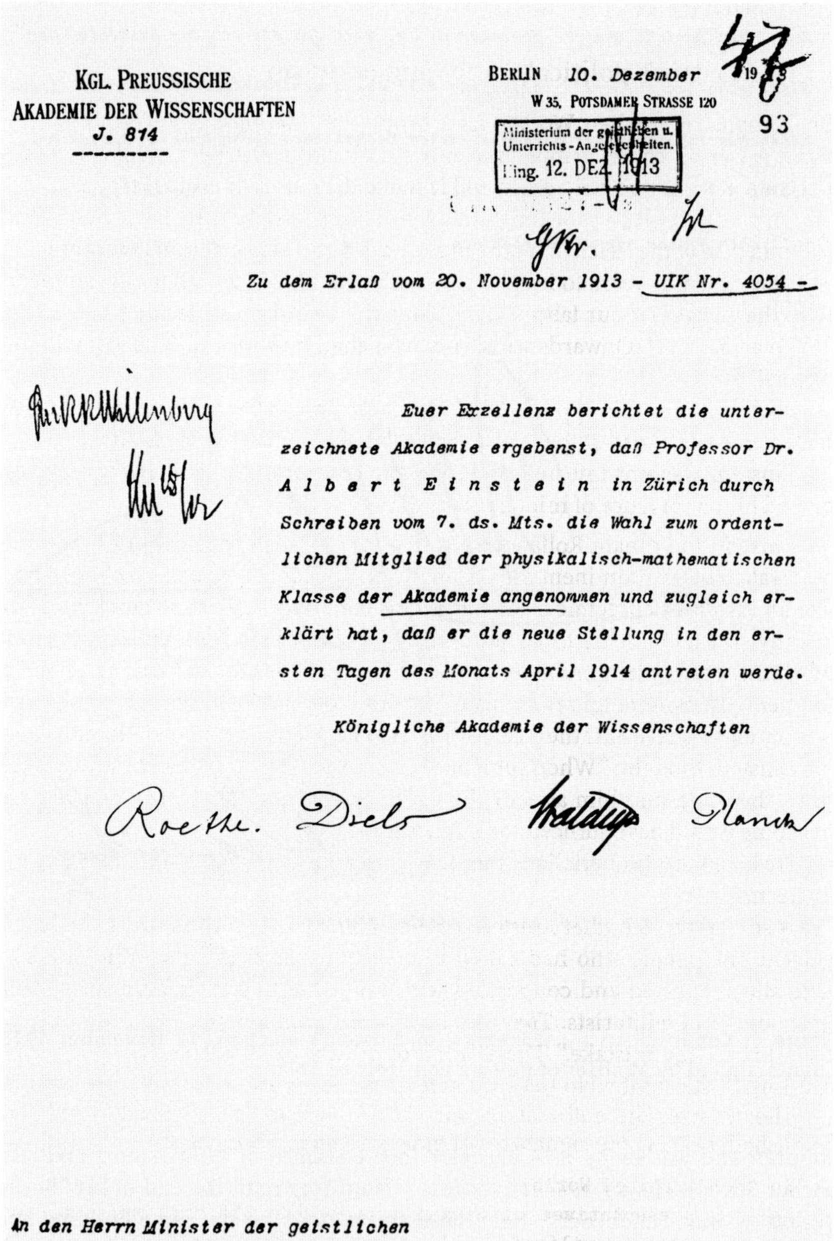

KGL. PREUSSISCHE
AKADEMIE DER WISSENSCHAFTEN
J. 814

BERLIN 10. Dezember 1913
W 35, POTSDAMER STRASSE 120

93

Ministerium der geistlichen u. Unterrichts-Angelegenheiten.
Eing. 12. DEZ. 1913

Zu dem Erlaß vom 20. November 1913 - UIK Nr. 4054 -

Euer Exzellenz berichtet die unterzeichnete Akademie ergebenst, daß Professor Dr. A l b e r t E i n s t e i n in Zürich durch Schreiben vom 7. ds. Mts. die Wahl zum ordentlichen Mitglied der physikalisch-mathematischen Klasse der Akademie angenommen und zugleich erklärt hat, daß er die neue Stellung in den ersten Tagen des Monats April 1914 antreten werde.

Königliche Akademie der Wissenschaften

Roethe. Diels Waldeyer Planck

An den Herrn Minister der geistlichen
und Unterrichts-Angelegenheiten

**Figure 7: Official notification of Einstein's acceptance – the Academy of Sciences to the Prussian Minister of Culture, 10 December 1913. Handwritten marginalia: administrative notes by Hugo Andres Krüss (Kr)**

## 1.3 World War I

### 1.3.1 Einstein's political stance and activism

On 1 August 1914 when World War I broke out, the masses jubilantly went to arms. A war psychosis gripped the entire nation as never before. All classes and all strata of society. The kaiser could feel like the kaiser of every German.

On 7 August 1914 a proclamation by the kaiser was published. It stated: "Since the founding of the Reich, it has been the passionate effort of mine and my ancestors these past 43 years to keep the peace in the world [...]. But our opponents envy the success of our labors [...]. Thus the sword must decide [...]. So rise up! To arms! [...]. Onwards with God, who shall be with us, as he was with our fathers!"[87]

A coauthor of this appeal, Adolf Harnack,[88] was able to capture the kaiser's sentiments and those of the masses so well, because they were *his own*.

Einstein was not infected by it. On the contrary. The goings on only confirmed his abhorrence of militarism and war.

Einstein to Romain Rolland, 22 March 1915:[89] "May your magnificent example awaken other prominent men from the inexplicable delusion that has gripped like an insidious epidemic even competent men of otherwise steady reason and healthy sentiment! After three hundred years of assiduous cultural activity, must our Europe really be remembered in coming centuries for not having advanced further than from religious mania to patriotic mania? Even scholars of the various nations behave as if their cerebrums had been amputated eight months ago."

Einstein thought: "When someone can only march in rank and file to a single tune, then I despise him already; he got his cerebrum only by mistake, because his spine would have sufficed for him entirely."[90]

He hated the "imperialistic mentality, which is dominating influential circles in Germany."[91]

It seems, though, that Einstein pushed a few unpleasant thoughts under the carpet: The people who had called him away to Germany and who had energetically supported and cooperated with him were to a large part out-and-out "patriots" and militarists. *They* – to borrow Einstein's words – had certainly got their brains "by mistake." These "men of otherwise steady reason and healthy sentiment" succumbed to the nationalistic madness – that "insidious epidemic."

Shortly after the outbreak of war, the "Appeal to the Civilized World" was published, bearing the signatures of ninety-three intellectuals.[92]

To the Civilized World![93]

We as representatives of German science and the arts raise protest before the entire civilized world against the lies and blasphemies with which our enemies are seeking to soil Germany's pure cause in the dire battle for survival that has been forced upon it. [. . . ].

*It is not true* that Germany is to blame for this war. Neither the nation wanted it nor the government, nor the kaiser. On the part of Germany, the utmost has been done to prevent it. [. . . ]. It was only

**Figure 8: The same signature appears on appeals to strengthen military strength and on certificates appointing the "diligent friend of peace" Albert Einstein. Wilhelm II: "I no longer know any parties, I know only Germans."**

when a superior force that had long been lurking at the borders fell upon our nation from three sides, that it rose up as one.

*It is not true* that we maliciously violated Belgium's neutrality. France and England were evidently resolved to violate it. Belgium was evidently in collusion. It would have meant self-destruction, had we not intervened.

*It is not true* that a single Belgian civilian's life and property fell before our soldiers without desperate self-defense so dictating. [. . . ].

*It is not true* that our troops brutally attacked Louvain. They shot a portion of the city's population, with a heavy heart, in punishment because a raging mob had perfidiously assaulted them in their barracks. [. . . ] reluctant as we are to concede a greater love of the arts to anyone else, we categorically refuse to preserve a work of art at the price of a German defeat.

*It is not true* that our military leadership is ignoring international laws. It will have nothing to do with undisciplined atrocities. [. . . ]. Those who make pacts with Russians and Serbs and offer the world the disgraceful scenario of setting Mongols and Negroes loose on the white race have the least right to pose as defenders of European civilization.

*It is not true* that fighting our so-called militarism is not fighting our culture, as our enemies hypocritically contend. Without German militarism the German culture would have been obliterated from the face of the earth long ago. [. . . ].

Believe us! Believe we shall fight this battle to the end as a civilized nation, that holds such legacies as Goethe, Beethoven and Kant as sacred as it does its own hearth and its own soil.

This we vouch for with our names and on our honor!

There follow ninety-three signatures, among them signatures of people who had strenuously sought and organized Einstein's election and transfer to Berlin:

- *Max Planck,* professor of physics, Berlin,
- *Walther Nernst,* professor of physics, Berlin,
- *Fritz Haber,* professor of chemistry, Berlin.

But the signatures also included

- Prof. *Adolf von Harnack,* managing director of the Royal Library, Berlin,
- His Eminence Ernst Haeckel, professor of zoology, Jena,
- Philipp Lenard, professor of physics, Heidelberg,
- Wilhelm Ostwald, professor of chemistry, Leipzig,
- Richard Willstätter, professor of chemistry, Berlin,
- His Eminence Wilhelm Röntgen, professor of physics, Munich,
- Wilhelm Wien, professor of physics, Würzburg,

as well as many other leading members of the sciences and the arts and culture, including:

- His Eminence Wilhelm von Bode, managing director of the Roy. Museums, Berlin;

- Prof. Wilhelm Dörpfeld – Berlin;
- Max Liebermann – Berlin;
- Gerhart Hauptmann – Berlin;
- Hermann Sudermann – Berlin.

The *most renowned representatives of German culture and science* were amongst the undersigners of the "Appeal to the Civilized World." In structure this document was defensive, but *in content* it was the unconditional legitimization of a policy of aggression. "No other manifesto during the World War discredited the moral standing of the German intelligentia abroad [...] more than this 'appeal' [...]. No manifesto so provoked chauvinism on the opposing side without it even obtaining as much as a hint of its intended effect."[94]

"It is not true ...," "It is not true ...," "It is not true ...." People who in their own field accepted nothing that could not stand up to the most rigorous test, willingly let themselves be used as warmongers. Blind hatred against "Russians, Mongols and Negroes" was their defense as representatives of a civilized nation.

That was not all. The leaders of German academia placed themselves and their services completely at the disposal of warfare.

Fritz Haber is probably the most striking example.

"On 28 July 1914 Haber requested leave of absence for six weeks to take a health cure at a spa in Karlsbad followed by a period of relaxation. He wanted to treat his 'gallstones and melancolia' ('Gallensteine und Gemüth'). But he added in his application: 'If the political situation were to take such a form as to draw our country into a conflict, I intend to return from this vacation.' Three days later World War I started. Everyone was swept up in an inexplicable wave of patriotic enthusiasm. Deputy Sergeant Fritz Haber also reported himself for duty. But he was turned away for age reasons. Other duties more crucial to the war were waiting for him than marching with the field artillery to the front line to counter the enemy."[95]

Haber conceived the fundamental scientific and technical ideas for the synthesis of nitric acid, an important component of explosives and fertilizers. "Without the crucial development of the Haber-Bosch process, the German fighting power would have been destroyed already in spring 1915 and the German population would have starved from a shortage of fertilizer."[96]

Because Haber's contributions had been so important, Planck thought he might be able to convince Hitler to revise his anti-Semitic policy in 1933 – to no avail, of course. "Following Hitler's seizure of power, I had the responsibility as president of the Kaiser Wilhelm Society of paying my respects to the *Führer*. I believed I should take this opportunity to put in a favorable word for my Jewish colleague Fritz Haber, without whose invention of the process for producing ammonia from nitrogen in air the previous war would have been lost from the start."[97]

Haber had, after all, been the father of chemical warfare. He had been the consultant chemist for top ranking leaders of the Army and head of gas attack and gas defense. The Kaiser Wilhelm Institut of Physical Chemistry and Electrochemistry in Dahlem, near Berlin, exchanged "its statutory duty of promoting pure science for activities in the area of gas warfare and anti-gas defense."[98]

At Haber's initiative other talented scientists were taken on board as well: James Franck, Gustav Hertz and Otto Hahn. But it was possible to decline. Max Born is an example, who later wrote: "During the war he and I broke off our relationship. He wanted me to join his war gas team, which I bluntly refused to do."[99]

According to international law, what Haber did was a *war crime*.

The "Regulations concerning the Laws and Customs of War on Land" of The Hague Convention of 18 October 1907 explicitly forbade the use of poison gas in a war. Article 23 reads:

In addition to the prohibitions provided by special Conventions, it is especially forbidden
(a) to employ poison or poisoned weapons;
(b) to employ arms, projectiles, or material calculated to cause unnecessary suffering;
(c) to employ projectiles whose sole purpose is to spread poisonous or asphyxiating gases. The splintering effect must always exceed the toxicity.[100]

This convention had been signed by Germany as well. Consequently, what Haber had done was a war crime even according to *German* law!

Thanks to Haber, gas warfare had its beginnings on 22 April 1915 near Ypern. The result was agonizing death for thousands of people. For this service to the fatherland, Haber was proud to receive, among other distinctions, the Iron Cross 2nd and 1st class.

Einstein chose a quite different position. He became co-signer of a counter-manifesto to the appeal by the ninety-three intellectuals. In the middle of October 1914 Dr. Georg Nicolai, professor of physiology at the University of Berlin and doctor of the imperial household, drafted a manifesto "to the Europeans"[101] together with Prof. Wilhelm Foerster (a repentent co-signer of the foregoing "Appeal to the Civilized World").

Manifesto to the Europeans

While technology and commerce inexorably drive us toward factual acknowledgment of international relations, and thus toward a common global civilization, never before has a war so intensely interrupted the cultural communalism of cooperation as the present one. Perhaps it strikes our conscience so conspicuously only because so very many mutual bonds existed, the breaking of which we now feel so acutely.

Even if this state of affairs should not surprise us, those to whom this common world civilization is in any way dear, would be doubly obligated to fight to uphold these principles. However, those of whom one ought to expect such convictions - that is, primarily scientists and artists - have thus far uttered statements that would almost exclusively suggest that coincident with the breach in relations their desire for further continuance of these real relations has waned. They have explicably spoken in a martial spirit - but have yet to utter words of peace.

No nationalistic passion can excuse such a mood; it is unworthy of the word culture, as it is understood throughout the world. If this mood gained universality among the educated, it would be a misfortune indeed.

It would be a misfortune not just for civilization but for the national composition of the individual states - we are firmly convinced of this. It would be a misfortune for the very cause that ultimately unleashed all this barbarity.

Technology has made the world *smaller;* the *states* of the large peninsula of Europe appear today as close to each other as did the *cities* on each of the small Mediterranean peninsulas in ancient times. And by virtue of the multifarious interrelations between the needs and experiences of each individual, Europe - one could almost say the world - already comprises a single unit.

Thus surely it would be the duty of educated and well-intentioned Europeans at least to make an attempt at preventing Europe - as a consequence of its deficient total organization - from suffering the same tragic fate as ancient Greece once had. Must Europe too exhaust itself with fraternal strife and go to ruin?

For today's raging conflict is not likely to produce any victors; only the vanquished will be left. That is why it seems not just *a good thing,* but a dire *necessity, that educated men of all nations* direct their influence in such a way that the *terms of peace not become the wellspring of future wars* - uncertain though the outcome of the war may now still seem. The fact that this war has plunged all European relations into an equally *unstable* and *plastic state* should rather be put to use to create out of Europe an organic whole. - The technological and intellectual conditions for this are already given.

How precisely this new European order may be possible is not the topic here. We only want to emphasize in principle our firm conviction that the time has now come for *Europe to act as one body to safeguard her lands, her peoples, and her culture.*

We believe that this will exists latently among many, and by our concerted enunciation of this will, we hope to transform it into a driving force.

To this end, it would seem that everyone who cherishes European culture and civilization, that is, those whom *Goethe* had presciently called ''*good Europeans,*'' must first gather together. For we must not lose hope that our collective voice will be heard - even above the din of arms - especially if those who now enjoy the esteem and authority of their educated peers number among these ''good Europeans of tomorrow.''

But it is essential that Europeans first assemble, and if - as we hope - we can find enough *Europeans in Europe,* that is to say, enough people for whom Europe is more than a geographical concept, being an important ideal guarded close to their hearts, then we shall try to

**Figure 9: Walther Nernst (center) "as Staff Scientific Advisor on the way to his post" (original caption) – 1915**

rally such a league of Europeans together. - It should thereupon become our voice and guide.

Our only intention is to inspire and stimulate; thus, provided you share our convictions and our determination to *lend resounding expression to the European will,* we urge you to send us your signature.

But the response was disappointing. The document was disseminated privately and was "sympathetically received by many." But "even those who agreed with it ultimately refused to allow their names to be published for formal reasons or out of principle. Since the 'appeal could only be of value if it bore the names of recognized authorities, however,' its authors abandoned their plan, 'deeply concerned' about their 'isolation'."[102] It was only *later,* thanks to Einstein's world renown, that the appeal caused any furor. At its conception it made no impact whatsoever. It was not even published.[103]

The appeal nevertheless occupies a central place in Einstein's political biography. It marks the beginning of his activism. Einstein's later views are an extension of convictions expressed in 1914 (and certainly not any break with them). These convictions include, for instance:

- that, as a consequence of developments in science and technology and the associated internationalization of life, war should stop being a means toward solving disputes;
- that a European community of states – a "European League" – was necessary as an instrument toward settling or preventing conflicts.

The statement that "the terms of peace should not become the wellspring of future wars" was prophetic. This initiative was also typical of the later period: it concurred with Einstein's beliefs but the idea to respond to the 'Appeal to the Civilized World' with a 'Manifesto to the Europeans' did not originate with him. *Someone else* thought of the idea and saw in his "esteemed friend and mindmate" Einstein a *partner,* whose achievements outside the great arena of politics could serve as a mouthpiece. It was only in his science that Einstein was truly original, creative and innovative. Even so, Einstein was not an apolitical person, an "apolitical pacifist," as Castagnetti and Goenner have suggested with reference to the period of the Great War.[104] He knew only too well where and when he should say yes or no. His idealism did not prevent him from being a realistic, reasonable person. (Since when – incidentally – is idealism the opposite of "politics"?) Despite its inconsistency and internal strife, the organization that Einstein soon joined as an active member[105] prepared the ground for what was to come: the November Revolution of 1918. In this respect Count zu Eulenberg was right when he contended: The pacifist movement "in Germany helped [...] prepare the ground for the revolution [...]. The crippling of the will to fight as sought by the pacifist movement was a precondition for the initial and subsequent success of the revolution."[106]

It was but a small step from the 'Manifesto to the Europeans' to membership in the pacifist New Fatherland League (NFL, Bund Neues Vaterland). Even though the 'Manifesto to the Europeans' was not published, Einstein's signature was no secret among the founders of the league. Their social connections were diverse. Einstein, on the other hand, had no reason to decline an invitation to join. He entered the league sometime between 21 March and 24 April 1915 (no precise date is given in the minutes of the NFL).

The New Fatherland League had been founded in the fall of 1914.[107] "After Hellmut von Gerlach, Kurt von Tepper-Laski and Otto Lehmann-Russbüldt had agreed, already in August 1914, to counter the first occasion of growing pan-German imperialist capitalist power by forming, alongside the democratic socialist mass movements, a decidedly pacifist and democratic syndicate of like-minded men from public life be formed, in November 1914 the N.F.L. was founded under the presidency of Kurt von Tepper-Laski."[108] (As far as these "men" are concerned, incidentally, the league had remarkably many women among its membership!)[109]

The minutes of the NFL mention Einstein's name repeatedly. (It is not always clear from the proceedings, however, and sometimes only indirectly indicated, whether and when Einstein was in attendance.)

From the minutes of the New Fatherland League:

Meeting of 21 March 1915:[110]
Attending [. . . ] guests: [. . . ] Prof. Einstein, Frau Einstein

On this very day – it was a Sunday – the pacifist Walther Schücking visited the office of the NFL. He noted afterwards:[111]

In the afternoon of 21 March I visited the office of the New Fatherland League in Berlin for the first time. I had to walk into the courtyard of the building on Tauentzienstrasse [. . . ]. In the office I was very cordially greeted by Miss Lilli Jannasch, a very nice lady in her thirties [. . . ], and by Mr. Lehmann-Russbüldt [. . . ]. That afternoon a meeting of the league was taking place in the conference hall of the German sports association on Schiffbauerdamm, which was very interesting. [. . . ]. Besides Mr. von Tepper-Laski [. . . ] a number of other famous people were there. They included Privy Councillor of Commerce Arnhold from Dresden, a man from high finance [. . . ]. Professor Einstein was also there; it was the first time I heard his name. Through a law he had discovered regarding the universality of time he is supposed to have accomplished a scientific feat of the very first order and has for that reason been drawn away from his Swiss home country by the Ministry of Culture in order to devote himself to research in Berlin with no teaching obligations. [. . . ]. The 'New Fatherland' League is supposed to organize [. . . ] the opinions of those who view the goal to strive after not as phantastic conquest but lasting peace. From the outset this circle had advocated the idea that a contract for world peace needed to be organizationally laid down so that in future such catastrophies are eliminated. [. . . ].

7th meeting on 24 April 1915:

Lilli *Jannasch* reports furthermore that negotiations are being undertaken with the 'Forum,' the 'Deutschen Briefzeitung-Gesellschaft' and the 'Staatsbürger' about the takeover of the League's Proceedings [Mitteilungen]. At the suggestion of *Einstein*, a suitable agreement should be subjected to a vote beforehand at a regular meeting.[112]

9th meeting on 10 May 1915:[113]

On the basis of this new §2, at the proposal of Professor *Einstein*, the engineer Georg von *Arco* is unanimously elected as vice-president.

11th meeting of 31 May 1915:[114]

In the appendix to the minutes of the meeting, the confidential and not alphabetically arranged list of the members of the NFL, Albert Einstein is nominated as 29th in order (of a total of fifty-nine members). The nomination of Elsa Einstein follows at place 58.

12th meeting of 14 June 1915:[115]

The ''Appeal by intellectuals of nationalist and internationalist persuasion is transferred to a committee, for which the following gentlemen are suggested: Count *Arco, Einstein, Goldscheid,*[116] *Herzog.*''

13th meeting of 28 June 1915:[117]

> ''Concerning the planned 'Appeal by intellectuals' it is reported that the gentlemen Count *Arco*, Prof. *Einstein*, *Goldscheid* and *Kestenberg* have taken up preliminary work at a meeting on 16 June. The appeal is not to originate from the league itself but from individual persons and have an international character.''

The "Minutes of a meeting of German interested persons on 30 August 1915"[118] held at Lilli Jannasch's apartment may also be considered as a session of the NFL. It involved consultations about the formation of a German delegation for the "Great Council of the international Central Organization for a Durable Peace" and in particular about cooperation with the Dutch peace movement (Anti-Oorlog Raad). Pursuant to a Dutch proposal, the aim was to recruit about ten people from each participating country ("who should represent different political orientations and social groups as far as is possible"). A list was compiled of persons to be enlisted for the cause. Prof. Einstein and Prof. Rade were supposed to be approached through the NFL.

When and how Einstein was approached is not reflected in the files. But his acceptance is recorded on 22 October 1915.[119] Professor Schücking was presumably the recipient of his reply (it is among the Schücking papers).

*22 X 15.*

*Highly esteemed Colleague!*

*From Miss Jannasch I hear that you would like to have me included in the Anti-Oorlog Raad's Gr. Council. I have absolutely no experience and am not an efficient person in political affairs. Nonetheless, I am quite willing to support this splendid cause. So do include me, if you consider it fruitful. I must inform you, however, that I am Swiss and consequently cannot figure as a German.*

*With all due respect, Yours very truly,*

*A. Einstein*

*Wittelsbacherstr. 13, Berlin*

*A brother of the unfortunate ''Fatherland''*

The autumn of 1915 was generally a time of much pacifist activism on the part of Albert Einstein.

On 16 September he visited Romain Rolland in Vevey (Switzerland). Although this was their first meeting, they had known each other for some time already. On 22 March 1915 Einstein had written him: "From the daily paper and through my connections with the highly creditable 'Fatherland' League I learned of how courageously you placed your life and person at risk toward eliminating the so ominous misunderstandings between the French and German peoples."[120] He offered his services "as a tool," by virtue of his "connections with German and foreign individuals in the exact sciences." Rolland was informed of Einstein's impending visit by the secretary of the NFL, Lilli Jannasch, on 29 June 1915.[121] The day finally arrived on 16 September 1915. In his 'Diary of the War Years' Rolland described his meeting with Einstein in detail.[122]

As a direct consequence, Einstein became the object of an investigation by the German police – no different from the NFL in general.

A police investigator was set on Einstein's case. This Mr. Göring reported the results of his work on 5 January 1916.[123]

The report mentions, among other things that Einstein moved on 10 October 1914 from Ehrenbergstrasse in Dahlem to Wittelsbacher Strasse 13 as a tenant.

Göring wrote down everything he could find out about Einstein's biography. The facts he got wrong were that Einstein (among other things) had studied *at the University of Munich,* that he had become professor at the German University in Prague in *1910,* that he had "done his doctorate" in *1905,* that Einstein's election had "been confirmed by His Majesty the German Kaiser" on *12* November 1913. In short: virtually nothing about his life prior to 1913 is correct. Einstein's current position was described basically correctly, however.

The following point was of political relevance:

He is not recorded here as politically suspicious and nothing of a negative nature could be found here in this regard. Nor are proceedings regarding his character [. . . ] available.

Einstein is known here as a member of the 'New Fatherland League,' but has not yet drawn any attention to himself through any agitatory behavior within the pacifist movement. From a moral point of view he enjoys the best reputation conceivable and there are no penal records on him.

He has a subscription to the paper *Berliner Tageblatt.*

A transcription of the registration page is attached.

*Nowadays,* a comment on the reference to the *Berliner Tageblatt* is necessary; at that time it was superfluous. This paper had been founded in 1871 by the Jew Rudolf Mosse and was being managed by another Jew Wolff (chief editor). It generally reflected the views of the liberal middle class and was an organ of pacifist sentiments. For that reason it was particularly despised by the military authorities. Without the personal intervention of the "Kaiser-Jude" Albert Ballin with the head of military censorship, the *Berliner Tageblatt* would have gone bankrupt as a result of a lengthy ban imposed in August 1916.[124] So, from the point of view of war advocates, the mere fact that Einstein was a subscriber to this paper was politically suspicious.

The New Fatherland League became the target of "growing restrictions [...] by the military authorities."[125] The deciding factor for this treatment was a memorandum that the NFL submitted to the Reich Chancellor and the members of Parliament in the summer of 1915. It was entitled: 'Should We Annex?'[126] and was a direct response to a proposal by the six major trade associations from 20 May 1915 demanding large-scale annexations. On the justification that dissemination of this memorandum had broken censorship rules, the High Command of the Official Seals had already threatened to close down the NFL on 29 July 1915.[127] Later the secretaries Lilly Jannasch and Elsbeth Bruck were given prison sentences. Finally, on 7 February 1916 the High Command of the Official Seals imposed a complete ban on all activities of the NFL. The league obeyed the order but recommended to its members to continue the activities as private persons. For an idea of the historical context in which this was going on, this happened around the time of the beginning of the Battle of Verdun (January to July 1916), but a long time *before* the bitter "cabbage" winter of 1916/1917, *before* the founding of the Independent Party of Social Democrats (IPSD) in April 1917, and *before* the mass demonstrations against starvation and war. This means that the NFL had stopped being a dangerous element for the hawks in government long *before* fortune turned the tables in the war.

Uncertain about how the league members would conduct themselves, the police continued their surveillance. On 8 March 1916 the Commander's Office of the Capital Berlin issued a complaint to the academy that Einstein had left on a trip without officially applying for a travel permit from Police Headquarters.

Letter from the Commander's Office of the Capital Berlin to the Royal Academy of Sciences, 8 March 1916 (excerpt):

> [. . . ]. The Swiss citizen Dr. Albert Einstein, residing in Wilmersdorf, Berlin, at Wittelsbacherstrasse 13, alleges that he occupies a salaried position as a member of the Academy of Sciences. Information is most humbly requested whether his information agrees with the facts; in the affirmative case, the request is made that the pertinent personnel files be made available. Dr. Einstein has left repeatedly on travels without personally registering his departure here in Berlin at the Police and without personally registering his arrival with the Police at his travel destinations, which he as a neutral foreigner is obligated to do.

On 25 April 1917 the president of the Berlin Police informed the High Command of the Official Seals about eleven known "representatives and promoters of pacifist activities." The "Swiss citizen Professor Dr. Einstein [...] member of the 'New Fatherland' and of the Dutch 'Anti-Oorlog Raad'," was named among them as a "very avid friend of peace."[128] His name also appeared on the 'List of notable pacifists residing in the State Police District and Environs of Berlin,' dated 29 January 1918.

But soon there was no more reason for the military authorities to monitor Einstein's activities. His labors on the general theory of relativity sapped all his energy and in early 1917 he became seriously ill. Rolland still found reason to note on 17 July 1917: "Einstein arrived from Berlin quite ill, he has lost 30 kilos. He needs to go on a cure to recover [...]. He is quite exhausted."[129] Einstein himself wrote him on 22 August: "I would definitely not neglect to visit you if my health were a bit more stable; but the smallest undertaking often takes its toll afterwards."[130] Besides reflecting a weakened state of health, Einstein's correspondence from this time also indicates political resignation. He had lost confidence that an organiziation like the NFL was capable of doing anything decisive toward putting an end to the war. He was "firmly convinced that this straying of minds can only be steered by hard facts." It was hopeless fighting against power-hungry sabre rattlers "with intellectual weapons."[131] "Hard facts" meant nothing less than military defeat for Germany. For the first time in his life, his pacifist persuasion had to bow before other points of view.

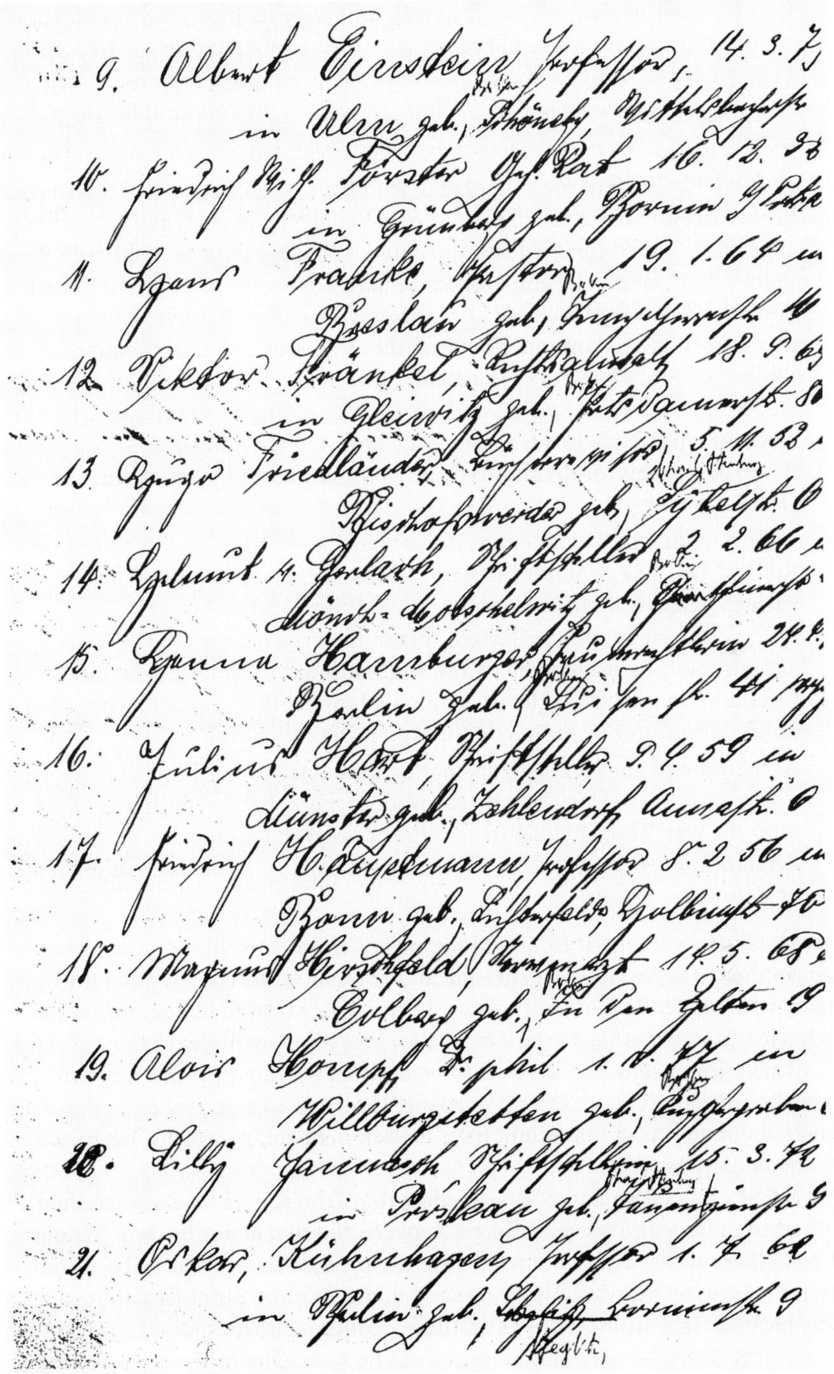

**Figure 10: 'List of *notable* pacifists residing in the State Police District and Environs of Berlin,' dated 29 January 1918. "*9. Albert Einstein*" (ninth on a list of thirty-one names.)**

Despite this deeply pessimistic assessment with its inconsistent conclusion, Albert Einstein's experiences of the war and his membership in the New Fatherland League had a lasting effect on his values. His membership placed an accent on his later political views and activities. The war set the course of Einstein's later life.

- At the NFL Einstein met the people who later – as members of the German League of Human Rights (GLHR, the renamed NFL in 1923) as well as in other capacities – accompanied him or chose a similar path in politics:[132] Ernst Reuter, Emil Gumbel, Lehmann-Russbüldt, Eduard Fuchs, Eduard Bernstein, Toni Mendel (one of Einstein's later lovers), Friedrich Wilhelm Foerster, Magnus Hirschfeld. Many members of the NFL were future members of the League of Friends of the New Russia (Eduard Fuchs, Lehmann-Russbüldt, Hugo Simon, Helene Stöcker and others – including Einstein). And, like Einstein, Magnus Hirschfeld, Count Arco and Arthur Wolff later served on the board of trustees for the children's homes maintained by Red Aid in Germany (RA, Rote Hilfe).
- Important international connections had their origin in the NFL. Romain Rolland was one, who on his part did his utmost to include Einstein in his various pacifist activities (if only by employing Einstein's fame). A *co-founder of the NFL*, Hellmut von Gerlach, made the arrangements for Einstein's reappointment as a member of the International Committee on Intellectual Cooperation of the League of Nations. Many further contacts with scholars on the committee or indirectly through it were the result.
- As mentioned above, Einstein's connections with the NFL had a lasting effect on his worldview. He picked up many ideas there that he continued to uphold after the war. They include issues of peace and the promotion of mutual understanding among nations – including the call for the creation of the League of Nations.

Besides Einstein's retreat from politics for health reasons, there was another factor conducive to his political survival and preventing an escalation of his differences with his professional colleagues: his tendency toward irratic, contradictory behavior, in part perhaps even a certain degree of opportunism.[133]

Anti-crime agent Göring could not find anything politically suspect to report, because there *was nothing* to report (Einstein was careful enough to keep quiet about anything that might have broken his neck politically, perhaps even professionally as well).

Haber once commented in retrospect that "the war years drew us apart."[134] Whether there was a proper quarrel between the two is not known. Even if so, a complete rupture could not have been the result. Einstein knew about Haber's activities during the war. It was no secret. Einstein's first officially paid residence was located right inside Haber's institute. Nothing about what was going on in the rooms downstairs could go unnoticed by him. One of the tests being conducted there on 17 December 1914 even caused the death of one of Haber's coworkers. Such conflicting political views and even Haber's active complicity in the butchery of the war did not destroy their friendship.

Einstein's philosophy of survival is summarized in notes Romain Rolland made after a conversation with Einstein on 16 September 1915 in Vevey (Switzerland).[135]

From Romain Rolland's diary:

16 September. Professor A. Einstein, the ingenious physicist and mathematician at the University of Berlin, who has written me during the course of the past winter, comes to visit me from Zurich [. . . ]. Einstein is unbelievably frank about his opinion of Germany, where he lives. No German has such frankness. Any other person would have suffered from the sense of isolation as a thinker during this terrible year. Not he. He laughs. He has managed to write his most important scientific work during the war. I ask him whether he mentions his views in the presence of his German friends and discusses them with them. He says no. He contents himself with posing many questions - as Socrates did - to upset their peace of mind. Everything I learn from him is not particularly encouraging; for it reveals the impossibility of arriving at a lasting peace with Germany without first smashing it to pieces. [. . . ]. The will for might is spreading everywhere, an admiring faith in force and a firm resolve for conquest and annexations. The government is much more moderate than the people. It would like to evacuate Belgium but it cannot do so. The officers threatened a revolt. Major bankers, industrialists and trade associations are omnipotent; they are demanding reparations for their sacrifices; the kaiser is just a tool in their hands and in the hands of the officers: he is good, weak, in despair about this war, which he had never wanted, which he had been forced to want [. . . ]. Tirpitz and Falkenhayn are the protagonists of this bloody deed [. . . ]. The socialists are the only (somewhat) independent element; it is still only a small minority of that party who are rallying around Einstein. The New Fatherland League is making very slow progress but is not expanding. Einstein does not anticipate any renewal of Germany of its own volition: it lacks the energy, the bold initiative. He is hoping for a victory for the Allies, who would destroy the power of Prussia and the dynasty. [. . . ]. Einstein and Zangger are dreaming of a divided Germany: Southern Germany and Austria on the one hand, Prussia on the other. [. . . ]. It is not food that could be most scarce but chemical products essential for the war. It is true that the truly admirable resourcefulness of German scholars offer new substances as substitutes for the missing products. Einstein says, you could not imagine the organizational energy that has been expended, including all the best minds [. . . ], the University of Berlin [. . . ] is more loyal to the government and imperialistic than any other university: the professors are appointed expressly with this purpose in mind. [. . . ].

These were not just momentary exclamations of bitterness. Einstein expressed similar views repeatedly. In his letter to Romain Rolland from 21 August 1917, he wrote: "Only facts can dissuade the majority of the misled from their delusion that we were living for the State and that its intrinsic purpose was the greatest power possible at any price.

The finest alternative to lead us out of these deplorable circumstances seems to me to be the following: America, England, France, and Russia conclude a military arbitration agreement for all time [...]. Any state that has a democratically elected parliament and whose ministers depend on a majority within this parliament ought to be allowed to enter into this pact. [...]."[136] On 11 September 1917 Rolland also notes: "Einstein visited virtually no one during his stay in Switzerland. Yet in a letter to a friend he beseeches the Allies to carry on to the very end, to the destruction of Germany."[137]

Even if only half of what Rolland recorded was right, Einstein's statements would have been regarded, from the official German point of view, as treasonable.

The initiator of the 'Manifesto to the Europeans,' Georg Friedrich Nicolai was a warning example. The only difference was – aside from Einstein's incomparably higher achievements as a scientist – that Nicolai announced what he thought of the war *in public* and *repeatedly*. Shortly before the war started, he was one of the head doctors at the Tempelhof military hospital. He was at the same time permitted to pursue his research at the University of Berlin. Shortly after the signing of the 'Manifesto to the Europeans,' however, he was demoted to a lowly medical assistant in the military hospital for contagious diseases in Fort Graudenz. There – thanks to an indulgent fort commander – Nicolai wrote his book 'The Biology of War.' Shaken by the sinking of the British passenger ship the 'Lusitania' by a German torpedo, he exclaimed to a few of his coworkers at table: "the violation of Belgian neutrality, the deployment of poison gases, or the sinking of commercial shipping are not just moral atrocities but boundless stupidity which sooner or later is going to ruin the German kaiserreich." Someone reported him to the authorities and he got transferred. He was ordered not to make any further political statements. Nicolai thereupon refused to take an oath of loyalty and obedience toward the government. He was degraded to a common soldier, harassed and humiliated. His tasks were reduced to the simplest of office work. Nicolai was sentenced to imprisonment. At liberty again later and at work in the sanitary service, Nicolai eventually fled by plane across the border.[138] After the war, Nicolai was, as a consequence, one of the most hated "traitors to the fatherland" for the political right. Prompted by protests organized by extreme-right students against lectures by a "deserter-professor," the senate of the University of Berlin, under the leadership of the ultranationalist president Eduard Meyer, unleashed an unprecedented smear campaign against a member of its faculty.[139] Nicolai was deemed unworthy of continuing his teaching duties. His inability to practice his profession eventually led to his decision to accept a professorship in Argentina in 1922, where he stayed for the rest of his life.

Einstein would not have been spared a similar fate of confinement and subsequent expulsion if he had demanded the subjugation, division and destruction of Germany *in public*. Major scientific feats would not have been able to protect him. Max Planck, Walther Nernst and Fritz Haber would not have been able or even have wanted to help him anymore. But Einstein took care not to make such radical confessions out in the open. His statements had been made in confidentiality to Romain Rolland, a *foreigner and pacifist*. Einstein did not discuss such things with his *German* colleagues. "He contented himself with posing them many questions." But in most cases, opportunity for that never arose. Haber, one of the people who had created with "admirable resourcefulness" the necessary "chemical products essential for the war," were usually at the front line. Against his better judgment, Einstein had often taken a protective stance before members of his field of expertise and contended that scientists and mathematicians were "strictly internationally minded" – in contrast to historians and philosophers, who were "for the most part chauvinistic hotheads."[140]

Although Einstein's pacifist activities distinctly differed from Nicolai's, the two remained closely associated with each other – both politically as well as personally – after the fiasco of the Manifesto to the Europeans. Probably Einstein's most covert and spectacular deed as a pacifist was assisting the fugitive Georg Friedrich Nicolai in escaping the country on 20 June 1819.[141] The War Ministry had demoted its rebellious serviceman Nicolai yet again and transferred him from Danzig to the provincial backwater of Eilenburg near Leipzig. The only consolation was its closer proximity to Berlin, making it much easier for his friends to visit from Berlin. One of these friends was Ilse Einstein – an ardent admirer of Nicolai (and daughter of Albert Einstein's later wife Elsa) and another was the Pole Fanja Lezierska – a member of the Spartacus League (out of which the CPG emerged on 31 December 1918). The commander of the barracks in Eilenburg thought severer measures were necessary for Nicolai. It was decided that he be forced to undergo training in the handling of a gun. But before it came to that Nicolai managed to desert. Where could he run to? He found a hideaway among the Einsteins. On 25 May 1918 – "after 4 days of calm and most earnest reflection" – he wrote to the minister of war to inform him of his motivations and then waited 4 weeks in vain for a reply. His letter had also instructed the minister about where to send the response: "Prof. Albert Einstein, Berlin W. 30, Haberlandstr. 5." Nicolai did not fail to also point out that once the minister's decision was issued he "would find ways and means to have this decision made public."[142] So as not to reveal his hideaway, Nicolai made it look like he had fled south. His letter arrived at the Ministry of War in Berlin by registered mail from Munich – through the help of others. On 20 June Nicolai finally escaped to Denmark – in an airplane procured by the Spartacus League and piloted by one of its members. As soon as he was safe, Albert and Ilse Einstein were each variously and promptly informed.

So Einstein was complicit to desertion. Both Einstein and Nicolai carried this secret to their graves.[143] Otherwise Einstein – like Nicolai – may well have been expelled from Germany right after World War I. Nicolai had gone too far; Einstein had not yet crossed that line of political tolerance.

Einstein was occasionally *courageous;* but he was never really *dangerous.* The 'Manifesto to the Europeans' is not the only document bearing his signature. His name is also found on another document demanding peace without annexations but it does not give the least impression that co-signer Einstein was wishing the demise of Germany. Einstein joined others in putting his name to the statement that Germany had not gone to war "with conquest as its aim." It even contained the affirmation "that the German people would be able to claim a prize of victory."[144] Einstein did not object. He signed his name under what he himself could not believe.

In *this case,* Einstein was willing to compromise. In signing, he joined erstwhile co-signers of the 'Appeal to the Civilized World," *Adolf von Harnack* and *Max Planck.*

Einstein's response to a plea by Elisabeth Warburg, wife of the director of the Bureau of Standards (PTR), Emil Warburg, is also characteristic of his attitude during the war. Elisabeth Warburg asked for Einstein's intercession toward having her only son discharged early. Einstein and his cousin Elsa (whom Elisabeth Warburg presumed was his wife) enjoyed the hospitality of the Emil Warburg household as particularly welcome occasional guests.

Elisabeth Warburg to Albert Einstein:[145]

*Elisabeth Warburg* *Charlottenburg, 21 March 1918*
*Marchstr. 25b*

*Esteemed Professor,*

*How kind of you to want to write to my son. I am so very worried and anxious, you know. Would that your letter still reached him in good health. Living hell must have broken loose over there. It is a sin against human beings, an outrage, that these mass murders are being committed on and on. Do we have a right to this?*

*My husband was always so overjoyed with his son. He once told me in confidence that he was one of the very greatest. These hopes we placed in him, must all of it be for naught? Is he not to fulfill his destiny? I fear he is too much involved over there to extricate himself. Why also must it be that his division, in particular, be deployed right at the very front? This is a boundless misfortune for us. Forgive these desperate pleas; I am not resigned, I still rebel against fate and would like to save what can be saved. How good that you want to help me with this. I am enclosing two letters for you, written to describe him. With cordial regards, also to Mrs. Einstein, ever yours,*

*Elisabeth Warburg*

Einstein immediately came to the rescue. But he first had to be sure that the person concerned would accept his help. This feedback was essential, for Otto Warburg had enlisted as a volunteer in 1914. He had been stationed with the 202nd Uhlan Guards Regiment which was eventually deployed at the Russian front. His reminiscences of this period reveal that he even liked being on active duty for his fatherland at the time.[146] But his loving mother saw it very differently. In her view, what was happening out there was nothing short of *mass murder*.

Einstein to Otto Warburg:[147]

*23 III 1918*

*Highly esteemed Colleague,*

*You are probably surprised at receiving a letter from me, because until now we have only walked past each other, without actually ever getting to know each other. I must even fear rousing some form of indignation in you by this letter; but it has to be.*

*I hear that you are one of Germany's most talented younger biologists of great promise and that presently representatives in your particular field are quite mediocre here. I also hear, though, that you are stationed over there at a very vulnerable outpost, so your life is constantly hanging by a thread! Now, please step out of your own shoes and into those of another discerning creature and ask yourself: Is this not madness? Can this post of yours not be filled by an unimaginative average person of the type that come 12 to the dozen? Is it not more important than all that big scuffle out there that valuable people stay alive? You know it yourself very well and must admit that I am right. Yesterday I spoke with Prof. Krauss, who is also entirely of my opinion and is also ready to arrange to have you recalled for another assignment.*

*My plea to you, which arises from what has been said above, is therefore that you support our efforts in securing your personal safety. Please, after a few hours of serious reflection, write me a few words so that we here know that our efforts will not founder on your attitude.*

*In the ardent hope that in this matter, for once, reason will exceptionally prevail, I am, with cordial regards,*

*Yours truly,*

*A. Einstein*

*Haberlandstr. 5, Berlin W.*

Otto Warburg's response is not known. He presumably acquiesced. Later his father Emil Warburg, his boss Carl Corren – director of the Kaiser Wilhelm Institute of Biology – and Minister of Culture Schmidt-Ott also applied for his early discharge.[148] They were successful. In October 1918 Otto Warburg was back at the institute. Einstein's intervention had proven useful. But the fact that this Warburg did not become a pacifist and managed to pursue a successful career as a Jew – serving as institute director – even during the Third Reich is not Einstein's doing.

Einstein's position was indisputably reasonable. While coming to the aid of a worried mother, he was also serving science (the object of his assistance later became a Nobel laureate). But his arguments also reveal a generous helping of elitism and condescension.

Einstein was a *pacifist.* But his pacifism had limits. He did nothing that might jeopardize his exclusive working conditions and eventually drive him into exile. It was all very well for Einstein to criticize the militarists in general. He tolerated the warmongers among his colleagues all the same. And vice versa. He was forgiven or his opinions were simply dismissed as secondary, perhaps even as eccentric whims. One did not need to know what political views exactly Colleague Einstein entertained. His *usefulness* to science far outweighed any problem he as an individual posed his colleagues and the Prusso-German Kaiserreich. He respected the limits of what was acceptable behavior: he stayed useful.

There was another reason why Einstein survived the war period without suffering any serious political damage. Many of his fellow scholars and science politicians with a say in his fate were not quite as aggressive and intolerant in their normal dealings with each other as one might expect of co-signers of the Appeal of the Ninety-three. As mentioned above, the open letter to Reich Chancellor von Bethmann-Hollweg of 27 July 1915 bore the signature not only of Albert Einstein, Adolf von Harnack, Max Planck and David Hilbert, but also of leaders in commerce and finance: Prince Henckel von Donnersmark, Franz von Mendelssohn (president of the Chamber of Commerce of Berlin and Member of the Upper Chamber), S. Bleichröder, Carl F. von Siemens and many others besides. With the votes of the Social Democrats and the Center Party, a *majority* was reached in parliament for passage of a resolution on 19 July 1917 to sign a peace without annexation claims!

Ultimately, other laws prevailed at the Prussian Academy of Sciences than at universities. The academy could tolerate a foreigner and pacifist among its ranks, especially as moderate a one as Einstein was, despite its official belligerency. This staid attitude had not been easily won. It was mainly due to the efforts of Max Planck, secretary of the physico-mathematical class.

The academy's position on its membership from "enemy countries" as well as on international scientific relations in general is an example.

In a letter from 17 July 1915, the minister for intellectual and educational affairs, von Trott zu Solz, informed the head of the Imperial Cabinet of Civilian Affairs, von Valentini, that he "shared His Excellency's view that for the present there was no reason to attach such great importance to the wartime conduct of the scholars and artists in question." It is pointed out that the Berlin Academy had hitherto upheld this view but would be deliberating on 22 July on the question of "whether steps ought to be taken against foreign members who had made particularly derogatory and hateful statements about German science."[149]

So it was by no means sensational or even any indirect conscious promotion of a "traitor to the fatherland" to appoint Albert Einstein to important posts during the war. These posts were as:

- director of the Kaiser Wilhelm Institute of Theoretical Physics and
- as trustee (Kurator) of the Bureau of Standards (PTR).

In both cases the kaiser signed the appointments *in person*. A declared *opponent* of the system could not have expected such an honor. That would have exceeded the forbearance and tolerance of "His Most Gracious Kaiser, King and Lord." We may exclude the possibility that the authorized officials concealed what they knew about Einstein. Nor was the Secret Service as active as it later was to become.

### 1.3.2 Appointment as director of the Kaiser Wilhelm Institute of Physics

The idea of creating a Kaiser Wilhelm Institute of Physics dates back to 1913.[150] The idea took on more concrete form in early 1914. After sounding out the situation during a visit with Head of Department Schmidt-Ott at the Ministry of Culture, to find out whether such a proposal had any realistic chance, Nernst submitted to him a petition on 4 February 1914 he had co-signed with Haber, Planck, Rubens and Warburg.[151]

In the 'Applicaton re. justification for a Kaiser Wilhelm Institute of Physical Research'[152] the signers set forth reasons for the necessity of establishing such an institute and suggestions on how to materialize this plan.

Petition, regarding justification for a Kaiser Wilhelm Institute of Physical Research:

We propose to establish a Kaiser Wilhelm Institute of Physical Research. The *purpose of this institute* is supposed to be to assemble at once or one after another particularly suitable researchers in physics to solve important and urgent problems of physics. The relevant problems would thus be tackled in a systematic manner both by means of mathematical and physical considerations as well as, and especially, through experimental analyses conducted in the laboratory by the researchers concerned toward obtaining the most comprehensive results possible.

The *seat* of the institute we would imagine to be in a small building in Dahlem [. . . ].

For the *costs* of maintaining the institute and, primarily, for the experimental investigations conducted in fulfillment of the institute's purpose, we apply for an annual allocation of 75,000 marks. [. . . ] Although cooperations of this type among divers experts is virtually new in physics, in other sciences it has long since become commonplace; [. . . ]. Modern physics in particular poses at present such important and far-reaching problems, and in all probability will continue to do so increasingly in the future, that large-scale treatment surpasses the powers not only of any individual but even of a larger institute of physics. [. . . ] The bodies of the institute are:

1. the board of trustees [. . . ].

2. the scientific board, constituting a Permanent Honorary Secretary and 8 physicists, among these 4 Berliners and 4 non-Berliners. As Permanent Honorary Secretary we propose Prof. Einstein, the other members are reelected every three years.

3. the working committee. [. . . ]

Haber Nernst Planck Rubens Warburg

Although the proposal pointed to other sciences as its model, its issuers were envisaging a completely new form of organizing and promoting science. Genuinely modern *big science* was the plan. The responsibilities of the future governing bodies and officials were correspondingly heavy. Speedy materialization of this project was to secure for Germany lasting leadership in many areas of physics. The approach this institute was to take served – as it turned out – as a pilot project for Schmidt-Ott's later major undertaking for the promotion of research: the *Notgemeinschaft der deutschen Wissenschaft.*

The man proposed by these gentlemen as honorary secretary of the board, Albert Einstein, turned out to be completely unsuitable, as we shall see in other contexts. The applicants were apparently of the opinion that a great discoverer was necessarily also a great organizer of science. The "prize hen" they brought back with them to Berlin had laid many a golden egg in the area of physical research but none really in the area of science policy-making. *Organization* was not one of Einstein's fortes and he had little taste for the bureaucracy inevitably attached to it. The time-consuming tasks involved in coordinating the envisaged novel interdisciplinary collaborations were more than Einstein was either willing or able to do. He had opened the door to a new world-view in physics, but as far as science policy was concerned, he was not yet ready for the twentieth century.

Planck must have had his doubts about the direction of the institute as early as January 1914. The minutes of a conversation recorded on January 1914 reads: "*Planck.* Wait with institute for Einstein until he has familiarized himself thoroughly with Berlin circumstances [...]."[153]

In a retrospective handwritten message from 10 March 1924, Planck informed the Ministry of Culture that the KWI of Physics had thus far remained provisional, first of all, for lack of funding, "on the other hand and principally, for lack of a suitable person." Planck thought he had (now) found in Max von Laue the right man for the job.[154]

Janos Plesch, a medical doctor who had befriended Einstein in the 1920s, was probably right with his recollection that Einstein "always knew how to arrange things so that 'de facto he had no institute on his back.' He simply wanted 'to keep his mind clear'; nor did he want to dispose over the activities of others – thus he was neither 'leader' nor subordinate."[155]

The approvals by the two institutions mentioned in the petition, the Koppel Foundation (signature: Leopold Koppel) and the Kaiser Wilhelm Society (signature: Adolf von Harnack) were quickly obtained. In a joint submission dated 5 June 1914[156] they applied to the minister of intellectual and educational affairs for the creation of such an institute. The arguments set forth in the petition submitted by Haber, Nernst, Planck, Rubens and Warburg were reiterated, adding the latter also in attachment. The focus of the petition by the Koppel Foundation and the Kaiser Wilhelm Society was on the *financial* arrangements, however. The minister was informed that Councillor of Commerce Koppel would have the institute building built "at his own cost" and that annual contributions of both the Koppel Foundation and the Kaiser Wilhelm Society of 25,000 marks each were secure. A matching sum was anticipated and solicited from the ministry.

At the same time (on 6 June 1914) the Leopold Koppel Foundation applied to the kaiser, in accordance with its by-laws, "to please to graciously approve" a materialization of the institute (signatories: Leopold Koppel as president, Schmidt[157] as recording clerk and Diels as member).[158] This document also reveals that the funds Koppel promised (at that time a gift in the amount of 540,000 marks) had originally been intended for charitable purposes "especially in the area of social welfare." The board of trustees requested permission to use this sum to found a Kaiser Wilhelm Institute of Theoretical Physics.

Quite an irony of history! A charitable enterprise is sacrificed up for a man with a passionate concern for social welfare. But all this happened without Einstein's knowledge or involvement.

The interplay of Kaiser Wilhelm Society and Koppel Foundation, of science and capital, once again worked outstandingly well by the efforts of Haber, Nernst, Planck, Rubens, and Warburg, through the liasons between von Harnack and Koppel, and between Koppel and Schmidt-Ott. The network of interests was so tightly knit, everything was so well prepared, that nothing really could go wrong.

War was not part of the original plan. On 3 October 1914 Leopold Koppel and Adolf von Harnack humbly informed the minister of intellectual and educational affairs of their presumption that the planned state funds could probably not be expected under the present circumstances and therefore recommended that the plan of creating a Kaiser Wilhelm Institute of Physics "temporarily be shelved" and "further pursuit of the matter be abandoned for the time being."[159]

The institute is founded nevertheless during a time of war – with the approval of the State but without public funds. This was due to a major private contribution by the manufacturer Franz Stock. Following the example of Koppel and others, the source of this gift was treated with utmost discretion. It was a matter of course that this gentleman be welcomed into the Kaiser Wilhelm Society as a new member. Von Harnack's written reminder and information of this fact to the members of its senate is dated 18 January 1917.[160]

The institute was founded on 1 October 1917. As expected, Albert *Einstein* was appointed its director. *Schmidt-Ott* (who had risen from head of department to the top office), represented the Ministry of Culture.[161] *Leopold Koppel* represented the Koppel Foundation. The industrialist *Wilhelm von Siemens* became chairman of the board of trustees.[162] Einstein – in league with major industry and finance.

On 3 December 1917, Harnack informed the kaiser – evidently somewhat belatedly – of the newly established institute, and asked for his blessing. He also gave the kaiser a brief history of its creation and its working approach. Its director was to be *Albert Einstein*. In profound respect, and even more devotedly than in other petitions, Harnack begged permission to call the institute "*Kaiser Wilhelm Institute of Physics*."

Adolf von Harnack to Wilhelm II, 3 December 1917 (excerpt):[163]

Most Serene, High and Mighty Kaiser and King! Most Gracious Kaiser, King and Lord!

I respectfully ask leave to inform Your Imperial and Royal Majesty that two new research institutes of the Kaiser Wilhelm Society for the Advancement of the Sciences have just recently been established, the Institute of German History and the Institute of Physics. [. . . ].

Following a meeting held in the summer of 1917, at which Berlin physicists were also in attendance, the [society's] senate approved the founding of an institute, [and] elected as director Professor Einstein, member of the Academy of Sciences [. . . ].

In utmost awe I therefore beg Your Imperial and Royal Majesty to most graciously please to approve that the Institute of German History bear the name ''Kaiser Wilhelm Institute of German History'' and that the Physics Institute bear the name ''Kaiser Wilhelm Institute of Physics.''

The kaiser was gracious and permission was granted. Thus the supreme war-lord sanctioned a pacifist as director of a *Kaiser Wilhelm* Institute. It is not known whether Einstein was in any way disturbed by this. Nor is there any indication that he might have found it insufferable to rub shoulders with such capitalists and patriots. He was much less inclined to perform the organizational tasks expected of a director. A trip abroad served later as a welcome opportunity to cede the managerial duties to his deputy Max von Laue. And that is how things remained.[164]

### 1.3.3 On the board of trustees of the Bureau of Standards

Einstein's appointment on the *Board of Trustees of the Bureau of Standards* had greater *political* relevance than his directorship of the Kaiser Wilhelm Institute of Physics.

First of all, the Bureau of Standards (Physikalisch-Technische Reichsanstalt, PTR) was *directly answerable to the Reich.* (Its duties lay "in the area of pure and applied physics and partly encompassed general analyses aimed at solving scientific problems of far-reaching scope and significance for theory and technology.")[165] The PTR was very different from the Kaiser Wilhelm Society, which settled its affairs as a society independent of the imperial state.

Secondly, the issue of appointments on the board of trustees of the Bureau of Standards arose *during the war* – in the case of the Kaiser Wilhelm Institute, the way had already been paved *beforehand.*

Thirdly, the *deciding authority* for the PTR was the *Reich Office of the Interior,* not the Ministry of Culture. Scientific interests assumed a comparatively minor role.

In 1916 two positions on the Bureau of Standard's board of trustees became vacant. Both Professor Schwarzschild, director of the Astrophysical Observatory (deceased on 11 May 1916 as a consequence of war injuries) and Professor Dorn had passed away.

The presiding chairman of the PTR's Board of Trustees (Lewald), in agreement with the president of the PTR (Warburg), intended one of the vacancies for Royal Bavarian Building Officer, Oskar von Miller. The other was destined for Albert Einstein. A corresponding proposal was submitted to the state secretary of the interior on 24 October 1916.[166] In his submission, Theodor Lewald (simultaneously undersecretary of state at the Interior Ministry), explicitly pointed out that Einstein was a *citizen of Switzerland,* and upheld the view that this ought not pose any more of an obstacle toward his appointment than it had for the Academy of Sciences.

There is no indication in the files whether Lewald was merely performing his official duties or whether he had any personal interest in Einstein's appointment. Whatever the case may have been, he energetically promoted Einstein's candidacy. The career he later followed is remarkable.

**Theodor Lewald** was born in Berlin on 18 June 1860. After completing his studies of law at Heidelberg, Leipzig and Berlin, he entered the Prussian civil service as an administrator in 1885. In 1894 he became a government advisor at the Reich Office of the Interior and was assigned Reich commissioner for the World Fair at St. Louis in 1904. In 1910 he advanced to head of department at the ministry, 1917 to undersecretary of state, and 1919 to secretary of state at the Reich Ministry of the Interior. In 1921 Lewald retired from state service but continued to be involved in the negotiations with Poland as plenipotentiary until 1927. From 1922 he was listed in the 'Handbuch für das deutsche Reich' as chairman of the Historical Committee at the Reich Archive.

In the early 1920s Theodor Lewald embarked on a career in an entirely different field. In 1919 he became chairman of the Committee for Gymnastics of the German Reich. 1921 he organized, together with Carl Diem, the creation of the German College for Physical Education in Berlin and was appointed member of the International Olympics Committee in 1924. As a member of the IOC Lewald submitted in 1930 Berlin's application to host the Olympic Games in 1936. It was approved 1931 in Barcelona. 1933 Lewald became president of the Olympic organizing committee for the Summer Games. It was a calculated move on the part of the Nazis. As a "half-Jew," Lewald was able to counteract attempts to boycott the games in Berlin or to reassign them to another host country. With a half-Jew at the helm of Germany's delegation, the anti-Semitic statements made there could not possibly be deemed of any great importance. Only after the games had been held, in 1936, did Hitler pressure Lewald to resign his membership in the IOC. With his business finished, he was dispensable.

The secretary of state at the Reich Ministry of the Interior advised the minister of intellectual and educational affairs on 26 November 1916 – after a month's consideration – of these appointment plans,[167] requesting to know whether there were any objections. The signature is identical to the one under the submission to the Reich Office of the Interior by the president of the PTR's board of trustees: Lewald. That means: on 24 October 1916, *President* Lewald plied *Ministerial Department Head* Lewald for assistance in appointing Einstein. In the letter to the minister of culture, another argument is brought forward: the "installations at the institution made for experiments in the fields of military and marine technology are kept secret." But not even the preservation of military secrets was, in the view of the secretary of state, any argument against appointing Einstein (who at *that* point in time was *not* regarded as a citizen of the Reich but as a *foreigner*).[168]

The minister of intellectual and educational affairs (by proxy: Schmidt-Ott) likewise had no objection.[169]

On 22 December 1916 the Reich Office of the Interior approached the Reich chancellor as well as the kaiser for approval of Einstein's appointment. Once again it is noted: "Einstein is a citizen of Switzerland, however, this fact ought not to present an obstacle for his appointment to the board of trustees."[170]

The kaiser provided his signature. Place: main headquarters.[171] Once again, war-lord promotes pacifist. Einstein was supposed to work, not drill. But it also proves that his pacifism remained within the limits of acceptability. Otherwise the kaiser could not have been so gracious. Such an appointment petition could never even have come that far.

Per the report of 22 December 1916 I wish herewith to appoint the member of the Royal Prussian Academy of Sciences, Professor Albert A. Einstein, as member of the Board of Trustees of the Imperial Bureau of Standards. *Main Headquarters, the 30th of December 1916.*

Wilhelm I.R.

## 1.4 Upshot

Einstein's appointment to the Berlin Academy of Sciences, indeed his presence in Berlin, served *mutual* interests. Einstein's achievements and reputation served the ambitions of the German Kaiserreich as a great power. Einstein was supposed to lend German science new fame and influence in the world. His coming to live and work in Berlin were good preconditions for the capital of the German Kaiserreich to remain a research metropolis and consolidate itself as a world leader in the field of physics. The procedure used in appointing Einstein confirms the tight interlacing of scientific, economic and political interests. The appointment to the Berlin Academy of Sciences and the associated move to Germany and Berlin also satisfied *Einstein*'s own interests. Glamorous working conditions, including close physical and intellectual proximity to leading physicists of the day, obscured any political reservations Einstein might have entertained. Able to devote himself entirely to research, and freed from material worries, the young Einstein could regard the omnipresent militarism in the German Kaiserreich as a tolerable burden.

*In the Kaiserreich* Einstein was more an object than a subject of politics. He did what everyone expected of him. The gentlemen from Berlin had speculated with him "like with a prize laying hen,"[172] and their investment was a good one. Not even two years had elapsed since the creator of the special theory of relativity had commenced work in Berlin when, on 4 November 1915, he was able to lay before the Academy his second offering, the "General Theory of Relativity." Thus came imperishable fame not only to its father but also to its host, the Berlin Academy. After the outbreak of war, Einstein was among the very rare exceptions in Germany to be able to continue to expend his efforts exclusively on pure – non-military – research. He exploited this chance to the fullest and, as expected, did not come away with empty hands.

"Besides the 'annus mirabilis' at the Patent Office in Berne in 1905, the period of a little over a year from November 1915 until February 1917 was the most productive stage of Albert Einstein's creative life. During these fifteen months he wrote fifteen scientific papers. Two of the most important contributions to quantum theory were among them as well as, primarily, the brilliant pinnacle of his success: the general theory of relativity. These Academy reports were the culmination of his ideas conceived eight years before on the principle of equivalence. In his tentative extensions of it, he laid the foundations of a new scientific cosmology within the limitations of recent findings. Only Einstein's own famous year 1905 can provide a comparative gauge for the profundity and intensity of this research."[173]

Einstein's Swiss citizenship was not only respected during this period; it also afforded welcome justification for his continued concentration on pure research. The distinctions and promotions from the hand of the kaiser demonstrate the expectations and confidence placed in Einstein continuing to bring fame to Germany and German science. The aim was to strengthen his ties to Germany and not, as was later to be the case, to isolate and drive him out of Germany. These distinctions and honors are also proof that Einstein's political actions were, overall, that of a loyal subject. His behavior was not perceived as damaging to German interests.

Einstein was a decided opponent of war, dictatorship and officious subordination. His membership in the New Fatherland League was clear proof of this. But Einstein never belonged to any particular political party, either then or later in life. "Party discipline" and adherence to any specific party platform contradicted his impulsive need for independence of mind and action. If one must localize Einstein's political position along the scale of political parties, it would fall "tendentially toward the left wing of the Social Democratic Party." Einstein's political views and actions only became more generally known and gained political relevance after the war, however. A certain reserve in making public statements was one aspect of his loyalty toward the Kaiserreich. He made no political moves that might have endangered his own working conditions and his collaboration with Planck, Nernst, Haber and others. This was the price for Planck's, Nernst's and Haber's loyal forebearance toward the occasional uncomfortable political statement by Einstein. Besides, he was a scientist, not a politician. He had other things to do. While the general theory of relativity was in the making, he was a man possessed with work. There was little time left over for anything else. Afterwards, by the end of the war, his health had been undermined. He was, as he himself wrote, exceedingly pleased with his scientific achievements, "but quite shattered." In early 1917, he eventually became seriously ill. It took more than a full year for his health to be completely restored – with many ups and downs. That was another reason why little time was left over for any major political feats.

Einstein's position along the political spectrum was generally toward the left; he was a pacifist and a democrat. Nevertheless he vacillated on many things. A deed did not necessarily always follow his word. And his deed sometimes even went against his professions. Einstein condemned the deployment of poison gas as a weapon of war. He was nevertheless a pleasant acquaintance (some even contended: a good friend) of Fritz Haber. In secret, Einstein hoped for the demise of Germany; in public, however, he would occasionally say the opposite. Einstein the pacifist even participated in designing the tools of war – airfoils for aircraft.[174] The obvious contradictions in Einstein's thought and behavior are, on one hand, an expression of a certain opportunism; on the other, however, a comprehensible insecurity in the process of forming one's own political judgment. Irrespective of his moral development and its reasons, Einstein remained a useful political tool for Germany.

## 1.5 Einstein in private – not quite private

As chance would have it, the period of Einstein's life during the German Kaiserreich was not destined to remain generally identical with the period of the first World War, with its focus on his labors on the general theory of relativity. It also presented breaks and junctures with lasting consequences for his private life.

Einstein debarked from the corridor train from Aachen at the zoo station in Berlin on 29 March 1914.[175] In mid-April his wife Mileva also arrived in Berlin with their two boys (Hans-Albert, born on 14 May 1904, and Eduard, born on 28 July 1910). It is hard to explain why Einstein thought such a move a good idea. His marriage since 1903 was, by that time, in fact, already in pieces. Did he think he could combine the liason he had developed with his divorced cousin Elsa by continuing to be a married man? He wrote his friend Besso in Switzerland that "the extremely agreeable, really fine relationship with my cousin, the permanent nature of which is guaranteed by a renunciation of marriage," was doing him tremendous good.[176] As he soon found out, this was a serious mistake. His wife was not willing to assume the role of housewife and his cousin was not only affectionate, but a clever and practically-minded woman – she wanted more than just to be his lover, cook and nurse.

In the summer of 1914, Mileva left again with the children for Switzerland. It took a while longer, however, for her to become convinced of the senselessness of the marriage with all its desperate hopes. She finally gave her approval to the divorce early in 1918.[177] Einstein also needed time. Wanting a divorce meant giving in to his lover's more or less gentle pressure. Perhaps gratitude was involved, perhaps love. Whatever the case may have been, his complete recovery from his many bouts of illness were indisputably due to Elsa Einstein's devoted efforts. Choosing an apartment closer to where she lived was thus the practical thing to do. So Albert Einstein left Wilmersdorf for Schöneberg. He moved – or shall we say, more precisely: he was moved (although, officially, Albert was the sole tenant, his cousin took care of the practical transfer of his things during the summer of 1917, while he was away on a health cure in Switzerland – not just into her neighborhood but into her very building, at Haberlandstrasse 5). His move into *her apartment* soon followed. Finally, on 14 February 1919 he was divorced from Mileva, by cause of "adultery," "character differences" and other serious reasons. As the guilty party, Albert Einstein was interdicted from remarriage for a period of two years.[178] On 2 June 1919 he and his cousin were standing before the justice of peace, all the same.

Following the guidelines of the present book, no more than a brief sketch of this stage of Albert Einstein's life would have sufficed. As a strictly private matter, his divorce and remarriage could be dispensed with. This was not the case, however.

Einstein was not as inconsiderate as some have thought. He was willing to sign over to his former wife Mileva the full amount of his anticipated Nobel award. He also wanted to make sure the divorce would not affect her claim to widow benefits from his employment contract in Berlin. So he wrote a letter about this matter to the responsible official at the Prussian Ministry of Science, Arts and Culture: *Hugo Andres Krüss*. In it he only reiterated what resolute Elsa Einstein had already stated in person at the ministry.

Einstein to Krüss:[179]

*Prof. Albert Einstein* *Berlin{-Dahlem}*
*{Faradayweg 4} Haberlandstr. 5.*
*(Post Lichterfelde West III)*
*Highly esteemed Professor,*

*With reference to the interview you recently granted my cousin relating to the request concerning my terms of employment, I take the liberty of presenting the situation to you again in writing. The following is involved.*

*Linked with my employment is the right to a widow's pension. I now intend to be divorced in order to remarry. Through this, my present wife would lose her pension entitlement to the advantage of the second wife. This situation would entail a hardship for my present wife, however, and is at odds with my sense of justice, especially since my financial circumstances would not make it possible for me to compensate for this loss sufficiently otherwise.*

*My request now entails having my employment contract modified so that in the case of my death my current wife become the sole beneficiary of the pension, even though she be divorced from me.*

*I entreat you not to look upon my request as presumptuous, and I add in its favor that the granting of it would not create any troublesome precedence case; for my engagement is of a thoroughly unique nature and does not fall under any government-position category. Furthermore, this concession would result in absolutely no disadvantage to the State Treasury.*

*I would be greatly obliged to you for your support in this matter.*

*In utmost respect, I am*
*Yours truly,*
*A. Einstein*

Consideration for Mileva's situation was perhaps not the only motivating factor. Cold calculation may well also have been at play: The better the offer, the greater the probability that Mileva would not propose alternatives and slow down the official divorce process. Her successor in matrimony was magnanimous to relinquish such rights.

Before replying, Krüss consulted Official Advisor Trendelenburg, who wrote him that Einstein could not be helped in this matter. Einstein could only support his first wife "as an arbitrary third person."

On 15 April Krüss drafted his reply, the fair copy of which was sent out on the same day (below is the handwritten draft submitted to the clerks for transcription).

Krüss to Einstein, 15 April 1918 (excerpt):[180]

[. . .]. *I am deeply sorry not to be able to give you a more favorable response to your inquiry. The regulation on widow's pensions is based on general legal foundations from which exceptions cannot be made in individual cases by special agreement. Pursuant to it, the widow's pension can be released for disbursement only to the current lawful widow; a transferral is explicitly ruled out pursuant to the law. Accordingly, I can think of a settlement only in the form of your taking out a life insurance policy to the benefit of your first spouse, in the case of your death, and in the amount of the lawful widow's pension. In this way, your first spouse would be provided with the full equivalent of a state widow's pension.*

*What annual premium payments will be required for this I cannot estimate, yet it does not seem to me out of the question that the Minister would be prepared to help you, in case you were to regard it as too great a burden on your annual outlay. [. . .]. I shall be glad to assist you in this matter, as far as it lies in my power.*

Thus Krüss settled the business to the best of his knowledge and conscience. More he could not do. The *mere fact* that he had approached the minister about it and drawn a positive response was a lot already. Whether or not Einstein agreed with his view of the legal situation, it was *Krüss*'s duty to make sure that the *applicable legal regulations* were being respected while serving the applicant as best as possible. It was not a question of whether Krüss *personally* shared a similar legal interpretation to Einstein's. Besides, it would have been too burdensome to others, if not just to Krüss himself. A civil servant cannot live with a constantly bad conscience; the stress would finish him off emotionally. He either had to conform or leave the service.

This exchange of letters between Einstein and Krüss about this delicate subject is interesting for another reason as well. Krüss handled *all* matters concerning Einstein at the ministry from as early as *1913* until his retirement in 1925 (in that year Krüss became managing director of the Prussian State Library). His initials abound in the relevant files. So through his official duties he was perfectly aware of the *private* affairs of the man he was dealing with. When Albrecht Fölsing writes that Einstein's remarriage (hence also his divorce) was an event "that probably no one beyond Haberlandstr. 5 knew about,"[181] we must bring forward at least *one* exception: *Krüss* was informed of it. This brings us to another important point. After 1925 Krüss was an important confidential advisor of Albert Einstein and his proxy on the Committee on Intellectual Cooperation of the League of Nations. Thus civil servant Krüss accompanied Einstein throughout his *entire* "Berlin period." Einstein and Krüss – personifications of the relationship between scholar and official or between science and politics.

Chapter 2

# During the Weimar Republic

## 2.1 Boycott of German science

On 11 November 1918 a truce was signed with the Allies at Compiègne. The war was over at last. Germany had lost. It was weak and isolated. What followed was, in the words of the prime minister of victorious France, merely a continuation of the war by other means. *Clemenceau* wanted to eliminate Germany as a serious contender for political power. Henceforth, *France* should determine what happens in Europe. On 28 June 1919 the treaty of Versailles was signed.

The treaty of Versailles cost Germany over 10 percent of its population. The consequences of the war combined with the peace terms settled at Versailles essentially equated a collapse of the German economy. Germany's share in industrial productivity around the world shrank from 16 percent in 1913 to 8 percent in 1919. In that year German manufacturers generated only 38 percent of the output of six years before. The territorial losses imposed by the treaty have to be counted in this figure, of course. But an area of identical size produced in 1919 at best 45 percent of its prewar output. The territorial losses cost Germany 74 percent of its resources of iron ore and bituminous coal. After the hand-overs prescribed by treaty, German commercial shipping was reduced to 10 percent of its prewar volume. Total productivity in industry, trade, agriculture and communications only reached the prewar level again in 1927, surpassing it by 5.7 percent two years later.

Those who once shared Adolf von Harnack's view that the armed forces and science were the "two sturdy pillars of Germany's might" must now have regarded science as the only thing left to vanquished Germany. Although somewhat biased, it was not completely wrong. The war had changed science but not seriously damaged it. On the contrary, science had shown how indispensable it was to the military and the economy. It had benefited from its subjugation to a common cause.

This explains the pride with which the Academy of Sciences self-confidently declared: "Following Germany's collapse, it is most appropriately the responsibility of German science to rebuild the respect in other countries for German endeavors and to reestablish the necessary relations abroad."[182]

Fritz Haber expressed himself along the same lines in a speech on 10 December 1921: "What our nation accomplishes in the 30 billion hours of work it affords, year by year, for the world economy, will define our future. Raising the practical efficiency of our working hours will define our future. But such practical efficiency depends entirely upon our scientific capabilities and the economic efficiency built on this capability!"[183] In 1925 he called the surrender of German inventions and manufacturing know-how to the former enemy "gateways into economic fortresses" abroad, and "the area in which no international obstacles exist" for forging ahead beyond the borders.[184]

Looking back on that period, Georg Schreiber – a man of influence on the cultural policy of the 1920s and a member of parliament[185] – remembered many years later (1952): "Once again, with an eye to the military necessities, it seemed appropriate to support with science the modest minimum of military defense granted by the treaty of Versailles."[186]

These were not just empty promises. During the 1920s astonishing energy was expended on maintaining and building upon the renown of German science. Available resources, limited though they were, were applied toward the common cause in a unique sense of solidarity that put aside existing rivalries between individual scientists. The creation of the Emergency Association of German Science (*Notgemeinschaft der Deutschen Wissenschaft*) in 1920 was one expression of these efforts. The last Prussian minister of culture of the Kaiserreich, Friedrich Schmidt-Ott, was an important person behind this drive.

Before an emergency gathering of technicians on 13 March 1920 (coincidentally on the day of the Kapp *Putsch*)[187] Schmidt-Ott and Haber drafted plans for an organization to distribute state and private grants in support of scientific research. All the German scientific associations were to become affiliated with it.[188] In mid-April, at the proposal of Fritz Haber, the Academy of Sciences officially invited their honorary member Schmidt-Ott to take over the establishment of an Emergency Association of German Science. After such careful preparations, the next steps followed speedily. Secretary of State Theodor Lewald at the Reich Ministry of the Interior was especially involved. He was a close friend of Schmidt-Ott since the days of the Kaiserreich. By the end of May the most important professional associations had announced their willingness to join: the cartel of academies, the Union of Universities (*Hochschulverband*), the KWS, the Society of German Scientists and Medical Doctors, and the German Federation of Associations in the Technological Sciences (*Deutscher Verband technischer wissenschaftlicher Vereine*). In June Schmidt-Ott invited an exclusive gathering of leaders in politics and scholarship in the academy wing of the Prussian State Library[189] to resolve the creation of the Emergency Association. In the hall of the library the association was officially constituted and its by-laws, drafted by Schmidt-Ott, were ceremoniously adopted before a large gathering of forty presidents of universities, the academy and professional federations, in the presence of their host Harnack – head of the library and president of the KWS. Harnack was by that time honored as the Nestor of science policy-makers in Germany. Berlin was chosen as the seat of the association.

Schmidt-Ott was elected president of the Emergency Association. The mathematician from Munich, Walther von Dyck served as first vice-president; Haber and Harnack assumed the other vice-presidencies, with Harnack presiding over the main committee. The funding was provided by the State as well as by private business. Appeals for donations from industrialists, bankers and businessmen in trade and retail attracted considerable yet not nearly sufficient sums, because only the interest accrued on the contributed capital could be distributed as grants. The lion's share of the financing had to come for the State budget, in particular from the Reich Ministry of the Interior and the Prussian Ministry of Culture.

Within the association, numerous professional committees were formed for the various fields of science. Renowned scientists examined grant applications and proposals for the allocation of the funds. The following types of applications were considered for financial assistance:

1. research stipends, especially for young scientists,
2. educational travel,
3. scientific publication (printing costs for articles or complete books, assistance for longer-term enterprises, such as thesaurus and dictionary compilations, or financial support for publishers), leading to the formation of a publishers' subcommittee,
4. acquisitions of scientific literature (also from abroad),
5. experimental apparatus (through loans of instruments or equipment, donations of supplies or laboratory animals), also leading to separate subcommittees.

Schmidt-Ott generally reached the final decisions himself. He moved into the Berlin palace with his small group of coworkers and passed judgment on applications for financial assistance and on the overall orientation of the Emergency Association. He was no less authoritarian than his mentor Althoff. He was just more conservative and continued to entertain good relations with the former kaiser at Doorn, despite the changed circumstances. But although conservative, he was very much a realist as well. He would not be driven by emotional fanaticism. He did what rational necessity dictated.

Science, the intellectual basis of modern industry and weapons technology, became an instrument of politics. It was used *indirectly as well as directly*. Ideology was deployed as a means to break through Germany's isolation and to gain new influence.

The Prussian minister of culture, Haenisch, a Social Democrat, wrote in a letter to the presidents and senates of all the universities of Prussia on 24 November 1919: "Peace is at our door. With its return, the barriers that have separated our nation from the community of other nations for far too long will, we hope, fall. During years of enmity, with every means used against us in battle, the bond of intellectual communality among nations was broken. It is now time to retie this bond. No social station has a more profound duty to do this than the world of German science. Untouched by humiliation and persecution, German science has retained its value. Now that we are standing on the ruins, in a struggle over our very survival as a nation, long condemned to economic powerlessness or dependence on self-serving foreign aid, we rely to a special degree on the powers of our own intellect. Destiny has assigned a huge task here to German science."[190]

**Haenisch, Konrad** (14 Mar. 1876–28 Apr. 1925). Politician (SDP).

| | |
|---|---|
| 1898 | Newspaper editor of the *Pfälzischen Post* (Ludwigshafen) and the *Mannheimer Volksstimme*. |
| 1899 | editor of *Sächsischen Arbeiterzeitung* (Dresden). |
| 1900–1911 | director of the *Dortmunder Arbeiterzeitung*, during this time also editor of the *Leipziger Volkszeitung* for a year (1905/1906). |
| 1911–1915 | director of the leaflet headquarters of the SDP cadre. |
| 1913–1920 | board member of the professional organization of socialist writers (Verein Arbeiterpresse) and editor of its professional publication. From 1913 member of the Prussian house of parliament. |

**Figure 11: Friedrich Schmidt-Ott, an influential man – *during the Kaiserreich:* Head of department at the Ministry of Culture 1911–1917, minister of culture 1917–1918; *during the Weimar Republic:* 1920–1934 president of the Emergency Association of German Science (renamed German Research Association in 1929); *during the Third Reich:* 1920–1945 chairman of the supervisory board of the dye concern F. Bayer & Co. (in 1925 it joined the dye consortium IG Farbenindustrie AG)**

1915–1921 chief editor of the Social Democratic weekly *Die Glocke*.

1918–1921 Prussian minister of science, arts and culture.

1923–1925 president of the local government in Wiesbaden.

1924 member of the Reich committee of the republican Reichsbanner Schwarz-Rot-Gold.

The official cultural policy pursued during the Weimar Republic was developed by C. A. Becker, at that time still secretary of state. This future Prussian minister of science, arts and culture published a memorandum on the 'Politico-cultural responsibilities of the Reich' in 1919. He drew it up on behalf of the Reich government and submitted it to the constitutional committee of the national assembly.

**Becker, Carl Heinrich** (born 12 Apr. 1876 in Amsterdam, died 10 Feb. 1933 in Berlin). His father was a banker and consul. University studies at Lausanne, Heidelberg and Berlin.

1900–1902 trips to Spain, Egypt, Sudan, Greece and Turkey.

1902 certification for academic teaching in Semitic philology at Heidelberg, where he was appointed extraordinary professor 1906.

| | |
|---|---|
| 1908 | appointment as professor of oriental history and culture at Hamburg, followed by full professorship in oriental philology at Bonn 1913. |
| 1916 | joined the Ministry of Culture as rapporteur. |
| 1919 | secretary of state at the Ministry of Culture, 1921 minister of culture; that fall again secretary of state. |
| 1925 | Prussian minister of culture; |
| 1930 | retirement from office. |

Becker defined cultural policy as a "conscious deployment of intellectual values in the service of the nation or the State, for internal consolidation and external dealings with other nations."[191]

In Becker's opinion, the end of the German Kaiserreich was "unfortunate."[192] Realizing that the former ideals had been shattered and discredited, he sought something new for Germans to hang on to. He thought that nationalism ought to be given greater emphasis than before. He called the recent conflicts between the social classes an "indignity" and a "shameful demonstration of weakness" before foreign countries.[193] True to his opinion that the demise of the Kaiserreich was unfortunate, hence that it was essentially a good system, Becker arrived at a conclusion that may be described as the quintessence of his whole conception of cultural policy: "We only have to decide to idealize the traditions of the Bismarckian and Wilhelmian era. The Reich, for want of military power of its own, needs ideological power."[194] In other words: Becker wanted a continuation of the former dominant policy of war, but by other, more subtle means.

His appeal was that "the Reich [...] conduct foreign policy with ideas."[195] He was fully aware that these "apolitical" methods were defined by the circumstances of his time. He wrote: "Through its political and economic exclusion, the German nation has only its ideology left as a weapon in the struggle among nations."[196]

Nothing fit Becker's cultural policy better as a tool than *science*. On one hand, it was possible to use it in a "politically innocuous form." On the other hand, owing to the growing economic importance of scientific knowledge since the war, its deployment had anyway become a necessity. This dual usefulness turned it into an unusually important factor in German cultural policy. Members of the German government with any realistic perception of the situation thus had every reason to praise the value of German science.[197]

The conviction that science counted among the most precious goods left to the German Reich served as a uniting bond. This insight was not the result of a change in mentality. It was merely the result of a cold assessment of the war's consequences. "The conviction that German commercial export needed its ideal reputation to be supported and fostered" had been "a commonly held view among industrial circles" before the war already.[198]

The "enemies abroad" were also aware that downtrodden, but not yet extinct, Germany still had *science* to its name. So the decision was made to *boycott German science,* in full knowledge that science cannot flourish well in confinement.

By severing international ties, the victors were going after the Germans' vital nerve. Success would mean victory once again.

But first, some justification for the boycott had to be found in order to be able to secure the moral high ground. The behavior of the Germans soon gave them that justification. The arguments put forward were: One cannot cooperate with the Germans because they are generally to blame for the war and its reprisals. Without their scientific squads Germany could neither have begun nor fought the war. Germans are morally despicable and their culture is at a historical low point. International organizations of science had hitherto only been a "springboard for Germanism." The most influential and frequently cited evidence presented to legitimize the boycott of German science was the notorious appeal by the ninety-three German intellectuals 'To the Civilized World.' This appeal was as useful as a target as it was objectionable. So having the German scholarly community repent would only be disadvantageous. New hurdles had to be put in place. Whoever was willing to concede would have to distance himself emphatically and in every way (preferably in writing) from the *Appeal by the Ninety-three* (and refrain from uttering the slightest criticism of the opposing side).

Thus science too became embroiled in "a continuation of the war by other means." It was only an extension of what had been begun during the war. In February 1915 the Académie des Inscriptions et Belles Lettres and the Académie des Sciences in Paris expelled signers of the 'Appeal by the Ninety-three': von Wilamowitz-Moellendorf, Harnack, Dörpfeld, de Groot, K. Robert, von Baeyer, E. Fischer, F. Klein and Waldeyer-Hartz. The Chemical Society in England soon followed suit by expelling nine of its German members from among their ranks. That same year, the annual convention of the British Association of the Advancement of Science at Manchester passed a resolution to boycott German science and proposed to deflate German dominance in science by founding new international review journals.

The change in mood over Germany and German science could hardly have been more abrupt. *Right up to the outbreak of war* in 1914, "Germany was considered the 'land of science,' not just by Great Britain's relatively small 'scientific community,' but by large swathes of the British public as well."[199] Scientific life in Germany was lauded in most exalted terms. *After the war,* however, ignorance and contempt were supposed to wipe away what had not long ago been an entirely different attitude. The heftiness of the boycott was thus also an expression of a guilty conscience.

The change in attitude among the French was not *quite* as drastic as among the British. Franco-German history had cast a different mold. Their relationship with German scientists was less reverential and more matter-of-factual and restrained. France was the traditional enemy of the Germans, no less than Germany was for the French. Nonetheless, World War I marked the end of former exchanges between German and French scientists. On 3 September 1915 the president of the French academy, Piccard, declared, henceforth "personal relations between scholars of the two belligerent parties was impossible."[200] Then the French prime minister, Clemenceau, read out before the French senate the "manifesto of the so-called intellectuals" and described it as "the greatest crime of Germany; a worse crime than all the deeds we know of."[201]

Evidently, German intellectuals were not any more belligerent than their opponents. They just happened to be the losers.

At the conference of the Union of Academies in London, from 9 to 11 October 1918, an *official boycott* of German science was passed by the Entente states. The vote was presided over by the British statesman, Balfour. At the conference from 26 to 29 November 1918 the Union of Academies decided a plan of action for excluding German science and at the convention from 18 to 28 July 1919 in Brussels the statutes of the International Research Council (*Conseil international des recherches*) were ratified. Another important player in the boycott of German science beside the International Research Council, which was a union of scientists, was the Union académique internationale, a union of scholars of the liberal arts. Assuming that this would be the deathblow to German science, invitations to cooperate were issued to *neutral* countries as well.

As was later revealed, the "mollifiers" were thereby welcomed among their ranks. Indeed, many, such as Einstein's friend Hendrik Antoon Lorentz from the Netherlands, understood their entry into the International Research Council and the Union at the outset as an initial step toward eventual inclusion of the Germans. The United States of America pursued their own interests and never were involved in the boycott of Germany. The same applied to Japan. As more powerful rivals and heirs of Europe, they were anyway not vulnerable to arm-twisting.

Articles 24 and 282 of the treaty of Versailles finally carved the boycott of German science in stone, as the official common policy among western European nations.[202] Although it is not explicitly referred to as a *boycott of German science,* this goal is nevertheless clearly articulated. With the exception of the agreements on the *International Committee of Weights and Measures* and the *International Institute of Agriculture,*[203] the victors at Versailles annulled all official treaties between Germany and other states concerning the fostering of international scientific relations.

The boycott against German science was primarily expressed in the following forms:

1. Deletion of German scholars from membership rolls in international organizations, notably presidents.
2. Refoundings elsewhere of central bureaus then located in Germany; founding of scientific unions.[204]
3. Cessation of international collaborative efforts in which German involvement was indispensable, e.g., the international chronometric service.[205]
4. Campaigns against internationally reputable scientific journals published in Germany and the creation of substitutes; omission of German publications in international bibliographies.
5. Exclusion of Germany from international scientific conferences.

From among 106 conferences organized in Allied countries between 1922 to 1924, Germany was excluded eighty-six times,[206] that is, 81 percent of the time. If you also take the conferences into account that were held in Germany, Austria and the neutral countries, hence a total of 135, the rate of Germany's exclusions comes to 66 percent. This percentage was reduced by 1925 to forty-seven (34 out of 72 conferences = 34 %) and by 1929 to seventeen (17 out of 99 conferences = 17 %). The effect of the boycott of German science by the Entente powers was that the number of bureaus of international associations located in France rose from 18 to 37, in Belgium from 13 to 21, in England from 9 to 14.[207] The organizers were therefore also the profiteers of the boycott.

To find out the real scope of the boycott, the German Foreign Office solicited reports from its foreign embassies on 14 June 1921. The reports that came in confirm, on one hand, *that* boycott measures were being taken. On the other hand, they demonstrate that it was not a *global* boycott. So some German complaints turned out to be stage thunder.

*One* foreign institution did in fact resist the boycott from the very beginning: The Nobel award committee. In doing so it even occasionally acted against the principles of neutrality.[208] How else do we explain the award of only three of the four outstanding Nobel prizes in 1919/20, all three of them destined for *German* scholars: Planck, Stark and Haber? The award to Fritz Haber for his invention of poison gas was received with dismay in other countries as well as among the Swedish public. Some despised the choice as downright provocative. How do we interpret the abstention by the Swedish Academy to award two Nobel prizes in literature otherwise than as a "breach of neutrality"?[209] This "tribute to German science," as Ambassador Nadolny put it in his report about the award of the prizes for 1919 on 1 June 1920, annoyed France and other countries with good reason.

Within Germany there was general consensus on the strategy to follow to alleviate the plight of German science.[210] But the question of *how* to address the boycott of German science was the subject of heated debate. A major proportion of the German "intelligentia" wanted to respond in kind and pressed for the severance of any remaining international ties. In the view of this group, those who wanted to make use of remaining or restored avenues toward a return to closer scientific exchanges with other countries were "traitors to the German front."[211] They were people suffering from "a deficiency of patriotic reserve"[212] and completely "unappreciative, apathetic and mentally lethargic";[213] they were "obedient lackeys" of the French government.[214]

With reference to membership in international organizations, including the Committee on Intellectual Cooperation of the League of Nations, Georg Karo wrote: "Everyone who keeps a place in his heart for German culture should make sure that no naïve, credulous and unreliable persons take those seats and thus help French cultural hegemony along to victory. For each and every German scholar has a part to play in protecting German honor and dignity. Such a noble office obligates!"[215] The secretary of the Academy of Sciences, Roethe thought: "Germans have every reason to wait staunchly in a 'splendid isolation' not of their own choosing."[216] The Reich Center for Scientific Reporting issued the following assessment in 1925: "The stance generally held here is that a boycott has to be answered with a counterboycott."[217]

Not everyone shared this view by Kerkhof, the head of the Reich Center for Scientific Reporting. Minister of Culture Becker, the directors of most of the academy institutes as well as numerous politically liberal or leftist scientists, writers and artists took pains to use every opportunity that presented itself to help restore normal relations with intellectuals in other countries – provided there was no loss of face. It was not a matter of confrontation but of patience, before the gap that had formed would eventually be bridged again.

The so-called "subtler means" that Bethmann Hollweg had suggested as a policy as early as 1913 were deemed of some importance. It is reflected in the minutes of a meeting on 17 June 1920 between Minister of State Dernburg, Minister of State Schmidt-Ott, Privy Councillor Krüss and the director of the America Institute, K. O. Berling.

One of Dernburg's points was that the America Institute offered "a welcome opportunity for resuming some contacts with influential groups in the United States and for seeking out new ones as well within a politically inoffensive context. From such *apolitical relations* one could ultimately anticipate quite substantial profit. A thriving resumption of cooperation now would be advantageous not only in intellectual respects but in economic respects overall as well."[218]

*"Apolitical"* ... "advantageous in economic respects..." A closer look at the biography of *Schmidt-Ott,* the most experienced and deft politician among these consultants, would be worthwhile.

**Schmidt-Ott, Friedrich** (4 Jun. 1860–28 Apr. 1956). Schmidt-Ott was as cultivated a man as he was a staunch conservative. He was a school friend of Wilhelm II in Kassel. He also had personal ties to the head of the Imperial Cabinet of Civilian Affairs, von Valentini ("my friend the head of cabinet, von Valentini"),[219] to Minister of Culture von Trott zu Solz (whom he succeeded) and General Ludendorff. 1911–1917 department head at the Ministry of Culture. 6 Aug. 1917 appointed minister of state and minister of intellectual and educational affairs ("I did not think I might side-step the kaiser's expressed wish").[220] When he was forced to resign during the revolution in November 1918, he called his coworkers together to deliver his "farewell speech, to which I declined to invite my successor."[221] After the revolution, Schmidt-Ott "voluntarily took part in the nightly patrols, with steel helmet and gun at the ready."[222] His sons Eduard and Albrecht were involved in the Kapp *Putsch* as volunteers of a *Freicorps* brigade. "Nothing, unfortunately, came of this [attempt] to found a national government, despite Ludendorff's involvement."[223] His heavy work schedule did not prevent Schmidt-Ott from visiting Kaiser Wilhelm II in exile from 4 to 30 June 1921. Throughout his entire stay at Doorn, he wore "only my field-gray captain's uniform, [...] together with the Iron Cross on its white ribbon and the Commander's Cross of the Royal Hohenzollern Order hanging from my neck."[224] 1920–1934 President of the Emergency Association of German Science (*Notgemeinschaft*). Schmidt-Ott was an influential representative of German science policy right up to the Nazi era. Despite his conservative, indeed reactionary, political views, he was not a fanatic. He was a true *Realpolitiker,* who did not shrink from fostering scientific relations even with the Soviets during the 1920s. 1920 Schmidt-Ott became president of the Society for Eastern European Studies (founded 1913), the same year as his appointment

> to the Emergency Association of German Science. On 1 November 1920 he assumed the chairmanship of the supervisory board of the Dye Factories F. Bayer & Co. (which later joined the consortium of major dye manufacturers in 1925 to form IG Farbenindustrie AG). "I consider it not only a privilege to have been on the supervisory board of this enormous enterprise until 1945 [...]; it afforded me personal contacts in industry that were of utmost value to me in carrying out the scientific responsibilities entrusted to me."[225] Schmidt-Ott's autobiography contains many details. He had a good memory until very old age. But he does not mention that IG Farben produced the poison gas for the Nazis' extermination camps, nor that it was brought before the war-crimes tribunal at Nuremberg for that reason. It is also strange that the name *Albert Einstein* does not appear anywhere in his autobiography, even though he was involved in many aspects of Einstein's personal affairs. His direct dealings with Einstein as recording clerk of the Leopold Koppel Foundation, as minister of culture and as presiding member of the board of trustees of the Kaiser Wilhelm Institute of Physics were numerous. So this lack of mention in his memoirs was certainly not a matter of chance. Perhaps at the end of the day Schmidt-Ott regretted having once helped this leftist and Jew so very much in the past. On the other hand, it was risky to publish any critical reminiscences about Einstein after World War II.[226]

In a letter discussing ways to restore relations between Germany and America and the associated financing, the America Institute suggested to His Excellency, Legation Councillor Schmidt-Elskop on 18 August 1921: "According to our experience here, collections designed to inform adverse elements in America only have any prospect of success if, from the outset, the sponsoring organizations are *beyond suspicion of any pro-German bias.*"[227]

In short, the alternative to Kerkhof's method was wishing an end to the boycott of German science without making any vociferous demands. Fewer manifestos and appeals, but for it, more concrete efforts toward drawing nations closer together again. The preferred approach was through lectures by prominent and politically unobjectionable researchers about problems in science and the results of their research. Unassuming, patient and modest appearances that do not reveal the true underlying motive. They had to be as "apolitical" as possible and completely devoid of patriotic flare. Much fanfare for Goethe – with profit in mind (best is when the praise is sincerely directed at *Goethe,* in ignorance of its service to the *national* cause.)

And who else was better suited to the task than *Albert Einstein?* He was politically unobjectionable, to say the least. Indeed, he was a professed opponent of war at a time when, for others, war had defined every action. Unassuming and unconventional as he was in appearance, and somewhat other-worldly besides. *He* did not have to put on an act. *He* was no German nationalist. He was honestly convinced of the necessity for peaceful collaboration among scientists and among their native countries. This coincided with German interests perfectly. Added to that, this was *Einstein:* one of the best – if not the very best – that the world of science had ever produced. How better to break through the boycott?

This was precisely why Einstein was needed so much. This was also precisely why he was attacked and hated and abused so much as well.

In a memorandum dated 11 January 1923 demanding that Germany withdraw from the *International Committee of Weights and Measures* and the *International Institute of Agriculture,* its author Kerkhof argued:

[. . . ]. I would not like to omit to comment, however [. . . ], that in countries of the Entente there are many scholars who in the name of equality honestly wish that scientific relations with Germany be restored [. . . ]. But the case of *Einstein* demonstrates best that caution is advisable here too.[228] At the *invitation* of French friends he went to Paris and his trip is now being quite cleverly exploited by French propaganda against Germany.

On this basis it would only serve the dignity and interests of German science to decline any involvement in the above-mentioned international congresses and [. . . ] to announce Germany's withdrawal from the conventions [. . . ].

In a document dated January 1924 Kerkhof mentions no names but it is still obvious *whom* he means:

Because peaceful international scholarly exchanges [. . . ] would have foiled French plans for war, extermination and hegemony, the French government has been conducting the *boycott of German science* since 1918; and through its obedient lackeys has tried to lure not only scholars mainly from the Allied and neutral countries into supporting the boycott but once even a very highly regarded, *more or less German scholar* (as ''separatists'').[229]

Who else could fit the description of a "more or less German scholar"? Einstein the "obedient lackey" of the French government, the separatist and traitor. Einstein, as a foreign agent and a supporter of the boycott of German science.

The man who was later to substitute for him on the Committee on Intellectual Cooperation of the League of Nations and ultimately become his successor was ... Hugo Andres Krüss.

The Foreign Office invited a select few gentlemen with "lengthy experience in relations with foreign scholars" (Einstein was not among them) to a meeting on 6 February 1925 in order to discuss how to address the boycott of German science in a manner that would assure "that German science maintain a single, unified front."[230] At this meeting Krüss (at that time still a department head at the Prussian Ministry of Science) stated:

Einstein, who himself constantly emphasizes his supranationality, could not be a valid representative of the German nation at the League of Nations Committee on Intellectual Cooperation. Considering that, under the circumstances, it is out of the question to reject an invitation to Germany to participate at the Parisian Institute,[231] we should now focus on choosing really very good competency for election to this post with the backing of the whole of German science. Abstention on this issue would be worse, because the other side would never have difficulty finding willing outsiders as representatives of Germany, who would then lack the crucial characteristics they ought to have in guarding the interests of German science.

The message was crystal clear: Einstein was *not a valid representative of Germany, neither fully competent nor fully backed by the German scientific community, and an outsider too compliant toward the French!*

**Hugo Krüss** (11 Jan. 1879–28 Apr. 1945). Studies in physics and doctorate by 1903.

| | |
|---|---|
| 1907 | appointed assistant at the Prussian Ministry of Culture. 1909 professorship. On 10 Mar. 1918 appointed privy councillor and rapporteur at the Ministry of Culture. |
| 1920 | department manager at the ministry. On 1 Oct. 1920 Krüss was appointed by Minister Becker (decree of 23 Sep. 1920) as successor to Naumann to head the University Division (Division U I) at the ministry of Culture. 1922 department head at the ministry. |
| 1925 | managing director of the Prussian State Library (after the departure of Harnack's successor Milkau). |
| 1929–1930 | chairman of the League of Nations Subcommittee on General Scientific Affairs and member of the International Library Conference at the International Institute of Intellectual Cooperation in Paris (1931 elected for five years). |
| 1936 | chairman of the Reich Advisory Board for Library Affairs. |
| 1940 | member of the Nazi party (application for membership on 26 Mar. 1940, approval already on 1 Apr. 1940). |

We shall meet this Mr. Krüss many more times in the role of Einstein's adversary – a very restrained but determined one. This was the man who – after a longer period of mutually beneficial collaboration – eventually drove Einstein off the Committee as representative of Germany at the International Institute of Intellectual Cooperation. So *he* was the "really very good competency" he had been thinking of. When the Nazis rose to power he remained a loyal supporter. Antonina Vallentin reported: When the German Army marched into Paris, Dr. Krüss was there with them. He demanded the surrender of the library of Strasbourg and threatened in the event of a refusal "to take compensation from the Bibliothèque nationale."[232] *Other* sources do not corroborate this version *exactly,* but they do document that Krüss was appointed "Reich commissioner for the protection of libraries and the supervision of book supplies in the western area of operation" and in this capacity had close ties to Reich Minister Goebbels, Reich Leader Alfred Rosenberg and the ambassador in Paris, Otto Abetz. Krüss certainly was not the initiator of the confiscations of foreign property but a meticulous and diligent aid. His notes on conversations with Rosenberg (7 August 1940) and Goebbels (17 September 1940)[233] do not provide the slightest hint that he might have had qualms about the confiscations of "libraries of the Freemasons and [of] certain Jewish and church libraries," not to mention the seizures of "certain private libraries of undesirable persons" or "library holdings desirable for the national cause." It was not Goebbels, *but Krüss* who drew into the discussion the idea that "Freemasons, Jews and emigrés must be regarded as antagonists of Germany and that their intellectual weapons should be treated accordingly as with other implements of war." Was there any room left for a sense of justice and conscience in the otherwise so fussy Hugo Andres Krüss? He *only* "prepared a survey of the existing valuable artefacts of German culture in the enemy states" in order to "facilitate the Führer's decisions on eventual future reparations claims on the basis of these surveys."

Krüss, "a dutiful, 'correct' civil servant bound to his oath of office," had conformed "rapidly, both terminologically and in substance," and rigorously satisfied "the directives required of him in the areas of library business and staffing," which incidentally also involved "the removal of politically disagreeable and most of the Jewish personnel (1933–35)."[234]

Similar things could be said about Karl Kerkhof's later walk of life. *His* response to Hitler's victory over France reveals his fury at French cultural policy and the boycott against German science. In compliance with a request by the Foreign Office from 5 July 1940 ("in acknowledgment of your exceptional experience"), Kerkhof submitted a memorandum dated 22 July 1940 on reforming Franco-German intellectual relations and international cultural exchanges in general.[235] It settled his old scores with "French cultural imperialism" and demanded "a reorganization under *Germany's leadership* in the fields of the sciences."

Kerkhof's fury at French vengeance after World War I was merely an expression of his annoyance that Germany had *lost* the war. Kerkhof was anything but a champion of justice and international intellectual cooperation. The boycott of German science by the French provided a convenient opportunity to stir up hatred and call for revenge.

Kerkhof and his mind-mates obstructed reconciliation with the boycotters of German science, but they could not prevent it. By 1925 the boycott had fallen to pieces, and on 10 September 1926 Germany finally became a member of the League of Nations.

This was partly the work of *Albert Einstein*. His important and lasting contribution toward international cooperation and peace revolved around his *merit as a scientist.*

## 2.2 World renown

Toward the end of World War I, the Royal Society of London equipped two scientific expeditions to observe the solar eclipse on 29 May 1919 in Brazil and West Africa. Their purpose was to test predictions of Albert Einstein's theory of relativity.

On 17 October 1919 Einstein was able to submit the following preliminary notice to the scientific journal *Naturwissenschaften:* "According to a telegram sent by Professor Lorentz to the undersigned, the English expedition under Eddington to observe the solar eclipse on 29 May has observed the deflection of light at the limb of the solar disk required by the general theory of relativity." On 23 October he wrote to Max Planck: "This evening at the colloquium Hertzsprung showed me a letter by Eddington, according to which precise measurement of the plates has furnished the exact theoretical value for the light deflection. It was gracious destiny that I was allowed to witness this."[236]

The official announcement of the expedition results was made on 6 November 1919 at a joint meeting of the Royal Society and the Royal Astronomical Society under the chairmanship of Joseph John Thomson. Einstein's prediction that starlight passing by the sun, i.e., light traveling in the vicinity of large masses, is deflected, was acknowledged as correct. The expedition results were considered convincing proof of the validity of the theory of relativity.

A few days later, relativity theory had become a political issue. Its creator was already a well-known figure within his field. But now the whole world heard of him. Einstein became the topic of the day. Everyone wanted to see the greatest living scholar. He was depicted on the front page of major newspapers; his theory soon became the subject of a film.

Never before (nor afterwards) in human history did a scientific theory grab the interest of the public at home and abroad to such a degree. The print media and soon also radio broadcasting became powerful amplifiers of Einstein's fame. "Anyone who could read, knew 'something' about relativity and modern atomic physics during the 1920s."[237] Interest in the theory was a mass phenomenon, in the truest sense of the word. It was so intense that the twenties could well be described as a "relativistic decade." Even after the initial excitement in the media about lectures and publications on relativity theory had died down, it remained an inspiration to the muses: it reappeared, for instance, in Thomas Mann's 'The Magic Mountain' and Robert Musil's novel 'The Man without Qualities' as well as in Josef Goebbels's novel 'Michael' published in 1929.[238]

The physicist Ernst Gehrcke studied the question of relativity theory as "mass suggestion" in great depth. Although his arguments leave many open questions, he collected "historical documents on cultural psychology" with unique thoroughness.[239] His collection contained over 5,000 newspaper clippings. His publication from 1924 cites 309 periodicals that covered Einstein and relativity theory, 152 of them were from Germany and 71 from France. The most frequently quoted newspapers are *Vossische Zeitung* and *Berliner Tageblatt.* His reproductions of documents from the period *before* the Great War are also valuable. With occasional reference to statements made by Einstein, Gehrcke produced proof that the theory of relativity certainly did have a hypnotic effect on the public at large, even though most people did not understand it. (Gehrcke failed to offer a reason for its attractiveness.) Gehrcke was an outright opponent of the theory, appeared in the disreputable company of the anti-Semite Weyland and was *personally* rebuked by Einstein. Nonetheless we have to concede him this much: he took great pains to keep to *the substance* of the issues. He let *the facts* (documents) speak for themselves. Gehrcke was apparently fascinated by the idea that "a mathematical theory [...], hence a purely academic matter far removed from practical application, could find such broad coverage in the daily press." His work is, itself, a "historical document on cultural psychology."

When Albert Einstein traveled to Paris at the end of March 1922, he was "the most popular personality in Paris" – thoroughly "à la mode." Academics were not the only ones wanting to know when and where Einstein was scheduled to deliver his lectures: politicians, artists, the petty bourgeois, security personnel, cab drivers, waiters, pick-pockets ... Even the "cocottes at the Café de Paris grill their escorts about whether Einstein wears glasses or is a chic sort of guy." While touring some war-torn regions, he was recognized by a couple of French officers ("in full uniform") in the company of a very dignified lady. "All three of them rose from their seats at table to bow respectfully to the great physicist."

That year in Tokyo during the annual chrysanthemum festival on 22 November, a traditional celebration of the imperial family in communion with its people, the focus of attention was not on the empress nor the prince regent nor the imperial princes, but Albert Einstein. The audience of some 3,000 forgot the importance of the day. "All eyes were on Einstein; everyone wanted at least to have shaken the hand of the most famous man of the present day. One admiral in full uniform pushed his way through the crowds only to say to Einstein 'I admire you!' and retreat again."

Einstein was worshiped like a new messiah.

How do we explain this rare (if not "unique") phenomenon in the history of science and of civilization? Why did Einstein become a "media star" – one of the first – within a matter of days? Why did the press and radio broadcast Einstein's renown around the world and reach every social stratum? Why the applause of millions, who could not really understand the substance of his theory?

First of all: In the aftermath of a lost war and somewhere between revolution and counterrevolution, many Germans were swept up in a roller coaster of powerful emotions. All former values were being questioned and doubted. Many saw the theory of relativity as a confirmation of their experiences and premonitions. As Einstein himself put it: "It is a peculiar irony that many people believe that the theory of relativity offers a support for the anti-rational tendency of our times."[240] They were looking for something new to hang on to. If there was nothing left to believe in, there was at least the consolation of Albert Einstein's glamorous fame and "scientific" confirmation that all values were relative.

Second: It was a sheer stroke of luck for vanquished and boycotted Germany to have a *native son* be the originator of a theory that revolutionized our physical conceptions of the world. Max Planck already knew long ago that the theory's "scope and breadth is probably only comparable to the one defined by the introduction of the Copernican system of the universe."[241] It did not take politicians long to realize this. So the liberal newspaper *Volksblatt in Spandau* had every reason to cheer, in its issue dated 5 February 1920: "[...] even if pan-German professors dragged Germany's reputation and its scientific renown down more than a little, Einstein raised it up again, filling the outside world with so much German learnedness that one could describe the inventor of the relativity principle, which revolutionized technology and physics, as a *financial share in the German valuta*." Einstein had become a *German hero*. The recent humiliating defeat was hardly behind them; Germans could stand tall again. The losers could come out victors of the battle of battles. Hail Einstein, hail German supremacy of the world.

Third: The German Jewry also had plenty of reason to celebrate and make the most of Einstein's accomplishment. The greatest scholar of all time was *one of their own*. Jews, and especially German Jews, had suffered much recently. Even those proud of being "Germans" and imbued with a patriotic spirit to do "more than the call of duty for the fatherland,"[242] had to submit when the Ministry of War ordered a "Jew census" on 11 October 1916 (in reaction to rumors that Jews were evading their compulsory military service). The sheer humiliation of being relegated among a stigmatized race! Thanks to Einstein, that could soon be brushed aside. They could hold their heads up high again. Was it so objectionable for them also to join the proud chorus and put Einstein's name on a pedestal?

But why did Einstein's fame not stop at the German borders? Why so much fanfare and reverence even in "enemy" countries and by millions of people who could not assess the scientific merit of his theories? Why so much honor for a person who, despite all his side-stepping ("I am Swiss, not a German"; or "not a typical German") had to be respected as a "*German* scholar"?

World War I was a turning point not just for Germany. Many countries in the world were in a state of ferment. There was poverty, striking and situations resembling civil war even among victors of the World War. There were the first signs of the desintegration of the colonial empires. The October Revolution in Russia and the Bolshevik victory over the intervening French and British forces were a warning beacon for many. The collapse of prewar values was not restricted to Germany. Elsewhere, too, Einstein's theory seemed to be a confirmation that the wheel of history was continuing to turn. What was "at the bottom" did not have to stay "at the bottom." Elsewhere too, people thought they had found in Einstein and relativity theory something to believe in during such unsettled times.

A second reason for the world-wide admiration was that Einstein certainly was not a "typical German." He distanced himself from the war while millions of others, not only Germans, rushed to the battle field to exterminate their "hereditary foes." Not being a "typical German," it was safe to claim allegiance with Einstein. Moreover, he was a "*better* German" – so there was every reason to do so with enthusiasm. His theory seemed to confirm the grim experiences of the war. Einstein's views of world politics agreed with the deep yearning for peace shared by many people throughout the world. The response would have been very different if Einstein, the great scholar, had been a warrior (he would inevitably have been demoted to a "supposed" great scholar). Foreign admiration of Einstein was thus also a demonstration of disdain for the "other," "old" Germany.

A third reason for Einstein's international image is an extension of his importance to German Jews. Jews around the world were proud to put forward one from among their own ranks as the greatest scholar of mankind (and also capitalize on it).

Much as there is to say about the political, social and psychological reasons for Einstein's fame: His *scientific* accomplishments remained the basis of this fame.

**Figure 12: Title page of the *Berliner Illustrirte Zeitung* dated 14 December 1919** ➠
**Commentary on the following page of the newspaper:**
***A New Authority in the History of the World*** **The year 1543 marks the beginning of a new period in the way we think of and view the world, indeed it is a new period in the development of mankind in general. The publication of a simple book was what started it all, the work by Nicolaus Copernicus: 'On the Revolutions of the Heavenly Spheres' ('De revolutionibus ordium coelestium'). This book, written by the introspective thirty-three-year old canon of Frauenburg had a revolutionary effect. Many a leading contemporary fought assiduously against it. Even the likes of Luther wrote: "This fool Copernicus wants to turn the whole of astronomiam upside down." Even so, the Earth was dethroned from its place in the universe; man was no longer at the center point of creation. What Copernicus had set in motion, Kepler and Galileo built upon, and Newton completed. The enormous celestial bodies obey the laws this English natural philosopher laid down; tiny molecules follow their orbits according to them, the diameters of which often do not reach the size of a millionth of a millimeter. For over two centuries Newton's laws were used to gain mastery over almost all the natural processes in the universe. But as the inquiries delved ever deeper into electrical processes, to develop what is called electro-mechanics, and progressed ever further into the area of radioactive phenomena, contradictions arose about the consequences one had to draw from Newton's theories. In vain did such illustrious physicists as Heinrich Hertz, the conceptual father of wireless telegraphy and telephony, seek to eliminate the contradictions in order to maintain the apparently so stabily built framework of Newton's classical mechanics. Other physicists also tried their acumen and looked around for saving experiments, but the contradictions only multiplied. Then a relatively youthful scholar, Professor Albert Einstein, the youngest member of the Prussian Academy of Sciences, came boldly and resolutely forward – he is 41 years old today – to test our current conceptions of space and time. With penetrating reason, with the subtlest of tools from higher mathematics, he tackled our ideas about motion, he analysed the concepts of mass, inertia and gravity. By showing that we cannot determine any absolute motion, that we are compelled to apply our concept of time in all our physical measurements, he eliminated the gaping contradictions between electrodynamics and Newtonian mechanics. But his theory of relativity, the result of years of profound contemplation, yielded much more. It showed astronomers how to calculate hitherto unexplained deviations in Mercury's orbit. It predicted that under certain conditions a ray of light would have to be deflected away from its straight course in the direction of gravity. The solar eclipse on 29 May 1919 became the touchstone of Einstein's conceptions; and the English scientists, who were studying their consequences, had to concede that Einstein had bested Newton. Newton's laws of gravity are just a special case of Einstein's theory of relativity. Our conceptions of space and time have to be modified on the basis of Einstein's theories; and just as in Copernican times, a change in the picture we have of the world is taking placing. A new epoch in the history of mankind is dawning and it is inseparably linked with the name Albert Einstein**

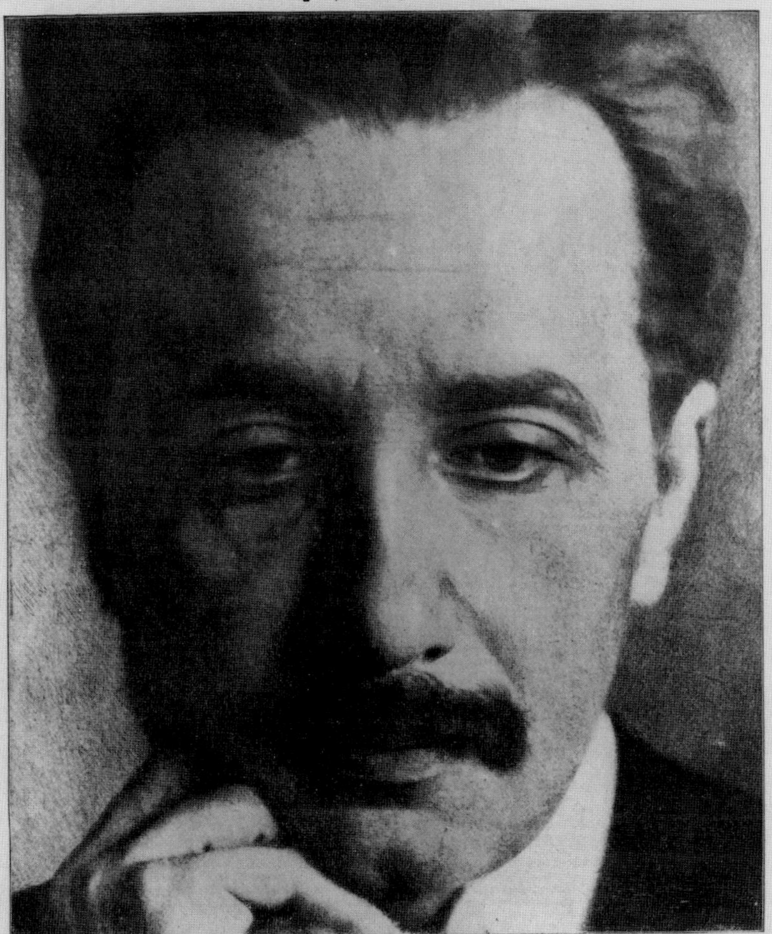
14. Dezember 1919
Nr. 50
28. Jahrgang

Berliner Illustrirte Zeitung

Einzelpreis des Heftes 25 Pfg.

Verlag Ullstein & Co, Berlin SW 68

Eine neue Größe der Weltgeschichte: Albert Einstein, dessen Forschungen eine völlige Umwälzung unserer Naturbetrachtung bedeuten und den Erkenntnissen eines Kopernikus, Kepler und Newton gleichwertig sind.

Phot. Suse Byk.

**Figure 13: "Der Einstein Film," 1922. Excerpt from the film program:**
**The foundations of Einstein's Theory of Relativity.**
**Film adaptation by Professor G.F. Nicolai and Hanns Walter Kornblum in collaboration with Dr. Buek (Berlin)/Professor Otto Fanta (Prague) and Dr. Rudolf Laemmer (Zurich). Explanatory script: Hanns Walter Kornblum**
**Film producer: Colonna-Filmgesellschaft**
**Introduction**
**Virtually never before has a hypothesis from an abstract science agitated the passions of its contemporaries as much as Einstein's theory of relativity. [...]. How can this interest by the general public [...] be explained? Surely because Einstein has – whether consciously or not – shaken the very foundations of our habitual old views. He has upset and stirred us all so profoundly that this little speck called Earth, which had hitherto seemed so firmly rooted, seems suddenly insecure [...]. The unease caused by this momentous event in world history prompts us to want to be able to judge for ourselves, to know exactly what this revolutionary idea involves and to try to regain a firm foothold. [...]**

## 2.3 Sponsorship – the Einstein Tower

### 2.3.1 The appeal for the "Einstein Donation Fund"

Einstein was a coveted personality: The universities at Zurich, Leyden and elsewhere had already extended invitations to entice him away. Germany had to do something if it wanted to maintain its leadership in physical research. It was warning enough that a *British* expedition had delivered the experimental proof of a theory conceived by a *German*. Einstein *had* to be retained.

So it was not coincidental that a short time after 6 November 1919 a proposal was submitted to the State Budgetary Committee of the Prussian Constituent Assembly regarding the budget of the Ministry of Science, Arts and Culture for the financial year 1919 aimed at promoting Einstein's research.[243] Not long before that fateful day of 6 November 1919, a memorandum submitted by Gustav Müller and Erwin Freundlich in 1918 requesting a grant of 30–50,000 marks had got inextricably caught up in bureaucratic red tape.[244]

Prussian Constituent Assembly.

S t a t e B u d g e t a r y C o m m i t t e e
P r o p o s a l № 1 5 1

in re
Budget of the Ministry of Science, Arts and Culture
for the Financial Year 1919

The Committee would resolve:

to petition the State Government, in agreement with the Reich Government, to release the necessary funds for Germany's continued collaboration with other nations toward developing Albert Einstein's fundamental discoveries and toward furthering his own research.

Berlin, the 26th of November 1919

Dr. Schlossmann, Dr. Friedberg, D. Rade, Otto, Dr. Thaer
Dr. Fassbaender, Gottwald, Dr. Hoetzsch, Frau Dr. Wegscheider
Hennig, Lüdemann, König (Frankfurt), Dr. Weyl

The signers of this proposal represented a broad political spectrum. Their party memberships ranged from the conservative German National Peoples' Party to the Independent Socialists.[245]

This document bears sheet number 1 of the file *Einstein's Theory of Relativity* compiled by the Prussian Ministry of Science, Arts and Culture. It was soon realized that the ministry would be spending much time on this process. The official responsible for this matter was the previously mentioned Prof. *Hugo Andres*

*Krüss* – initially as a regular employee at the ministry and from 1920 as department manager (responsible for universities, polytechnics and the promotion of science).

Einstein was pleased to hear about the resolution from 26 November. While expressing his thanks he also added his own reservations about passing such a resolution in a time of "extreme need," because it could prompt discontent among the people. He even gave suggestions about how to promote research on the theory of relativity with much more modest means. It would suffice, Einstein thought, "if the country's observatories and astronomers would simply place a portion of their apparatus and labor at the service of this cause." Einstein also used this opportunity to recommend Dr. E. Freundlich at the Astrophysical Institute in Potsdam as "the only German astronomer (aside from Schwarzschild) to advance the field [...]" and to suggest that "in accordance with a proposal by Director Müller, this astronomer [...] receive an observer position very soon at the Potsdam institute with the objective of working on testing the general theory of relativity."

Einstein to Minister Haenisch, 6 December 1919:[246]

Prof. A. Einstein

W 30, the 6th of Dec. 19.
Haberlandstr. 5

To
the Minister of Science, Arts and Culture
W 8, Unter den Linden 4

Highly esteemed Minister,

A day or so ago I received a report by the State Budgetary Committee of the Prussian Constituent Assembly, according to which it is planned that 150,000 marks from the State Treasury be placed at my disposal in support of research in the area of the general theory of relativity. However great my joy and feeling of gratitude for this truly generous gesture, I nonetheless cannot withhold painful reservations. In these times of extreme need, would such a resolution not justifiably trigger bitter feelings among the public? I believe that we can promote research effectively in the area of the general theory of relativity even without use of special public funds, if the country's observatories and astronomers would simply place a portion of their apparatus and labor at the service of this cause.

Hitherto Dr. E. Freundlich at the Astrophysical Institute in Potsdam was the only German astronomer (aside from Schwarzschild) to advance the field. It would be of great serve to the cause if, in accordance with a proposal by Director Müller, this astronomer were to receive an observer position very soon at the Potsdam institute with the objective of working on testing the general theory of relativity.

Finally, I would like to mention a minor but nonetheless urgent request that regards my own working conditions and for which I earnestly petition immediate resolution. By arrangement of the Housing Department of the town of Schöneberg, a room in our building was granted to me to accommodate my terminally ill mother and a nurse. The building owner objects and is making all sorts of misrepresentations to the authorities to postpone or obstruct handing over the room, so that I am, as a matter of fact in the predicament of having to make room in my study for the ill person if the affair is not speedily decided in a favorable direction. A word by the Ministry of Culture directly to the *Rental Settlement Board* of the town of Schöneberg (Einstein-Eisfelder case)[247] would bring to a prompt decision this matter of such importance and urgency to me. I entreat you to forward the requested communication directly to the authority mentioned, since I am being deluged with telegrams by the patient's doctor reporting that the transfer of the patient is becoming more and more difficult by the day.

In utmost respect,

A. Einstein

Einstein could not do anything about the proposal to the State Budgetary Committee, of course. By the time he expressed his reservations the petition had already been submitted and the consultations by the Prussian Constituent Assembly on the budget of the Prussian Ministry of Science, Arts and Culture for 1920 were already underway, having started on 3 December 1919. On 12 December the petition was "approved [...] without a separate vote."

From the minutes of the 94th meeting of the Prussian Constituent Assembly, 10 December 1919:

Dr. Schlossmann (D[emocratic] Party): [. . . ] among the gems entrusted to the care of the Minister of Arts, Science and Culture, *our academies, universities and polytechnics* are at the very top and this ranking is true now more than ever. For, today we are living in chaotic circumstances; so much of what we had once looked up to has collapsed, so many of our hopes, our dreams for the future are shattered. We are living in a time in which all values are undergoing new scrutiny [. . . ]. Only things considered of value abroad still mean something [. . . ] even here at home, and our universities and colleges are among the things still enjoying unconditional esteem even now.

Welcome proof of this fact is, of course, provided [. . . ] in that just recently *three German scholars, three Prussian scientists have been distinguished with the Nobel prize:* Messrs. Planck and Haber in Berlin and Professor Stark in Greifswald. [. . . ]. In addition to these three Nobel laureates, I mention a man who has elevated the reputation of German science abroad extraordinarily: Mr. *Albert Einstein, the man who pointed out new courses and new paths for our inquiry of Nature and who rightfully assumes a place next to the greatest minds of all time.* It is a major feat of our administrators of culture that we have succeeded in acquiring this man for our German science and thus for the University of Berlin while he was still working out the basic scheme of his fundamental ideas.

With this view in mind, the State Budgetary Committee has unanimously approved the petition:

To apply to the State Government: [. . . ].

Toward this end it will be particularly necessary that the astronomical instruments of our observatory be maintained at the highest level and continually improved upon. [. . . ].

These are shafts of light in our so somber times. These men *guide us to the genuine value of international ties; for true and lucid science knows of no border posts.* The community of all scientists is at the same time the community of humanity, and such a cosmopolitan point of view, a view that certainly finds room next to patriotism, we should like to safeguard. He who injures such relations, relations between German science and scientists of other countries, is thereby destroying the last lifeboats that may well be able to lead us into the range of a happy future. [. . . ].

Member of Parliament Dr. Schlossmann provided the arguments for the proposal to the State Budgetary Committee. His pathos and faulty knowledge led him to assert some untruths, for instance: Einstein would ("if I am not mistaken") be made honorary speaker at the inaugural celebration of the new academy building located on Unter den Linden (i.e., the "academy wing" of the State library.) The important element of Schlossmann's argumentation was that it captured the mentality and circumstances of defeated Germany. Unobtrusive cosmopolitanism and an appeal for the conciliatory role of science now served as the groundwork of original *national* interests. Schlossmann was certainly not a wolf in a sheepskin. It was not necessary for him to disguise his true feelings. His words reflected what he thought. Times had changed, consequently also the people who now had a say.

"Without a separate vote" – that means *all* parties were agreed on this point. Not a single representative objected to the proposal. The budget for the Ministry of Culture was otherwise hotly debated, bordering even on personal insult. The conflict between the "spirit of Weimar" and the "spirit of Potsdam" aroused passions. On *one* issue they were in agreement, however. As Representative Kloss from the German National People's Party put it, the new Germany needed "engaging, vigorous ideas."

One of these ideas was Einstein's. It certainly proved its worth as a national advertiser! This gift out of the blue during such hard times had to be made the most of.

Another reason for this unanimity across the political board may have been that in December 1919 Einstein was still a relatively unknown figure. The errors in Dr. Schlossmann's speech are indicative of this. A few months later, when more was known about the Jew and pacifist Einstein, consensus among the parties was virtually out of the question, when it was a matter of funding experimental tests and further research on the theory of relativity.

Einstein's letter to the minister of culture reveals that *even he* only found out about the plans to collect and make available major funds for experimental research to verify his theory *after the fact*.

It goes without saying that Freundlich obtained the observer position at the Astrophysical Observatory very soon afterwards (early 1920). In the margin of Einstein's letter there is a handwritten note by the official in charge – Dr. Krüss:

"The idea to promote Dr. Freundlich via the Astrophys. Obs. came from me after the repudiative attitude of P[rivy] C[ouncillor] Struve made his further presence at the Babelsberg Observatory impossible. P[rivy] C[ouncillor] Müller has now, most welcomely, taken an interest in the matter. I hope that it will be possible to make Fr. into an observer at Potsdam." The fact that earlier *Privy Councillor Müller* had also opposed Freundlich's hiring at the Potsdam Observatory ("Freundlich [...] is not in any way suited [...] for our observatory")[248] was suddenly forgotten. It was also forgotten that Einstein had long regarded it as "desirable" that "Mr. Freundlich be employed at the astrophysical institute" (even though, in Einstein's words, he was "quite considerably less talented than Schwarzschild").[249]

The Ministry of Culture did, incidentally, take the steps Einstein wished in the matter of Einstein vs. Eisfelder. Krüss submitted a document to the minister,[250] that petitioned the Housing Department for the Schöneberg district of Berlin "to kindly see to the handover of the room, if possible, to Prof. Einstein,"[251] in order to assure that "Prof. Einstein's research work in the area of mathematical physics, which is now currently in the focus of scientific and public interest" not have to suffer any "interruption that might reduce the scientific reputation of Germany" and (this in second place) because "Prof. Einstein's mother is in a life-threatening condition." The letter was sent off *"today"* and *"immediately"* on 12 December 1919.

It is highly probable that *Dr. Erwin Finlay Freundlich* was the *initiator* of the proposal to the State Budgetary Committee of the Prussian Constituent Assembly.[252] "During a discussion between Dr. Freundlich and Undersecretary Becker in November 1919, the plan first emerged of granting Mr. Einstein special funds to pursue this problem [...]" of testing the theory of relativity against experience and investigating other consequences of it within Germany. "The efforts of Undersecretary Becker succeeded in uniting all parties of the Prussian Assembly in the passage of a resolution (State Budgetary Committee Petition No. 151) granting a sum to Mr. Einstein for his researches.[253]

Freundlich was certainly also the instigator of a private fund-raising campaign for the benefit of experimental verification and "elaboration" of the theory of relativity: the "Einstein Donation Fund."[254]

Albert Einstein Donation Fund

Albert Einstein's work on the general theory of relativity is an important turning point in the development of science, comparable only to that connected with such names as Copernicus and Newton. Experimental verification of its observable consequences, to prove the applicability of the new theory, must go hand in hand with further elaboration of the theory. Only astronomy seems suited to take up this task at the present time. It thus faces a responsibility of enormous significance.

The academies in England, America and France have recently set up a commission, excluding Germany, to busily lay the experimental foundations of the general theory of relativity. Persons concerned about Germany's cultural standing are honor-bound to provide what funds they can to enable at least one German observatory to work on the theory in direct collaboration with its author. These funds are intended for procuring the observational equipment necessary for the Astrophysical Observatory, which is placing itself at the service of this cause, to work successfully on this problem.

Approximately 500,000.- marks are needed.

The Prussian Ministry of Culture has promised to support this endeavor insofar as the funds granted by the State Assembly allow.

We request that contributions be addressed to the financial institution, Mendelssohn & Co., Berlin W. 56, Jägerstrasse 49-50. Bank account: Albert Einstein Donation Fund.

Priv. Coun. Prof. G. Müller
Director of the Astrophys. Observatory, Potsdam

Priv. Coun. Prof. Fr. Haber
Director of the Kaiser Wilhelm Institute of Phys. Chemistry

Priv. Coun. Prof. H. Struve
Director of the Babelsberg Observatory, Berlin

His Exc., Prof. A. von Harnack
President of the Kaiser Wilhelm Society
for the Advancement of the Sciences

Priv. Coun. Prof. W. Nernst
Director of the Institute of Phys. Chemistry
at the University of Berlin

Priv. Coun. Prof. M. Planck
Professor Ord. of Theoret. Physics at the University of Berlin

Priv. Coun. Prof. H. Rubens
Director of the Institute of Physics at the University of Berlin

Priv. Coun. Prof. E. Warburg
President of the Reich Bureau of Standards

The justification for this funding-raising campaign is typical of the postwar period: "The academies in England, America and France have recently set up a commission, *excluding Germany,* busily to lay the experimental foundations of the general theory of relativity. *Persons concerned about Germany's cultural standing are honor-bound* to provide what funds they can to enable at least one German observatory to work on the theory in direct collaboration with its author." *Freundlich* knew which lever to pull to open the cash registers of the State and German industry. He purposefully beat the national drum to make Einstein into a *German.* (He did not fail to refer to his own accomplishment, by pointing out that the *first* tests of relativity theory "against experience [...] were first undertaken *in Germany*).[255]

It also later seemed appropriate to put down accomplishments made elsewhere – particularly those by the English. "The theory was posited exclusively by Einstein," Freundlich wrote on 5 April 1921, "but even the method toward its verification was worked out completely by us already in 1914 with a comprehensiveness [...] that makes the method used by the English two years ago look downright dilettantish." He continued by saying that the English approached the experimental verification of relativity theory "without the necessary expertise."[256] Freundlich felt free to discredit the results by the British observers even though he knew at very latest from an article Einstein had published in the *Berliner Tageblatt* of 27 August 1920 that Einstein himself was of the opposite opinion, referring there to the "masterfully conducted English measurements of the deflection of light rays by the Sun."

That is how Einstein the *internationalist* became Einstein the *nationalist,* despite his personal preferences.

It is interesting that the people behind the Einstein Donation Fund included all the instigators of his initial appointment to Berlin: *Planck, Nernst, Warburg,* and *Rubens.* The list is completed by the influential promoter of science in Germany Adolf von Harnack and the by no means less influential institute director, *Fritz Haber* (whose influence – the reason for his inclusion on the list – was not primarily due to his official function). The astronomers Müller and Struve were necessary embellishments, not owing to any particular personal inclination for the theory. Their institutions, the Astrophysical Observatory in Potsdam and the Babelsberg Observatory in Berlin, were simply the closest ones suitable for experimental verification of relativity theory.

This is where the political story of the Einstein Tower or the "Einstein Institute" at the Astrophysical Observatory in Potsdam starts. The appeal for the "Einstein Donation Fund" is, in fact, its founding document.

The tower, the "Einstein Institute," never did fulfill its original purpose.[257] But in the present context, this is of secondary importance. Funds were raised and donated on the assumption that the research conducted inside the tower would meet every expectation.

### 2.3.2 Donations for the Einstein Tower

The initiator of the Einstein Donation Fund was an excellent organizer. Freundlich's talents and dedication contributed crucially to the fund-raising success and ultimate construction of the Einstein Tower in Potsdam. But this success also had two strong pillars of support:

1. Einstein's endorsement and fame,
2. the interests of the State, science and industry.

It also helped, of course, that the grand time had at long last arrived that Freundlich had been waiting for. In 1920 he was appointed observer, in 1922 main observer and in 1921 also manager of the Einstein Institute. He was not just on the board of trustees, the board even "conveyed [...] all the work to its one member, Prof. Freundlich. The other members noted the developments with interest but without exercizing any clear authority."[258] Einstein had given Freundlich a full power of attorney on matters concerning the construction of the building on 24 April 1920 (in the handwriting of the architect, Erich Mendelsohn, incidentally).[259]

All this agreed with Freundlich's ambitions. Practically speaking, he also had the say on the capital of the Einstein Donation Fund.

Freundlich's ambition and talent as an organizer were not the sole drivers of the project's rapid progress. It benefited primarily from Freundlich's foresight. He had started the construction plans for the tower spectrograph in very good time. His idea was to give the tower an *architectonically* challenging design that nevertheless stood up to the task from a *technical* point of view. He was soon able to interest his friend, the then still young and unknown architect Erich Mendelsohn (1887–1953), in the project. Thus the vision became a plan, and when the plan was mature, chance granted it materialization. It was due to Freundlich's and Mendelsohn's careful preparations and commitment and the interest of those commissioning the work that the time needed to complete the tower was not only sensationally short but also that the original vision took shape without too many alterations (with some minor defects in the construction).[260]

On 29 June 1920 the Ministry of Culture hoped the construction would be finished "still this year."[261] Although this turned out to be wishful thinking, the building phase was still short. On 28 October 1920 the Prussian Building Surveyor's Office inspected the building. Freundlich was able to report that the construction was essentially finished.[262] The main structure was inspected on 19 August 1921; and the final inspection took place on 14 February 1922. 1922/1923 the laboratory, equipped with its electrical furnace, was set in operation. 1924 the telescope with its 300-mm lens was mounted and the main spectral apparatus installed. Finally, in 1925 the 600-mm lens also arrived. The research could begin.

During a period of severe economic shortages and scarce research funding, in the middle of the inflation, substantial amounts of public and private money were readily expended to promote research on Einstein's theory of relativity. If you add together Einstein's salary, the maintenance costs of the Einstein Tower, and the costs incurred by Freundlich's solar eclipse expeditions, this was probably the most expensive research project of the 1920s.[263] The tower became a symbol. Thanks to its architect Erich Mendelsohn, it is "what is perhaps the most important building of the immediate postwar period."[264] No one could foresee that the tower would not be able to keep its promise to science.

The Prussian state contributed much more than their original commitment of 150,000 marks. In 1920 it approved the purchase of a tower spectrograph for 200,000 marks. The annual grant for running costs came to 20,000 marks. The driving force behind this generosity was the Prussian Ministry of Culture (at that time under Minister Haenisch). The intention was to prevent that "further development of an extremely important theory conceived by a German scholar be left exclusively to foreigners." That was precisely why the Einstein Donation Fund was exempted from gift tax that would otherwise have applied.[265] Haenisch supported Erwin Freundlich's petition "to entrust the construction of the tower to the certified private architect Erich Mendelsohn [...], who has been acting in the most selfless manner as a consultant on the design of the new facility and has prepared all the necessary static calculations and drawings without any compensation at all, while it still seemed completely unlikely that these plans could be materialized."[266] The Ministry of Finance agreed.[267] The petition was granted.

Private individuals, industrialists and bankers also opened their purses. An excerpt from the list of donors was published in April 1920 to announce the success of the campaign and invite further sponsors.[268]

This list documents how much major industry and banking supported the construction of the Einstein Tower for the glory of Germany and German science. Famous personalities like Walther Rathenau were also among them.

By 16 April 1920, 296,634.40 marks had been received. A renewed appeal – explicitly pointing to the exemplary contributions by the Reich Federation of German Industry and the optical companies Carl Zeiss and Schott & Partners – stimulated further donations. By 30 December 1921, a total of 1,395,280 marks donations plus interest in the amount of 2,486.46 marks (less taxes) could be entered in the books.[269]

On 19 May 1922 Freundlich could report to the board of trustees of the Einstein Donation Fund receipts of 1,416,766.46 marks.[270] The Zeiss Company installed the tower spectrograph at cost, a total of 200,000 marks (in early 1921 the apparatus was on the normal market for one million marks).[271] The Zeiss Company also made available, free of charge, "a large lens worth many hundreds of thousand marks" in the form of a loan for ten years.

The notable names among the sponsors, the considerable sums contributed and the deposit dates all indicate energetic and purposeful fund-raising activity in the beginning months of the campaign. This was largely *Freundlich's* accomplishment.

Hans Ludendorff, director of the Astrophysical Observatory since 1921, praised his subordinate, Erwin Freundlich's commitment in a letter dated 27 May 1921 to the Minister of Science, Arts and Culture.

Hans Ludendorff to the Prussian Ministry of Culture, 27 May 1921:[272]

[. . . ] During the elapsed year Dr. Freundlich expended much personal effort toward organizing the Einstein Donation Fund and managing the construction of the tower telescope. Since all the negotiations with the sponsors and the firms involved in the construction were in his hands, he was often compelled to take part in conferences in Berlin that often took up the whole day. Because the design of the telescope is being carried out by the Zeiss Company in Jena according to his specifications, frequent trips there are necessary. [. . . ]. In the coming month an additional trip to England is necessary, since Dr. Freundlich has been invited by the Universities of Manchester and Oxford to accompany Professor Einstein on his visit, in order to facilitate with his command of the English language and expertise in the new theory discussions about relativity theory with the English physicists. [. . . ].

*All* the founding firms of the dye consortium *IG Farben* were contributors to the Einstein Donation Fund.[273] Six contributions were made in close temporal succession (between 8 and 18 May 1920). Three times, sums in the amount

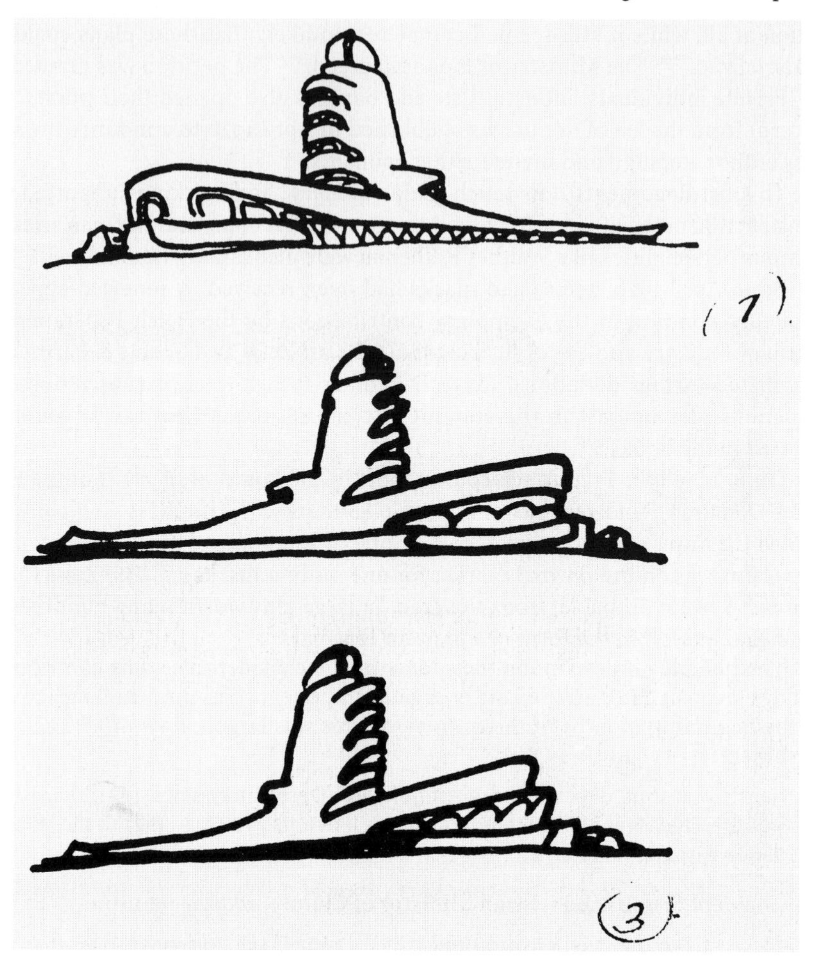


**Figure 14: Three sketches by Erich Mendelsohn, from different angles of perspective (from the northwest and northeast) in different versions (1920). Erich Mendelsohn to his wife, 18 June 1920:**
**The tower blue-prints are ready [...] Freundlich is only taking them to the ministry on Monday morning [...] Tomorrow the major construction plans start. Concerning the technical drawings that I am sending out to you tomorrow, here are 3 sketches.**
**1. The former version.**
**2. The last one. You can see the final step after removal of the corner stones on either side of the entrance. Thus the ring concept develops at the most prominent spot into a very bold sweep of ferroconcrete. I am going to use concrete completely, if ever possible, for the upper structure and the ring side of the tower.**
**3. A variant of 2. The ring of windows protrudes from the surface of the wall. What do you think? There is still time. As soon as the ministerial permit is here (private architect on government property – impingement upon sacrosanct rights) – Freundlich is in good spirits, I am skeptical to the teeth – I hope to further absolution very promptly.**

4. September 1 9 2 1
Nr. 36
30. Jahrgang

Berliner

Einzelpreis des Heftes 75 Pfg.

Illustrirte Zeitung

Verlag Ullstein, Berlin SW 68

Der neue Einstein-Turm auf dem Telegraphenberg bei Potsdam,
der zur experimentellen Prüfung der Relativitätstheorie aus Mitteln der Einstein-Spende errichtet worden ist.
Entwurf von Erich Mendelsohn - Berlin.

Phot. Heddenhausen & Weiß.

**Figure 15: Certainly headline news: "The new Einstein Tower on Telegraph Hill near Potsdam…" Front page. *Berliner Illustrirte Zeitung,* no. 36, dated 4 September 1921.**

of 24,000 marks were among them. This was no coincidence. But the files do not reveal who was the driving force behind it. The most likely candidates are: Fritz Haber, Walther Nernst and perhaps also Friedrich Schmidt-Ott (since 1920 chairman of the Baeyer Company's supervisory board). The case for Haber is his connection to the chemist and BASF man, *Bosch,* who developed the manufacturing process based on Haber's idea for synthesizing ammonia. Their "Haber-Bosch process" was a technological feat of historical dimensions. The close collaboration between Haber and Bosch and the Baden Aniline and Soda Factory (BASF)[274] just a few years earlier had been a crucial element of German arms production and war policy. The chemical industry reaped such great profits as a result that it was a mere triviality to make any contribution Haber might have wished. The *chemical* industry was thus a major promoter of research on relativity theory even though it could not expect to gain a thing from it.

*Jews* were another significant category among the sponsors: Rudolf Mosse, J. Friedländer, Hermann Gerson, Franz von Mendelssohn, Isodor Cohn, Max M. Warburg, Georg Tietz, the Grand Lodge of Germany, Delbrück, Walther Rathenau and many others.

It is also noteworthy that a large proportion of the private donations came from *foreign* sources, including the USA: Albrecht Pagenstecher, New York (one of the largest contributions of all: 200,000 reichsmarks) and Henry Goldman, New York (25,000 reichsmarks).

"The deterioration of the mark in 1923 necessitated immediate expenditure of the funds for instruments and requisite laboratory equipment, so the Foundation's capital was depleted by the end of 1923. The research could continue to proceed essentially by a grant of the Ministry of Culture."[275] The foundation received additional funding for compensation of the staff from BASF and the Emergency Association of German Science.

The "Einstein Foundation," which grew out of the "Einstein Donation Fund," was its legal administrator and dispenser of the contributed sums.

The governing by-laws of the Einstein Foundation went into force on 4 January 1922.[276] According to Article 1, the non-profit "EINSTEIN Foundation [...] shall make use of the funds provided by voluntary contributors exclusively for the purpose of promoting EINSTEIN's scientific research and related areas of research [...]." Article 2 names the members on the board of trustees:

1. Professor Dr. Albert Einstein, Berlin, member of the Academy of Sciences,
2. Dr. E. Finlay-Freundlich, Potsdam,
3. Professor Dr. H. Ludendorff, director of the Astrophysical Observatory, Potsdam,
4. Professor Dr. C. Bosch, Ludwigshafen on the Rhine,
5. Privy Councillor, Professor G. Müller, Potsdam, member of the Academy of Sciences,
6. Dr. R. Schneider, Berlin, managing director of the Reich Federation of German Industry.

Later trustees were: Priv. Coun. Prof. Krüss for the minister of culture; Dr. Jeidels, managing director of the trading corporation Berliner Handels-Gesellschaft; Dr. L. Ruge, lawyer. In 1925 the board decided that implementers of decisions taken by the board could not be among its members. Einstein informed the Ministry of Science, Arts and Culture in November 1925 (the letter is undated) that Freundlich was no longer on the board on these grounds.[277] In summer 1931 the board of trustees of the Einstein Donation Fund consisted of: Einstein, von Laue, Nernst, Schrödinger, Paschen, Franck, C. Bosch, Jeidels, Ruge, Berliner, Krüss and Ludendorff.

Article 3 of the by-laws specified the Astrophysical Observatory as the foundation's official place of business. According to Article 4, the facilities and apparatus acquired by the foundation's capital and any liquid assets were to become the property of the State as of 1 January 1932.

The inclusion of Messrs. Bosch and Schneider on the board meant that two influencial members of German industry had a say in its internal affairs.

By the time the tower was completed, however, Einstein, the person most directly concerned, had already partly lost interest in it. One reason may have been that growing tensions between Einstein and Erwin Finlay-Freundlich since 1921 had finally gotten beyond the point of tolerability. Shortly after the construction work on the Einstein Tower came to an end, the personal relations between Einstein and Freundlich broke off completely. There is no record of what broke the last straw. But *Einstein* came belatedly to this painful realization in a letter to Ludendorff dated 15 September 1925. It needs no further comment.

Einstein to Ludendorff, 15 September 1925 (excerpt):[278]

With regard to Mr. Freundlich, you know my opinion, of course. In any case I have broken off personal relations with him and could have added a few very fine ''specimens'' to the list of sins you reported. He counts among the very few with whom I consider such a rigid attitude necessary. But I respect his organizational achievements and act accordingly, as you have also most commendably done on the occasion of his appointment. Thus we both serve the cause, even though we value the man and the scientist little. He is not worth getting upset about.

Einstein's assessment of Erwin Freundlich as a *scientist* does need further comment. Was it justified? Others were of a different opinion.[279] We cannot go into the specifics here. Personal qualities are one thing; scientific qualities quite another. The history of science has proven time and again that a scientist's character and attitude are not appropriate yardsticks for his professional merit.

If we must take a closer look at Erwin Freundlich's behavior, it is primarily in order to show that Einstein's evaluation was justified and that the personal animosity between Hans Ludendorff and Freundlich, with its many consequences, cannot be explained as primarily due to Ludendorff's political stance. It is also demonstrated that the conduct of Einstein's former collaborator even partly damaged Einstein's international success (if only to a minor extent). Many people inside as well as outside Germany viewed Freundlich as a close and trusted colleague of Einstein.

Einstein obviously was on quite amicable terms with his colleague Ludendorff, despite their opposing political positions.[280] Einstein proposed *Ludendorff* in his stead when he declined an invitation by the Mexican government to join an expedition to observe the solar eclipse there in 1923. (He did not even mention Freundlich.) Note in the files of the Ministry of Culture, 1923:[281]

> Professor Einstein informs us that he cannot accept the invitation by the Mexican government because he intends to stay in Germany for the time being. He requests that the Mexican government be relayed his thanks and would very much appreciate it if the planned expedition took place anyway. He proposes Professor Ludendorff as its leader.

The Ministry of Culture informed the Foreign Office of this on 15 March 1923.[282] This is not to say that Ludendorff and Einstein always saw eye to eye on everything.

For instance, Ludendorff complained to Einstein in mid-1928 that "under your chairmanship the board of trustees of the Einstein Foundation, whose member I am, has [not] sought my opinion" when a proposed budget was up for approval by the trustees. Ludendorff perceived this as a "great unkindness."[283] "It pains me," Ludendorff continued, "that you should stoop to such an unkindness toward me and it surprises me, particularly since not long ago in a letter still in my possession you acknowledged my objectivity and at the same time expressed in unmistakable terms your opinion of Prof. Freundlich as a person and a scholar."

This time it was *Einstein* who was in error, not Ludendorff. Einstein had to bear the consequences of his own negligence in administrative questions and Erwin Freundlich's unauthorized actions.

Compared to Freundlich, Ludendorff seemed to Einstein "considerably the clumsier of the two, [...] nonetheless [...] much the more decent."[284] Verification of the "Einstein effect" was, in fact, a part of the agenda of a solar eclipse expedition led by Ludendorff in 1923 to Mexico.[285] Even *after* the fascists took over government, and *after* Freundlich had left, the Astrophysical Observatory – to which the Institute of Solar Physics was now attached – still ordered new apparatus from the Zeiss Company in Jena in February 1934 with the aim of continuing "on-going analyses on light deflection in the gravitational field of the sun."[286] There was no more mention of the "Einstein effect," but the actual topic of inquiry had not changed. In 1934 it was not Ludendorff who distanced himself from the theory, but *Freundlich*.

The bad feelings between Freundlich and Ludendorff reached the point of no return. Reconciliation was out of the question.

Einstein was probably never so thoroughly mistaken about a person's character as in the case of Erwin Finlay-Freundlich.

All the same, the Einstein Donation Fund and the construction of the Einstein Tower in Potsdam are proof that interwar Germany took a live interest in promoting Einstein's research. Albert Einstein's research certainly was "national valuta" yielding generous interest. Shortages and inflation posed no obstacle.

On 1 January 1932 the Einstein Institute was handed over to the Prussian state in accordance with its by-laws, thus becoming a department of the Astrophysical Observatory.[287] This was just about the time when Freundlich began to question what he had been working on for so long: "After the scientific community

**Figure 16: *Berliner Illustrirte Zeitung,* no. 36 of 4 September 1921: "Prof. Einstein and Prof. Ludendorff during the astronomers convention in Potsdam"**

had finally come around to acknowledging and even admiring the theory of relativity in 1919, Freundlich began to express doubts in 1931 about the validity of the theory and to interpret the experiments (light deflection and gravitational redshift) that everywhere else were being considered as triumphant vindications of Einstein differently from the majority of physicists, astrophysicists and astronomers."[288]

A short time afterwards, the Nazis rose to power. Promptly, on 28 March 1933, a demand was made to change the name of the Einstein Institute. Freundlich agreed.

Excerpt from a letter by Senior Civil Servant Theodor Vahlen at the Ministry of Culture to Reich Commissioner B. Rust dated 28 March 1933:[289]

> The name 'Einstein Tower' must disappear, and I suggest that you issue an order as soon as possible to assign it the name 'Institute of Solar Physics.' I have already spoken with the director of this institute, Professor Freundlich, who has agreed to this change.
>
> Berlin, the 28th of March 1933.

It was only a matter of time before Freundlich was expelled from the institute and forced into exile.

## 2.4 A target for right-wing propaganda and violence

The period after the war produced a tower to promote and celebrate Einstein but it also produced the fuel for the bonfires on which his works would later be burned.[290] While some revered him, others hated him. Einstein experienced the Third Reich before the Weimar period even came to an end. The only difference was: the elements who grabbed the reins of power in 1933 – with the assistance of the State machinery, the economy and the military – were still in the gutter in 1919/20.

Einstein's fame was also a burden. *Because* his success was the topic of the day, everyone wanted to know what else this man thought and did. Einstein would hide himself away or leave town unexpectedly to escape this publicity. All he wanted was to devote himself to his science. But the time he could do that was past. There was no going back for him into a world free from politics. Henceforth he was the object and soon also the subject of major politics.

His fame was an enormous sounding board for any pronouncement he made about the world or politics. To Hedwig and Max Born, Einstein wrote on 9 September 1920: "Like the man in the fairy tale who turned everything he touched into gold – so with me everything turns into a fuss in the newspapers: suum cuique [each to his own]."[291] Whether he laughed, looked annoyed, or said nothing at all, the newspapers continued to publish reports about Einstein. Articles and manifestos of political relevance bearing his signature only amplified his political influence. His commissions and omissions gained an importance that, as Fritz Haber pointed out on 9 March 1921, "were formerly only attributed to the dealings of princes."[292]

Einstein was euphoric about the end of the war and applauded the "November Revolution" of 1918 and the kaiser's abdication, still unaware of the sacrifices to come. "The great event has taken place!" he wrote his sister and brother-in-law in Lucerne two days after the revolution. "That I could live to see this!! No bankruptcy is too great not to be gladly risked for such magnificent compensation. Where we are, militarism and the privy-councillor stupor has been thoroughly obliterated."[293] On the *very same day* – yet another sign of his enthusiasm – he wrote his mother: "Don't worry. All has been going smoothly, impressively even, up to now. [...]. Now I'm beginning to feel really comfortable here. The bankruptcy has done wonders. [...]. Among the academicians, I am some kind of high-placed Red."[294] He reiterates this last point shortly afterwards in a letter to his friend Besso in Switzerland: "I am enjoying the reputation of an irreproachable socialist [Sozi]; as a consequence, yesterday's heros are coming fawningly to me in the opinion that I could break their fall into emptiness."[295]

This reputation as a "*socialist*" remained with him for the rest of his life. Even so, under more concrete circumstances Einstein rejected this title. The term socialist was too restrictive for him. Moskowski quotes Albert Einstein: "If I had to choose a term, I would rather say: a democrat of liberal thinking, in the broadest sense of the word."[296] In a letter dated 19 April 1920 he mentions that Ulrich von Wilamowitz-Moellendorf had described him as an "independent socialist."[297] But he did *not* add any comment about this categorization (from which we may conclude that it could not have been entirely misplaced). It was out of the question for him to join a party. Indeed, Einstein wondered, "How an intelligent man can subscribe himself to any particular party is simply a puzzle to me."[298] "Party discipline" in any form would have been torture for Einstein. He never understood how truth could be deposited with any single party. The underlying logic of the political confrontations was: People on the extreme right assigned Einstein among the extreme left-wing, and the leftists soon thought that he was fully and completely on their side. Thus no distinction was made between Einstein and Liebknecht, or between Einstein and the Bolshevists.

Around the time of the revolution and even later, Einstein sympathized with the communists in Russia and Germany. It only justified the hatred the conservative right had for him.

His reputation of being a "*communist*" also originates from the period of the November Revolution. The fact that he never was one, changes nothing. But the fact that he sympathized with their cause is just as indisputable. "Politics are actually developing consistently in the direction of Bolshevism," he wrote on 27 January 1920. What he meant was: this development was desirable and good.

Excerpt from Einstein's letter to his friend Max Born, 27 January 1920:[299]

The political situation is developing consistently in favour of the Bolsheviks. It seems to me that the Russians' considerable external achievements are gathering an irresistible momentum in relation to the increasingly untenable position of the West; particularly our situation. But before this can happen, streams of blood will have to flow; the forces of reaction are also growing more violent all the time. Nicolai is being attacked and insulted so much that he is no longer able to lecture, not even at the Charité. [. . . ]. France is really playing a rather sorry role in all this (all the same it is to their credit that they have rid themselves of the Tiger [Georges Clemenceau]). Victory is very hard to bear. [. . . ]. By the way, I must confess to you that the Bolsheviks do not seem so bad to me, however laughable their theories. It would be really interesting just to have a look at the thing at close quarters. At any rate, their message seems to be very effective, for the weapons the Allies use to destroy the German Army melt away in Russia like snow in the spring sun. Those fellows have gifted politicians at the top. I recently read a brochure by Radek - one has to hand it to him, the man knows his business!

A few days before this statement was made Radek had taken part in the founding meeting of the Communist Party of Germany – on 30/31 December 1919 – (together with Karl Liebknecht, Rosa Luxemburg, Paul Levi and Wilhelm Pieck) as a member of the Russian delegation.

Konrad Wachsmann, the architect of Einstein's summer villa in Caputh, later remarked that Einstein was "a confirmed democrat, who had earlier himself flirted with anarchism."[300] Rolland gives a similar report: "Bolshevism is spreading in a strange way among German intellectuals. Even those who had founded anti-Bolshevist societies let themselves be infected by it after a few months. The physicist A. Einstein has also succumbed to this influence."[301]

Einstein was enthusiastic about the revolution in Germany. Only now was he beginning to feel really "comfortable" in Germany. But this euphoria did not last long. Actual events soon made him sense that the "reactionaries were becoming increasingly hot-headed." The "reactionaries" seemed to have been beaten, but they were not powerless – just injured and therefore much more dangerous. Culprits, scapegoats had to be found. It simply could not be that Germany was inferior on the battlefield. That is how the *stab-in-the-back* myth [*Dolchstosslegende*] was born. The "Sozis," and more so the "Bolsheviks," became, as a consequence, the archenemies of the reactionary right. On 15 January 1919 the leaders of the recently founded Communist Party of Germany (CPG) were murdered: Karl Liebknecht and Rosa Luxemburg. (Einstein: "She was far too good for this world.")[302]

Besides *leftists, Jews* were another favored object of the blood-thirsty forces of reaction, and especially so when both these qualities coincided – as was true of Luxemburg. This coincidence happened to be quite frequent as well. Jews played a decisive role in the history of the labor movement: Marx, Lassalle, Bernstein. It was not just fortunate but in a certain way also unfortunate that, as Stefan Zweig wrote to Romain Rolland on 24 February 1920, "all slightly liberal and independent minds of German literature are Jews. And the fact that the heros of the revolution were also Jews or half-Jews (consider Liebknecht,[303] Haase, Luxemburg, Eisner, Toller, Nicolai) has provoked such a fury among the 'patriotists' that anything sponsored by Jews is suspicious from the start. You know how Einstein was hooted down in Berlin, the greatest scientist of Europe of our day."[304]

Never – in the view of Golo Mann – "had anti-Semitic passions raged as heftily as in the period between 1919 and 1923. It was much more vicious then than between 1930 and 1933 or between 1933 to 1945."[305] Jews became the target of hate not just by nationalists but also by much broader strata of the population.

Under such circumstances Einstein quickly realized that he too was a *Jew*. Not that he particularly *wanted* to do so. It was not the Jews, but *anti-Semites* who taught him this lesson. Unlike others in his milieu he did *not* buckle under the pressure and try to deny his Jewish heritage. Contrary to Walther Rathenau, Fritz Haber and many others, Einstein was of the opinion that Jews ought to live according to their own ways and stand up for their own traditions. Einstein already believed then what he later put down on paper: "One can be an upholder of European culture, a good citizen of a state and at the same time be a faithful Jew."[306] He became a Zionist without abandoning his cosmopolitan views.

In conformance with these views he spoke out for the interests of Jews, particularly those who had the hardest lot in Germany: immigrants from the East, the so-called *Ostjuden*. Theirs was the hardest lot also because many Jews who had been living in Germany for a long time and had managed to build a comfortable living for themselves disdained immigrant Eastern Jews and wanted to avoid being identified with them in any way. Ostjuden were shunned not just because they were *foreigners*, but also because their mere presence posed an obstacle to assimilation efforts. "Strongly assimilated German Jews, the majority of whom were from the middle class, perceived the mostly orthodox refugees from the East as an outright danger. They feared that the immigrants – by virtue of their foreignness – would provide anti-Semites with 'objective reasons.' Driven by this fear they themselves adopted anti-Semitic prejudices and applied them to their fellow brethren in faith. They tried to avoid any kind of identification with them [...]. Thus German Jews often regarded Russian and Polish Jews as inferior and occasionally even applied for State assistance in preventing their immigration."[307]

Einstein's efforts for his "fellow clansmen," as he liked to say, mainly consisted in traveling and issuing appeals for the benefit of the World Zionist organization (which included his trip to America in 1921) and offering courses for immigrant "young foreigners studying in Berlin from Russia, Poland, Bulgaria, Rumania, and Lithuania, mostly Eastern Jews, who lack the opportunity of matriculating here at the university."[308]

Einstein and L. Landau submitted a petition to Prussian Minister of Culture Haenisch on 19 February 1920 on this justification, requesting permission to describe them as courses "offered by Berlin University professors accredited by the State."

**Leopold Landau** (born on 16 July 1848 in Warsaw, deceased on 28 Dec. 1920 in Berlin). From 1876 private lecturer, from 1895 titular professor at the University of Berlin, from 1902 extraordinary professor. Member of the General Medical Council of Berlin-Brandenburg. 1906 privy councillor of medicine. On 23 Jan. 1920 oath of allegiance to the constitution of the Reich.

Einstein offered to give lectures on the topic 'Introduction to Theoretical Physics.' Other proposed lecturers were: Prof. Dr. James Franck (experimental physics), Prof. Katzenstein (surgery), Prof. Landau (gynacology). Haenisch had no objections. On 4 August Einstein and Landau reported to the minister about how the first semester had faired. 182 persons had registered to attend, 175 among them were Russian or Polish, 2 were Bulgarian and 5 Hungarian. Einstein had taught a course on experimental physics and on theoretical physics together with Dr. von Horvath for 10 auditors.

There was energetic resistance at the ministry, however, to allowing the participants to attend regular lectures at the university: "unheard of" and "That was not how the matter was intended. Please intervene energetically. Flat circumvention" were marginalia at the ministry. On 16 August the minister protested.[309] But that did not shield Haenisch from hefty attacks by the press.

On 14 April 1920 the *Deutsche Zeitung* published an article under the heading 'Haenisch and the Galician Jews. Special professorships for Einstein and Nicolai-Löwenstein?'[310] It closed its sketch of the enrollment procedure with the following commentary:

Mr. Minister is putting every conceivable hurdle against university study by our German students. But these foreigners, who are, of course, almost exclusively Russian Jews, for whom money is not an issue, are given the possibility of doing so by the circumvention described above, and are even granted State accreditation to boot. What does our German student body think of this? It certainly will not be long before elements like the Jewish traitor Professor Nicolai-Löwenstein obtain another chair in this way.

And what about the German student body? - - - K.

Now it was publically out in the open for all to see. Einstein was a *Jew* and an advocate of *Jews*.

**Figure 17: Einstein at a conference of Jewish students (around 1921)**
**Einstein: "Despite my declared international mentality I still do always feel obliged to speak for my persecuted and morally oppressed fellow clansmen, as far as it is in my powers. [...] after seeing recently from countless examples how perfidiously and unkindly fine young Jews are being treated here, in an attempt to deprive them of educational opportunities." ... "I do what I can for my fellow Jewish brethren, who are being treated so meanly everywhere [...]."**

So at the beginning of the 1920s Einstein, too, was publically viewed as *a leftist and a Jew*, or as he himself wrote: "a Jew of liberal, international bent."[311] Without this association the controversy over relativity theory would not have been nearly as bitter. The bone of contention in the *political* disputes at that time was actually not the scientific content of relativity. The criterium dividing the parties was rather their proximity to the political stance of the theory's creator. This was true elsewhere as well, although not quite as extremely as inside Germany. In France, Dreyfus supporters declared that Einstein was a genius and Dreyfus opponents called him a donkey.[312] The association "Einstein – relativity theory – Judaism" stuck in the public mind, even though thousands of *non-Jewish* physicists in Germany and abroad, political opponents of Einstein among them, acknowledged the theory of relativity as one of the greatest achievements in the history of science and accepted it as a natural basis of their research.

Einstein became the victim of blind hate outside the country as well.

A notice published by the Paris edition of the *Chicago Tribune,* no. 1059 on 8 June 1920 essentially means: Einstein is a Hun and a pig, because he is a *German.*

From: Paris edition of the *Chicago Tribune,* no. 1059 dated 8 June 1920 (from the French translation):

Voices from Americans in Europe. Protest against the award of a medal to a *German.*

The Hague, 3 June. To the editors. I read in the American edition of the Tribune that the University of Columbia has awarded a medal to a Boche named Einstein. This event takes place a few days after Decoration Day, when all Americans stand with bared heads before the graves of men who had died in battle against the Boches.

Does the University of Columbia forget that a Hun is still the same pig he was in 1914? Does the University of Columbia forget that the Boches have proven themselves unworthy of the lowest signs of distinction from us? And does Columbia forget that we are actually still at a state of war with Germany?

Can't newspapers exercise any influence in this regard?

An important American university honors a Boche! That's enough to make one *sick!*

H.

Things looked very different as soon as it became known that Einstein was a Swiss citizen, had not signed the Appeal to the Civilized World during World War I, and actually wasn't even "German." As the foreign world became friendlier toward him, the hate *inside* Germany only worsened. Well before he was officially expatriated by the Nazis, Einstein had long since been banished emotionally.

A report by the German envoy in Shanghai, Thiel, from 28 November 1922 to the Foreign Office reveals how quickly the mood could swing around:[313] "It was very amusing to see how the French tabloid L'Echo de Chine capitalized on the arrival of a new agent of Boche propaganda, whereupon a white-washing by the chief editor Dr. Vallet followed today, saying that he wasn't a Boche at all, considering that he was born in Ulm – that is to say, in Czechoslovakia, and had furthermore become Swiss, etc. Then, naturally, it is recounted how shaken Einstein had been at the sight of destroyed regions, in particular the cathedral of Reims."

Few have experienced such a relativity of values as Albert Einstein.

Einstein was the object of hatred *even abroad,* but it was the strongest in his *German* fatherland. Circumstances that mitigated his case or even reaped applause outside the country, thus making it possible for him to become an emissary of Germany, were persuasive arguments for hatred at home, particularly among those who could not stand seeing Germany a defeated nation and needed to find an internal enemy.

The arguments Einstein's opponents used against the theory of relativity were essentially the following:

1. An appeal to "healthy common sense." An attempt was made, using formal logic and rudimentary mathematics, to disprove the theory of relativity. The unintuitive nature of a scientific theory in its beginning stages, particularly for the

layman, was taken as proof of its invalidity. Just this already exposes the demagogic mendaciousness of the anti-Einstein militants. Einstein's political opponents did not intend to elucidate the topic but to confuse and mislead the masses. The working class was demagogically played off against a segment of the intelligentia, specifically against Einstein. "Our poor workers don't realize that their yearning for a better lot in life is being shamefully abused," Roderich-Stoltheim wrote in his publication from 1921, 'Einstein's theory of deceit.'[314]

This Roderich-Stoltheim was one of the most vituperative Jew haters after the war. His book 'The puzzle of Jewish blood' [Das Rätsel des jüdischen Blutes, Leipzig, 1919] – sold in a hundred thousand copies – fanned anti-Semitic passions. He stated that Jews mainly seduced German virgins to bastardize the Aryan world. Nor did he have any qualms about accusing Jews of child rape. This, Roderich-Stoltheim contended, was one of the Jewish plots to conquer the political and economic leadership of Germany.

2. Einstein was accused of turning physics into a dialectic affair and thus drawing it within the range of Marxism. Einstein was characterized as a Bolshevist and his theory was termed "Bolshevist physics." Fricke, head of the so-called German Society for Research on Universal Aether and Intuitive Physics (founded in 1933), provided what might be termed as the "classical" formulation of the anti-Einstein camp: "Modern physics, as it appears in Einstein's theory of relativity, as well as in Lorentz's electron theory, Bohr's new atomic physics and Planck's new theory of energy quanta, [...] offers a strange chaotic parallel to Marxism. Both are based on 'materialism' [...]."[315]

3. This hate campaign was taken so far as to hold Einstein responsible for the revolutionary political developments. The content of the theory was made to sound as if Einstein had struck down every last brick in his destruction of the foundations of physics. The moral implications were purposefully drawn in. The *Weser-Zeitung* of 22 June 1921 : "Relativism is one of the symptoms of decay and decomposition." The diatribe mentioned above on 'Einstein's theory of deceit' stated: "Professor Einstein posited a principle of relativity that destabilizes all our conceptions of space and time and is supposed to convince us that nothing solid, reliable or true exists in the world."[316] A reader unaware of the slanderous nature of the text would conclude with its author: "Nothing ... of permanence is supposed to exist anymore [...]. The moral effect [... should] not be underestimated [...]. It [is] tailored to diminish the value of life itself and to prevent trust in the values of civilized life."[317]

4. Einstein was accused of carrying out a publicity campaign, scientific plagiarism and "throttling his opponents."[318] The famous formula $E = m \cdot c^2$, a consequence of the special theory of relativity, is renamed the "Hasenöhrl law." The Hungarian professor Palagyi alleged that Einstein, Minkowski and Lorentz had stolen his theory of 1901 and made it substantially worse. As early as 11 February 1920, the chauvinistic paper *Völkische Beobachter* declared that Einstein's theory actually did not have any scientific merit. It was "Jewish scientific plunder" or a "pettifogging bluff."[319] On one hand, the theory of relativity was supposedly false, on the other hand, Einstein was charged with plagiarism.

5. Finally, all the attacks on Einstein had an anti-Semitic flavor. The connec-

tion between anti-Semitism and the charge of publicity-making and plagiarism is expressed, for instance, in the *Völkische Beobachter* from 11 February 1920: "First comes the official beating of the drum about Mr. Einstein's invention, so that his name, and above all, his racial origin become sufficiently known, because what little of supposed worth that Jews have adeptly managed to achieve at the cost of others, naturally has to be shouted from the rooftops."[320] Only a Jew could purportedly have conceived relativity theory. 'Jewish thinking' was incapable of recognizing the essence of things; it was "forever deprived of profound secrets of the essence of things."[321] The *Deutschvölkische Monatshefte* (1921, issue 1) boldly asserted: "Einstein demonstrates precisely by this destructive relativity theory that he is of the Jewish race, who are certainly good critics but not beneficial creators like the Germanic people." Etc.

6. For want of real arguments, assault came as a frequent substitute, along with its quintessence: the call for his assassination. It did not necessarily always have to be so blatant. The following cynical suggestion was given to anyone having to defend himself at court for an act of violence against Einstein: "If something similar should happen to Mr. Einstein [...] he could scarcely seek his rights before a court. His opponent would argue, with good reason: All motion is relative. It cannot be proven that I hit Mr. Einstein. I stretched my arm out and Mr. Einstein's cheek suddenly came flying in relative motion against the palm of my hand."[322]

So much in simplified synopsis. Klaus Hentschel's doctoral thesis goes into the details of interpretations and misinterpretations of relativity theory by Albert Einstein's contemporaries.[323]

Let us turn now to the chronology of the attacks on Albert Einstein.

In December 1918 the highpoint of the November Revolution was already in the past. The forces of reaction were preparing for retaliation. The spring of 1919 was a bloody one. The labor movement lost its best two leaders, Rosa Luxemburg and Karl Liebknecht, on 15 January 1919. On 21 February, Kurt Eisner became another victim of violence and at the beginning of May the heads of the Soviet Republic of Bavaria Eglhofer, Leviné and Landauer, also lost their lives. During this time Einstein also became a target of the animosity. During a public meeting that spring, a student bellowed out: "One ought to slit this Jew's throat!"[324] Georg Nicolai, a co-signer of the Manifesto to the Europeans – the countermanifesto to the Appeal by the Ninety-Three – was harrassed and tormented to the point that he finally left Germany for Argentina. Einstein preferred to leave on an extended trip abroad. He returned to Berlin that summer, but the reactionaries were still on the advance. Einstein was particularly affected by anti-Semitism prevalent among the "educated" class.

When Einstein rose to extraordinary fame toward the close of 1919, a systematic campaign of persecution took off as well at the beginning of the following year. It culminated in the official establishment of a Syndicate of German Scientists for the Preservation of Pure Science, reg. assoc. (*Arbeitsgemeinschaft deutscher Naturforscher zur Erhaltung reiner Wissenschaft e.V.*) in August 1920. This association has gone down in history as the "Anti-Einstein League."

The Berlin engineer Paul Weyland was its leader under the conceptual authority of the Nobel prizewinner Philipp Lenard.

**Philipp Lenard** (born 7 Aug. 1862, deceased 20 May 1947). Nobel prize 1905 (physics) "for his work on cathode rays."[325] He worked under Helmholtz in Berlin and under Bunsen in Heidelberg. Lenard was assistant to Heinrich Hertz and published a posthumous edition of Hertz's 'Principles of Mechanics.' At Heidelberg since 1907. He was the first noteworthy German scientist to take up the Nazi cause. On 8 May 1924 an article he coauthored with Johannes Stark appeared in the *Großdeutschen Zeitung* announcing their sense of duty to proclaim publically their discovery in Hitler and his cohorts of the spirit they had always been looking for.[326] In 1928 Hitler and his deputy Hess visited Lenard at his apartment in Heidelberg. On 9 July 1933 Lenard applied for membership in the Nazi party.[327]

The Anti-Einstein League set itself the following goals: "To liberate science from freebooters. Among certain circles a method has emerged to reduce science to an object of commerce or political publicity for a particular race [...]. The world of research must close a united front against such mucking up of science [...]. We seek to purge German science of Jews."[328]

This was a clear declaration of war against Einstein. Ridiculous though these people were in wanting to "save" science and "keep it clean," they were dangerous nonetheless.

Twenty lectures were announced for the late summer of 1920 with the aim of culling the "nonsense" and "phantasms of a philosophical dilettante."[329] Their topic was to be the refutation of the theory of relativity. Thus the persons charging Einstein with publicity campaigning set about doing so themselves. The first lecture was scheduled for 24 August. The Jewish philosopher Kraus in Prague was invited to participate as a kind of figurehead to demonstrate that, for heaven's sake, no political intentions were involved. But this speaker realized what they were planning to do with him and withdrew just in time.

Weyland tried to bait Einstein's scientific opponents into attending. Some fell for it, others did not. Weyland lost many followers as soon as they saw through the swindle. A letter Weyland wrote to a physicist – not publically identified by name – on 23 July 1920 reveals his luring tactics.

A letter by Weyland dated 23 July 1920:

Esteemed Professor!

Now that unanimous agreement has been reached among serious members of the exact sciences about rejecting Einstein's research, we are planning also to present the educated lay public with counterarguments, after it has long enough been fed, to the point of vomiting, with Einstein's ideas. I, as secretary of the Einstein opponents, inquire of you whether you are willing to participate in these lectures against Einstein and could provide you with more details upon receipt of your acceptance.

For reasons of urgency I ask you kindly to reply by wire. The business ought to yield a profit of about 10,000 to 15,000 marks for you.

In great respect, yours very truly,

Weyland

This letter was published in the *Berliner Tageblatt* of 4 September 1920, and later also in other papers, to expose Weyland's scheme.

An article in the *Deutsche Zeitung* of 15 September 1920 came to Weyland's defense and added apologetically that the sums involved were not between 10 and 15,000 marks for *one* lecture but for *ten*. Any further financial gains would be applied toward a compilation of "rare works." Weyland's letter had, it continued, therefore truly reflected the facts and because he personally had no major financial resources, he had sought out sponsors.

On 24 August the first lecture evening of the series took place in the great hall of the Berlin Philharmonic. At the entranceway disciples of "pure science" sold swastikas and the mentioned "rare works": agitatory anti-Semitic pamphlets and newspapers. Einstein had also been invited and came with Nernst to listen to the arguments against his theory. Weyland was the first to take the podium. He bombarded Einstein with charges of "publicity seeking," "scientific Dadaism" and "plagiarism." According to his description, all academies and universities that had distinguished Einstein, foremost the Berlin Academy of Sciences, could only be regarded as publicity establishments run by bunglers and idiots. All the while, Weyland busily hawked the pamphlets on display in the entranceway. An intermission of 15 minutes was arranged specifically for sales purposes. Loudly protesting shady intentions by Einstein and publicity-making, Weyland demonstrated how virtuosic he was in doing precisely that. Finally, the audience started to call out "Let's get to the point!" Weyland's "point" in reply consisted in declaring that appropriate measures had been taken to turn out scandal mongers. Weyland was succeeded by Professor Ernst Gehrcke, who took pains to keep to the facts. The version of his talk published in 1924 contained neither political statements nor even anti-Semitic insinuations.[330] Gehrcke may have been on the wrong side, scientifically speaking, but he could not be faulted with base ignorance of the literature or the theory's course of development (as far as this is possible of an opponent). But under the given circumstances his appearance was an indisputable sign of solidarity with Weyland's excesses.

Professors von Laue, Nernst and Rubens (none of them Jews) protested against the defamatory remarks made against their colleague in an article published on 26 August 1920 in the *Täglichen Rundschau*.

Declaration of solidarity by von Laue, Nernst and Rubens:

At the meeting in Philharmonic Hall, which was supposed to elucidate Einstein's principle of relativity, objections were raised of a derogatory nature not just against his theory but also, to the deep regret of the undersigned, against his *character as a scientist*. It cannot be our task to give closer details here on the unprecedented profound conceptual feat that led Einstein to his theory of relativity. Astonishing confirmations have already been made; further verification must, of course, remain the business of future research. We would like to emphasize something, however, that had gone unmentioned yesterday. Irrespective of Einstein's research on relativity, his other research has already earned him a permanent place in the history of our science. Thus his influence on

## Zum Geleit!

Die deutsch-völkischen Monatshefte, die hiermit zum ersten Mal in die Welt ziehen, sind das Ergebnis sorgsamer Ueberlegung. Der Herausgeber ging von der Erkenntnis aus, daß alle bereits vorhandenen Zeitschriften dieser Art das deutsch-völkische Ziel wohl im Auge haben, dieses aber mit ganz wenigen Ausnahmen nicht mit genügender politischer und sachlicher Präzision zum Ausdruck bringen. Wenn wir Deutschvölkischen etwas erreichen wollen, so müssen wir ein Programm haben und nicht nur ein ideelles, sondern ein politisches. Das ideelle ist vorhanden. Der „Bund der

Es ist selbstverständlich, daß sich der Herausgeber im schärfsten Gegensatz zur D. N. V. befindet, solange diese in ihrem Partei-Marx (wie sie wenig originell ihre Leitsätze nennt) immer nur vom jüdischen Geist statt vom Juden selbst redet. Entfernt die Juden aus Deutschland, und deren zersetzender Geist geht von selber! Es ist selbstverständlich, daß nach den Gesichtspunkten der Finanzreform, die in diesen Blättern verfolgt werden, ein Zusammenarbeiten mit den Herren Hergt, Delbrück und Helfferich ausgeschlossen ist, die unter der Aegide Bethmanns tätig waren und natürlich unmöglich umlernen können. Insbesondere muß gegen die sonst ausgezeichnete Persönlichkeit des Herrn Helfferich aus prinzipiellen Gründen Stellung genommen werden.

Antisemitismus der D. N. V. vorbei! Viel hat Helfferich geleistet, aber er ist kein neuer Mann. Er ist belastet mit den Allüren des ancien régime und wird nie einer Reform in unserem Sinne nützen können. Er gehört zu denen, welche die neue Zeit nie verstehen werden und stets nach den alten Zuständen schreien! **Wir wollen aber nicht mehr**

4 Heft 1

**das Deutsche Reich, das von einem Dutzend Juden, die sich in die Reichsbank teilen, an der Nase herumgeführt wird! Wir wollen nicht mehr das alte Deutsche Reich, in welchem neujüdischer Adel dominierte. Aber das neue Deutsche Reich wird uns Helfferich nicht bringen!**

Seine ganze Vergangenheit ist eine unausgesetzte Kette von Beweisen seiner großkapitalistischen Gesinnung. Als Helfferich sein Hindenburg-Programm aufstellte, hat er nicht den staatspolitischen Weitblick besessen, den Lloyd George

**Figure 18: The first (and last) issue of the *Deutsch-Völkische Monatshefte* edited by Paul Weyland, with excerpts from the introduction**

scientific activity, not just in Berlin but throughout Germany, could hardly be overstated.

Whoever has had the pleasure of knowing Einstein more personally knows that no one can outdo him in acknowledging the work of others, in personal modesty and in his distaste for publicity. It seems to us an obligation of justice to express this conviction of ours without delay, since no opportunity was given us to do so yesterday evening.

von Laue. Nernst. Rubens.

On 31 August the *Berliner Tageblatt* published another declaration by other persons. Their message to Einstein was: "In indignation about the pan-German incitement against your exceptional character, we affirm to you, in truly international spirit, the sympathy of all liberal persons, who are proud to know you rank yourself among them and number among the leaders of science worldwide." Oskar Bie, Alexander Moissi, Max Reinhardt and Stefan Zweig were among the signers of this letter.[331]

The papers are full of reports and commentaries. On 27 August 1920 a humorous rhyme appeared in *Morus*, poking fun at what the meeting at Philharmonic Hall was all about:

| Die Einstein-Hetz | The Einstein Flap |
|---|---|
| Germanen, uns wagt man zu bieten, | Germans! Someone dares to offer us |
| die Theorien des Semiten. | the theories of a Semite. |
| Da macht sich so ein Mauschel breit | So some Yiddish jabber gets around, |
| und läßt die Zeit im Raum verschwinden, | and lets time vanish into space, |
| Verleugnung ist's der ‚großen Zeit', | It's a denial of the 'grand era,' |
| ihm fehlt das Nationalempfinden. | He lacks a sense of patriotism. |

Having attended the conference at Philharmonic Hall, Einstein sent his response on 27 August 1920 in his own paper, *Berliner Tageblatt*.[332]

Einstein's reply in the *Berliner Tageblatt*, 27 August 1920 (excerpt):

My Answer re: Theory of Anti-Relativity Inc.
by *Albert Einstein*

Under the unassuming name of ''Arbeitsgemeinschaft deutscher Naturforscher (Study Group of German Natural Philosophers)'' a motley crew has joined together whose present purpose of existence is to lower the theory of relativity and its originator in the eyes of nonphysicists. Recently the Messrs. *Einstein* and *Gehrke* gave a first lecture along these lines in the Philharmonic (auditorium). I am very well aware of the fact that these two speakers are unworthy of an answer by me but I have good reasons to believe that every motive, [beside] the striving for truth lies at the bottom of this enterprise. (If I were a German national with or without swastika instead of a Jew with liberal, international views then.....)* [Footnote in the original translation: This statement ends abruptly here and is given exactly as in the ''Berliner Tageblatt.''] I am answering for the sole reason that this has been repeatedly desired by well-wishers so that my conception might become known.

First, I would like to say that, as far as I know, there is hardly a scholar today who has done noteworthy investigations in the field of theoretic physics who would not admit that the entire theory of relativity in itself has been built up logically and is in harmony with those experimental facts which have been gained up until now. The most outstanding theoretic physicists - I will name *H.A. Lorentz, M. Planck, Sommerfeld, Laue, Born, Larmor, Eddington, Debye, Langevin, Levi-Civita* - are linked with the foundation of the theory and have contributed most valuable facts to the theory. As outspoken opponents of the theory of relativity, I would know only *Lenard* among the physicists of international importance. I admire *Lenard* as a master of experimental physics: in theoretical physics, however, he has accomplished nothing, so far, and his objections against the general theory of relativity are so superficial that I have not considered it necessary to answer them in detail. I am considering doing this now.

I am being accused of pushing propaganda for the theory of relativity which is in bad taste. I can surely say that I have always been a friend of the well-considered, plain word and of the concise presentation. High-sounding phrases and words give me goose pimples whether they are related to the theory of relativity or anything else. I have often made fun over effusions which now have been attributed to me. Furthermore, I give these gentlemen of that association this privilege. [. . . ].

Abroad, it will make a peculiar impression, particularly upon my Dutch and English colleagues, Mr. H.A. *Lorentz* and Mr. *Eddington*, who have been deeply engrossed in the theory of relativity and who have lectured on it repeatedly, when they see that the theory as well as its originator in Germany himself is thus being slandered.

All the facts in Einstein's text may have been correct but his reply does have a sharpness to it that could only offend. Telling a Nobel laureate in *public* that he had not yet done *a thing* in the field of theoretical physics and did not have the least idea about relativity theory meant driving the polemics to a climax and tearing open wounds that would never be healed. Instead of staying in dignified retreat, Einstein threw oil onto the firey passions. He was right that *many* of his opponents were not really interested in the subject per se, just in politics and ideology of the most primitive kind. But *not all* of them. Ernst Gehrcke[333] may well have been very naïve (as Einstein himself was on many occasions). But there was no reason to throw him into the same pot with people whose attack was less aimed at relativity theory than at "Jews of liberal international bent." Gehrcke had published his rejection of relativity theory long before it made such waves. Seven papers appeared on the subject between 1911 and 1913.[334] By mixing science with politics Einstein himself set the tone of subsequent discussions of his theory. Gehrcke did have a point when he wrote in 1924 that "this telling closing confession of Einstein's" aroused "political and racist passions" and diverted "the attention of the public away from the substance of the theory of relativity" and unleashed "a flood of essays in the same tone in various papers."[335] In contrast to Lenard, Gehrcke did not become an ardent advocate of the Nazi party line. It may also have been for *that reason too* (not just because of a laxity toward his teaching obligations) that his superior, the Dean of the University of Berlin saw "no need to continue his permission to teach" there in 1937.[336] He never became a member of the Nazi party (like Lenard) or the Nazi Storm Detachments known as the SA (like Weyland).

It was also expecting too much of the readers of the *Berliner Tageblatt* to discuss *scientific issues* in a regular newspaper, as Einstein seemed to think necessay (his numerous lectures before audiences with an inadequate knowledge of science show that this was not the only instance).

Well-intentioned contemporaries also heftily criticized "that rather unfortunate reply in the newspapers."[337] Einstein regretted it as well: "Everyone has to sacrifice at the altar of stupidity from time to time, to please the Deity and the human race. And this I have done thoroughly with my article. This is shown by the exceedingly appreciative letters from all my dear friends."[338] But he did not have the courage to apologize, as Sommerfeld expected of him.[339] Einstein could have worked on limiting the damage, but it would not have prevented the open conflict with Lenard, who soon afterwards, in 1924 openly pledged his allegiance to Hitler. He did not even respond to Gehrcke's reply in the *Berliner Tageblatt* of 31 August 1920. In it Lenard had surmised that Einstein would have difficulty "producing the evidence correlating with political and personal motivations my many years of factual rebuttals to the theory of relativity."

It is worth mentioning here that Einstein's comet-like rise years before overshadowed Lenard's place in science and science policy. Before there was any thought of appointing Einstein to the Berlin Academy and nominating him director of the Kaiser Wilhelm Institute of Theoretical Physics, Lenard had drafted at Althoff's behest a 'Memorandum and plan for a German Institute of Physical Research.' It was submitted in *December 1906*.[340] So he certainly could hope to become its director. Einstein's appearance spoiled everything for him. Lenard probably now also thought he had a personal account to settle with him. Personal envy of Einstein, scientific and political opposition all contributed toward his allowing his hatred to escalate.

News that Einstein had the intention of leaving Germany spread fast. Democratically minded persons abroad were shocked about the conduct in Germany. Einstein's initial reaction was indeed to that effect and the public believed it.

On 2 September 1920, the German chargé d'affaires in London reported to the Foreign Office in Berlin that the attacks on Einstein were extremely damaging to Germany's reputation abroad.

Report by the German chargé d'affaires in London, 2 September 1920[341]:

In re: Professor Einstein.

The English papers have passed on the news about the hefty attacks against the famous physicist Professor Einstein. Today's 'Morning Post' even publishes a notice that Professor Einstein was intending to leave Germany and emigrate to America.

Although it is known here in England too that a prophet finds no appreciation in his own land, the attacks on Professor Einstein and the incitement against this famous scholar does make a very bad impression here. Professor Einstein is just at this moment a cultural factor of the first order for Germany, because Einstein's name is known among the widest circles. We should not chase a man out of Germany, with whom we really could do some cultural propaganda.

Should Professor Einstein in fact have the intention of leaving Germany, I would consider it desirable, in the interest of Germany's reputation abroad, if the famous scholar could be retained in Germany.

sig. Sthamer.

The German Foreign Office forwarded this report to the minister of science, arts and culture requesting his opinion. The minister's answer was detailed. He emphasized: "If the German chargé d'affaires in London points out in his letter from the 2nd inst. that the prophet finds no appreciation in his own land, this statement could not apply to Professor Einstein." He then itemized what distinctions Einstein had received since 1913. In conclusion he played down the attacks on Albert Einstein: "Against this, slight importance should be attached to the attacks of a few unqualified persons, who certainly lack professional legitimacy to pass judgment on Einstein's theory."[342]

Nonetheless, Haenisch did recognize the *danger*. On 30 August he was already toying with the idea of expressing his dismay to Einstein and asking him to stay in Germany.[343] He sent a letter to this effect on 6 September 1920; on 7 September it was released to the press.

From: *Wolff's Telegraphisches Büro*. Morning edition, 7 September 1920[344]:

Berlin, 6th of September. As we have been informed, in response to the recent events the Prussian minister of education has sent the following letter to Professor Albert Einstein:

Highly esteemed Professor!

With pained embarrassment did I gather from the press that the theory you champion has been the object of malicious public attacks extending beyond the bounds of factual judgment and that even your professional integrity did not remain safe from disparagement and libel. It is especially gratifying to me that scholars of acknowledged reputation, even prominent members of the University of Berlin among them, are standing by you with regard to this affair, repulsing the worthless attacks against your person, and pointing out that your research has guaranteed you a permanent place in the history of our science. As the best qualified are supporting your cause, it will be that much easier for you not to give any more attention to such ugly activities.

Thus I may surely also express the resolute hope that there be no truth to the rumor that due to these nasty attacks you wish to leave Berlin, which has always been proud, and always will be proud to count you, highly esteemed Professor, among the most brilliant jewels of its science.

In voicing my very particular esteem, I am yours very sincerely,
Haenisch

Einstein answered the minister's letter immediately. He thanked him and reassured him that he intended to stay in Germany because "Berlin is the place in which I am most deeply rooted through personal and professional ties."

Einstein to Minister Haenisch, 8 September 1920:[345]

Prof. A. Einstein

Berlin W 30, 8 IX 20.
Haberlandstr. 5

To the Minister of Science, Art and Culture, Mr. *Haenisch*
*Berlin*

Your Excellency's letter of the 6th inst. fills me with a sense of sincere gratitude. Quite apart from the question of whether I deserve so much benevolence and high regard, in these last few days I realized that Berlin is the place in which I am most deeply rooted through personal and professional ties. I would follow a call outside the country *only* in the case that external circumstances force me to do so.

In utmost respect,
Your Excellency's loyal servant,
A. Einstein

Efforts had been made to keep Einstein in Germany well before September 1920. On 23 October 1919, for example, the academy petitioned the ministry for a raise in Einstein's salary from 12,000 to 18,000 marks retroactive to the first of the month and the ministry approved it on 9 December 1919. When the ministry suggested a salary raise for the Professor of German Studies Burdach, who held the same kind of independent position without teaching obligations as Einstein's at the academy, the academy rejected it at first on the justification that "it only applies for an improvement of budgeted compensation of an individual member in cases where there is a danger that the academy would lose him."[346] Einstein's salary was increased repeatedly in the following months (from 1922 with reference to inflationary pressures as well as to Burdach's salary).

The academy did not want to lose Einstein. But when it came down to standing behind its famous member after the attack on 24 August 1920, the presiding secretary, Roethe declined on the argument that the declaration by Nernst, Rubens and von Laue was sufficient. He thought it advisable that the academy stay in the background, especially considering that not all its members were inclined to issue a certificate of approval to Einstein.[347] Roethe continued: "The even-tempered declaration released by Nernst, Rubens and Laue seem to me to suffice entirely." Nor did he see any reason to call a special meeting of the academy. Planck generally agreed, especially since he knew that Einstein actually did not want to leave Berlin.[348]

Minister Haenisch's attempt to down-play the seriousness of the situation to the Foreign Office in his reaction to the letter by the German chargé d'affaires of 2 September 1920 even makes out Einstein's opponent as "matter-of-factual."

The minister of science, arts and education to the Foreign Office in Berlin, 19 September 1920 (handwritten draft):[349]

*Immediately.*

To the Foreign Office [. . . ]. If the German chargé d'affaires in London points out in his letter from the 2nd inst. that the prophet finds no appreciation in his own land, this statement could not apply to Professor Einstein. Even before the war prominent German scholars drew attention to this important man and his unparalleled accomplishments for science. By imperial decree of 12 Nov. 1913 he was confirmed as a member of the Prussian Academy of Sciences and financial means were granted him to allow him to live completely for his science - freed from the worries of seeking a livelihood. His merits have also been fully acknowledged by the Prussian State Assembly; a proposal in

**Figure 19: Conrad Haenisch of the Social Democratic Party. Prussian minister responsible for science, education and cultural affairs from 14 Nov. 1919 to 21 Apr. 1921.**

which the State government was petitioned to make available funds in conjunction with the Reich government was accepted by the State Assembly, in order to make possible Germany's continued successful collaboration with other nations toward developing Albert Einstein's fundamental discoveries, allowing him also to pursue his own research.- [. . . ]. - Scholars such as Lenard in Heidelberg, who are opponents of Einstein's theory, have thus far presented their opposing scientific convictions in a matter-of-factual manner. Against this, slight importance should be attached to the attacks of a few unqualified persons, who certainly lack professional legitimacy to pass judgment on Einstein's theory. The reason why they managed to receive such attention is that they were heralded in the press by a less than dignified peddler's publicity for Einstein that had nothing to do with him personally.

The Minister pp.

The Anti-Einstein League had miscalculated. Instead of the twenty announced lectures, only two such evenings actually took place.

The first attack on Einstein had been rebuffed, but the second was already in the making. Lenard made an appearance before the scientists convention in Nauheim (19 to 25 September 1920) and attempted to debunk Einstein's theory of relativity. He had every right to do so. But in view of Lenard's past and the way he confronted Einstein, his performance was politically tainted.

Lenard did not stand a chance at the convention, though. He simply could not measure up to Einstein in a scientific debate. Lenard's main argument was that Einstein's theory was unintuitive. Einstein replied: "Intuitive ideas have their weaknesses, just as does the so oft-cited human common sense."[350] It only spurred Einstein's opponents on to press the "healthy common sense" issue. They used demagogic means to inflame public sentiment against Einstein.

The defeat in Nauheim enraged Einstein's political opponents. This anger came down on the organizer of the convention, Max Planck, and other prominent German physicists. Under the headline 'The Scientists Convention in Nauheim. Throttling of Einstein Opponents,' Paul Weyland wrote in the *Deutsche Zeitung* of 26 September: It is "high time [...] that this rats' nest of scientific corruption get a fresh airing."

Who was this Paul Weyland? A swindler, slanderer and imposter with heroic ambitions, a demagogue, fascist and anti-Semite.

Numerous documents can be found in the file 'Einstein's Theory of Relativity' on his criminal activities during the 1920s. Andreas Kleinert's truly meticulous research and other documents draw a complete portrait of Weyland.[351]

**Paul Wilhelm Gustav Weyland** born 20 Jan. 1888 in Berlin, deceased 6 Dec. 1972 in Bad Pyrmont. He presumably attended preparatory high school; good foreign language skills (English, French, Spanish). According to Weyland's own (retrospective) account, he was a chemist by training. He is occasionally described as an engineer during the 1920s. 1919 first emergence as a novel writer. 1920 marriage, 1933 divorce. In issue 1 (1921) of *Deutsch-Völkische Monatshefte,* edited by himself, he published articles on revolutionary poets, sculpture, a tale and a poem. 1920 organizer of the anti-Einstein campaign. Since fall of 1921, extended trips abroad, including a trip to the United States. 1922 plans for a trip to Norway. Weyland applied to receive financial support from the German consulate in Hammerfest for his purported astronomical expedition there. (Since no consulate existed at Hammerfest at that time, his letter was forwarded to the German embassy in Christiania).[352] It forwarded the paperwork to the Interior Ministry, which contacted Prof. Ludendorff, director of the Astrophysical Observatory in Potsdam, for his opinion. Ludendorff replied that Weyland had organized the lectures in Philharmonic Hall, "the purpose of which was to incite the wider public against relativity theory." He returned Weyland's application "with the comment that I urgently advise against any support of Mr. Weyland."[353]

In 1923 – that is just before Einstein's trip there – Weyland traveled to Scandinavia. A commentary appears in the *Svenske Dagbladet* under the heading 'A suspicious figure: false dove of peace or dangerous customer? An eminent authority of pan-German anti-Semitism stays in Stockholm. Propaganda visit on occasion of Einstein's imminent Nobel lecture?' The Prussian minister of science, arts and culture was concerned and urged the German embassy: "In my opinion everything possible ought to be done to prevent him from doing anything more in this direction. It would otherwise be expected with great certainty that sooner or later incidents that are exceedingly detrimental and embarrassing to the reputation of Germany will occur."[354]

Weyland's later career in brief: On 5 May 1933 application for membership in the Nazi party. At this time Weyland was director of the SA chorus. On 28 Sep. 1933 Weyland was "expelled permanently from the SA [...] because of his criminal record and conduct damaging to the party as well as owing to unauthorized absences from duty." In October 1933 the squad leader and "SA man, Weiland," fled to Prague.[355] 1933 and subsequently, further fraudulent activities abroad. Weyland posed as an opponent of National Socialism. On 22 July 1936 emigrated from Germany. On 31 Mar. 1939 returned to Germany; immediately detained and from 1940 to 1945 interned in a concentration camp. After the liberation he worked for the American occupying forces, later for the CIA. On 11 Jan. 1954, Weyland became a citizen of the United States of America and continued there what he had been doing in Germany during the 1920s: he denounced *Einstein* (this time to the American criminal authorities, the FBI). Thus Weyland returned to the stage where he had started his political career.

In 1922 there was a new wave of virulent aggression against Einstein. All efforts were directed at denigrating his trip to Paris as a "dishonorable" representative of the country and "traitor," etc.

For lack of arguments the political reactionaries turned to other means. Walther Rathenau was assassinated. On his funeral day, an officially declared day of mourning, Philip Lenard demonstratively held his lecture at Heidelberg. Within this political context, a series of other political murders took place and Einstein, too, was named as a desirable target. He was forced to cancel the summer term lecture he was in the middle of teaching at the University of Berlin, because he had been "urgently advised, in the interest of [his] personal safety, not to perform any public functions at this time."[356] On 6 July 1922 Einstein wrote from Kiel to cancel a talk he was scheduled to deliver before the Society of German Scientists. His justification in this letter to Max Planck was: "For I am supposedly one of a group of people against whom nationalists are plotting attacks [...]. So there is nothing else one can do but be patient and go abroad."[357] He first simply avoided making any kind of public appearance. He wrote Solovine: "I, too, am constantly being warned, have abandoned my lecture and am officially absent, while being, in fact, here. Anti-Semitism is very strong."[358] Finally he left for a three-month stay in Holland.

At the eighty-seventh convention of German scientists in September 1922, signatures were even being collected for a "protest" by a group of doctors, mathematicians and philosophers against relativity theory. How right Einstein was, when he wrote Hermann Weyl as early as 23 November 1916: "Although the theory still has many opponents at the moment, I console myself with the following circumstance: The otherwise established average brain power of the advocates immensely surpasses that of the opponents!"[359]

The blood-thirsty hate campaign against Einstein continued. In 1924 the fascist philosopher Dietrich Eckhard demanded outright that Einstein be assassinated, in his book 'Bolshevism from Moses to Lenin. Dialog between Adolf Hitler and me.'[360] Eckhard quotes Hitler's opinion: "The physicist Einstein, who let himself be revered like a second Kepler, declares that he has nothing to do with Germanism. He thinks that the habit of the Central Association of German Citizens of the Jewish Faith to present themselves only as a religious community of Jews, not as nationalists is 'dishonest.' A black swan? No, just someone who believes his people are out of the wood and therefore no longer needs to make false pretenses."[361]

This fascist leader, Adolf Hitler, did not want to help Jews out of the wood but into the grave. So any distinction between Jewish voices pro or contra the Central Association was moot. He was clear enough about that in *Mein Kampf*. The hysterical Jew-baiting 'Protocols of the wisemen of Zion' was a suitable tool for that purpose. It appeared in 1919 and was sold in large numbers in various editions. "It goes without saying that the readers of the 'Protocols' could hardly do otherwise than identify the idea of 'disguised' Jewish confusion theories of science specifically with Einstein and relativity theory – and with a few other ambivalent formulations besides."[362]

Despite all this, in 1924/25 the attacks on Einstein subsided again for a while. "Outwardly, at least," Max Born wrote in his memoirs. "Under the ashes the animosity toward him continued to smolder on until it flared up again openly in 1933."[363]

## 2.5 Emissary and emigré – Einstein's foreign travels

### 2.5.1 Reasons and purposes of Albert Einstein's travels abroad

In the years 1920 through 1925 Einstein was abnormally frequently on lengthy travels. He spent the period from April 1922 until March 1923 almost continuously outside the country. Altogether, his foreign trips from 1920 to 1925 lasted more than seventy-five weeks, so Einstein was away from Germany for at least one and a half years.

Einstein naturally gave very many lectures on relativity theory. Their purpose was in most cases not primarily scientific. (Very few of his listeners could even understand what he was talking about.) His travels and lectures served *other* purposes and *other* functions.

Einstein himself later stated that the main purpose of these trips had not been scientific. Einstein's response in 1932 on a questionnaire re. "VI. Scientific travel" on the occasion of his election to the Imperial German Academy of Scientists in Halle (Leopoldina):[364]

> VI. Occasional lecturing trips to France, Italy, Japan, Argentina, England, the United States, which - with the exception of the travels to Pasadena - did not really serve research purposes.

So why so much traveling?

We find an initial answer by looking at the kinds of people who invited Einstein – as far as they are known – and funded his excursions. These inviters and funders were:

- the local 'Astronomical Society' for the trip to Copenhagen in June 1920;[365]
- American Jews and the University of Manchester for the trip to the United States and England[366];
- the Collège de France (or, if the *Berliner Tageblatt* of 24 March 1922 was correct with reference to the French *Matin:* the "Michonis Foundation") for the trip to France in 1922;
- Mr. Yamamoto, editor of the *Kaizo*, a popular monthly magazine, for the trip to Japan;[367]
- the Jewish cultural 'Asociación Hebraica' and Institución Cultural Germano-Argentina' and a few wealthy Jewish businessmen for the trip to Argentina;[368]
- the sojourn in Palestine in 1923 (on the way back from Japan) was doubtlessly financed by Jews or a Jewish organization;
- the plan of attaching a visit in Spain to the homeward trip from Japan and Palestine arose from a request by the German ambassador of Spain, Ernst Baron Langwerth von Simmern in September 1922 and forwarded by Heilbron and Solf.

This does not mean that Einstein's interests coincided with the goals of his hosts or were *identical* with the purpose of the trips. As a closer look at the trips will reveal, at least three influential factors were involved:

**Figure 20: A new file: 'Lectures by Professor Einstein Abroad' compiled by the Foreign Office since March 1922.**

1. The hosts and funders, including the World Zionist organization.
2. The German state and, in particular, the scientific community in Berlin.
3. Einstein personally.

Einstein was free to decline and he occasionally did so. He declined invitations to Batavia in 1922, to China in 1923, and to Mexico in 1923 and 1926. He was already "firmly resolved" around the time of his Japan trip "not to drive around the world so much."[369] But he continued to do so nonetheless – despite the considerable time it cost and the associated physical stress.

Why did he stay abroad so long? Why did he leave so often precisely around the beginning of the 1920s?

A more sweeping answer will be possible once we have taken a closer look at the trips themselves. For the moment, let me list the three most important reasons:

First. The trips served to consolidate his reputation in the scientific world and disseminate his theory. For all his modesty, he was fully aware of his own merit. He was obviously not ambivalent about whether adherents or opponents of his theory had the upper hand. He had no need for blatant propaganda. But a complete retreat from the public eye would have been seen as a sign of extreme arrogance (or at least interpreted as such). He would have hurt his own image.

His sudden loss of interest in traveling in 1926 shows that an understandable personal interest in it was not the only or most important reason for his frequent voyages.

Second. As a social animal, Einstein shared certain *interests* with other persons and groups. Where necessary, he was ready to serve "as a famed bigwig and decoy-bird."[370] He did so for the German state, the interests of his Berlin colleagues, the World Zionist organization and other institutions both inside and outside the country. With his travels Einstein made a significant contribution toward terminating the boycott of German science. When the boycott broke down, Albert Einstein's period of extended travels ended as well. This temporal parallel was no coincidence.

Einstein was a suitable emissary precisely *because* he could not be fit into any particular political mold. His unconventional deportment made him at once a complicated and pleasant guest (a genius and likeable oddball). His unconventional manners, apparently at variance with any diplomatic form, and his political opinions caused the Foreign Office trouble often enough. Anyone else would have been brought into line. Not Einstein. He had to be taken as he was, acceptable or not. Outside the country, people often got the impression that he had no inclination for Germany. But *afterwards,* when the diplomatic sum was added up from a foreign trip, it turned out that he had been *just* the right person for it.

Einstein was particularly suited as a solicitor of German interests because this admired authority of his field and most internationally renowned scholar came *from Germany* and was seen as a *German* without being viewed as a *German nationalist.* The controversy over his citizenship so irritating to many of his contemporaries could not alter this fact. Indeed it ultimately even proved useful, because it seemed clear that he could not have been sent by the German government. His not having been a co-signer of the 'Appeal to the Civilized World' but rather of the 'Manifesto to the Europeans' turned out to be exceedingly useful for his diplomatic mission. He could not be counted among the Germans that foreigners distanced themselves from. This was simultaneously a protection for Einstein's dialog partners. They were spared the accusation of betraying national interests to enemy Germany. It was still acceptable to speak with *Einstein.* That was hardly the case with *Haber.*[371]

For this reason I would like to contradict the thesis advanced by Brigitte Schroeder-Gudehus that *Fritz Haber* played a *pioneering role* in the renewal of international collaboration in science. It is incorrect to assert that Einstein's many foreign travels, "construed as an important step toward the restitution of German scholarship in international scientific relations [...], is unjustified, however."[372] *Einstein* was the pioneer. *Einstein* softened the opposing fronts. Haber had a chance to do so on his own initiative and with his organizational talents only once the ice in international scientific relations had already been broken.

After World War I governmental authorities purposefully tried to draw political capital out of past refusals to sign the Appeal by the Ninety-three. Confirmation of this is a report by the German embassy in Paris dated 28 September 1920 to the Foreign Office. It reports that W. Foerster, the German chairman of the International Committee of Weights and Measures, had compromised himself by having signed the Appeal by the Ninety-three. The German ambassador regretted that he was not one of the scholars who had withheld their signatures, "because then it would have been certain, that the neutral parties and even some of the Allies would have joined Germany's side."[373] After the war, "traitors to the fatherland" could evidently be of some use to the fatherland.

The German government, and particularly the Foreign Office, were greatly interested in Einstein's trips abroad. The sheer number of reports concerning Albert Einstein and inquiries and requests for information directed to German embassies outside the country is indicative of this.

Third. Finally I arrive at a point that is given too little notice in the existing literature. Einstein's foreign travels were a form of *emigration*. He did not go into exile but went away on long voyages. Thus he avoided people who did not want to have him or see him in Germany. There was reason enough to do so. The early years of the Weimar Republic, as has already been mentioned, were fraught with assassinations of leftists and Jews.[374] As a Jew of the political leftwing, Albert Einstein, too, was in jeopardy. Foreign countries offered him much more security than the German "fatherland." The government was thus relieved of much worry. At the same time, it profited politically from Einstein's growing international fame on his tours around the world. (Einstein's frequent absences abroad were also financially beneficial for Germany. His travel costs were taken over by others during an economically difficult period, and his salary payments could be suspended.) This "emigré" was at the same time an "emissary." Because the situation in Germany drove Einstein beyond its borders for a *long time,* he served as a German emissary for a *long time* as well.

Numerous statements by Einstein, some of which have been quoted above, reveal how aware he was of the dangerous situation at home.

Einstein's letter to Max Planck of 6 July 1922:[375]
rec. 7 Jul. 22. Planck.[376]

Kiel, 6 July 22.

Dear Colleague,

This letter is not easy for me to write; but it really does have to be. I must inform you that I cannot deliver the talk I promised for the Scientists' Convention, despite my earlier definite commitment. For I have been warned by some thoroughly reliable persons (many of them, independently) against staying in Berlin at present. I am supposedly among the group of persons being targeted by nationalist assassins. I have no secure proof, of course! But the prevailing situation now makes it appear thoroughly credible. If it had been an action of substantial professional importance, I would not let myself be swayed by such motives, but this involves a merely formal act that someone (e.g., Laue) could easily do in my place. The whole difficulty arises from the fact that newspapers mentioned my name so often and thereby mobilized the riffraff against me. So there is no helping it besides patience and travel. I ask you one thing: please take this little incident with humor, as I myself do.

With amicable greetings,
Yours, A. Einstein

An excerpt from A. Einstein's letter to W. Orthmann, chairman of the study group of mathematicians and physicists at the University of Berlin, in reply to Orthmann's letter of 18 September 1922,[377] that requested a continuation of his lecture course:

I had to discontinue my lecture in the summer semester, because I had been urgently advised, in the interest of my personal safety, not to perform any public functions at this time. I unfortunately cannot lecture next semester, because I am absent from Berlin. I was pleased to see such an interest in the subject as expressed by this request of yours; I shall try to make good on what is still outstanding as soon as possible.[378]

From a report by the German embassy in Tokyo from 3 January 1923 to the Foreign Office:[379]

During his absence, to be precise on 15 December, while he was already staying in the south of the country, the Adyertiser[380] reprinted a Kokusai-Reuter telegram according to which Maximilian Harden in Berlin stated before a court of law: ''Professor Einstein went to Japan because he did not consider hinself safe in Germany.'' - As this news seemed liable to damage the extremely favorable effect of Einstein's visit for the German cause, I asked Einstein by telegram for the authority to declare it untrue. Initially I received a telegram that the matter was too complicated for an answer by wire and that a letter would follow. In the subsequent letter from Miyajima of the 20th of the mo. Einstein writes the following verbatim:

I hasten to forward to you the more detailed sequel to my reply by telegraph. Harden's statement is certainly awkward for me, in that it makes my situation in Germany more difficult. It is not completely right, but neither is it completely wrong. For, people who can survey the situation in Germany well are indeed of the opinion that there is a certain threat on my life. I did not assess the situation the same way before Rathenau's murder as I did afterwards, however. A yearning for the Far East, led me, to a large part, to accept the invitation to Japan. On the other hand, it was the need to get away from the tense atmosphere in our homeland for a while, which so often has placed me in difficult situations. But after the murder of Rathenau, however, I was very relieved to have an opportunity of a long absence from Germany, taking me away from the present heightened danger, without my having to do anything that could have been unpleasant for my German friends and colleagues.

There were other reasons for his travels too. Even so they were a welcome opportunity to avoid the threats on his life and limb. He already knew in April 1922, for instance, that he would be traveling to Japan and Spain that October.[381] But he "very much welcomed the opportunity of a lengthier sojourn outside of Germany."[382]

## 2.5.2 The first excursions after the war – trips to neutral lands

The destinations of Einstein's postwar travels were the cities Prague, Vienna, Leyden, Oslo, and Copenhagen.

It is not the aim of the present book to retrace the exact itineraries of these trips. The *political function and effect* they had is what is important.[383]

Whether the German diplomatic embassies reported on *every* one of these trips abroad remains open to future research.

As it appears (judging from the files), the first report originated from the German embassy in The Hague (Holland), dated 25 May 1920. The mere fact that the ambassador submitted a report on it reveals the importance attached to this trip. Moreover, the Foreign Office passed the report on to the Reich minister of the interior, with the request that the Prussian minister of science, arts and culture also be informed. The trip was politically useful for Germany primarily because, as the ambassador thought, it "contributes substantially toward drawing the scientific communities of Germany and Holland closer together." Disregarding Einstein's own ambitions, this trip served toward dismantling the boycott of German science and generally fostering a return to mutually conducive relations between Germany and other countries.

From the report by the German embassy in The Hague, May 1920.[384]

Professor Einstein has been staying in Leyden since a few days, where he has connections with one of his most important forerunners in the field of science, the former professor at the University of Leyden - now Utrecht - and renowned Dutch physicist Lorenz.[385] Professor Einstein delivered a talk there on 20 May on 'Space and time in modern physics.' The function, which took place in the university's memorial hall, bore the character of a special honor for Professor Einstein, lent particular expression by the fact that the speaker was asked to speak from the raised podium, which is otherwise reserved for the Rektor Magnificus [university president]. To heighten the memorability of the moment, the dean also recalled in his introductory speech that such famous men as Descartes and Huygens had sat at the foot of this podium. [. . . ]. Without regarding the scientific importance of these lectures, Professor Einstein's activities here can only be described as very favorable, because they contribute substantially toward drawing scientific groups in Germany and Holland closer together.

In attachment I have the honor of submitting an article from the 'Nieuwe Rotterdamsche Courant' in original, which reveals the interest in Professor Einstein's expositions shown by the local press.

sig. Rosen

A few days later, on 22 June 1920, the German legation in Christiania (now Oslo) reported on three talks by Einstein on "the so-called theory of relativity propounded by him."

Report by the German legation in Christiania, 22 June 1920[386] (excerpt).

The professor from Berlin, Dr. Ernst[387] Einstein, held three lectures on the so-called theory of relativity here on the 15th, 17th and 18th of this mo. at the invitation of the local student association. The lectures were unusually well received among the public and the press. Admiration for the scholar was extraordinary. After the lectures, Professor Einstein was named an honorary member of the student association. I have the honor of enclosing herewith the commentary on the lectures by the local paper 'Aftenposten,' in the original.

sig. Mutius

Four days later, on 26 June 1920, the German legation in Copenhagen also reported home.[388]

Report by the German legation in Copenhagen, 26 June 1920 (excerpt).

Following an invitation by the local ''Astronomical Society,'' Professor Albert *Einstein,* who arrived from Christiania, delivered a lecture here as well. The topic was ''Gravikation[389] and geometry.''

The press of every orientation recently indicated Professor Einstein's importance in long articles and interviews as the ''most famous physicist of the present day.'' The reviews of the lecture emphasize the genius of Einstein's theory of relativity and the great clarity of his presentation. The lecture was acknowledged with enthusiastic applause. [. . . ].

Even though Einstein is Swiss by birth and supposedly Jewish by heritage, his research is nevertheless a part of German science. His many years of working in Berlin and the multifarious links between his theoretical conceptions of the world and those of German scientists did not prevent the English from welcoming his theory with warm and unqualified admiration, as soon as the results of the solar eclipse of May 1919 had confirmed it decisively. - There are many signs that some English scientists welcomed this opportunity with pleasure as the beginnings of a reconciliation in the world of science. [. . . ].

sig. Neurath.

A couple of phrases in this report are of interest besides the contemporary praise. The essential emphasis is: "Even though Einstein is a foreigner," and – worse still – "supposedly of Jewish heritage, his research is nevertheless a part of German science." The opinion expressed here was that of *Konstantin von Neurath.*

*In 1920,* German Envoy *von Neurath* was still of the view that Germany could very well make use of Einstein. As Reich foreign minister *in 1934, von Neurath* looked after Einstein's expatriation. One may assume that it was not Mr. Neurath's views that changed in the interim but *temporal circumstances. Later,* when Einstein had lost his direct access to the wielders of power in Germany, this Mr. von Neurath, in his capacity as Reich protector of Bohemia and Moravia, sent Jews and other opponents of the Third Reich to concentration camps and gas chambers. In 1920, however, with Germany still a weak and defeated nation, the very same Neurath was willing to employ foreigners and Jews like Einstein for the good of the fatherland.

**Konstantin von Neurath** (born on 2 Feb. 1873, deceased on 14 August 1956); 1901 began service at the Foreign Office. 1903 vice-consul in London. 1909 legation councillor. 1914–1916 ambassador in Constantinople. 1917–1918 head of the Civil Cabinet of the king of Württemberg. 1919 envoy in Copenhagen. 1921 ambassador in Rome. 1930 ambassador in London. From 1932 to 1938 Reich minister of the Foreign Office. 1937–1945 member of the Nazi party. 1937 award of golden party badge and personal appointment by Hitler as SS-Gruppenführer (lieutenant general). 1943 promotion to SS general (Obergruppenführer). 1938 head of the Secret Cabinet Council. Member of the Reich Defense Council. 1939–1943 Reich protector of Bohemia and Moravia and upon appointment also distinguished with the Order of the Eagle. On 1 October 1946 during the proceedings by the International Military Tribunal against main offenders (charges: (1) conspiracy, (2) crimes against peace, (3) war crimes, (4) crimes against humanity) von Neurath was sentenced to 15 years imprisonment. On 6 November 1954 he was released prematurely from Spandau Prison.

What Hitler wrote to Neurath when he asked leave to retire by reason of his advanced age proves that, contrary to some postwar myths, he was his loyal paladin, despite a few differences of opinion: "During the five years of our joint collaboration, your council and insights have become indispensable to me. If upon releasing you from the ongoing business at the Reich Foreign Ministry I thus appoint you as President of the Secret Cabinet Council, this is so that I may retain an advisor at the pinnacle of the Reich in future as well, who has stood by my side in trusty loyalty during five of the most difficult years. It is my deep heartfelt desire to thank you today for that."[390]

Not long afterwards, when the politically motivated, but purportedly "scientifically" based anti-Einstein campaign *in Germany* reached its first highpoint, a trip by Einstein into "enemy England" was still out of the question. But the German chargé d'affaires in London thought the same as his colleague Neurath in Copenhagen (see the letter by the German chargé d'affaires Sthamer from 2 September 1920, reproduced above).

*Thenadays* Einstein was considered "*a cultural factor of the first order*" – albeit only "at this moment." *Thenadays* the advice of the chargé d'affaires was: "We should not chase a man out of Germany, with whom we really could do some cultural propaganda." Einstein's function could not have been more explicitly defined.

### 2.5.3 The voyage to the United States and England

The first major trip Einstein set off on was in 1921 to the United States, with a stop-over in England for a brief visit on the way back.

This trip is remarkable for two reasons. First, it involved two countries that had both been enemies of Germany during the recent World War. Second, its overall function was as a fund-raiser for Zionism and the future Hebrew University of Jerusalem.

After World War I, Germany was courting the favor of the United States and the latter was not adverse to assigning it a special role in Europe. Germany was to be a counterbalance to England and France at the same time as serving as a future anti-Soviet drop hammer. This American policy was not immediately apparent. That would have led to friction with the countries of western Europe. But the refusal by the USA to ratify the Versailles treaty and its decision to sign a separate peace treaty with Germany on 25 August 1921 were clear signs that the United States chose to follow its own path and develop its own policy with Germany. But much time had to pass and much effort had to be expended on either side before a closer relationship could develop and before reservations in the USA could be overcome and public opinion swayed in Germany's favor.

In pursuit of these basic political goals, Germany's cultural policy was decisive. During the exploratory stage, what mattered was an improvement in the dialog between the two countries. Great importance was attached to proceeding in the least politically "objectionable" manner possible. The true goals had to stay in the background so as to be the more effective. The America Institute in Berlin was particularly active in this respect. Its president was Minister of State Schmidt-Ott – the last minister of culture of the Prusso-German Kaiserreich and president of the Emergency Committee of German Science. Its directors included such influential people as the future head of government, Cuno, the bankers Koppel and Max Warburg, Minister of State Dernburg and Minister of Culture Becker (as vice-president). Hence all entirely "apolitical people"!

Germany spied this chink in the enemy front and knew how to exploit it. Intense scholarly exchanges between America and Germany without obvious political goals helped draw the states closer together and were suited to putting pressure on British or French policy towards Germany. Consequently, whatever the motives were behind Einstein's trip to America and regardless of the interests it was intended to serve, the fact that Einstein arrived from *Germany*, was a *German*, and repeatedly came forward explicitly as such, made his trip also serve the interests of the *German* Reich. Whether Einstein was conscious of this is of secondary importance. Secondly, if Einstein *was* conscious of this implication, his trip was an important contribution toward *breaking the boycott of German science*. By making an apolitical appearance, Einstein was not hampering the goal but helping it on.

There is no causal link between these national goals and the other remarkable aspect of Einstein's trip: its function as a promoter of *Zionism*. It complicates the matter, though. Right after World War I, anti-Semitism and anti-Zionism were developing into a widespread and dangerous factor of political life in Germany (likewise even in USA). The goal of Zionism was to create an independent Jewish state in Palestine. After the war the idea drew its force primarily from two sources. First, it enjoyed the particular support of the British government (and for that reason alone was distasteful to Germany's reactionary political right). This was no secret even then. Weizmann declared in his memoirs: For some influential French and Italians "Zionism was nothing more than British imperialism in disguise."[391] Second, many Jews joined this nationalistic movement for whom life in Europe and Germany after the Great War was becoming virtually intolerable owing to discrimination and persecution. Eastern European Jews who had immigrated to Germany (the "Ostjuden") were a major pool of support for the movement.

Under these circumstances Einstein's voyage to the United States had a contradictory nature. On one hand, it was useful to Germany interests, on the other hand, it served a cause that was highly controversial – even among German Jews.

Einstein joined the Zionist movement only after World War I. Before coming to Berlin he did not take the Jewish issue seriously, even though he had personally got a taste of anti-Semitism during his school days in Munich and even during his stay in Switzerland. In the opinion of Kurt Blumenfeld – since 1909 propaganda head of the Zionist Association for Germany and since 1920 manager of the "Keren Hajessod" (Palestine Foundation fund) for Germany – Einstein never considered it an important life's decision when he exited or entered the Jewish community officially.[392] He rejoined the Jewish community in Prague only because it was not permitted for dissidents to be employed at universities of the Austrian empire. But Max Brod and Hugo Bergmann tried in vain to win him over to the Zionist cause at that time. When he was confronted with the situation in imperial Germany after 1914, Einstein's attitude changed. He wrote about this in a letter from 19 March 1929: "It may well be that my conduct has had an influence on Jews and others, but that was not from my own doing. I became 35 years of age, as it were, without knowing that I was a Jew. Only in retrospect do I realize that – despite living in a Christian environment and being married to a Greek-orthodox woman – my closest friends were Jews. It fell to my Berlin milieu to enlighten me about my belonging to the Jewish nation, is it therefore surprising that I understood the lesson? [...] so use me as you see fit [...] but know that I am entirely aware that in this capacity I am more an object than a person."[393]

The prehistory of Einstein's enlistment for the Zionist fundraising voyage begins in 1919.[394] At that time the Zionist, Felix Rosenblüth, compiled a list of famous Jewish scholars to be persuaded into joining the Zionist cause. The first discussions with Einstein took place soon afterwards. Less than two years later, on 10 March 1921, Weizmann telegraphed the head of the World Zionist organization, Kurt Blumenfeld, to ask Einstein to go on a promotional trip to America. The purpose of this trip was first to call to life the American Keren Hajessod, and second to collect money to establish the Hebrew University of Jerusalem. At first Einstein declined participation. He thought the Zionist agenda lent exaggerated importance to agriculture and initially did not think much of a Jewish university either. Einstein thought no better of the founding of a Jewish state. He was for a *homeland for Jews* in Palestine but not for an independent state. Besides, he did not like the role he was supposed to play on the trip. "One is only using my name, which is now the talk of the day," he replied to Blumenfeld.[395] But Blumenfeld would not give up. He knew how to serve the matter up to Einstein, put it in a favorable light and draw links to Einstein's political views. On 29 March 1921 he reported to Weizmann: "In the discussion of Keren Hajessod, the idea that all work would be carried out according to strictly economic aspects first hit home with Einstein. Especially the question of Jewish labor and the Jewish laborer was effective on Einstein, who has a strong socialist streak."[396] Finally, Einstein agreed but less out of sympathy for Zionism than out of compassion for Jews subjected everywhere to severe persecution. He had no illusions about the role cut out for him. On 8 March he wrote to Solovine: "I don't enjoy going to America at all but am doing so on behalf of the Zionists, who have to beg for dollars for the educational institutions in Jerusalem, for which I must serve as famed bigwig and decoy-bird. [...]. On the other hand, I am doing whatever I can for my clansmen, who are being treated so meanly everywhere."[397] At the same time Blumenfeld tried to convince Walther Rathenau also to come along on the trip, with Einstein's assistance, who was a friend of Rathenau. But Rathenau opposed Zionism and categorically declined.

Thus Einstein's trip to America was the most controversial of them all. Many Jews were adamantly against it. Fritz Haber, for instance, tried to dissuade Einstein, whom he referred to as "the most important of all German Jews," in a letter from 9 March 1921. His urgent warning to Einstein reportedly was that "it would damage his career and the institute of which he was a respected member, if he got himself involved with Zionists, especially with such an outspoken one" as Weizmann.[398] Haber admonished him that serving Zionism and going on the trip to America would hurt the interests of Germany and German Jews. It would be an act of disloyalty, of betrayal even, in the view of this nationalistically minded Jew.

Fritz Haber's letter to Einstein, 9 March 1921:[399]

Dear Albert Einstein,

Our many years of friendship compels me to write you today.

I am told that you are soon departing for the United States on behalf of efforts to establish a Zionist university in Jerusalem and that you are undertaking this voyage in the company of English proponents of this idea and are afterwards also going to stay in England at the invitation of the government over there. From the papers I found out in addition that this plan prevents you from attending the Solvay conference in Brussels, to which you have been invited as the only representative of our country.

Wherever your steps take you, my affection for you will follow. Everything you do, as long as I have known you, has always originated from noble human nature and the kindness of your heart. I feel this so strongly that it is difficult for me [. . .] to speak to you in your liberal humanity about the obligations that everyone of us has. I would not do so if I was not convinced that by it I was helping you. But [. . .] I do not want to have a man who is dear to me later say that he embarked on a path inevitably strewn with painful conflicts, without having been warned by his friend.

If you travel to America *at a moment when* the new president is delaying consultations on the law defining the declaration of peace between the United States and Germany; if you travel with English friends of Zionism, *at a moment when* the sanctions are exposing more starkly than ever the contrast between England and us; then you are publically proclaiming before the whole world that you do not want to be anything but a Swiss, who by coincidence has his residence in Germany. I beg you to consider whether you really want this proclamation now. Now is a time when belonging to Germany is an act of martyrdom. Do you really want this demonstration of personal alienation at this moment? People's need for profundity has deservedly coronated you and has lent significance to your commissions and omissions as were formerly only attributed to the dealings of princes. What you do, you do not just for the effect you yourself intend.

The English and Belgians want to divest the name Albert Einstein of the German character hitherto attached to it. If you permit them to do so, German Jews will have to suffer for it.

For the whole world you are today the most important of all German Jews. If *at this moment* you ostensibly fraternize with Englishmen and their friends, then the people here at home will regard this as evidence of disloyalty by the Jews. So many Jews fought in the war, died, or became impoverished without complaint, because they considered it their duty. Their lives and deaths did not eradicate anti-Semitism but reduced it to being regarded among those who care about the honor and greatness of our country as maliciousness and ignobility. Do you want to sweep aside so much blood and suffering by German Jews by your conduct? [. . . ].

Fritz Haber

Einstein answered immediately. He did not deny that Weizmann was just making use of the name Einstein for propaganda purposes. Nor did he dispute having deviated in this concrete case from his consistently antinationalistic mentality. He emphasized that the way Jews were being treated forced him to take Jewish solidarity more seriously than before. He rebuffed the charge that he was acting disloyally and thanklessly toward his German friends. Despite a lack of "affinity for the political structure of Germany," he certainly did make "considerations of tact." He defended himself and said he repeatedly stressed his being a *Berliner* and a *German* while he was abroad. There was no going back: "To America I must go."

Einstein to Haber:[400]

*9 March 1921.*

*My dear friend Haber,*

*What happened with this America trip, which cannot be changed under any circumstances, is the following. A couple of weeks ago, when no one was thinking of the political developments, a Zionist whom I value highly came to me with a telegram from Prof. Weizmann, the content of which was that the Zionist organization was inviting me to travel to America with a few German and English Zionists to consult about the educational affairs of Palestine. I was not needed for my skills, of course, but only for my name. They anticipate from its promotional power considerable success among our rich fellow clansmen of Dollaria. Despite my declared international mentality, I still do always feel obliged to speak for my persecuted and morally oppressed fellow clansmen, as far as it is within my powers. So I gladly accepted, without more than five minutes' consideration, even though I had just declined all American universities. So this much rather involves an act of loyalty than one of disloyalty. The prospect of establishing a Jewish university pleases me particularly much, after seeing recently from countless examples how perfidiously and unkindly fine young Jews are being treated here, in an attempt to deprive them of educational opportunities. I could tell you about other incidents last year that would compel any self-respecting Jew to take Jewish solidarity more seriously than had been necessary or seemed natural in former times. Think of Röthe, Wilamowitz-Moellendorf and the celebrated Nauheim guard, who only shook off the fool Wieland*[401] *in the end for opportunistic reasons.*

*No reasonable person could accuse me of disloyalty toward my German friends. I declined many attractive calls to Switzerland, Holland, Norway and England, without even thinking of accepting any of them. This I did, by the way, not out of attachment to Germany but to my dear German friends, among whom you are one of the most exceptional and well-intentioned. Any affinity for the political structure of Germany would be unnatural for me as a pacifist. Now, however, there are tact considerations that arise from the moment; at the present moment they produce a conflicting situation that could not have been foreseen.*

*The situation is exacerbated by my having accepted an invitation a few weeks ago to give a talk at the University of Manchester. The choice of exactly when to give it has, incidentally, been left largely to me. A few weeks ago, no sensible German would have approved of my rejecting it. Now my acceptance looks like a provocation toward Germany, but that was quite certainly not my fault. If the ominous political situation should continue, I would perhaps be able to dispense with the visit to Manchester. Our colleagues there would understand, if I explained the reasons in full friendship and honesty. Scientific cooperation is, incidentally, by no means the State. If scholars would take their profession more seriously than their political passions, they would guide their actions more according to cultural than political aspects. It must even be said that in this regard the English have behaved much more nobly than our local colleagues. They are to a large part Quakers and pacifists. How magnificent their attitude has been toward me and relativity theory in comparison! You may perhaps not have followed this so closely, but I can only say: Hats off to the fellows! For the English I am, by the way, a Berliner, through and through, whose international mentality is known to them. So their kind invitation really ought to be acknowledged. They recently sent an inquiry via the German envoy in England whether I would visit London if I received an invitation of an official nature. It is fortunate that this invitation has not yet been issued. In any event, this fact also shows that English scholars do not want any enmity.*

*All this is cura posterior. To America I must go, however, because*

*I have given my firm acceptance and the steamship seats have already been booked. I am just carrying out a natural obligation.*

*I would have gladly gone to the Solvay conference and I declined the visit only with a heavy heart. Nernst was furious, by the way, when he heard that I had been invited there and was intending to go. You regret - likewise for nationalistic reasons - that I had to decline. Doesn't this remind one of the nice ancient fable of father, son and donkey?*

*Dear Haber! An acquaintance recently labeled me a 'wild animal.' This wild animal likes you and will look you up before departure, if at all feasible in this hustle and bustle. Until then, my warm greetings,*

*Yours,*

*Einstein.*

The small delegation (of five members) embarked on their voyage from Rotterdam. *Elsa Einstein* was one of the party.

**Elsa Einstein** was, according to Armin Hermann, irritating company.[402] Einstein's derogatory remarks do exist, such as referring to her as "the only piece of furniture [...] along on the trip" or complaining that "my old lady [... could] not refrain from hovering over me the whole day long and keep finding something to put right about me."[403] Whether Elsa was in fact burdensome remains an open question. The answer would belong in a biography of Einstein. The relevant issue *here* is whether she *harmed or helped* the undertaking. The latter is more likely and would probably also apply to Einstein's next extended trip – the voyage to Japan. Precisely *because* she frequently found "something to put right" about him, Elsa caught in time many of Einstein's laxities that would certainly not always have been greeted with applause. It is also worth mentioning that Elsa Einstein's knowledge of human nature was not worse than her husband's but apparently rather better. In this respect her presence was a *political* factor and therefore deserves mention. Whether Einstein was of a different opinion is, in *this* context, a secondary issue.

A balanced assessment is provided by Konrad Wachsmann from personal experience. He was the architect of Einstein's summer villa in Caputh. "Later I experienced the constant appearance of beggars, tramps, solicitors or enthusiastic tourists at the door, who just wanted to catch a glimpse of Einstein. Because he listened to whoever approached him, he would never have got any work done if his wife, daughter or maid had not already separated the chaff from the wheat."[404] Elsa Einstein complained that "Einstein paid attention only to his own research and, at best, to matters that interested him for humain or political reasons. Everything else he left to her, and she was often even admonished because she was not prepared to take obligations into consideration that a man of his station happened to have. She was always the buffer between the public, with its constant demands on Einstein, and him, who turned his attention to the public only when he happened to think it right to do so for some reason."[405]

"His wife had to relieve him of every decision, indeed everything that did not at that instant concern him or was a nuisance."[406] Elsa Einstein was a "kindhearted and devoted wife"; she relieved him of "almost all problems and shielded him from excessive impositions from the outside world, which was constantly besieging him, even though she herself would probably have preferred to have been seen in this world much more often. Her love was indivisible and she simply could not understand how her husband could occasionally take an interest in other women."[407]

With such a wife at his side, very much was made very much easier for the great scholar. What use would it have been if the spouse had thought and felt the same way about everything as Albert Einstein himself? Without this wife, the somewhat reclusive scholar would very many times have foundered in public. So we have good reason to assume that Elsa Einstein exerted a beneficial influence also in *political* and diplomatic matters.

Einstein's work in the USA for Jewish charity was multifacetted. For example, the *Jüdische Rundschau* from 20 April 1921 reported that the famous millionaire Rosenblum from Pittsburgh immediately contributed a larger sum for the university in Jerusalem after having been received by Einstein.

As anticipated, Einstein served as a "famed bigwig and decoy-bird." Weizmann preferred to do the talking himself. The head of the World Zionist organization did not want to have his plans ruined by any undesirable *political* statements that escaped from Einstein.

Among the lectures that Einstein held on *scientific* issues, the ones he delivered at Princeton University are most prominent. They went down in history as the classical treatises on relativity theory. On 9 May 1921 Einstein was conferred an honorary doctorate at this university.

Another incident worth mentioning is Einstein's meeting with Charles Steinmetz (1865–1923), a very famous American engineer, on 23 April 1921.[408] By then Steinmetz had already positioned himself among the political left. On 16 February 1923 he eventually wrote to Lenin to offer his assistance in the rebuilding of Soviet Russia. When Einstein and Steinmetz met, "they talked together for hours about Russia, Lenin and the new red-tinted dawn of mankind."[409]

The scheduled climax of Einstein's trip was a visit to the White House. There was no exchange of ideas, though. President Harding could understand neither German nor French, Einstein could barely understand English. But both of them were men of standing, so even a handshake with a friendly smile were politically significant acts.

Elsa said to a reporter for the *Nationalzeitung* (19 June 1921): "My husband also met with President Harding. But this meeting was only pantomime-like, since President Harding and Professor Einstein could not converse with each other. For the president speaks neither German nor French, and my husband no English. So the meeting was only relatively short. It consisted in mutual friendly handshaking and posing together for a photograph."

President Harding's reception of a *German* scholar confirms that this trip was certainly not dedicated exclusively to the cause of the World Zionist organization. Contrary to Fritz Haber's fears, it also served *German* interests.

Einstein remained in the USA until 30 May. All in all, the trip was a success. First, it furthered the set goal. Substantial contributions were made for the Keren Hajessod.[410] Second, it enhanced Germany's reputation beyond its borders and helped put an end to the boycott of German science. Looking back, Einstein wrote to Ehrenfest on 18 June 1921: "My trip was also good for the restoration of international relations. Everywhere I went there was good will, a cordial reception and a mentality of peace."[411]

The *Vossische Zeitung* (15 June 1921) concluded in satisfaction that the trip "was a successful advertisement for Germany," especially since Einstein "never denied being German." "Although a few still belligerent papers denoted the scholar as a 'Swiss professor,' Dr. Einstein himself never denied being German." The trip, it continues, "contributed not little [...] toward repairing friendly relations with Germany." The same paper also noted that the use of the *German language* in Einstein's lectures was of extremely great *political* importance (forgetting to mention that Einstein could not have done otherwise). In the aftermath of the war, the paper states on 2 July 1921 ('Prof. Einstein's travel impressions'), public opinion in the USA was so agitated "that even use of the German language was as good as prohibited. Currently a definite turn around is taking place. Einstein described how cordially he was received by scholars and scholarly corporations. People everywhere were pleased to speak German and thought of German scientists and institutes with genuine sympathy."

On the return trip from America, Einstein also visited England. His meeting with Lord Haldane and Prime Minister Lloyd George were of particular political significance.

The philosopher and politician Haldane had more than his work on relativity theory (1921: 'The reign of relativity') in common with Einstein. He – like Einstein in Germany – was an *outsider* among his countrymen, and, certainly after the war, was not a typical representative of Britain. Shortly before the war had broken out he had advocated reconciliation with Germany. This "germanophil attitude" cost him his position as Lord Chancellor in May 1915. This did not affect his views and he refused to join the chorus of germanophobes. For this reason he was a very suitable messenger of the new policy of more favorable relations between Germany and Great Britain after the war.

Einstein delivered a talk at Kings College. The mixed audience initially received him with cautious reserve. Einstein was a famous scientist but he was also a *German*. At first there was no applause. Einstein spoke about the international role of science, mutual exchanges between scholars, the role of the English in the development of physics and about Newton. He thanked his English colleagues and emphasized that without their help relativity theory could scarcely have been verified so quickly. His talk was a eulogy on the international community of science. The result was a transformation of the mood not only in the auditorium but among scientific circles in the whole of England as well.[413]

The weekly publication *The Nation* noted a definitive "turning point in postwar sentiments." The German ambassador and Haldane himself were of the following opinion: Einstein's reception in England would help smooth the path toward improved relations.[414]

*Die Nationalzeitung*, 19 June 1921 (excerpt):

''*Professor Einstein's Voyage to America and England*

*A Conversation with Mrs. Einstein* [. . . ]. Because my husband used the German language there (as he does everywhere) in delivering his talk, it was *the first time* since the war *that German was spoken again at an official occasion in London*. [. . . ].

**Figure 21: Group photo on board the "Rotterdam" on the way to America, 21 March 1921. From left to right: Ben-Zion Mossinson, Albert Einstein, Chaim Weizmann, Menachim Ussishkin.**[412]

The welcoming speeches by Lord Haldane and the president of the university [. . . ] pointed out that with Einstein's visit in London *scientific relations between Germany and England were once again officially resumed*. A similar declaration about *the official resumption of Germano-American scientific exchange* was made at a New York dinner. Everywhere, Einstein's visit and talks were given an importance extending beyond the scientific sphere to general relations between hitherto hostile nations.''

Even the *Daily Mail*, one of the most anti-German papers, had to admit:

He is a German and delivers his lectures in the German language, a fact that does not particularly recommend him to an English audience. He is a Jew. [. . . ]. He is a revolutionary. And yet the reception he got was not just friendly but downright enthusiastic. [. . . ]. It does not matter whether one understands his language or not. One knows one is under the spell of a forceful personality, a powerful mind.[415]

Another remarkable aspect about the trip to England is that use of the German language – Einstein's native tongue – was politically extremely significant. Einstein's talk was the first to be held in German in London since the outbreak of

324 Berliner Illustrirte Zeitung Nr. 22

Prof. Dr. H. Müller-Breslau, der bekannte Brückenkonstrukteur und Lehrer der Statik, der den 70. Geburtstag feierte. Phot. Dührkoop

Olga Wohlbrück.

In dem schönen Freiburg, wo Olga Wohlbrück jetzt wohnt, ist das neue Bild der Verfasserin der „Athleten“ aufgenommen worden. Dort hat sie auch diesen Roman beendet, der wiederum ihre starken Erzählereigenschaften zeigt und ihre Kenntnis der internationalen Typen. Denn Olga Wohlbrück hat wie wenige Autoren Europa gesehen. In Wien ist sie geboren und kam dann nach Moskau und Kiew, wo sie das Gymnasium besuchte. Sie wurde Schauspielerin in Paris und Schriftstellerin in Berlin. Hier ist sie mehrfach erfolgreich an Bühnen aufgetreten. In dieser Zeit hat Olga Wohlbrück sich mit dem Komponisten Waldemar Wendland vermählt, dem Schüler Humperdincks, von dessen Opern mehrere mit großem Erfolg über die Theater Deutschlands gegangen sind.

Prof. Einstein mit seiner Gattin als Gast beim Präsidenten Harding in Washington.

Frau v. Hindenburg †, die Gattin des Generalfeldmarschalls.

**Figure 22: Einstein's visit with American President Harding (*Berliner Illustrirte Zeitung,* no. 22/1921, p. 324.)**

war. "The stormy ovations that the intellectual elite of London gave the German scholar could not have been more hearty and sincere than if the audience had been celebrating an English intellectual hero. There was no distinction here between German and English. [...] the auditorium sensed that this stranger, who was speaking in a foreign language before an English public had outdone the teachings of this English scholar [Newton]."[416] Lord Haldane and the president of the university *also* spoke in German!

In compliance with a request by Max Planck, Einstein reported before the Prussian Academy of Sciences on 23 June 1921 about the results of his trip. At the subsequent meeting of 14 July, however, Wilhelm Schulze (an academy member and full professor of linguistics at the University of Berlin 1901–1932) protested vehemently against the report being accepted as an official topic of discussion in the minutes if the meeting on 23 June, because it involved "a trip in the service of Zionist propaganda."[417] A majority of the academicians stood behind Einstein, however. Max Planck pointed out that the academy obviously was interested in hearing about experiences that its members had with foreign scientific institutions during trips abroad. Einstein emphasized that in any communications with foreign scholars he had always introduced himself as a *German* and as a professor from *Berlin.*

Upon the invitation of the Red Cross, Einstein also presented a report on 30 June 1921 to an exclusive audience. *Reich President Ebert* attended, along with a

**Figure 23: "When one is famous: Prof. Einstein in London as a guest of Lord Haldane." (*Berliner Illustrirte Zeitung,* no. 26, 26 June 1921, p. 386)**
**"The most photographed man of the present day" (*Frankfurter Zeitung,* 1 July 1921).**

few cabinet members.[418] Thus the highest representatives of the Reich also attached much higher importance to Einstein's trip than just as propaganda for Zionism.

All in all, by this trip Einstein accomplished much toward improving Germano-American and Anglo-German relations in culture and science. It was still too early to speak of an "official resumption of Germano-American and Germano-British scientific exchanges" – as the *Nationalzeitung* exaggeratedly hoped. Nevertheless the trip was an important step in overcoming the boycott of German science. Fritz Haber's fears had proven unfounded.

### 2.5.4 Visiting the French, 1922

In 1922 Einstein received three invitations to visit France. The first came from the League of Human Rights, the second from the French Philosophical Society and the third from the Collège de France. He declined the first two right away. The third, extended by a friend of his, the physicist Paul Langevin, also received a refusal at first, but Einstein changed his mind and accepted it a few days later.

On the justified assumption "that the academy is interested in all events that concern international relations," he informed the secretary's office of the academy on 13 March 1922.

Albert Einstein to the secretary's office of the Academy of Sciences, 13 March 1922:[419]

Prof. Dr. A. Einstein Berlin W 30, 13th of March 1922.
To the Secretary's Office Haberlandstr. 5.
of the Prussian Academy of Sciences

Berlin

Esteemed Colleagues,

It is now the third time this year that I have received an invitation of a public nature from Paris; the first time from the ''Ligue pour les droits des hommes,''[420] the second time from the French Philosophical Society, the third time from the Collège de France, which last invitation asked me to give a few guest lectures. The invitation is issued by the teaching faculty of the Collège de France and was conveyed to me by my friend and fellow colleague in the profession Langevin. The letter of the latter explicitly points out that this event should serve the restoration of relations between German and French scholars. The relevant passage states: ''L'intérêt scientifique veut que les relations soient rétablies entre les savants allemands et nous. Vous pouvez y aider mieux que personne, et vous renderez un très grand service à vos collègues d'Allemagne et de France et par dessus tout, à notre idéal commun en acceptant.''[421]

I answered my friend's letter initially with a polite refusal, indicating that my main reason was out of a sense of solidarity toward my colleagues here.

I could not rid myself of the feeling, however, that with my refusal I had followed the path of least resistance rather than my true duty. A conversation with Minister Rathenau had the result of turning this feeling into firm conviction. Thus a few days after my first refusal I wrote a second letter to Prof. Langevin in which I retracted my refusal and stated my willingness to accept the invitation if other measures had not in the meantime been taken respective the relevant guest lectures. Thus I agreed with Langevin to travel to Paris at the end of the month, in order to deliver the lectures.

In view of the circumstance that the academy is interested in all events that concern international relations, I consider it appropriate to submit the above information to the academy.

In utmost respect,
A. Einstein.

On 28 March Einstein traveled to France. He was warmly greeted at the border by the physicists Langevin and Nordmann.

This trip took place at a time of heated controversy between Germany and France. In January 1922 the addition of Raimond Poincaré to the French cabinet brought a militant nationalist to the top of government. Poincaré spoke out stridently in favor of an inflexible stance on the issue of German reparations. His attitude was equally unconciliatory toward the Soviet Union. At that time in Germany the cabinet was being led by the so-called "compliance politicians" Wirth and Rathenau (Rathenau as foreign minister), who were seeking reconciliation with the western powers. Their less rigid stance also toward the Soviet Union left its mark primarily in the treaty of Rapallo. Rathenau was vehemently opposed by the so-called "catastrophe politicians" led by Stinnes, who chose an unforgiving stance on every point, wanting to provoke open conflict with France.

As Einstein's letter of 13 March 1922 to the secretary's office of the Academy of Sciences documents, the German foreign minister, *Walther Rathenau,* was personally involved in the materialization of the trip. Rathenau advocated dialog and negotiations in order to break the front of Germany's opponents: negotiate, argue and gain time; on the way to the negotiation talks, reveal the differences between the opponents and exploit them in Germany's favor. In Rathenau's view, negotiations were Germany's best weapon and the most effective tool of its resurrection. This was the only way to weaken and eventually diffuse French foreign policy, which was oriented toward dictates and boycotting.[422]

The conference of Genoa with its important implications for postwar history happened to begin on the day Einstein returned from France – on 10 April 1922. On the sidelines of this conference – on 16 April 1922 – the Russo-German treaty, mentioned above, was signed in Rapallo. It agreed the return of diplomatic and economic relations between Germany and the Soviet Union and relinquished all claims to reparations between the two countries. The signatories of the treaty were Soviet Foreign Minister Chicherin and German Foreign Minister *Walther Rathenau.*

**Walther Rathenau** Born 9 Sep. 1867 in Berlin; assassinated on 24 June 1922 in Berlin. Son of the founder of the AEG. 1884–1889 studies in mathematical physics under Helmholtz, chemistry under Hofmann, philosophy under Dilthey. 1889 doctoral thesis on 'The absorption of light by metals.' Following military service and practical training in Switzerland 1893–1899 director of the Electrochemical Works Co. Ltd. in Bitterfeld, 1902–1907 manager of the Berliner Handelsgesellschaft. 1915 President of the board of AEG. 1914/1915 head of the Raw Materials Department of the Prussian War Office. Expert on economic policy at the conferences in Versailles (1919), Spa (1920) and London (1921). Minister of reconstruction in the first cabinet under Wirth (10 May to 26 Oct. 1921) and foreign minister in the second Wirth government (31 Jan. to 24 June 1922). Participated in the International Economic Conference at Genoa. Separate signing of the Rapallo treaty between Germany and the Soviet Union. Assassinated on 24 June 1922 by radical right officers of the "Organization Consul."

Einstein and Rathenau met each other in 1916. We already know what Einstein's position was on the war which was then still raging full force. Rathenau's stance *at that time* will be sketched here briefly as head of the Raw Materials Department at the Prussian War Office. He was one of the people who "did everything in their power to smooth the way for [General Ludendorff] to the highest position of command in the Army."[423] In a letter dated 16 September 1916 to Ludendorff he approved the deportation of seven hundred thousand Belgian laborers to Germany to work in heavy industry as a part of the "Hindenburg program."[424] When Einstein and Rathenau first met in 1916, it was a meeting of dove and hawk.

The two men had many basic points in common, despite their opposite opinions on the war. Einstein and Rathenau were both of Jewish origin (although Rathenau would have preferred to forget that), about the same age, and both had studied physics...

Einstein's "feelings [...] for Rathenau were [...] those of delighted admiration and gratitude for his giving me hope and consolation in the current dismal situation of Europe and for his sharing with me unforgettable hours as a clairvoyant and warm person." Einstein regretted it when Rathenau was appointed minister.[425] Given "the attitude of a large majority of the educated class in Germany toward Jews, his conviction of proud reserve among Jews in public life would have been the natural thing."[426]

On the other hand, neither was Rathenau prepared to share all of Einstein's positions – as his attitude on Zionism shows. Chaim Weizmann recalled: "This visit has remained vividly in my memory. It was a conversation with Walther Rathenau, whom I met at Einstein's home one evening. In a sudden gush of words he immediately launched an attack on Zionism [...]. The quintessence of what he presented was: he was a Jew but felt like a German and devoted all his energy toward building up German industry and restoring Germany's reputation in the world."[427]

Nevertheless the influence Einstein and Rathenau had on each other was considerable. Fritz Haber confirmed this in a letter to Einstein from 28 June 1921: "Your word weighs heavily with him."[428] And Einstein likewise was open to arguments Walther Rathenau brought forward. The trip to France is the clearest proof of this.

This, incidentally, was not the first trip Einstein undertook at Rathenau's behest. He was on the road in February 1921 already for the same reason. He went to Amsterdam together with Count Kessler "to find out the views of the International League of Unions on the Parisian resolutions."[429] The request presented there was to draw the problem of reparations "away from the very limited Franco-German framework and make it part of the bigger problem of the global economic recovery or at least of the recovery of Europe including Russia [...]. For this, organized workers should also be included in a decisive way."[430] Kessler's recollections do not reveal what Einstein said or did on that occasion. Presumably he was used – as he often was later on – only as a "famed bigwig."

The extent of Rathenau's influence on the *trip to France* can also be gathered from the internal files of the Foreign Office.

An excerpt from a memorandum drawn up for Legation Councillor Dr. Soehring on 13 March 1922,[431] reads as follows:

According to information by the president of the University of Berlin, Professor Nernst, Professor Einstein has been invited by the Société de Physique de France in Paris to deliver a lecture but has declined for now, because he is overworked and does not feel his command of French is adequate. Political reasons played no part, the invitation and refusal were issued in a completely amicable form among colleagues. There can be no question of any prohibition by the Berlin Faculty of Philosophy toward Professor Einstein.

Recently Professor Einstein has also been invited to give a lecture in Brussels and he is considering whether he should not combine both trips and therefore travel to Paris after all.

Professor Nernst is happy to forward any wishes by the Foreign Office in this matter to Professor Einstein, which should only be regarded as suspended.

Berlin, 13 March 1922.

**Figure 24: Walther Rathenau, German Minister of Foreign Affairs in the vehicle he would be assassinated in on 24 June 1922.**

There is a marginal note in Walther Rathenau's handwriting dated 16 March 1922: "I see no objection – Should there be any, I request notif. R." Below it the reply from the culture department of the Foreign Office: "No objection! On the contrary!"

Rathenau's mind had already been made up *beforehand,* as the date of Einstein's letter to the academy reveals. The marginal note was just a reiteration of his standpoint.

A publication by Einstein-opponent Ernst Gehrcke from 1924 shows that Walther Rathenau's personal involvement did not remain secret. Gehrcke wrote: "A source close to Einstein assured me that he only decided to travel at the special request of Rathenau, then German foreign minister. Rathenau supposedly had placed hopes of a reconciliation between Germany and France on the relativity trip to Paris."[432]

Einstein's fellow academicians in Berlin warmly welcomed this opportunity to restore normal relations with French scientists, especially considering that, formally, the motivation for the trip had been purely *scientific*. The invitation that Einstein described to the academy explicitly mentioned the intention that *scientific* relations between Germany and France be resumed. The current president of the University of Berlin, *Walther Nernst,* also emphatically recommended that Einstein travel. His offer to relay any wishes by the Foreign Office in this regard to Einstein were not just in fulfillment of any official obligations. Nernst was very probably concerned that no irritations arise from Einstein's guilelessness in matters of diplomacy.

Einstein's colleagues and notable patrons of science regarded his trip as a social event of the very first order.

A week before the scheduled trip, on 20 March 1922, Elsa and Albert Einstein hosted a major dinner, "lending this truly dear, almost childish couple a certain naïvité." Harry Count Kessler recorded other guests in the party as "the extremely wealthy Koppel," the president of the Bureau of Standards, Emil Warburg, and the banker Franz von Mendelssohn. "Radiant kindness and simplicity transported even this typical Berlin society from the conventional and illuminated them with an almost patriarchical, fabulous aura."[433] The circumstance that Count Kessler had conversed with Walther Rathenau just a few hours previously about his own trip to France, about the preparations for the International Economic Conference in Genoa and other matters of foreign policy,[434] is likely to have influenced the conversation about Einstein's imminent trip. During that evening Einstein asked Kessler "to repeat more than once and verbatim" what the French mathematician Painlevé had communicated to him with regard to the trip to Paris. (In 1927 Painlevé was minister of war and briefly also prime minister.)[435] Einstein complained that "university people were probably holding [... the trip] against him." "But these people were truly awful. Just thinking about them made him feel nauseous. And he hoped to be able to achieve something in Paris toward the restoration of ties between German and French scholars."[436]

Even so, the departure had an almost conspiratory character. The German ambassador in Paris was at a loss at how to conduct himself. He did not want to put himself forward too much, but he certainly did not want to ignore Einstein's presence either. His biggest worry was finding out whether Einstein "would present himself as a German and a representative of German science." So a few hours before Einstein disembarked, he sent a coded telegram to the Foreign Office.[437]

Telegram by the German ambassador in Paris, 27 March 1922 (excerpt).

[. . . ]. Professor Einstein who, as we know, is going to hold lectures here at the Collège de France upon French invitation, arrives tomorrow morning in Paris. I intend to keep very much in the background, if only to spare him difficulties arising from the anti-German sentiment among some of the local scholarly world and not to place in doubt the purely scientific nature of Einstein's trip. I would, however, like to send him a welcoming message at the station, so as not to expose the embassy to the charge of having ignored the great scholar - provided that Einstein, who supposedly owns a Swiss passport, is traveling with a German passport.

Please reply upon receipt [. . . ] whether Einstein is carrying a German passport, whether his deportment of late indicates that he will present himself here as a German and a representative of German science, and whether any special statement by me to him is desired over there.

Mayer.

The Foreign Office replied immediately.

Berlin, 27th of March 1922

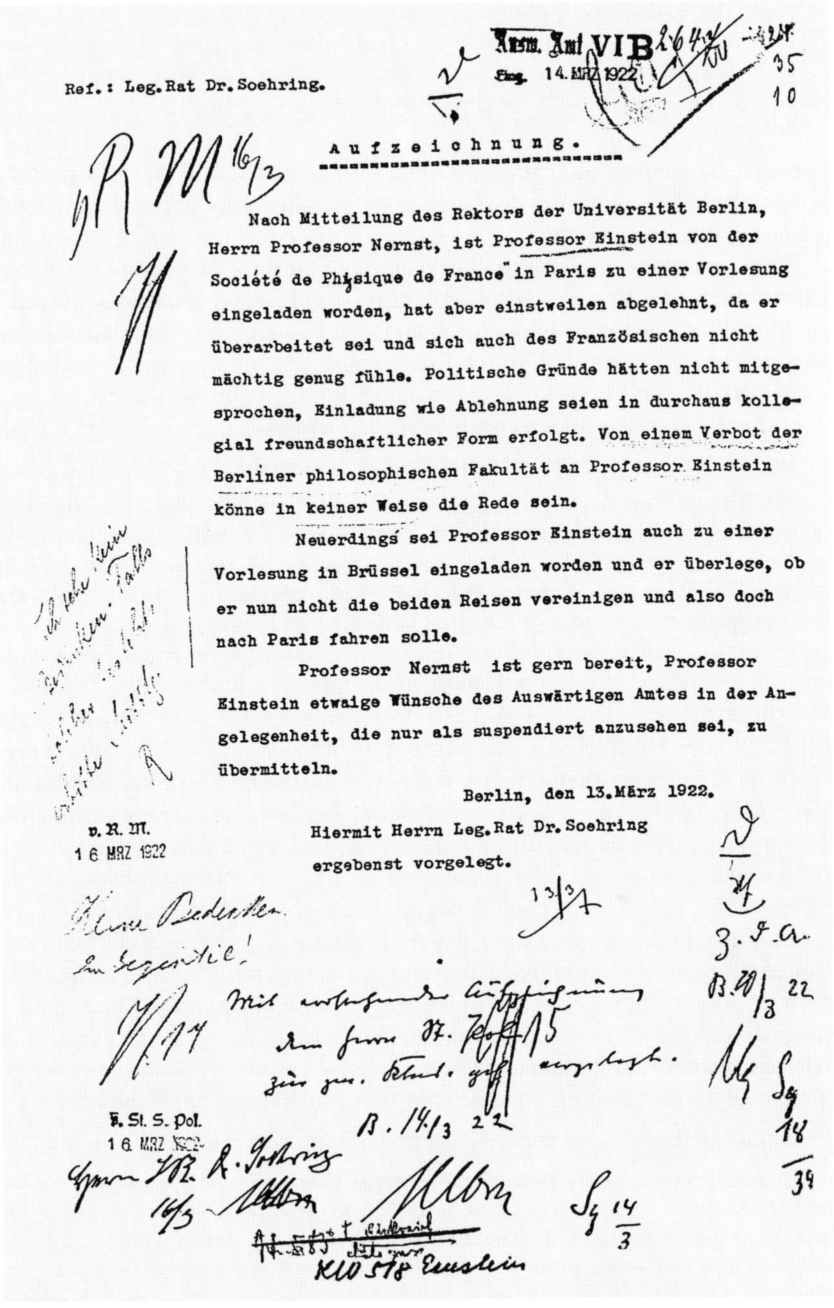

Ref.: Leg.Rat Dr.Soehring.

Ausw. Amt VIB 14. MÄRZ 1922

Aufzeichnung.

Nach Mitteilung des Rektors der Universität Berlin, Herrn Professor Nernst, ist Professor Einstein von der Société de Physique de France" in Paris zu einer Vorlesung eingeladen worden, hat aber einstweilen abgelehnt, da er überarbeitet sei und sich auch des Französischen nicht mächtig genug fühle. Politische Gründe hätten nicht mitgesprochen, Einladung wie Ablehnung seien in durchaus kollegial freundschaftlicher Form erfolgt. Von einem Verbot der Berliner philosophischen Fakultät an Professor Einstein könne in keiner Weise die Rede sein.

Neuerdings sei Professor Einstein auch zu einer Vorlesung in Brüssel eingeladen worden und er überlege, ob er nun nicht die beiden Reisen vereinigen und also doch nach Paris fahren solle.

Professor Nernst ist gern bereit, Professor Einstein etwaige Wünsche des Auswärtigen Amtes in der Angelegenheit, die nur als suspendiert anzusehen sei, zu übermitteln.

Berlin, den 13.März 1922.

v. R. M. 16 MRZ 1922

Hiermit Herrn Leg.Rat Dr.Soehring ergebenst vorgelegt.

Keine Bedenken. Im Gegenteil!

R. St. S. Pol. 16 MRZ 1922

K10 518 Einstein

**Figure 25:** ***"I see no objection – Should there be any, I request notif. R."*** **– Note by Reich Minister of Foreign Affairs Walther Rathenau (R) of 16 March 1922 on the planned "French visit" by Albert Einstein.**

Telegram (in plain text, urgent)
Einstein is traveling on a Swiss diplomatic passport. No objection to intended welcome there of famous member of our academy and university. Further action left to judgment there.
Heilbron

Although Einstein used a *Swiss passport* on his travels and was not particularly forthcoming about his itineraries, the Germans were actually less inconvenienced than the French.

The cause of the semi-conspiratory nature of Einstein's departure was strong opposition by French nationalists to the trip. They realized that it would create a gap in the front of opponents to Germany. *That* was also the reason for the German ambassador's restraint. At least outwardly, "the purely scientific nature of Einstein's trip" was not to be placed in doubt. Emphasis on the *scientific* character of the trip would most effectively serve *political* ends.

On 24 March 1922 the *Frankfurter Zeitung* reported:

The invitation by the Collège de France (not the Physical Society, as had been originally announced) to Einstein to deliver a series of lectures on his theory in Paris, gives a segment of the *nationalist* press welcome occasion for more agitation. 'Echo de Paris' states that the invitation had been passed *by a majority of only two votes* and that it had been expressly resolved to keep the invitation *secret*. Since this was not possible and the occasion of Einstein's presence raised fears of demonstrations particularly by student organizations, it would be preferable if Einstein did *not* come to Paris.

On 14 March 1922 (hence on his forty-third birthday) Einstein asked his old friend Maurice Solovine to keep the hide-away Langevin had arranged from him strictly secret. Only his friends were informed that he would be arriving in Paris in the evening of the 28th. It agreed neither with the date Nernst conveyed to the Foreign Office nor the date that the German ambassador gathered from press reports. "Einstein did not arrive in Paris at the presumably purposefully falsely indicated time but unexpectedly during the night of the 28th."[438] Einstein even "left the train station from a secondary platform,"[439] so the "reception I had planned was not possible." The ambassador did not know where Einstein was staying ("at a private apartment, apparently with the scholar Langevin"), and informed the Foreign Office that Einstein was "keeping very much out of public view."[440]

Report by the German embassy in Paris, 1 April 1922 (excerpt).

In conformance with your acquiescence there, I sent a member of the embassy to meet Professor *Einstein* at the train that the entire press had been informed he would disembark from at noon on March 28th, in order to hand him a letter from me. Einstein did not arrive in Paris at the presumably purposefully falsely indicated time but unexpectedly during the night of the 28th and he left the train station from a secondary platform, so the reception I had planned was not possible. Professor Einstein, who is accommodated at a private apartment, apparently with the scholar Langevin, and is evidently keeping very much out of public view, has not contacted the embassy yet either. [. . . ]. Perhaps only with the exception of the ''Action Française,'' Einstein's reception among his auditors as well as in the press has thus far been very amicable, often downright cordial.

Dr. Mayer

The *Berliner Tageblatt* devoted a long article on 12 April 1922 to Einstein's evasiveness from the public (evidently not even aware that Einstein had already returned to Berlin). It bore the headline: '*Hidden Einstein*' [Der verborgene Einstein]:

For the last ten days Professor Albert Einstein has been the most popular personality in Paris [. . . ] this German has conquered Paris. All the papers printed his portrait, an entire Einstein-bibliography has emerged [. . . ]. Einstein has become thoroughly à la mode. Academics, politicians, artists, the petty bourgeois, security personnel, cab drivers, waiters and pickpockets know when Einstein is delivering his lectures. [. . . ]. The cocottes at the Café de Paris grill their escorts about whether Einstein wears glasses or is a chic sort of guy. All of Paris knows everything and is telling more than it knows about Albert Einstein. Just one thing has still not been disclosed: where Albert Einstein is staying.

The mystery started with the arrival of the new Copernicus [. . . ]. He was supposed to arrive on a noon train [. . . ]. Albert Einstein really arrived in the dark of night, but even then no one saw him: because he disembarked from the other side of the passenger carriage, where solitude and darkness reign [. . . ]. So Einstein is being hidden away by his French friends and guarded like a rare treasure and this negative fame has contributed not insignificantly to his popularity. Because he has not said a single word that could be politically interpreted and reinterpreted, the serene glory of his scholarly renown is not marred by any garish flares. [. . . ].

Entrance tickets were issued in order to guarantee an audience limited to people with a genuine scientific interest. But everyone wanted to come and experience this Parisian "sensation." Einstein emerged victorious from the scientific debates after the lectures. The *Vossische Zeitung* reported with satisfaction that "the whole discussion took a turn that is thoroughly favorable to Einstein's theory." The Parisian Academy of Science also deliberated about whether to invite Einstein. But nationalist opposition was too strong. Thirty academicians declared that they would demonstratively leave the hall the moment Einstein entered, so a formal excuse was used in deciding against the invitation (Einstein was not a member of the academy and invitations were not normally issued to nonmembers). On 6 April Einstein did speak before the "French Philosophical Society" about philosophical issues of relativity theory.

On 28 March, still in the train to Paris, Einstein told his French companions about his great wish to view during his stay in France some of the areas devastated by the first World War. On 9 April this plan materialized. The motivation behind it was a profound sense of human compassion and abhorrence of war. He made no concessions to either side. He simply bowed before the graves of French and Germans alike, and protested against the accusation that Germans were a warrior nation.

Upon his return to Berlin the battle in the press was raging on in both countries, Germany and France.

The retaliators and militarists in Germany viewed Einstein as a shameless "defector," a person with no sense of patriotism. On the day that Einstein undertook this excursion into war-torn areas of France, on 4 April 1922, the *München-Augsburger Abendzeitung* and the *Deutsche Tages-Zeitung* both carried an article under the heading: 'Einstein's Frenchmen's trip': "In any event," it reads, "the authorized officials ought to have pointed out to him that for German citizens in official employ the time spent on scientific ingratiation with the French is completely inappropriate." The author of the article was Prof. Stark – like Lenard, a Nobel laureate and likewise a fanatic opponent of Einstein's theory just as of his political mission.

Even so, the general sentiment in Germany concerning Einstein's controversial trip was more positive than it was negative. The government appreciated the usefulness of his "Frenchmen's trip." It suited its purposes fine that Einstein "never made any secret out of having come to France as a representative of German science," as he put it in an interview (*Vossische Zeitung*, 18 April 1922). Einstein added: "You must also take into consideration that the invitation addressed to me by the Collège de France was addressed to a German scholar." What he said next, however, did not agree at all with the interests of the German government: The main obstacle toward closer reciprocal relations was, he thought, that biased opinions *on both sides* and prejudices "about the causal relations of the war" existed *on both sides*.

In France the press's reaction was identical, except that it was of the opposite sign. What many people in Germany saw as making him a "traitor to the fatherland" and a "defector" was his chance in France. Einstein's trip did not agree with the nationalists from the mere fact that he came from hated *Germany*. Most papers reacted in a friendly way, though. The *Matin* even thought Louis XIV would have called Einstein to his court – as the Danes did Rømer, the Italians Cassini, and the Dutch Huygens. The response by the *Action Française* was: "Whatever may have been Einstein's political stance, it is Germany itself that is being received in his person, it is German influence that is being consecrated. Under Louis XIV we had nothing to fear from Danish or Dutch influence. Today, however, we have everything to fear from German influence."[441]

"*Germany itself*" was Einstein's companion. Thus, thanks to his extraordinary accomplishments, Einstein had already become an institution. *Victorious* France feared nothing more than *German* influence – personified by *Albert Einstein*.

The tone was the same in *Le progrès de la Somme* (Amiens) from 30 March 1922: "Yes, I know that science has no fatherland [...]. But even though science has no fatherland, scholars certainly do, and it bothers me to know that Einstein's fatherland is Germany supreme! [...]. If I were Mr. Einstein, I would find it too boring, despite all my metaphys-

(a)

(b)

**Figure 26: Albert Einstein on the battlefields near Paris, April 1922**

23. April 1922
Nr. 17
31. Jahrgang

Berliner
Illustrirte Zeitung

Einzelpreis des Heftes 3.– M.

Verlag Ullstein, Berlin SW 68

Einstein,
der als Ehrengast in Paris an Disputationen über seine Relativitätslehre teilnahm und die Anerkennung errang, daß seine Theorie die bedeutendste Errungenschaft der modernen Wissenschaft sei.

**Figure 27: "Einstein, who as an honorary guest in Paris debated about his theory of relativity and drew the acknowledgment that his theory is the most important achievement of modern science."**

ical science, to come and pompously talk at length to people whose sons had destroyed my people."[442]

But it was precisely because he was so controversial on *both* sides of the border and had not openly taken sides with either one of them, that Einstein became a *harbinger* of reconciliation between Germany and France – but no more than a "harbinger" (the trip was, to quote Shakespeare, "the Nightingale, and not the Larke." It was for this same reason that he became the object of stinging attacks and slander.

At the request of the Foreign Office, the German ambassador submitted a final report on Einstein's trip on 21 April 1922. It might stand as the quintessence of the foreign policy surrounding his trip. "The case be as it may," he concludes, "there is no doubt that Mr. Einstein, who ultimately did have to be regarded *as a German,* has aroused much attention and new fame for the *German spirit* and *German science.*"[443] It even provides a reason for *why* it was possible: because Einstein had distanced himself from the "manifesto of the 93" and was anyway "just native to Germany."

Report by the German embassy in Paris, 29 April 1922 (excerpt):

[. . . ]. For the remainder of his stay in Paris also, Professor *Einstein* was much honored and acknowledged as a leading scientific mind. Attacks by the press remained isolated and elicited sharp rebuttals in the more liberal publications. From a scientific point of view it was a complete success. [. . . ].

Mr. Einstein did not visit the embassy. He said to a representative of the 'Berliner Tageblatt,' Mr. Block, who had suggested to him to visit me, that he was just a scholar and was a stranger to the ways of the world and its formalities. [. . . ].

After completing his stay in Paris he drove from here to the devastated area. His enthusiastic disciple Nordmann devoted an article to this excursion in 'Illustration' from 15 April, which I enclose.

It would be incorrect to assume that Einstein's favorable appearance in Paris demonstrated that on the scientific plane Germans could henceforth resume relations with French intellectual life and also foster personal connections in the old way without hindrance. Although Einstein's visit proceeded without any major discord, very satisfactorily, even, this can be ascribed to two causes. First of all, Einstein is a sensation that intellectual snobbery in the capital did not want to miss. On the other hand, Einstein had been most carefully made ''acceptable'' to Paris in that the press had declared, before his arrival already, that he had not signed the manifesto of the 93 but had, on the contrary, wanted to sign a countermanifesto; his opposing stance to the German government during the war was known, finally he was anyway Swiss and just native to Germany.

The case be as it may, there is no doubt that Mr. Einstein, who ultimately did have to be regarded as a German, has aroused much attention and new fame for the German spirit and German science.

Dr. Mayer.

### 2.5.5 Japan, Palestine and Spain

The idea of visiting Japan originated with Kitaro Nishida, one of Japan's most important philosophers.[444] The Japanese internist and researcher of tuberculosis Aihiko Sata was presumably also involved. The fact is, however, that Sata had been in Germany on a lecture tour after the war and *before* Einstein left for Japan. It was around the time when Albert Einstein had ascended to world fame. It is a fact too that Sata specially attended to Einstein during his Japanese visit.

Owing to reservations among the Japanese public toward the former enemy Germany, the invitation could only have come from *private persons* so soon after the war. The Japanese government could not have dared to do such a thing. At Nishida's suggestion, the magazine *Kaizo* extended the invitation and financed the voyage (more precisely put: prefinanced it; when Einstein was in Japan, he was very profitably marketed).

This time the Foreign Office wanted to be sure that during the tour it would play more than a mere walk-on part.

"It would be good if, this time, things do not go the way they went in Paris. Ambassador Mayer did not even have the benefit of glimpsing Einstein, although he had even taken the effort to have him met at the train station by an official from the embassy," Department Director *Heilbron* at the Foreign Office wrote to Germany's ambassador in Japan, *Wilhelm Solf*.[445] Heilbron was quite optimistic though: "In Japan Einstein will surely have less qualms about being seen in the ambassador's company than in Paris."

Solf and Heilbron were instrumental in assuring the diplomatic success of Einstein's trip.

**Wilhelm Solf** (1862–1936) was appointed ambassador to Japan in 1920, hence shortly before Einstein's trip (remaining in this capacity until 1928). So a restoration of fruitful relations between defeated Germany and its former enemy Japan would have very personal advantages. He could think of nothing better than having Albert Einstein go on a tour there. Solf realized that the colonial property lost to Japan could not be regained. The earlier colonial deportment of the former secretary of state at the Reich Colonial Office was henceforth no basis for any realistic policy. It was only through acknowledgment of the real state of affairs that had meanwhile developed and on a give-and-take basis that Germany (and Solf) could be successful in Asia. It was because Solf was an able and clever man that his work in Tokyo became "an era of liberal scientific exchanges between the two countries."[446] Solf was particularly suited to this post. During his youth he had studied Indian culture and he had great respect for Buddhist teachings. Besides law, he had also studied Sanskrit. He joined the Foreign Office with these skills. From 1900 to 1911 he was governor of German Samoa, 1911 to 1918 secretary of state in the Reich Colonial Office (colonial minister) and 1918 secretary of state at the Foreign Office (foreign minister). After the revolution Solf continued to serve in the same function at the behest of the Council of People's Representatives – very much to the displeasure of the kaiser. Solf is supposed to have once assured the kaiser that if the kaiser retired then he too would resign his state offices. He defended his (apparent) change of sides in a letter to Heilbron from 9 October 1922: "It is not a lack of direction for civil servants who had been loyal to the old government to be loyal collaborators in the new one [...]. Even those who do not like the new order should assist, for the German Reich is what is at stake here, not republic versus monarchy!"[447] Later, because the Reich government was not doing enough "to purge itself irreproachably before the world from the suspicion of plotting with the Bolshevik government" (in Soviet Russia), and because "nothing effective or resolute was being done against the openly proclaimed tyranny of the so-called Spartacus group," Solf eventually did tender his resignation, on 9 December 1918, and it was granted by Reich President Ebert on 13 December.[448] Solf left for Switzerland but did not stay long. Ebert's government very soon reengaged him and made him ambassador to Japan. His high regard for Ebert probably has the following foundation: "That man is a statesman, through and through. It is certainly to his credit that he guided us past the cliffs of ultra-radicalism and Bolshevism in 1918 and 1919. He could be one of the leaders in my utopian party of gentlemen."[449] Very soon Solf also realized, as he wrote in a letter to Privy Councillor Deutsch on 22 April 1922, "that a resumption of relations with Russia is a necessity of life for us as well as for Europe."[450] His political positions thus generally agreed with those of the foreign minister at that time, Walther Rathenau. They had been friends for years already.[451] He wrote furthermore in his letter of 22 April 1922: "To get a clear picture, I wrote Rathenau to please grant me brief leave in Germany. He thankfully did so very quickly and thus I shall be in Berlin in mid-July." (When he arrived, Rathenau was no longer alive: he had been assassinated by right-wing radicals on 24 June 1922). So Solf was in Germany when preparations for Einstein's trip to Japan were underway. This was very amenable to the cooperation between Solf and the authorized official at the Foreign Office, Department Director Heilbron, in re the Einstein affair.

Department Director Heilbron was a good friend of Solf and assured continuity at the Foreign Office. Heilbron had served as a rapporteur for Solf at the office and at the same time had been "a tactful advisor."[452] That was why Solf wrote him on 6 October 1922 that "my directorship of the Colonial Office had never given the kaiser cause for discontent. [...]. I held my peace in the face of criticism of my leadership during those ominous days of October in 1918. Not least, on your advice."[453]

On 24 April 1922 Einstein had informed Legation Councillor Soehring at the Foreign Office that he was traveling to Japan and China in October and was thinking of returning sometime in February. Soehring replied to Einstein on 24 July 1922 that "the German ambassador in Tokyo, His Excellency Solf, would soon be arriving in Berlin on leave" and that he contact Solf ("who will certainly take particuar interest in your trip") for any assistance he might need of Soehring.[454]

Einstein's "escape" to Holland may have prevented him from seeing this letter, or else he simply did not attach particular importance to Solf's suggestion. In either case, his enthusiasm to cooperate closely with the Foreign Office was limited. A note in the file from 28 August 1922 indicates that there was still "no statement by Prof. Einstein on file." Heilbron had reason to be disappointed: "In this case too Einstein has remained as aloof of the Office as possible." On 26 September – shortly before Einstein's departure – Heilbron finally had "the first opportunity [...] to find out details about his travel plans" when he joined Einstein at the Japanese embassy. He immediately reported what he found out during this interview with the Japanese councillor to Ambassador Solf, who was still in Berlin.

Ministerial Manager Heilbron to Ambassador Solf, 27 September 1922 (excerpt):[455]

[. . . ]. Professor Einstein intends to depart on 8 October on his trip to Japan. We were with him together with the Japanese embassy councillor, Matsubara, and this was the first opportunity that we had to find out about his travel plans. He had politely declined our offers of assistance already at an earlier stage. I did not omit to place our services at his disposal yesterday again, should he encounter any difficulties. At our passport office, the only one at the RM[456] that he contacted, we found out that he is traveling on a Swiss passport. [. . . ]. It would be good if, this time, things do not go the way they went in Paris. Ambassador Mayer did not even have the benefit of glimpsing Einstein [. . . ]. In Japan Einstein will surely have less qualms about being seen in the ambassador's company than in Paris. [. . . ].

sig. Heilbron.

Heilbron's letter also reveals that Einstein had meanwhile contacted Ambassador Solf – by-passing the Foreign Office.

Einstein and Solf were not as unacquainted as it might appear. In November 1918 Solf – at that time secretary of state at the Foreign Office – had asked to see Einstein, and Einstein had been happy to comply (Einstein's letter to Solf of 22 November 1918).[457] "As a consequence of the unstable political situation" Solf's reply had been delayed until 2 December. Solf hoped nevertheless to see Einstein "in the next day or so" and requested he contact him by telephone to make an appointment.[458] The files do not indicate whether the meeting actually took place; probably not. Shortly afterwards, on 9 December, Solf resigned his post. In any event, they had certainly been in direct contact with each other *long before* Albert Einstein's trip to Japan. There is reason to believe that this fact was also useful in the discussions that took place prior to the Japan tour.

Thus Einstein did *not* categorically reject all assistance. He may have circumvented the Foreign Office but he did inform the *ambassador* about his plans. So Heilbron could rest assured that "*this time*, things do *not* go the way they went in Paris."

A day later, on 28 September 1922, Heilbron wrote Solf once again to inform him that the German ambassador in Spain, Langwerth, wished to welcome Einstein and his wife as his guests during their planned stopover in Spain on their way back from Japan.[459] He asked Solf to use this opportunity to inform Einstein and say that Heilbron himself "had not been able to reach him."

The voyage began at the beginning of October 1922. Einstein had already informed the Academy of Sciences in advance, on 12 July 1922, to suspend his salary payments as of 1 October 1922 for an unspecified period of time.[460] The travel costs were assumed by Einstein's host, the publisher of the Japanese monthly magazine *Kaizo*. Thus he also dictated Einstein's travel itinerary and made full use of this lucrative marketing opportunity.

Einstein's wife was once again in attendance.

The original plan (in April 1922) was for an *extended* stay in China. In the end it was just a brief stopover.

The German ambassador in Peking, the German consul general in Shanghai, the chairman of the German Association in China, the Chinese Scholarly Societies and the School of Engineering in Woosung did their best to encourage Einstein to stay longer in China. The German ambassador in Tokyo was even asked to intercede. But it was all in vain. Einstein's manner of responding (i.e., of not responding) could no longer be excused by what Consul General Thiel called his "social awkwardness." It was a downright snub.

It is clear that Einstein had to decline all efforts during the *voyage out* to Japan – which were anyway on very short notice. Einstein told the consul general on board the steamship "Kitano Maru," "as far as Japan is concerned, I am under a very definite contractual relationship and could not assume any scientific engagements besides the ones agreed with his counterpart, which already now numbered 18."[461] The consul general's worries were not unfounded. Einstein was welcomed already in Shanghai by an emissary from Japan, and the "gentleman concerned presented an agenda for Shanghai that left Mr. Einstein not a single free minute throughout the entire period of his stay here."[462] The German consul general could only arrange for a slight trimming down of the hopelessly overloaded schedule.

This was only a foretaste of the taxing program awaiting Einstein in Japan. He did not "feel like sacrificing myself to such a degree," as he told Heilbron in September. But his time was in other people's hands.

On 17 November the Einsteins landed on Japanese soil for the first time at Kobe. Ambassador Solf was initially unable to greet him. He only returned from Germany on the 27th. On 28 November, the press (*Jiji*) "gave the following announcement about the return of His Exc.: The German ambassador, Dr. Solf [...] arrived on the 27th in the afternoon 6 o'clock via San Francisco and Honolulu in Yokohama [...]. Dr. Solf stated [...]: As I heard already on the high seas, the Japanese gave Professor Einstein an enthusiastic welcome. I am extremely pleased about that."[463]

The tour was "wonderful" but also "quite exhausting."[464] The lecture covered in the embassy report, on 5 December, lasted a full six hours (from 1:20 PM to 7:20 PM). Towards the end of Einstein's Japanese stay, Einstein had to get medical attention. On 26 December Einstein thanked his Japanese doctor Hayasi Myake for his helpful treatment.[465]

In the wake of Einstein's departure, Ambassador Solf sent a detailed report to the Foreign Office in Berlin. It comprehensively lists all the facts and – what is of interest to us here – it also provides an *evaluation* of Albert Einstein's trip to Japan.

This evaluation follows verbatim because it also provides the most important highlights of Albert and Elsa Einstein's Japanese stay.

Report by the German embassy in Tokyo to the Foreign Office in Berlin, 3 January 1923 (excerpt):[466]

Professor Einstein arrived in Japan on 17 November and departed again on 29 December. His trip resembled a triumphal procession. The visit of the Prince of Wales and Field Marshal Joffre had a courtly and military stamp, was programmatically prepared and found an officious echo in the Japanese press. There was none of that for Einstein's reception; the whole Japanese public participated, from the highest dignitaries to ricksha coolies, spontaneously, without preparation and without show! Upon Einstein's arrival in Tokyo, there was such a crowd of people at the train station that the police powerlessly had to tolerate the life-threatening throngs. The reception in the other cities, where he held lectures or came to view the land and its people, to relax from the strains of traveling, was the same as in Tokyo. [. . . ].

My personal relations to Einstein have developed into a friendship. Despite the superlative honors everywhere conferred upon him, he remained modest, friendly and unassuming. The highpoint of the distinctions of this famous man was this year's chrysanthemum festival! The focus of all attention was not on the empress nor the prince regent nor the imperial princes, but unconsciously and unintentionally, Einstein. The gentlemen at the embassy who had attended it - I arrived a couple of days afterwards - described to me how the audience of some 3,000 at this traditional celebration of the imperial family in communion with its people forgot the importance of the day because of Einstein. ''All eyes were on Einstein; everyone wanted at least to have shaken the hand of the most famous man of the present day. [. . . ].''

There were even caricatures of Einstein featuring his stubby pipe and his full, rebellious shock of hair, also hinting at his attire, which was not always quite suited to the occasion [. . . ].

By the time this report arrives in Berlin, it will be clearer when and where he will enter Germany. I would be grateful if the relevant border post be instructed to facilitate his entrance into German territory in every way possible, in conformance with the regulations. He has all kinds of honorary gifts in his luggage and his wife has requested that these objects be allowed through free of customs.

Solf.

A defender of German interests wrote this? A former *colonial* and *foreign minister* of the German Kaiserreich – now working under *new* premises! Such a man was able to appreciate Einstein's conduct despite their very different political opinions. Moreover, they had even become *friends*. The bit of criticism was subtle: Einstein's dress not always quite fitting the occasion. The only truly unpleasant thing for the ambassador was, because it damaged the "German cause," that the life of world-famous Einstein was no longer safe in Germany and word about that had spread abroad.[467]

And as far as their "friendship" is concerned, Solf did not say so out of the blue. It is certain that they got on very well with each other.

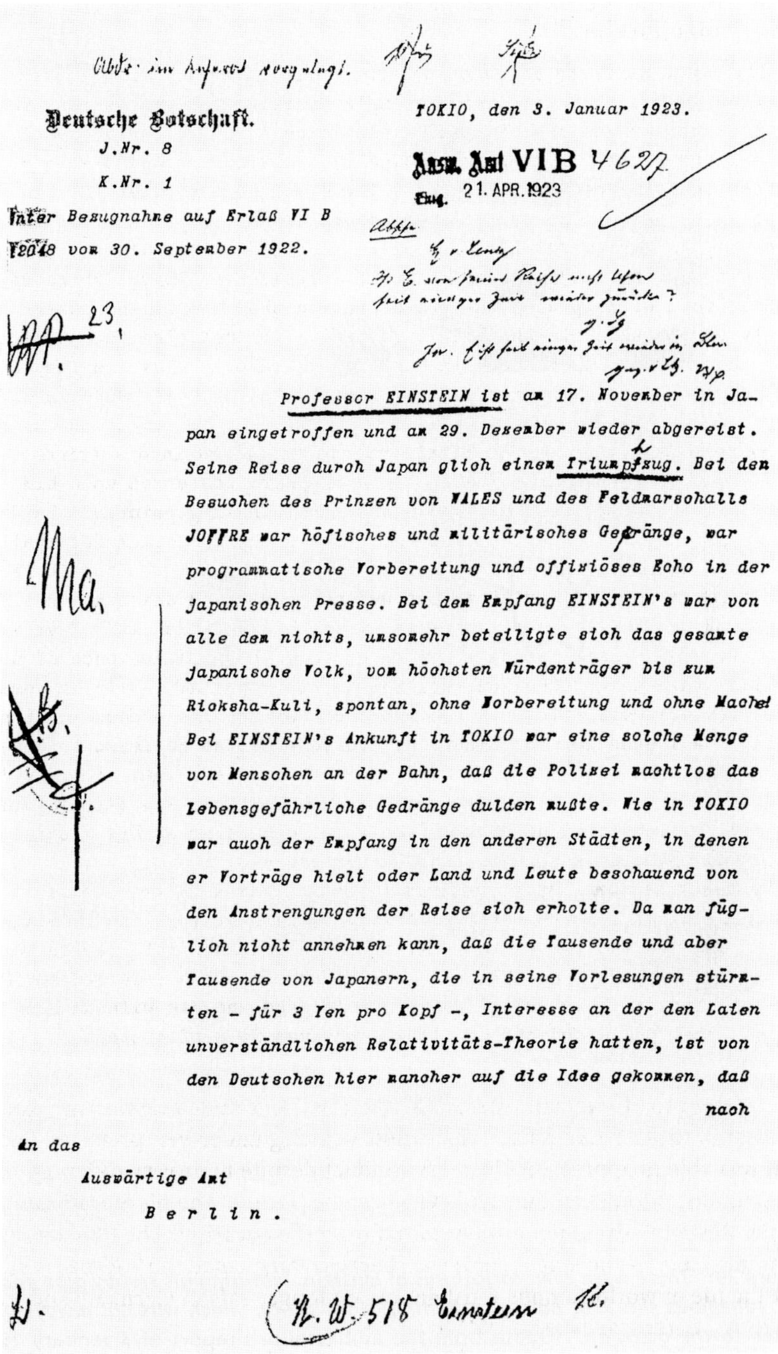

Deutsche Botschaft.
J.Nr. 8
K.Nr. 1
Unter Bezugnahme auf Erlaß VI B
12048 vom 30. September 1922.

TOKIO, den 3. Januar 1923.

Ausw. Amt VI B 4627
Eing. 21. APR. 1923

Professor EINSTEIN ist am 17. November in Japan eingetroffen und am 29. Dezember wieder abgereist. Seine Reise durch Japan glich einem Triumpfzug. Bei den Besuchen des Prinzen von WALES und des Feldmarschalls JOFFRE war höfisches und militärisches Gedränge, war programmatische Vorbereitung und offiziöses Echo in der japanischen Presse. Bei dem Empfang EINSTEIN's war von alle dem nichts, umsomehr beteiligte sich das gesamte japanische Volk, vom höchsten Würdenträger bis zum Rickscha-Kuli, spontan, ohne Vorbereitung und ohne Mache! Bei EINSTEIN's Ankunft in TOKIO war eine solche Menge von Menschen an der Bahn, daß die Polizei machtlos das lebensgefährliche Gedränge dulden mußte. Wie in TOKIO war auch der Empfang in den anderen Städten, in denen er Vorträge hielt oder Land und Leute beschauend von den Anstrengungen der Reise sich erholte. Da man füglich nicht annehmen kann, daß die Tausende und aber Tausende von Japanern, die in seine Vorlesungen stürmten – für 3 Yen pro Kopf –, Interesse an der den Laien unverständlichen Relativitäts-Theorie hatten, ist von den Deutschen hier mancher auf die Idee gekommen, daß

nach

An das
Auswärtige Amt
Berlin.

R.W. 518 Einstein

**Figure 28: "*His trip resembled a triumphal procession.*" Page 1 of the detailed ambassador's report of 3 January 1923 to the Foreign Office.**

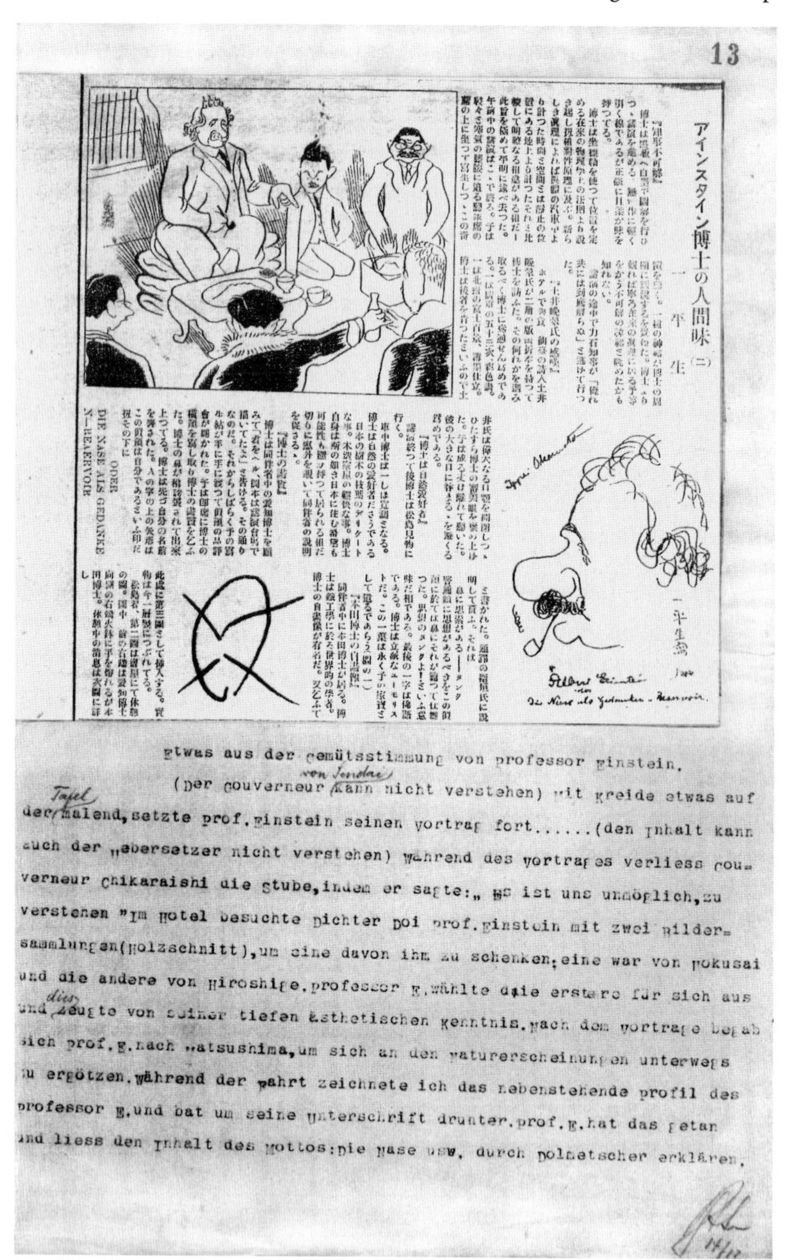

13

アインスタイン博士の人間味（二）

一平生

ODER
DIE NASE ALS GEDANKEN-RESERVOIR

Albert Einstein oder die Nase als Gedanken-Reservoir.

Etwas aus der Gemütsstimmung von Professor Einstein.

(Der Gouverneur von Sendai kann nicht verstehen) Mit Kreide etwas auf der Tafel malend, setzte Prof. Einstein seinen Vortrag fort......(den Inhalt kann auch der Uebersetzer nicht verstehen) Während des Vortrages verliess Gouverneur Chikaraishi die Stube, indem er sagte: „Es ist uns unmöglich, zu verstehen" Im Hotel besuchte Dichter Doi Prof. Einstein mit zwei Bildersammlungen (Holzschnitt), um eine davon ihm zu schenken; eine war von Hokusai und die andere von Hiroshige. Professor E. wählte die erstere für sich aus und dies zeugte von seiner tiefen ästhetischen Kenntnis. Nach dem Vortrage begab sich Prof. E. nach Matsushima, um sich an den Naturerscheinungen unterwegs zu ergötzen. Während der Fahrt zeichnete ich das nebenstehende Profil des Professor E. und bat um seine Unterschrift drunter. Prof. E. hat das getan und liess den Inhalt des Mottos: Die Nase usw. durch Dolmetscher erklären.

14/11

**Figure 29:** ***"There were even caricatures of Einstein, featuring his stubby pipe and his full, rebellious shock of hair, also hinting at his attire, which was not always quite suited to the occasion."*** **(Quote from the ambassador's report of 3 January 1923) Einstein's handwritten note under the portrait:** ***"Albert Einstein or the nose as a thought reservoir."***

Einstein gratefully told Wilhelm Solf that he had "experienced truly fine and uplifting hours [at your home ...] and gained the conviction that I have made the acquaintance of a kind and unassuming man."

As Einstein was about to leave Japan, Solf sent "Professor Einstein, passenger Hirano Maru" a telegram that was dispatched on 27 December 1922 at 4 PM.[468]

> Thanks for letter. Shall [. . . ]. Wish you both good trip. May the new year bring you as much fortune as your visit to Japan has brought the German cause. The good German cause, well beyond any chauvinism. Goodbye,
> Solf.

Einstein's personality, pacifist and internationalist stance, had apparently swayed the former German colonial minister in the same direction, at least momentarily.

As will be revealed later, Solf had more than one reason to be pleased with Einstein's trip. For Einstein had paved the way for *other* German scientists and politicians. One consequence of the trip was, for instance, that Fritz Haber took an interest in touring Japan. 1924 Haber traveled to Japan, and the idea was born to found a Japan Institute in Berlin. The 'Institute for the Advancement of Reciprocal Knowledge about the Intellectual Life and Public Organs in Germany and Japan (Japan Institute)' was founded on 18 May 1926. Its sister institute in Japan was inaugurated on 18 June 1927.

Haber, the incomparably better adept and enthusiastic organizer of scientific affairs, assumed its directorship. Nevertheless, it had been *Einstein* who had laid its intellectual and conceptual foundations.

The *personal* contacts between Einstein and Solf did not end with his departure from Japan. Solf's papers reveal that Einstein applied for his assistance at least twice in subsequent years. Once – on 7 June 1929 – Einstein asked him to help out the Lessing College in Berlin, which was floundering in a financial crisis. "My request is," Einstein wrote him, "that you receive Dr. Lewin [the director of the college, S.G.], who is a good acquaintance of mine, and at least allow yourself to be briefed on the matter [...]."[469] Solf satisfied Einstein's wish and received Mr. Lewin. After he took steps on behalf of the school, Einstein thanked him on 26 July 1929 "for thus doing a lasting service to intellectual life in Berlin.[470] The second occasion was on 18 October 1930, when Einstein requested that Solf provide a favorable evaluation for a planned charity drive for a famine-stricken area in China, and perhaps also involve himself in it. Einstein offered to sign an appeal in this matter.[471]

Solf himself remained a decided conservative. There was no question of any common political affinities with Einstein. But one thing united them: opposition to National Socialism. The "Aryan File" among Solf's papers[472] documents that he was committed to helping persecuted scholars, some of whom were Jews. In a strictly confidential letter his Japanese friend, Prof. Sata had requested that scholars, researchers, technicians and artists be appointed "if possible in our country, who have lost their positions through the political events in Germany." Solf replied with the necessary caution but still sufficiently definitely "many cordial thanks for the merciful compassion with the victims of the mentioned policy." On 7 August 1933 Solf sent off "in a separate envelope the names and addresses of a number of scholars and artists who would deserve finding employment in Japan and who will certainly perform useful service to the Japanese government and the Japanese people." He was "not of the opinion that the German government would have any kind of objection toward the Japanese government if the Japanese government tried to offer *these unfortunate victims* a helping hand."[473] Unfortunately, Sata broke the agreed confidentiality and published the list in the paper 'Mainichi' together with the proposed plan. Solf informed the German consul general, Ohrt on 12 October 1933, that he had thus jeopardized their plan "by at least 75 % ."

There was actually time now for a longer stay in China. The efforts by the embassy in Peking and the consul general in Shanghai offered opportunity enough for it. But Einstein was against it. On 22 December the embassy in Peking received a telegram that read: "Eivstein[474] does not plan to come to Peking Letter follows. Plessen.[475] By picture postcard Mrs. Einstein informed the chairman of the German Association, Berrens, at the end of December 1922 that her husband "could not accept the invitation, because he is continuing on to Singapore with the same steamship that he is arriving on from Japan."[476]

As already mentioned, the original plan had been to spend a number of days in China. As late as October 1922 he had asked Dr. Pfister in Shanghai to draw up, together with the host University of Peking, an itinerary of the lectures to be held in Shanghai and other cities.[477]

So it was that much more disappointing that Einstein had no time for China even on the way back, deciding instead to use the remaining time in a quite different way. Thiel grumbled that he "attended more to his brethren in faith than to Germans [...], even though everything had been done on the part of the Germans to impart to him the honors he could claim as a prominent representative of German science."[478]

The reason for Einstein's cancellation is not known. But it is probable that the German consul general for China had been right in fearing that the Japanese had "imposed on him an itinerary that would neither allow him to fulfill his national obligations nor leave him time to relax."[479] The illness at the end of the sojourn in Japan was the physical result of this overexertion. Even so, the manner in which he informed the German consulate general was far too short shrift. Not even "a certain social awkwardness" could excuse that. There was no apology, just the information that he would not be there. He did not even bother to write a letter himself but left it to his wife on a mere postcard. Demonstratively giving preference to others over people who had spent much effort on his behalf certainly is treatment contrary to common etiquette. It certainly was not "good breeding"!

Neither did he go to Batavia – also contrary to the original plan.[480] The German consulate general in Batavia reported to the Foreign Office on 29 January 1923: "The Einstein couple did not come to Batavia but continued their voyage from Singapore westwards. The reason for the change in the route is unknown to me."[481] Einstein even snubbed the Dutch, whom he had so much to be thankful for. In Palestine, however, Einstein disembarked again for twelve days. He had finally arrived where the homeland of the Jews and the Jewish university were to be founded.

During their stay in Palestine the Einsteins were the guests of the British high commissioner, Sir Herbert Samuel, at Government House (the building of the former German Empress Augusta Victoria Foundation) on the Mount of Olives.

Einstein delivered his first lecture on 6 February 1923 on Mount Scopus. This is still considered the first lecture at the Hebrew University of Jerusalem, even though its official inauguration took place on 1 April 1925. Other lectures were delivered in Tel Aviv and other cities. The German Foreign Institute of Stuttgart conveyed to the Foreign Office a press release about the lecture delivered on 13 February 1923. Despite annoyance that the invitation had been printed in Hebrew and "bore not a single European (let alone German) letter," the lecture was recorded as a complete success. It was an event conducted by a *German* professor in his own *German* language: "The thronging was so enormously strong that the doors had to be shut 1/4 of an hour early already. The audience was mixed: English, French, Americans, etc., Catholics, Protestants, Templars, and the majority: Jews. It was the first time after the war that Jerusalem saw such a large assembly come to listen to a lecture in German by a German professor."[482]

Einstein's mere *presence* in Palestine was a coup *for Germans*. It was only helpful that Einstein spoke *German,* not Hebrew. German, not Hebrew, was the language of the greatest scholar of the present day! It is rare indeed that a country could profit politically from a deficiency in foreign language skills by one of its citizens. That (along with zoology and botany) had in bye-gone days caused Einstein to fail an admissions test (to the Polytechnic in Zurich).

After a stopover in Marseilles, Albert and Elsa Einstein arrived in Barcelona on 22 February 1923. The German consulate general in Barcelona regretted in its report to the Foreign Office from 2 March 1923 that Einstein had only given his firm commitment to visit the day before. That was why "it was, therefore, not possible to prepare an immediate reception for him, especially in the press, as would have been in the best German interests."[483]

Einstein arrived in a country whose government was interested in restoring favorable relations with Germany. The Spanish resisted the boycott against German science that was being carried out under French leadership. In 1919 just after the founding of the International Research Council, 110 Spanish scholars called for a resumption of relations with German scientists. So the reception that the German scholar received in Spain was not just of great significance for Hispanic German relations but also for the dissolution of the political blockade of Germany in general.

The German consulate general in Barcelona had developed major initiatives toward converting visits by German scholars into political capital. It had been at this consul general's suggestion, on 4 October 1921, that the Foreign Office instructed the Reich authorities to inform him immediately of foreign travel plans by German scholars and artists, so that the German diplomatic missions could make appropriate preparations.

In Spain, too, Einstein was conferred extraordinary honors. He delivered lectures in Barcelona, Madrid and Saragossa. His stay in Toledo became one of "the nicest days" of his life. There he also met the philosopher Ortega y Gasset for a conversation and a walk through the city, which included a touring of the church Santo Tome containing El Greco's 'Burial of Count Orgaz.' "We are living in luxury in Toledo, in el Escorial," Elsa Einstein wrote, "only Spain could produce such a thing as Toledo. I'd give you all the Buddist temples of Japan and India in exchange. Spain is a country full of austere beauty. We are seeing much magnificent art and we are being pampered to the point of being ashamed [...] perhaps it's too much. But *he* remains untouched by it inside."[484]

The German consulate general in Barcelona and the German embassy in Madrid reported *at length* about Einstein's visits. This confirms how much importance was attached to Einstein's appearance in Spain. Differing, and sometimes even opposing political stances by the writers of these reports – Consul General *Hassell* and Ambassador *Langwerth* – from Einstein's own views were only marginal in comparison. *At that time,* Einstein was needed. *At that time,* when Einstein's pacifism was a sin, he was nevertheless thoroughly useful.

**Hassell, Christian August Ulrich von** (born 12 Nov. 1881 in Anklam, executed 8 September 1944 in Berlin). After completing studies in jurisprudence, legal expert in Tsingtao. 1909 Assessor at the Foreign Office. 1911 vice-consul in Genoa. 1914 seriously wounded at the battle of the Marne. 1916 official advisor in Stettin. 1918 entered the conservative German National People's Party (GNPP). 1919 embassy councillor and chargé d'affaires in Rome. 1921 consul general in Barcelona. Envoy 1925 in Copenhagen and 1930 in Belgrade. 1932 ambassador in Rome, in which capacity he was deeply involved in drawing up the treaties and agreements with fascist Italy. 17 Feb. 1938 dismissed from his post and placed provisionally in retirement (permanently retired on 10 Feb. 1943). Still retained close personal ties to Hermann Göring, one of the most powerful men during the Third Reich.[485] Through his position and contacts with various people of authority, he was well-informed early on about internal political affairs and the overall picture, including crimes committed by the Army, reprisals against the Polish intelligentia and the extermination of the Jews.[486] Close ties with the resistance movement (Beck, Goerdeler). His goal: resuming dialog with the Anglosaxon powers and protecting Europe from Bolshevism. Among the resistance he was regarded as a future foreign minister. On 8 September 1944 sentenced to death by the People's Court and hanged on the same day.[487] The relevant points of pertinence to his period in Barcelona, when he met Einstein, are: he was not a warmonger, being rather concerned about good western European relations, but a decided opponent of communism (and for that reason worried about keeping Einstein's reputation untainted by it).

**Langwerth von Simmern, Ernst Baron** (17 Mar. 1865–17 Nov. 1942). School leaving examination at a preparatory school in Neubrandenburg. 1884–1888 law studies at Heidelberg, Leipzig, Munich and Berlin. 1889 doctorate, 1896 certification for academic teaching. 1896/1897 private lecturer in Marburg. 1897–1898 travels to England, USA, France, Russia, Italy, Austria. 1901–1904 legation secretary in Athens. 1904 legation secretary in Lisbon. 1905 provisional chargé d'affaires and legation councillor in Tangier. 1908 first secretary in Berne. 1909 to 1919 in the Foreign Office (portfolio on Moroccan and Spanish affairs). 1919 placed in temporary retirement (tendered resignation in protest against the Treaty of Versailles). 1920 ambassador in Madrid. 1925 appointed Reich commissioner for the occupied territories (Coblenz). 1930 retired.

Both Hassell and Langwerth happily and emphatically informed the Foreign Office that Einstein's visit was good publicity *for Germany*. But Hassell was annoyed to have to report that the Spanish press had also pointed out Einstein's pacifism and his refusal to sign the Appeal by the Ninety-three German Intellectuals. A political turn had been given "which probably would have been best avoided." So this referee was particularly concerned to "rectify" statements and actions by Einstein that were too far to the left. Both of them – Hassell in his report and Langwerth in a telegram – informed the Foreign Office that Einstein had explicitly contradicted other allegations by stating that he "*was not a revolutionary in the field of science but rather an advocate of evolution and that he rejected the program of the communists.*" This passage was heavily underscored by the recipient in Berlin.

In conclusion, Hassell reported from Barcelona on 2 March: "All in all, Einstein's visit here, during which time he always presented himself as a German and not a Swiss, can be characterized as a complete *success for him and for the reputation of German science as well as for German/Spanish cultural relations*, thanks primarily to the personality of the savant, which inspired the keenest affection from all quarters."

Report by the German consulate general in Barcelona, 2 March 1923 (excerpt):[488]

[. . . ]. Professor Einstein, in the company of his wife, arrived here on his return trip from Japan on the 22nd of last month. It had been agreed well in advance [. . . ] that he would visit Spain, but his firm commitment was not received until a day before his arrival in Barcelona, and even then the exact date was left open. It was, therefore, not possible to prepare an immediate reception for him, especially in the press, as would have been in the best German interests. At the invitation of the Mancomunidad, Einstein held three lectures at the Institut d'Estudis Catalans and an additional lecture in the Real Academia de Ciencias. [. . . ] there was a large audience present, which listened attentively to every word, extended him an extremely warm welcome, and thanked him with enthusiastic applause. Professor Einstein was the object of numerous honors from all sides. [. . . ].

The press published detailed reports of his visit here and his lectures, emphasizing his significance and paying tribute to his scientific investigations. Most newspapers adopted a muted political tone, perhaps not without help from some local political circles, referring to Einstein's pacifist views while mentioning the fact that he had not signed the well-known manifesto of the 93 German scholars. A political overtone, which probably would have been best avoided, was sounded on another occasion. [. . . ].

The well-known socialist leader Angel Pestaña welcomed him, using the opportunity to mention the social struggles in Barcelona. According to newspaper accounts, Professor Einstein replied, among other things, that he, too, was a revolutionary, albeit only on the intellectual plane. [. . . ]. Subsequently, Einstein himself told an interviewer that this account was incorrect – in fact the opposite was true. Einstein maintained that he was not a revolutionary in the field of science but rather an advocate of evolution and that he rejected the program of the communists.

All in all, Einstein's visit here, during which time he always presented himself as a German and not a Swiss, can be characterized as a complete success for him and for the reputation of German science as well as for German/Spanish cultural relations, thanks primarily to the personality of the savant which inspired the keenest affection from all quarters.

Hassell

On 19 March Langwerth took stock of the visit.[489] The following passages were marked by the recipient at the Foreign Office:

Prof. Einstein was showered with every kind of honor, and one can surely say without exaggeration that *throughout living memory, no foreign scholar has been met with such an enthusiastic and exceptional welcome in the Spanish capital.*

At this occasion the Spanish minister of arts and sciences held a speech at the end of which he offered the scholar the hospitality of Spanish soil and financial support by the government in the event that conditions at the moment in his home country should make continuation of his researches temporarily impossible!

On 8 March he was officially conferred, in accordance with ancient traditions, an honorary doctorate during a special session of the university. Besides the university president, a professor and a student, I also delivered a speech, following the local custom, in the Spanish language. The topic I chose concerned the historical *development of cultural relations between Germany and Spain.* The speech was excerpted in almost all the papers of Madrid and seems to have been generally well received.

In summary, I can say about Einstein's visit that it was a *complete and unqualified success. By his Spanish tour, Professor Einstein has done an incalculable service to the reputation of German science* and esteem for its accomplishments, as the English press has also acknowledged. The simple and kind manner of this scholar contributed considerably to this success.

At the beginning of March Albert and Elsa Einstein returned to Germany. On 24 March he applied to the academy for release of his regular salary as of the next convenient date. On 1 April 1923 his monthly compensation in the amount of 854,310 marks resumed:[490] inflation in Germany was accelerating.

In those troubled times it was just another reason for Einstein to leave town again in 1923: to Sweden and for an extended period of time once again to Holland. There he was safe.

The many trips to Holland held an important place in that scholar's life. It is almost symbolic that when Einstein first took office in Berlin in April 1914, he did not arrive directly from his residence in Zurich but *from Holland,* where he had been offering lectures and staying with friends.

**Figure 30:** ***"On 4 March a festive session of the Academia de Ciencias took place under the presidency of H.M. the King, who handed Prof. Einstein a diploma for a corresponding foreign member. The Spanish minister of arts and sciences delivered a speech on this occasion."***
**(To Einstein's right, the Spanish King Alfonso XIII and the Spanish minister of public instruction.)**

There was no need for tedious preparations or formal invitations when going to Holland. In many ways, while Germany was Einstein's unbeloved fatherland, Holland was his *home*. Purportedly so rootless, with everywhere and nowhere to call home, Einstein felt safe there. These ties date back to the period before World War I and are anchored in his friendships with Hendrik Antoon Lorentz and Paul Ehrenfest. Lorentz (born 18 July 1853) was considerably older than Einstein and their relationship, extending back to before the Solvay conference in 1911, could almost be called that of father and son. Ehrenfest (born 18 January 1880), on the other hand, was about the same age as Einstein. They first met in Prague in 1912 after corresponding over an extended period. Lorentz promoted Einstein as much as he did Ehrenfest. (Through Lorentz's efforts Ehrenfest was appointed professor at Leyden as early as 1912.) *Both* Lorentz and Ehrenfest were early advocates and champions of relativity theory; *both* were among Einstein's most argumentative and inspiring debaters. In August/September 1919, before there was much interest in Einstein, they were already intent on luring him to a position in

Holland. When he declined, they arranged for a guest professorship at Leyden – and succeeded in convincing him to accept it as well.

Einstein traveled countless times from Berlin to Leyden and almost every time stayed there as a guest of the Ehrenfest family. The reasons for each trip were varied. In most cases, Einstein came in order to deliver courses and lectures, to discuss and debate with his friends. During the early 1920s he also went there to escape from the constant threat on his life in Germany. Thanks to the hospitality of his friends, as he wrote Planck on 6 December 1923, it was "at least a very merry banishment." He felt at home in Holland. This commuting to and fro (along with its flux of work) was another surrogate for migrating.

### 2.5.6 Sweden and Holland 1923

Einstein spent May of 1923 as guest professor in Leyden (Holland).

He went to Sweden for an entirely different reason. On 10 December 1922, while he was still in Japan, the Nobel prize in physics was conferred on him. The German envoy had represented him at the festivities in Stockholm and had expressed his thanks for the distinction.

Excerpt from German Envoy Nadolny's speech on the occasion of the award of the Nobel Prize to Einstein on 10 December 1922:[491]

> Since Professor Einstein is prevented by his travels to the Far East from receiving the high distinction of the Nobel prize in person, I have the honor and duty of receiving his prize from the hands of H.M. the King and expressing in his name my thanks for the fine words that Prof. Arrhenius expressed at the award of the prize and that Prof. Söderbaum just now devoted to the prizewinners in his address [. . . ]. [. . . ] again I would like to express the delight of my nation at being able once again to offer from among its own something of benefit to the whole of mankind, and finally the hope that Switzerland, too, shares in this joy, where the scholar had found his home and employment of many years. [. . . ].
>
> The Nobel laureate Einstein is known not just as a scholar and scientist but also as an enthusiastic missionary of international reconciliation. So it will certainly agree with him if I close my words with the wish that the noble and fine goals of Alfred Nobel [. . . ] come steadily closer to materialization and perhaps more quickly so in the future as well. [. . . ].

Einstein was handed the medal and diploma in Berlin. In order to receive the sum connected with the award (120,000 Swedish kronor), however, the statutes of the Nobel prize foundation stipulated that he come to Sweden and deliver a lecture in person. This happened on 11 July 1923 in Göteborg before an audience of 2,000 in the presence of the king.

On his way back from Sweden Einstein visited Niels Bohr in Copenhagen. *No* issues of political relevance are attached to this excursion (hence it is not covered here). In August 1923 he vacationed with his sons in Kiel as guests of Hermann

Anschütz. From Kiel he continued on to a physics conference at Bonn, which certainly *was* anything but a private affair. Einstein participated in expression of his solidarity with his German colleagues, who by choosing this location wanted to protest against the French occupying forces in the Rhineland. Following the conference he spent two weeks visiting Ehrenfest in Leyden. Three weeks later, on 7 November 1923, a six-week stay in Leyden began not for personal reasons but primarily for *political* ones. Einstein had been warned that his life was in danger. In view of the political unrest in Germany that was escalating further by the day, there certainly was reason to worry. On 9 November Hitler's Putsch, also called his "march on the Hall of Heroes," took place. Blood flowed on Germany's streets. Around this time the value of printed mark notes had plunged to unimaginable depths. On 20 November 1923 one gold mark was worth 1,000 billion mark notes (on 11 October it had still been worth 1 billion notes). On 21 November 1923, one dollar could buy 4,200 trillion marks.

Einstein now seemed to be inclined to stay permanently abroad. Max Planck begged him "not to take any steps that could make your return to Berlin permanently and forever impossible. Many offers and invitations of an enticing nature will now surely be made to you, for other countries have been envying us for a long time for this precious gem of ours. But do think of those who love and admire you here, and don't let them have to suffer for the groundless nastiness of a vicious mob, which we must under all circumstances get under control."[492]

Einstein came back. Months later, it appeared as if his period of travels had even come to an end. In 1924 Einstein was a quite sedentary person. He attended the academy meetings with a frequency and regularity not seen for a long while. "After the past three exceedingly eventful and lively years, Einstein was intent on guiding his life along a calmer course conducive to contemplation, and for whatever reasons, Berlin seemed to him to be the most suitable place. [...]. Einstein did not want to travel abroad again so soon [...]. He limited himself to a few trips to Leyden and to Switzerland as well as into his refuge by the fjord near Kiel in the Anschütz factory. From there he wrote in May 1924: 'The political situation has settled down and the far-too-many no longer worry about me, so my life has become more tranquil and undisturbed'."[493]

That was exactly it: "The *political* situation has settled down and the far-too-many no longer worry about me." He did not want to leave and did not have to; he *had performed his service for his fatherland.* Even so, he left on another longer voyage again in 1925 – to South America.

## 2.5.7 South America

There had been talk of a trip to Argentina long ago – in September 1922. Einstein only actually set sail for South America in spring 1925, however. The reason for the delay was political: opposition within the German colony there.

The reports by the German legation in Buenos Aires describe in great detail how the invitation came about and why so much time had to elapse in the interim.

On 11 September 1922 the Institución Cultural Argentino-Germana was founded in Buenos Aires – a society "whose purpose was to foster cultural relations between the two countries."[494] Nevertheless, "at the wish of Argentinian quarters, the German element of the society [...] was limited, with few exceptions, to the academically educated," resulting in a "considerably inferior number against Argentinians." The German envoy justified this as "in accordance with the aims and character of the society." The Argentinians did not want to harness themselves to *German* colonial interests and would not let themselves be dictated about the design of their relations with *Germany*.

The German envoy warmly welcomed the foundation of the society, because it was suited "to becoming an effective counterbalance to the annexation efforts initiated with far more substantial means by the French on the cultural plane." It had the advantage over French efforts "of not bearing, like these, the stamp of official propaganda through support by a foreign government." But the envoy also recognized the danger of collaborating with the "German elements." It was "not without some worry, since some German members have difficulty shaking themselves completely free of their habit of political propaganda and basing their communication efforts purely on the grounds of cultural motivations, the political effects of which are recognizable *only indirectly and not easily*."[495]

The German ambassador certainly regarded Einstein's trip as an element of cultural policy for the German Reich. But he did not want the political effects to be directly and easily recognizable. The cultural fanatics in the German colony did not realize that "finer means" were more effective tools toward furthering Germany's influence in foreign policy. This was particularly true of German cultural policy, aimed at weakening France's influence in particular. On the contrary, the fanatics wanted to transpose domestic conflicts onto the foreign arena. As will be shown, this actually had the effect of harming Germany's interests.

The proposal to invite Einstein to Argentina was made during the society's founding meeting. Afterwards the German envoy Pauli expressed the fear that "with the wish of the society's leadership to invite Professor Einstein to Argentina," the "ability of the German members to conform" was being subjected to a severe tolerance test and that it would be difficult "to make clear to the Germans involved [...]" that specific Argentinian interests could not be subordinated to "entirely irrelevant views" by a fraction of the German colony.[496]

These fears were confirmed just a few days later. At the the first meeting of the society's board of directors, on 25 September 1922, when a representative of the student association – just an invited guest, not even a member of the board – proposed electing Einstein as an honorary member of the society, it encountered such hefty resistance by the Germans that the petition was voted down.[497]

> The chairman, Dr. Seeber, responded to the petition that among German members of the Conseio Einstein was regarded as a 'défaitist,' for having conducted propaganda against Germany during the war, hence as a traitor to the fatherland. For this reason honoring the scholar would excite ill-will in the German colony of Buenos Aires. That is why it would be advisable to forego taking such a step.[498]

Besides, not a soul could understand Einstein's theories, so an invitation would have nothing to do with science and just serve as fashionable publicity. There were other scholars at least as famous, whose theories had even been scientifically confirmed. Such arguments were greeted with laughter by the Argentinians present. "They would like to meet Germany's great minds of synthetic thought," they scoffed, and pointed out that no one "had acted more strongly on Germany's behalf around the world than Einstein."[499] All the same, the proposal to invite Einstein was rejected and struck off the society's agenda. Pie-eyed, reactionary political positions had prevailed over the interests of international cooperation and even over Germany's political interests abroad.

The Argentinian university teachers would not stand down, however. As Dr. Pauli put it elsewhere, they "would not have nominations of German scholars be decided either by German political aspects or by the evaluations of local German academics and laymen." In other words: the Argentinians insisted on not allowing the Germans living in their country to prescribe whom they would invite. Since the avenue of the Institución Argentino-Germana was no longer viable, "the University of Buenos Aires then undertook to confer an honor on Professor Einstein and invite him here, if possible."[500] Toward the end of 1923, the press was eventually informed that the University of Buenos Aires had invited Einstein to deliver a course of lectures.

The university was not in a position to fund the plan on its own, and the financially powerful German colony was evidently not forthcoming. So other sources were sought. In the end the necessary sums were obtained from the Asociación Ibérica, a Jewish cultural club, a few wealthy Jewish businessmen and, finally, the Institución Argentino-Germana, after all – as the German envoy (now Karl Gneist) reported on 30 April 1925.

Einstein was evidently reluctant – as we have seen before – to involve the German official authorities in the preparations for his trip, because the Foreign Office was left in the dark about the specific departure date until 17 March 1925, shortly before the scheduled departure. They could not even find out anything through the grapevine. A handwritten note on a request by telegraph from Buenos Aires reads: "No official communication on Einstein's departure is on file at the Min. of Culture either."[501]

But it was not possible to by-pass the authorities completely. Einstein eventually was forced to report to the Foreign Office to settle his passport business. This time he had more cause to do so than usual. He came as a supplicant. When Ministerial Department Head Krüss met Einstein on 5 January 1925, he found out that he needed a valid passport. Einstein declared that he had "applied to the Swiss Legation as a Swiss citizen living abroad to have a diplomatic passport issued to him, pointing out his frequent travels abroad. His application had been refused."[502] The Swiss had finally had enough of this globe-trotter and border crosser. Einstein was left with the alternative of applying to the *German* authorities for a *German* passport. He finally had to admit where his home was.

Einstein made another appearance at the Foreign Office on 21 January 1925 in this matter. He reported that he was leaving on 5 March with the steamship

"Cap Polonia." His only request was to be issued a passport by the Foreign Office. His wish was granted, if only because "we always have considered it important that Prof. Einstein travel with a German passport rather than, as previously, with a Swiss passport."[503] On 5 February 1925 already the welcome news arrived of "forwarding herewith the promised ministerial passport for your trip to South America on the Argentinian visa."[504] On 20 March Ministerial Head of Department Heilbron telegraphed Buenos Aires: "Einstein is traveling on a ministerial passport of the Foreign Office. Has expressed no wishes concerning discretion or assistance. Heilbron."[505]

Einstein arrived in Argentina on 24 March; he departed again on 21 April. As usual, his timetable was full of a variety of engagements.[506] On 28 March Einstein delivered his first lecture in Buenos Aires. He also traveled to La Plata and Cordoba (where he was able to meet a good acquaintance of his again: Georg Friedrich Nicolai – coauthor of the 'Manifesto to the Europeans,' who had become full professor of physiology at the University of Cordoba since his escape from Germany in March 1922). Many conferences were being held in Buenos Aires and elsewhere, which Foreign Minister Angel Gallardo also attended. Einstein was received by President Marceto T. de Alvear. The Jewish community organized a conference on the situation of the European Jews. The National Academy of Exact Sciences, Physics and Natural Science appointed him as honorary member on 16 April. His itinerary also included a flight around Buenos Aires. Toward the end of his visit, Einstein joined students at a dinner. He felt very much at ease with them and performed some violin pieces by Schumann.

At the end of April the German envoy was able to report to Berlin: "One can say that the course of the events were crowned in every way with success. [...]. I do not hesitate to say that Professor Einstein's visit has promoted interest in our culture and thus also the German reputation more than any other scholar has succeeded in doing. [...]. *A better man could not have been found* to counter the enemy propaganda of lies and to destroy the myth of German barbarism."

This trip had another purpose besides "acquainting the Argentine public with the personality of the currently most frequently mentioned scholar of the Old World and thus, at the same time, promoting German–Argentinian intellectual exchanges." Einstein saw it as a welcome opportunity to advertise the Zionist movement.

Report by the German legation in Buenos Aires to the Foreign Office in Berlin, 30 April 1925 (excerpt):[507]

After a four-week sojourn, Professor Einstein left Argentinia again on 23 April. The trip had taken place on the basis of an invitation that the University of Buenos Aires had issued to the scholar to deliver lectures. The funders were primarily the local Jewish cultural club ''Asociación Hebraica''; the ''Institución Cultural Germano-Argentina'' as well as a few wealthy Jewish businessmen also made available substantial sums of money. [. . . ].

One can say that the course of the events were crowned in every way with success. The guest was most warmly received from every quarter and showered with distinctions, as probably no other scholar here before. An uninterrupted series of festivities, receptions, dinners and the like were prepared in his honor. [. . . ]. The minister of foreign affairs appeared for all his lectures and would not pass up on introducing the scholar personally to the President of the Argentinian Republic [. . . ].

Einstein's lectures were greeted with unusual enthusiasm by all his audiences, were booked to the last seat and regularly earned thunderous applause. [. . . ]. Hardly a day went by in which the press did not print column-length reports about all that had to do with the scholar's personality and his theory. The papers of every bent tried to outdo one another in doing so [. . . ].

Besides the actual purpose of this trip of acquainting the Argentine public with the personality of the currently most frequently mentioned scholar of the Old World and thus, at the same time, promoting German-Argentinian intellectual exchanges, the visit evidently also served another goal, namely, promoting and strengthening the Zionist movement in Latin America and establishing closer ties with European efforts working in the same direction. [. . . ].

Unfortunately, the local German colony shunned all the events, because some of its nationalistic members condemned an interview of Einstein's in the Nación as pacifist. [. . . ]. Public opinion in Argentina shrugged its shoulders at this tastelessness and continued on with everyday life; the Argentine press did not address this issue.

I do not hesitate to say that Professor Einstein's visit has promoted interest in our culture and thus also the German reputation more than any other scholar has succeeded in doing. [. . . ]. As the truth about the war-guilt lie and ''German barbarism'' is gradually beginning to be revealed also in Argentina, French cultural propaganda efforts are becoming more developed here, wielding great skill and generous resources. Until now we have not been able to counter these efforts with any decisive factor. Now for the first time a German scholar has come here, whose name is world famous and whose naïve and kindhearted, perhaps a little otherworldly manner appealed to the people here so extremely well. A better man could not have been found to counter the enemy propaganda of lies and to destroy the myth of German barbarism.

Gneist

The guest list for the reception hosted by the German envoy at his private home also reveals the official significance attached to Einstein's presence. This lengthy list was published in the *Deutsche La-Plata-Zeitung* of 18 April 1925. Notables among them included: the ministers of foreign affairs and of culture, the ministers of justice and public instruction, the minister of agriculture, the chief of protocol and the legation secretary (each, of course, with wife).

By its self-imposed boycott the German colony condemned itself irretrievably as a mere onlooker. "Public opinion in Argentina shrugged its shoulders at this tastelessness and continued on with everyday life; the Argentine press did not address this issue." Einstein, the purported "traitor to the fatherland," served his fatherland well, however, promoting "interest in our culture and thus also the German reputation more than any other scholar has succeeded in doing." More than anyone else, this German *pacifist* contributed toward extinguishing the "*myth of German barbarism*" during World War I (something that Einstein himself would never have even dreamt of refuting). Even his "naïve and kind-hearted, perhaps a little otherworldly manner" was transformed into a political factor – a tribute to his German fatherland.

From Buenos Aires Einstein traveled on to Montevideo (Uruguay), staying there for a week. He delivered three lectures at the university. As the German envoy reported to the Foreign Office on 4 June 1925, he was "received by the government, academic officials, the general public and the press in such a fashion as virtually no other scholar has ever experienced here before." As in Argentina, in Uruguay, too, Einstein was celebrated by president, foreign minister and sundry representatives of the State and economy. There, too, his "visit *for the German cause* was very useful." The highpoint was undoubtedly seen as the reception hosted by the *Confederation of German Associations* in his honor at the German Club.

Report by the German legation in Montevideo to the Foreign Office in Berlin, 4 June 1925 (excerpt):[508]

[. . . ]. Prof. Albert *Einstein* spent a week in Montevideo on his way from Buenos Aires to Rio de Janeiro at the invitation of the Faculty of Engineers of the local University and delivered three lectures on his theory at the University in the French language. The lectures were well attended and received generous applause even though only very few could understand it.

In addition, Prof. Einstein was received by the government, academic officials, the general public and the press in such a fashion as virtually no other scholar has ever experienced here before. He was the topic of the day in the city in conversations and in the papers for a week. The President of the Republic, the ministers of foreign affairs and public instruction received him, and in the Senate President *Buero,* whom Einstein had met in Paris, delivered a speech about him. The university conferred him an honorary professorship, the municipal administration placed a car at his disposal for the duration of his stay and invited him to take up quarters in the leading hotel of the city, which Prof. Einstein declined, because he preferred to live with a member of the Jewish colony, Mr. *Naum Rosenblatt.* Among the numerous functions in his honor, the banquet should be mentioned, which expense had been approved by the National Administrative Council and which the President of the Republic and various ministers attended.

The committee of the Confederation of German Associations unanimously resolved to greet Prof. Einstein with a welcoming delegation and give him a reception at the German Club. [. . . ].

Einstein left an excellent impression here with his simple, ami-

able manner. Because he is being celebrated everywhere as a ''sabio aleman,'' his visit has been very useful to the German cause.
Schmidt-Elskop

From 5 to 12 May 1925 Einstein then stayed in Rio de Janeiro (Brazil). He was a guest of the Academy of Sciences and other institutions, but also of the Israeli community and the "Germania" Society. (The ambassador's comment that the reception "*proceeded harmoniously*" suggests that he had feared otherwise.) The government did not take as great an interest as its counterparts in Argentina and Uruguay, but Einstein was "much celebrated as elsewhere" anyway. The envoy took stock: "With his unassuming way Einstein aroused personal sympathy in Rio as well, albeit his very obvious lack of concern about his toilette evidently was not held against him. His visit in Rio was doubtlessly also *of benefit to the German cause here*."

Report by the German legation in Rio de Janeiro to the Foreign Office, 20 May 1925 (excerpt):[509]

Arriving from Buenos Aires, professor Albert Einstein stayed here from 5-12 May. [. . . ].

Besides the honors bestowed on Einstein from Brazilian quarters, he was also especially celebrated by the local Israeli community; furthermore the ''Germania'' Society organized an evening that proceeded harmoniously. I myself had invited Prof. Einstein to a dinner of exclusive guests including personalities worthy of his honor. The foreign minister, whom I had likewise invited, had himself represented that evening by his head of cabinet. [. . . ].

With his unassuming way Einstein aroused personal sympathy in Rio as well, albeit his very obvious lack of concern about his toilette evidently was not held against him. His visit in Rio was doubtlessly also beneficial to the German cause here.

Before his departure from Rio, his Brazilian friends handed him a precious little box out of Brazilian wood in memory of his stay in Brazil. It was the result of a collection in which, among others, the local American ambassador was also prominently involved.
Knipping

At the beginning of June Einstein was back in Berlin. Thus ended the period of extended travels abroad. Einstein contributed *decisively* toward dismantling the boycott of German science. He took a leading role in guiding Germany back to normal relations with other countries. This was no small achievement. At the same time, he helped Jews around the world find house and home. Such was the work of a German Jew, political leftist, pacifist and – "traiter to the fatherland."

## 2.5.8 Foreign travels 1929–1933

Einstein's later excursions outside the country had an entirely different function from his earlier ones, but this is an appropriate place to give a brief synopsis of their political fallout. They also serve as indicators of the political changes that had since taken place in Germany. The relationship between the Weimar Republic and Einstein was not what it had once been (nor was this by any means the only thing that had changed). What Einstein could do to further Germany's influence in the world had been done.

In 1929 Einstein started to leave more frequently again on lecturing tours.[510] In August 1929, for instance, he participated in the constituent meeting of the Jewish Agency in Zurich and in November he stayed in Paris to give lectures at the Sorbonne and receive an honorary doctorate from its faculty of the natural and mathematical sciences. In late 1930 Einstein traveled to the USA in conformance with a contract to offer a lecture course at the Californian Institute of Technology (CalTech) every winter term. In toto, from 1929 until the rise of the fascists to power he spent about half his time abroad.

What conclusions can be drawn from the existing reports by the various consulates or embassies?

The almost precocious interest of the German government in Einstein's travels abroad between 1920 and 1924 subsided to a cool reserve. The enthusiastic reception abroad was now aimed much more at Einstein as an *individual* rather than as a representative of Germany. A much sharper distinction began to be made between Einstein and Germany. Einstein was a *scientist,* a *pacifist* and a *Zionist.* The time when he acted as an emissary of Germany had past. His international activities became more of a nuisance than a help to German foreign policy.

German Ambassador Hoesch's report from Paris on 22 November 1929 was – still – friendly.

Report by the German embassy in Paris, 22 November 1929 (excerpt):[511]

[. . . ]. Professor Einstein, who arrived here on the 7th inst. to receive an honorary doctorate from the faculty of the natural and mathematical sciences at the University of Paris, sojourned here until the 14th inst. [. . . ].

Professor Einstein was the focal point of this celebration, which I attended personally. While the other honorary doctors had to make do with loud applause, minute-long ovations distinguished Professor Einstein, setting his rank clearly above the other honorary doctors and indicating the unique esteem he possesses in the academic world of France.

I am not exaggerating when I say that, based on observations I have been able to make not just at this meeting but on other occasions as well, Professor Einstein enjoys as no other German scholar of the present day general esteem and respect in the local scholarly world and beyond. [. . . ].

On the evening of the 13th, I hosted a dinner in honor of Professor Einstein, who had taken up lodgings in my home, which was attended by the minister of education, former prime ministers Painlevé and Herriot, the former minister Professor Émile Borel, the president of the university, Professor Charléty, the deans of the faculties of the

```
mathematical and physical sciences and of philosophy as well as a num-
ber of famous physicists, mathematicians and other scholars [. . . ].
   sig. Hoesch.
```

The tone of this report is still very positive with the ambassador's hospitality even extending to having Einstein "take up lodgings in my home." This was still permissible because the future of the Weimar Republic had not yet been decided. Any evaluation of such reports must take its author's biography into account. Friendly reporting did not *yet* have any obvious disadvantages for the author. Hoesch was an advocate of international exchange and was particularly concerned about fostering neighborly relations between France and Germany. He did not support the change of direction in German foreign policy. He rather applied the brakes to it.

**Hoesch, Leopold von** (10 Jun. 1881 to 10 Apr. 1936). Law studies; 1895 1st state examination. 1907, as candidate for the diplomatic service, assigned to the German legation in Peking. 1909 doctorate, then attaché at the embassies in Paris and Madrid as well as at the Foreign Office. 1912–1914 legation secretary in London. 1915 legation in Sofia, 1916 embassy in Constantinople. Involved in the peace negotiations of Brest Litovsk and Bucharest. 1918 legation councillor in Christiania, 1919 chargé d'affaires in Madrid; from 1921 embassy councillor in Paris. From January 1923 chargé d'affaires in Paris, on 2 February 1924 appointed German ambassador in Paris. "In constant close contact [with Stresemann], H. adopted as his own, Stresemann's conception of a normalization of the shattered Franco-German relations and contributed significantly toward an improvement in the political climate." In autumn 1932 Hoesch was transferred to the ambassador's post in London. A biographical entry reads: "The initial National Socialist successes particularly in the area of foreign policy did not change his early scathing verdict on the law-breaking and war-making new regime. Ribbentrop wreaked his havoc in London without H. being able to prevent it. Having created under and together with Stresemann a chapter of trust in German policy in the West, he saw it so much diminished with the declaration of the Locarno treaty and the deployment of German troops in the demilitarized Rhenish zone that he called this action the first step to the 2nd World War. He died of a heart attack in London during this crisis."[512] Even so, it is doubtful whether Hoesch would have still been prepared to offer Einstein bed and board at his home *after* Hitler's ascension to power. It rather "appears as if Hoesch was not the most resilient of German diplomats. He made excuses for the German situation to his English counterparts more than was necessary. He pointed out to the Foreign Office [in Berlin] the positive and increasingly important trend for Germany, that Great Britain was instructed to keep out of any developments, above all those of a warlike nature, as much as possible [...] – advice that [...] must have been of great use to the new regime."[513] The resignation of Ambassador von Hoesch would have had a signal effect, especially considering that during the Weimar Republic he had counted as Germany's leading diplomat. But Hoesch was not capable of such resolute action as that of his colleague von Prittwitz in Washington. What von Prittwitz later wrote may well have also applied to Hoesch: "Whoever thinks he can use characters like Hitler always only notices too late that he himself has been used and is subsequently eliminated."[514]

A favorable judgment on Einstein, who agreed with Mr. von Hoesch in many things, was the logical consequence in 1929.

The last report that put a foreign trip by Einstein in a positive light was dated 21 March 1931. German Consul General Heuser in New York described it as "a *boon for the reputation of Germanism*, because he has been mainly named and celebrated specifically as a German scholar." This praise is relativized at the same time, though: "It does not seem out of the question that this effect would have been even stronger, had Einstein let himself be exploited less for Zionist purposes and by pacifist organizations during his sojourn here."[515] This way Consul General Heuser could safely distance himself personally from such pacifists and Zionists.

Another report dated 2 March 1931 from Los Angeles reveals the importance of the *character* behind the person doing the reporting. It sums up Einstein's stay in *Pasadena* (the main destination of his trip to America in 1931) as follows: "I regret to have to say that the visit by our famous contemporary has *remained without benefit for Germanism* as such or for our political and social standing."[516] The evaluation of Albert Einstein's stay in the United States was thus completely different (more "up to date," perhaps?) than the report by Consul General Heuser's from Paris.

Later diplomatic reports either comment *negatively* about Einstein or *not at all*. The reason for the latter is probably, as mentioned before, primarily that the network of power in Germany changed much more quickly than the staffing constellations in the foreign embassies. So they preferred simply to forego commentary and keep silent.

The report by Ambassador Prittwitz dated 19 March 1931 from Washington is one example. In discussing the "reaction of American Jewry to anti-Semitic trends in German National Socialism,"[517] he limited himself without comment to quoting from American newspapers, including one from the *New York Times* from 8 March 1931 which contains the comment: "It seems to me that very systematic attempts are being made by some English to destroy the relations between Germans and Jewish Americans. I conclude this particularly from various statements I happened to hear during the Zionist dinner in honor of Einstein that particularly refer to the rise of the Hitler movement in Germany. Many Jews here only see the Hitler movement from the anti-Semitic side and that is being thoroughly exploited against us."

Von Prittwitz, ambassador in Washington since 3 November 1927, evidently had his own opinion about this but left the commentary to the Foreign Office. In 1933 he distanced himself publically from the new regime in Berlin and resigned his ambassadorship. Evidently, not everything that happened in the German embassies initially agreed with the Nazis. Consul General Otto Kiep attended a banquet in honor of Albert Einstein in 1933, for example.[518] As a consequence he had to leave his post in New York, which he had been holding since 1930.

*Other* diplomats who distanced themselves from Einstein drew their reasoning in conformance with the new "spirit of the times." Restraint is not necessary or desirable where there is anything negative to report about that traitor Einstein.

What German Envoy Clodius reported from Vienna on 15 October 1931 is an adequate reflection of what was going on inside Germany. Officials wanted to have as little to do with Einstein as possible, "because he is a Jew and is considered a political leftist." *In the old days,* when opportunity arose to greet Einstein, royalty and government leaders were the first to come forward. Now it was better to keep far away from him. "*Neither the minister of instruction nor the presidents of the universities* attended Einstein's lecture," delivered in the Austrian capital on 14 March 1931. "Nor has Professor Einstein *been officially received or invited by any Austrian agency,* even though he has returned to Vienna for the first time since 9 years."

Report by the German legation in Vienna to the Foreign Office in Berlin, 15 October 1931 (excerpt):[519]

[. . . ] Confidential.

Professor Albert Einstein delivered a lecture on the current state of relativity theory on the 14th inst. at the Institute of Physics of the University of Vienna at the invitation of the Organizing Committee for Guest Lectures by Foreign Scholars.

It is characteristic of the way in which all things are treated, from the point of view of party politics, that the official Austrian authorities observe a particular reserve toward Professor Einstein, because he is a Jew and is considered a political leftist. Neither the minister of instruction nor the presidents of the universities attended Einstein's lecture, which otherwise drew very large crowds, of course. Nor has Professor Einstein been officially received or invited by any Austrian agency, even though he has returned to Vienna for the first time since 9 years [. . . ].

The right-wing press scarcely took note of Professor Einstein's presence. The papers on the left, by contrast, duly criticized the neglect by official Austria of one of the greatest living scholars.

sig. Clodius

German authorities turned a cold shoulder on Einstein for another reason as well. *He was no longer useful.* He had served weakened Germany very well indeed. But now that the country had regained its strength and influence, he was deemed dispensable.

## 2.6 Erstwhile Swiss, henceforth Prussian. Einstein's citizenship

Einstein's citizenship had never really been more than of secondary interest. This all changed when he was chosen for the Nobel prize. It suddenly became a *political issue*. Who was supposed to represent his country officially at the award ceremony on 10 December 1922 while he was away in Japan? A German envoy or a Swiss?

Inquiry by the German legation in Stockholm from 25 November 1922 to the Foreign Office (excerpt):[520]

German legation, Stockholm. Stockholm, 25 Nov. 1922

Telegram.

There is a difference of opinion here on whether Einstein is German or Swiss. In former case, I must accept Nobel prize, in latter, the Swiss envoy. Please wire info.

sig. Nadolny

At his birth on 14 March 1879 in Ulm, Einstein had acquired German (that is, Württemberg) citizenship.[521] When Einstein's parents moved away from Germany (Munich) in the summer of 1894 to Italy (Milan), Albert Einstein remained in Munich to finish the last three years of his preparatory schooling and take his final examination. The militaristic spirit at the school and an anti-Semitic attitude made life so unpleasant for him there that he spontaneously decided to drop the examination and followed his parents to Italy on 29 December 1894. Being less than sixteen years of age, he could still leave the country unhindered. He was released from Württemberg citizenship, upon application by his father, on 28 January, 1896.[522] But even this *Wunderkind* needed a school-leaving certificate in order to gain admission to college. So he moved to Switzerland and stayed there – with only a brief interlude in Prague as professor at the German University – until his appointment to Berlin. On 19 October 1899 Einstein applied to the "High Federal Council of the Swiss Confederation" for permission to acquire Swiss cantonal and municipal citizenship. Pursuant to a decree by the Federal Council, Einstein acquired Swiss citizenship on 8 March 1900.[523] This means that Einstein was stateless between 28 January 1896 and 8 March 1900.

Einstein knew what it meant to receive a call as professor at a German university (his appointment to one of the two full-time positions at the Berlin academy automatically included such university affiliation). Fritz Haber explicitly pointed out to him that one consequence of his appointment would be that Einstein would become a Prussian citizen.[524] This was not a secret either.[525]

Einstein decided, however, (as he recalled ten years later) to set as a condition for his acceptance of the appointment to Berlin that "absolutely nothing change" with regard to his citizenship.[526] He wanted to keep his Swiss citizenship, "because he feared discommodities – e.g., in obtaining a passport."[527] This fear was well founded, as very soon became evident. Being under forty years of age, he would not have been permitted to travel to neutral foreign countries during the war as a citizen of the German Reich. In addition, being only thirty-five years old in 1914, he would have been required to enlist himself in the military during the war.

He had supposedly been assured, however, that if he accepted the appointment, "he would *not* acquire Prussian citizenship."[528] The academy apparently made this concession offhand and, as the later furor about it reveals, without first carefully checking the legal situation and apparently without even making any note of it in the files. Einstein's appointment was not supposed to depend on the question of his national citizenship. The interests of the State were much more important than this apparently so trivial issue.

In actual fact, the academy and the German state treated Einstein as a *foreigner* and a Swiss citizen during the war. This special treatment was useful to the academy in two ways: Einstein was reserved for the purposes of pure research, his survival was guaranteed by keeping him safely away from the battlefield. On the other hand, it was never quite certain how a pacifist like Einstein would behave at the frontline. Letting Einstein stay Swiss was a way to avoid such unnecessary worries.

Henceforth the Reich authorities had to tolerate and acknowledge that Einstein was *Swiss* on many occasions, including Einstein's appointment to the board of trustees of the Bureau of Standards (PTR, which only became legally binding with the kaiser's signature).

The grounds for this signature were: "Einstein is a citizen of Switzerland, however, this fact ought not to present an obstacle for his appointment to the board of trustees. His former appointment to the Royal Prussian Academy of Sciences did not founder against it either."[529] At that time an inquiry about the matter had been submitted to the Reich Chancellor: "I have the honor of applying for Your Excellency's kind opinion on whether there be any objection to the appointment. Einstein is a citizen of Switzerland. The statutes of the Reich Bureau do not forbid the appointment of foreigners to the board of trustees, nevertheless installations at the institution made for experiments in the fields of military and marine technology are kept secret."[530] The supporting statement by the secretary of state of the Reich Office of the Interior and Vice-Chancellor K. Helfferich, dated 22 December 1916 were laid before Wilhelm II at main headquarters. On 30 December 1916, the kaiser decreed that Einstein be appointed on the board of trustees of the Bureau of Standards of the German Reich.[531]

Even after the war the Reich authorities were not initially disturbed by the fact that Einstein was regarded as *Swiss* and took a Swiss passport with him on his travels.

In March 1922 Ambassador Mayer wanted "to send him a welcome message at the station," only "provided that Einstein, who supposedly owns a Swiss passport, is traveling on a German passport."[532] He was also informed that: "Einstein is traveling on a Swiss diplomatic passport. No objection to intended welcome [...] of famous member of our academy and university. Further action left to judgment there."[533] The ambassador closed his report with the admission that although Einstein was formally treated as Swiss in France, he "ultimately did have to be regarded as a German."[534]

It was known that Einstein felt more comfortable traveling with a Swiss passport and thus was evidently more attached to Switzerland than to the German Reich when it came down to the question of national citizenship. His trip to Japan was a case in point.[535] Nevertheless, even his Japan voyage received good marks, having brought "much fortune" for the "*German cause.*"

The German position changed overnight when Einstein was to be distinguished with the Nobel prize. The German envoy regarded it even as his prerogative to attend the Nobel award ceremony on 10 December 1922 in Einstein's absence. That way, the whole world should see that Einstein was a *German* and belonged to the *German Reich.*

The Prussian Ministry of Science, Arts and Culture (represented by Dr. Krüss) was not certain, at first, whether Einstein was a Prussian citizen but took the position that Einstein's scientific work was intimately connected with Germany and that he was regarded as German all over the world. Consequently, the *German* envoy ought to represent him at the award of the Nobel prize – even though the only issue *not* being questioned at that point was that Einstein was a citizen of *Switzerland.*[536] The reply that the *Berlin academy* telegraphed to the German envoy was more definite: "Einstein is a Reichsdeutscher" (if not de jure, certainly de facto from his official place of employment and thus presumably also his affinity[537] – following the principle "ubi bene, ibi patria," where you are comfortable is your fatherland). This reply was a conscious half-truth: The academy knew fully well that Einstein was a Swiss citizen; hitherto, especially during the war, it had even been very convenient.

The Swiss envoy was bold enough to question whether Einstein actually was a Prussian citizen but the German envoy's brusk reply cut short his attempt to "claim" Einstein for his side. The German envoy thereupon "effectuated the representation on Einstein's behalf both at the award as well as at the subsequently held banquet and dinner hosted by His Majesty the King in honor of the laureates."[538] "In order to take the edge off any latent or perhaps still seething sensibilities," he thought it appropriate "to emphasize the Swiss share in Albert Einstein's person and work in his speech as well as at the banquet and in a telegram to the German press." A peculiar form of "courtoisie," indeed! Very "noble, chivalrous gallantry"!

It was all very embarrassing for the German envoy when the Foreign Office informed him – afterwards – that Einstein was in fact a *citizen of Switzerland.* The next essential step was to hush up the matter as best as possible, to avoid any diplomatic complications.

Thenceforth, the Germans, and particularly the Prussian Academy of Sciences, were intent on settling the matter to their advantage. The academy argued that, as a Prussian civil servant who had taken his oath of office, Einstein had assumed citizenship of the German Reich, "even if he had not already possessed it by birth." The academy did not explicitly contest his Swiss citizenship but waved it aside as a secondary issue.

On 29 December 1922 the Swiss legation sent the Foreign Office official notice[539] that Albert Einstein had "acquired Swiss nationality upon his becoming a citizen of the Canton of Zurich on 7 February 1901." The addendum was irrefutable: "According to statements he submitted to the legation, he had at the time of taking Swiss citizenship lost his German citizenship and had not regained it in the meantime."

After verifying the circumstances, the Academy of Sciences sent a full and final report to the Ministry of Culture on 13 January 1923.

Statement by the Academy of the Sciences, dated 13 January 1923 (excerpt):[540]

[. . . ] In response to the decree of 4 January of this year - UIK No. 3155 - the undersigned Academy humbly reports that the following aspects were decisive in her statement about Prof. Einstein's citizenship in the German Reich:

Prof. Einstein is a *full-time* member of the Prussian Academy of Sciences and as such occupies a regular budgeted position in the Prussian budget. [. . . ]

According to the congruent opinion of the juridical members of the Academy, Prof. Einstein is a direct *state civil servant*. As such he [. . . ] gave his *oath of office*, to be precise, on 1 July 1920 to the Constitution of the Reich and on 15 March 1921 to the Prussian Constitution. In his capacity as a direct civil servant of the Prussian state and from the fact that he has given his oath of office, the Academy concludes that Mr. Einstein has thereby automatically acquired citizenship of the German Reich, even if he had not already possessed it by birth. For only citizens of the Reich may be civil servants of the State [. . . ]. The Swiss citizenship formerly obtained by Prof. Einstein through naturalization is not affected by this, but in the Academy's view Prof. Einstein is, in any case, foremost a Reichsdeutscher.

A copy of this letter has been sent to the legation in Stockholm.

Prussian Academy of Sciences

Roethe Planck Rubner Lüders

The Ministry of Culture sent this notice to the Foreign Office on 26 February 1923 with the confirmation "that Einstein has Prussian citizenship."[541] However, "since Prof. Einstein is currently traveling outside the country, inquiries about his citizenship status could not be made of him personally." They were convinced, though, that through his election to the academy, Einstein had become a *civil servant and hence also a Prussian citizen.* (But they were still not quite sure whether he had ever lost his German citizenship.) It was a fact that "police records indicate him as Swiss."

In the end, with its letter of 31 March 1923, the Prussian Ministry of Science, Arts and Culture joined the Foreign Office in adopting the academy's view.[542]

The Office accordingly informed the Swiss legation in Berlin that Einstein was being regarded as a German citizen pending further clarification.[543]

But Einstein was not as easy to convince as the academy and the government had thought. In a letter from 24 March 1923 he stood his ground, pointing out that he had been promised otherwise when he accepted his appointment to Berlin. His referral to the files at the ministry for clarification of the matter prove that he had not asked for any written confirmation of this assurance at the time. Nor did he have any reason to doubt it before, especially during the war.

A. Einstein's letter to Secretary Lüders at the Academy of Sciences:[544]

```
Prof. A. Einstein Berlin W 30, 24 III 1923.
   Haberlandstr. 5.
Esteemed Secretary,
   With reference to your esteemed letter of the 15th of Feb. of this
yr., I take the liberty of informing you of the following. When my ap-
pointment to our Academy was being envisioned in 1913, my colleague
Haber called my attention to the fact that my appointment would have
the consequence that I would become a citizen of the Prussian State.
Because I consider it important that absolutely nothing change with
regard to my citizenship, I made my acceptance of an eventual appoint-
ment conditional upon satisfaction of this provision, which was then
also granted. I do not doubt that this matter can be verified in the
ministerial files; furthermore, I know that this situation is known to
my colleagues Haber and Nernst.
   In great respect, A. Einstein
   To the Pr. Academy of Sciences
   attn. Priv. Councillor Prof. Lüders
   Berlin N. W. 7.
```

When the ministry hunted through its files for documentation of this promise in 1913, nothing could be found. The trusting thirty-four-year old had been satisfied with a vague oral assurance.

Einstein requested that the certificate and medal of the Nobel prize be conveyed to him through the diplomatic representative of *Switzerland*.[545] To prevent further diplomatic scuffling, the *Swedish* envoy handed them to Einstein at the end of April 1923, at the request of the Nobel Prize Foundation.

Although Einstein did not contradict the verdict of the Prussian Ministry of Science, Arts and Culture during an interview there on 19 June 1923, he requested that "his eventual Prussian citizenship if possible not be made public knowledge."[546] So Einstein wanted exactly what the academy and the Reich authorities did *not* want: He wanted to keep quiet about his *Prussian* citizenship; the academy and government officials wanted to hush up his *Swiss* citizenship.

How little Einstein saw the need to comply with the academy's and the officials' preference is revealed in that he continued to use his *Swiss passport* on his travels.

On 7 February 1924 Einstein finally agreed to confirm that he had obtained Prussian citizenship in addition to his Swiss one. The academy requested this in writing – it was not so trusting as Einstein had been ten years earlier.

Affidavit by A. Einstein on the question of his citizenship for the files of the Academy of Sciences:[547]

At the Academy the question has arisen whether I have in addition to my Swiss citizenship also Prussian citizenship. At the Academy's suggestion I discussed this with Mr. von Rottenburg at the Ministry of Culture. He emphatically held the view that legal acquisition of Prussian citizenship was linked with my engagement with the Academy, since a contrary interpretation is not documentable in the files. I raise no objection to this interpretation.

A. Einstein

7. II 24.

It was not for Einstein, however, to decide the issue finally but *Switzerland.*

While arranging for his trip to South America, Einstein went to the Foreign Office to ask for a *German passport.* His reason was: "He had applied to the Swiss legation as a Swiss citizen for a diplomatic passport with reference to his frequent travels abroad. He had been turned away with the argument that such passports are not normally issued and the Swiss legation was anyway not responsible for him, Prof. Einstein, in this matter."[548]

Switzerland considered itself *no longer responsible* for Einstein. Its legation had informed him that it considered him a German and he should please act accordingly. He had no choice. He *had* to go to the German authorities if he wanted to accept the invitation to South America.

So Einstein had made his way to the Foreign Office on 21 January 1925. It had perhaps been a path to Canossa for him, but it had to be. The Foreign Office was evidently gratified; they graciously granted him the assistance he requested.[549]

On 22 January 1925 Einstein was duly and officially informed that the Office would take care of his business.[550]

He got his passport. A *German* passport.[551]

Thus the *Swiss* Albert Einstein became, against his will, the *Prussian* Albert Einstein.

The fact that this issue of national citizenship would not die down was partly through Einstein's own doing.

On 15 December 1926 police headquarters in Berlin inquired of the Ministry of Culture "for information, please, whether through his appointment he had acquired citizenship of the Prussian State on the basis of §14 of the Reich and State Citizenship Law of 22 July 1913."[552] The reason: "Prof. Albert Einstein, born 14 March 1879 in Ulm, residing in Schöneberg, Haberlandstrasse 5, is a citizen of Switzerland. Having been member of the Prussian Academy of Sciences in Berlin since 1913, he alleges also to have acquired Prussian citizenship."

The ministry responded on 1 February 1927 with a question: "for what reason should the citizenship of Prof. *Einstein* be clarified"?[553] The "humble" response was "that the citizenship of Professor Albert *Einstein* must be determined at the request of the Municipal District Election Office in Schöneberg, Berlin. It is required for the voter list." The police chief at the same time assured that he would handle "the matter with favorable discretion," considering that "the scholar has a claim to world fame."[554] On 28 March the police chief receive the desired information: "Prof. Einstein has [...] acquired citizenship of the Prussian State and therefore of the German Reich."[555] On 2 June 1927 the police chief informed the District Election Office of Schöneberg, Berlin, that Albert and Elsa Einstein "have Prussian citizenship."[556]

A few years later Einstein's German citizenship was retracted again. They were glad to be rid of him at last.

## 2.7 Einstein's membership in the International Committee on Intellectual Cooperation

When Einstein left on his travels in the first half of the 1920s, he was more an *object* than a *subject* of politics. But he also made a conscious effort over an *extended* period of time to be politically active at least in *one* capacity – as a member of the International Committee on Intellectual Cooperation.[557] As the tale of his collaboration will show, though, he never felt completely comfortable doing so. Many years later he gave a very characteristic resumé of it: "Despite its illustrious composition, it was the lamest endeavor I ever participated in, in my life."[558] The meetings of this committee of the League of Nations were, as Elsa Einstein wrote in a letter to Alfred Kerr in April 1926, "extremely boring" for Albert Einstein.[559] No activity caused him and others more trouble – from the beginning to the end of his membership. No other activity demonstrated so well how difficult it was to fit Einstein into any viable political structure. His membership in this committee of the League of Nations is an exemplary case of the relations between science and politics during the Weimar Republic.

Drawn in relation to history in general and to the history of this committee of League of Nations in particular, Einstein's membership can be divided into two parts: first, the period before his entry (1922) or reentry (1924) in the committee, until Germany's joining of the League of Nations (10 September 1926); second, the period afterwards. But no clear line separates them. In 1925 when Germany was working towards membership in the League of Nations, a new situation gradually emerged that produced an entirely different attitude toward membership in the committee.[560] From the point of view of *staffing*, the second period is marked by the appointment of Hugo Andres Krüss – managing director of the Prussian State Library – as Einstein's proxy on the committee. We shall see that the coincidental timing was not sheer chance.

### 2.7.1 The founding of the International Committee on Intellectual Cooperation

The Covenant of the League of Nations did not provide for a committee on intellectual cooperation. The Secretariat, under the directorship of the Japanese undersecretary general, Nitobe, only had an international section that, pursuant to Article 24 of the Covenant, dealt with the responsibilities and the statutes of international bureaus created by the collective treaties.

On the final day of the first meeting of the League of Nations, on 18 December 1920 a proposal advanced by Senator La Fontaine was accepted inviting the Council (of the League of Nations) to report to the Assembly at the next session on establishing a technical organization for intellectual cooperation. The proposal was accepted without actually knowing what specifically was meant by it (at least according to the recollection of then Undersecretary General Nitobe).[561] It would soon become apparent that this "*without actually knowing what specif-*

*ically was meant by it*" was in many ways characteristic of the entire work of the committee.

The Council discussed the matter on 1 March 1921 and concluded that the League of Nations' finances would not permit the founding of a technical committee and that at the moment best progress could be made in intellectual cooperation through intellectual efforts. The Council nevertheless suggested the Secretariat study the matter and report on its findings.[562]

At its meeting of 2 September 1921 the Council accepted a report on the subject[563] by Léon Bourgeois, a representative of France and president of the League's Council.

Léon Bourgeois stressed: "We are all convinced that the League of Nations needs nothing more urgently than to investigate the major factors of international opinion making – the systems and methods of education as well as research in science and philosophy. It would be inconceivable that the League of Nations strive to improve the means of exchanging material products without at the same time thinking of possibilities for the international exchange of ideas. A community of nations cannot exist without the spirit of mutual intellectual interaction." Bourgeois proposed the formation of a committee that should suggest ways to facilitate intellectual exchanges between nations, particularly respecting the exchange of information in the field of science, as well as respecting educational methods. First mention of the concept "intellectual cooperation" was made in this report.

Following Léon Bourgeois's report the Council asked the Assembly to suggest which steps the League of Nations could take to facilitate intellectual exchanges among nations.

On 21 September 1921, the (second) meeting, the Assembly of the League of Nations asked the Council to appoint a committee to study the problem of international intellectual cooperation. It was supposed to comprise a maximum of twelve members – "both men and women" – appointed by the Council. A report on measures that the League should take to promote the international exchange of ideas, particularly scientific information and educational methods, was scheduled for the next meeting. The "educational system" was subsequently struck from Bourgeois's original resolution because some members of the Assembly feared meddling by the League of Nations into the educational affairs of the individual nations. The original idea of including women on the committee with education in mind was retained, however, despite the expressed worry that the reputation of the committee could be hurt because of the extremely low proportion of women in science.[564]

On 14 January 1922, on the basis of a report by Hanotaux, the Council of the League of Nations finally ratified establishment of what was later called the "International Committee on Intellectual Cooperation" or "Commission pour la Coopération Intellectuelle."

The Council took this decision even though the available funds were extremely limited: "From the outset only a very small amount was allocated to the committee from the budget of the League of Nations, because it had not originally been envisioned in the Covenant. Although its later advocates, such as the Pole Halecki, pointed out that it had been the *intention* to add an article to the Versailles treaty about intellectual cooperation among nations, the Australian delegate took the absence of such an article to vote against allocating any funds to the committee."[565] With much effort, it was eventually possible to approve a small amount for the new committee from the budget in the section for international bureaus. Thus the Council was able at least to meet the travel and accommodation costs of the committee members.[566]

The conception of the Council of the League of Nations was that the committee initially concentrate on the following three points:

1. to promote international scientific conferences, associations, etc.,
2. to establish international ties between universities, particularly through exchange programs for professors and students as well as to organize an international university bureau and perhaps also an international university,
3. to organize an international scientific bibliography and the exchange of scientific publications.[567]

As far as its actual agenda was concerned, the committee was "granted the greatest liberty."[568]

No one seemed to notice, or else consciously ignored the fact that this goal directly contradicted the boycott by the French against German science.

### 2.7.2 Einstein's appointment to the committee

Much time necessarily elapsed between resolving to form the committee of the League of Nations, nominating its members and commencing actual operations, because the selection and nomination of its members was much more complicated than original thought.

The committee was supposed to consist of scholars "chosen with a view to their academic stature, irrespective of their citizenship."[569] Nationality was supposed to be ignored. Only the personal reputation of the individual members was supposed to count.[570] But such an ideal nominating principle was hard to follow. In actual fact *all nations of which the Council was composed* were eventually represented in the committee only excepting Japan and China. China was dispensed with because its universities were still too underdeveloped. Japan was omitted – contrary to original considerations – because it was already represented by Nitobe in the Council of the League of Nations.

In the end, proportional representation and diplomatic considerations were more important than lofty principles. On one hand, the fine dream, on the other, the reality of national rivalries. The committee was brought to life with this dilemma and that is how it stayed. Whenever the committee sought to obtain firm results from its efforts, it depended on national resources – institutions and financing – and correspondingly had to respect national interests as well.

A list of sixty candidates was drawn up for the Council from various sources from which twelve had to be selected. On 15 May 1922 the nominations were made public:[571] Banerjee (India), Bergson (France), Ms. Bonnevie (Norway), de Catro (Brazil), Ms. Curie-Sklodowska (Poland), Destrée (Belgium), Einstein (Germany), Gilbert Murray (England), de Reynold (Switzerland), Ruffini (Italy), de Torres Quevedo (Spain) as well as Robert Andrews Millikan and his proxy George E. Hale (USA).

The inclusion of *two* nonmember states was a decision of extraordinary significance. Although USA was not a member of the League of Nations, it was, at least, one of the victors of World War I. But Germany? Why choose a scholar from Germany? That loser of the war, thoroughly humiliated by the Versailles treaty (28 June 1919) and still an object of France's hate and desires. Its science was being systematically boycotted by international organizations under the guiding influence of France. Germany the pariah of the international community.

There was little inclination to choose a scholar from Germany. An Austrian might have been tolerable, but a German would be going too far. France applied the strongest resistance.

The final decision to choose a *German* centered around to main arguments.

1. The *initiators of the new committee* were seeking to restore fruitful cooperation among scholars from all countries. Boycotting the underdogs was not their aim. This was perhaps utopian and naïve in the given atmosphere of hate between winners and losers. It was a valiant struggle against the spirit of the times. One strong argument was that excluding Germans would have condemned the committee to failure from the start. German science had lost nothing of its world renown despite the war and despite the shortages. So the unanimous opinion was that "German science ought to be represented."[572]
2. All arguments in favor of the Germans would presumably have been ineffective without *Albert Einstein* as the candidate. His nomination caught the opposition off guard. Einstein's scientific accomplishments were so extraordinary and his antimilitaristic convictions so persuasive that his nomination could even break France's anti-German resistance (for many the saving justification was that Einstein wasn't really a proper German but a Swiss by citizenship).

It is to *Einstein*'s merit that in the prevailing atmosphere of hate an international institution could be established that broadly adhered to international principles (less through his activities in this institution than through his scientific prestige). "When Einstein's name appeared on the list," all considerations of nominating an Austrian were abandoned.[573] The French philosopher who later chaired this committee of the League of Nations, H. Bergson, told Undersecretary General Nitobe that Einstein was the only German whom the French could possibly accept.[574] By choosing Einstein, the Council of the League of Nations could afford "not only to honor Professor Einstein" but also "to acknowledge publically [...] that German science, too, is entitled to a seat on this committee."[575] Even though Einstein's nomination was ad personam, he was still regarded as a *representative of Germany.*

According to a report by the German consulate in Geneva, the first time Einstein was contacted about the membership proposal was while he was traveling in France in March and April 1922.[576] His admirer, Henri Bergson, professor of philosophy at the Collège de France and later president of the committee, would not let such a chance slip by. It was particularly favorable because Einstein's visit was being particularly positively received by the French public and most opponents of his nomination to the committee were bereft of their arguments.

On 17 May 1922, finally, Einstein was officially invited by the secretary general of the League of Nations, Sir Eric Drummond, to become a member of the International Committee on Intellectual Cooperation. Einstein acknowledged receipt of the letter on 30 May and announced his willingness to be elected to the "Commission pour la Coopération intellectuelle" but confessed "that I have no clear picture of the character of the work to be conducted by the committee" (and he was certainly not the only one – Marie Curie, for instance, also could not see what such a committee could really accomplish).[577] Despite these reservations, he felt it "a duty to answer this call, because no one in these times should refuse to collaborate in endeavors to attain international union."

Einstein to the secretary general of the League of Nations, 30 May 1922:[578]

Professor Dr. A. Einstein Berlin W. 30, 30th of May 1922
Haberlandstr. 5.

Highly esteemed Sir,

I acknowledge receipt of your letter of the 17th of this mo. and state that I gladly accept the nomination to the ''Commission pour la Coopération intellectuelle.'' Although I must confess that I have no clear picture of the character of the work to be conducted by the committee, I do feel it a duty to answer this call, because no one in these times should refuse to collaborate in endeavors to attain international union.

In great respect,[579]
A. Einstein

Société des Nations
aux bons soins de M. le Secrétaire Général Genève.

The Foreign Office was not enthusiastic about Einstein's nomination, but it could do nothing about it. It was *known* from a conversation between the German consul in Geneva and Nitobe[580] as well as from other information that Einstein was being regarded as a representative of *Germany* and *German* science. But because Einstein did not attach particular importance to becoming an official representative of Germany, a different tactic was chosen. A reply by the Foreign Office to the German consulate in Geneva from 7 July 1922 argues: since Einstein "is acting not as a representative of the Reich but as a private individual, the F.O. sees no reason to influence him, for instance, to decline the appointment; an opportunity for such does not present itself, considering that Prof. Einstein is not seeking any possibility to contact the F.O. about his appointment. The position held by the government about the League of Nations is not in any way compromised by Einstein's joining the committee."[581]

An internal memorandum in the file from 29 May 1922 already comments that it was "more than questionable whether any influence exerted by the F.O. on Prof. Einstein would lead to the desired goal." Elsewhere there is the statement: "In my view there was neither occasion nor opportunity to influence him toward declining the office.[582] He does not act as a representative of the Reich but as a private individual [...]. On the other hand, it would have also been a very touchy business to advise him to decline, especially since he was hardly likely to follow the advice."[583] Any influence on Einstein, it continued, could "at best [...] come through the intermediary of the Prussian Ministry of Culture."

That would have been the responsibility of Ministerial Manager Krüss. But he knew much better than the gentlemen at the Foreign Office, from years of experience, how hopeless such an attempt was. We may presume that he did not make the attempt.

Despite its reservations about Einstein's membership the Foreign Office attentively observed the activities of this committee from the very beginning.

On that 7 July 1922, when the Foreign Office distanced itself from Einstein's membership in the committee of the League of Nations, he had already retracted his initial acceptance, however.

Einstein to the secretary general of the League of Nations, 4 July 1922:[584]

Professor A. Einstein — Berlin W. 30, 4 VII 22.
Haberlandstr. 5.

Highly esteemed Sir,

In consideration of circumstances that became clear to me only since my acceptance of 30 V of this yr., I see myself unfortunately compelled to decline the nomination to the ''Commission pour la Coopération Intellectuelle'' retroactively, after all. I would not like to fail to take this opportunity, however, to also express my warmest sympathy with your efforts toward nursing international relations back to health.

In great respect,
A. Einstein.
Société des Nations
in care of the Secretary General
Genève

His reasons are set forth – in greater detail – in an enclosure as well as in a letter he wrote to Marie Curie. The situation in Berlin dictated that Jews keep a low public profile. He had resigned not only because of Walther Rathenau's assassination, but also because of the anti-Semitism reigning in Berlin and his impression that he was not the right person for the job.

Einstein's resignation annoyed and even offended the initiators of his nomination and others as well. Marie Curie replied: "I have received your letter and was very disappointed. The reason you offer for your withdrawal does not seem valid to me. It is precisely because dangerous and prejudicial opinions exist that it is necessary to fight against them; and you are in a position to exert an important influence in this regard, *if only by virtue of your personal reputation*, which permits you to fight for tolerance. I think that your friend Rathenau, whom I believe was an honest man, would have encouraged you to at least try to bring about peaceful international intellectual collaboration. I am sure you can retract your decision. Your friends here have fond memories of you."[585]

Nitobe and the president designate, Bergson, tried desperately to change Einstein's mind. The chief of the Information Section of the Secretariat of the League of Nations, Pierre Comert (also secretary of the committee), traveled to Berlin at the end of July.

Comert's letter a few months later, from 10 April 1923, puts on paper what happened during their meeting and what they discussed.

Pierre Comert to Einstein, 10 April 1923:[586]

> [. . . ]. I explained to you then that your sudden and unmotivated resignation could seriously prejudice the Committee on Intellectual Cooperation, because the public could misinterpret your sudden decision to retract your collaboration.
>
> With great honesty and in complete confidentiality you acquainted me with the particularly onerous reasons that led you to consider tendering your resignation. [. . . ]. I said to you that the precariousness of your personal situation in Germany seemed to me so considerable that, in my opinion, the members of the Council would absolutely not have dared to ask you, had they been able to suspect that this designation could make your situation in Berlin even more critical.
>
> We then analysed together, in complete good faith, whether under these - for me new - conditions it would be appropriate to confirm your resignation [. . . ].
>
> Before my departure from Berlin [. . . ] you had, however, informed me that you had abandoned the idea of resigning. The work of the League of Nations was so close to your heart, you said, that you were willing to accept certain risks for it rather than compromise the Committee's cause by an inexplicable resignation. [. . . ].
>
> Subsequent to these conversations, you wrote the Secretary General again, on 29 July. Your preparations for a trip to Japan would prevent you from attending the first meeting of the Committee on Intellectual Cooperation; you explained that upon your return your collaboration would therefore be the more zealous [. . . ]. It was with this kind letter that you took leave of us for the Far East.

The result of the two days of discussions with Comert (on 27 and 28 July 1922) was that Einstein retracted his resignation. On 28 July, at 2 o'clock in the afternoon, Comert sent a telegram to Geneva: "our friend withdraws resignation."[587] Einstein notified the secretary general of the League of Nations in writing the next day.

It soon became evident that it was more a matter of having been talked into it than having been convinced. He still felt he was "not the right person for this job." He had retracted his retraction but at the same time said that he could not attend the committee's first meeting. The public learned nothing about the resignation and reentry but the secretary general did publish the *second* part of his letter.[588]

First evening edition. [. . . ].
*Wolff's Telegraphisches Büro* [. . . ].
Berne, 5 August. Professor *Einstein* - Berlin addressed the following letter to the Secretary General of the League of Nations in Geneva at the end of July:
The necessity of settling a number of urgent matters before my imminent trip to Japan makes it, to my great regret, impossible for me to appear for the first conference of the Committee on Intellectual Cooperation in Geneva. In this situation I am only consoled by the circumstance that this inability to participate at the meeting will make any loss seem that much less from my being prevented by the above-mentioned trip from continuous collaboration for the coming half year. However, I do hope to repair this omission as quickly as possible after this period has come to an end.
In expressing the hope that the Committee's labors during this first half year will bear fruit, I remain, in assuring you of my great esteem,
A. Einstein.

The reference to the voyage to Japan that fall was not a valid reason but an excuse. Einstein could have participated. His schedule would have permitted it.[589] Einstein must have known that he would have disappointed the members of the committee once again and this was an alternative to participating in the committee's work – but it was also a lost opportunity to counteract the international boycott of German science.

The committee convened for its first session from 1 to 5 August 1922. With the exception of Einstein,[590] all the members elected by the League of Nations were present.

Bergson was unanimously elected its president, Murray its vice-president and the professor from Warsaw de Halecki its secretary. (A few years later, Bergson and Halecki withdrew from the committee; they were replaced in 1925 by H.A. Lorentz from Holland followed later by Murray; and Comert was secretary until 1924 when he was succeeded by Oprescu and finally de Montenach.)

The only thing they got from Einstein was friendly greetings by telegram. The secretary general's request, forwarded by Comert, that he inform the committee about his conception of his future collaboration,[591] remained unanswered. In the last minute he did try, without either much impetus or success, to find a substitute. (Comert was informed of this not by him personally but by his stepdaughter, Ilse Einstein, who was working as his secretary, on 1 October 1922 – Albert Einstein himself had already embarked on his voyage.)[592] His absence was regretted, "because especially his presence would have most brilliantly revealed the international character of the event." There was also "considerable doubt about whether the reason he gave for his hindrance – his imminent departure for Japan – was really true; questions were rather raised about whether his cancellation had been motivated by the not particularly favorable mood in Germany at that time for the League of Nations. Perhaps he had also believed with regard to the strong anti-Semitic current in Germany that he as a Jew had to be particularly careful and therefore could not appear as a representative of German science."[593] – This, at least, is what the German consul gathered a few days after the committee meeting from a conversation with Halecki and relayed to the Foreign Office on 3 September 1922. He did not abstain from adding that the opinion of many, including "quarters not antagonistically disposed toward the League of Nations," was that "the whole thing could become ridiculous, if the German scholar should stay away permanently." Halecki gave the promise "to take any wishes by the German government into consideration, as far as possible." He "seemed generally to find a certain cooperation with us important," the consul concluded.[594]

The secretary of the committee and the German envoy clearly regarded the committee's work as more than a private event by politicizing scholars. The consul also informed the Foreign Office that, for all its good will, the committee "lacked what is most important: money, which it is known the League of Nations does not have at its disposal but must always first beg off the individual governments for such purposes." He continued, "So we should not place any hopes on the scholars' committee initially but at least try to keep in contact with the positions of authority, namely the secretariat." With approval of the Foreign Office, the envoy volunteered to "keep an eye on" the matter.[595]

While awaiting his return, the people in Geneva received any news about Einstein's travels with particular interest. This explains why the Secretariat of the League of Nations was particularly consternated to hear that Einstein was planning to emigrate to Japan permanently.[596] It was a great relief to find out that it was only a false rumor.

### 2.7.3 Withdrawal of membership and retraction

Einstein had barely returned from his long trip to Japan when he withdrew his membership in the committee.

Einstein to Comert, 21 March 1923:[597]

Zurich, 21 III 23.

To the ''Commission pour la coopération du travail intellectuel.''

Dear Mr. Comert,

I have recently gained the conviction that the League of Nations has neither the energy nor the good intention to fulfill its great cause. As a serious pacifist I therefore do not consider it right to be connected with the same in any way. I ask you please to remove my name from the list of members on the committee.

In great respect,

Albert Einstein.

He wrote his friend Solovine at Pentecost in 1923: "I withdrew from a committee of the League of Nations, because I have no confidence in this institution anymore. This has earned me quite an odious reputation, but I am glad I did it. One has to maintain one's distance from hypocritical endeavors, even when they have a fine name."[598]

The immediate reason for this retraction was the occupation of the industrial Ruhr region by the French and Belgians on 11 January 1923. Einstein let himself be swept up – probably for the first time in his life – in the German outrage over the arbitrary exercize of foreign power. Einstein's protest coincided with the stance held by the German League of Human Rights, of which he was a member. His opposition to the occupation of the Ruhr region was then also the reason why he participated in the physics conference in Bonn in September. He wanted to demonstrate his solidarity with his colleagues who had moved the conference there in protest.

"I had accepted the nomination into the 'Commission de coopération intellectuelle' at the time, only to refuse it afterwards upon my return from Japan – in March last year – because of major reservations against the League of Nations," Einstein informed the university news editors of *Berliner Hochschulnachrichten*. (He also informed them then that he could not write more about the committee because he knew too little about it.)[599] He anyway "did not feel like [...] representing people who would surely not have elected me as their representative and with whom I [...] do not share the same views"[600] – thereby giving his probably *most immediate*, although unexpressed reason for his cancellation.

Comert's reply on 10 April 1923 was bitter, if not furious.[601] He had special reason to feel rebuffed and offended. He now realized that his arguments about half a year ago had only *persuaded* Einstein but not completely convinced him.

Marie Curie, Gilbert Murray and many others felt at least as snubbed. Murray wrote Einstein that his decision weakened the committee and destroyed the possibility of issuing a collective statement by the committee members against the French policy.[602]

Einstein's departure was obviously not welcomed beyond the committee, particularly among internationally minded leftist intellectuals. It was considered a breach of faith. But Einstein's opponents were just as thankless. They regarded this step as a sign of his fickleness.

The resentful reaction among the members of the committee of the League of Nations and elsewhere is an indicator beyond the plane of personal relations. It demonstrates that, despite its overrepresentation by the French, the committee was not an obedient instrument of French cultural imperialism. (Murray had criticized Einstein precisely because his resignation destroyed the option of the committee members collectively protesting *against the French policy* as a single voice!) The committee did *not* come into existence in Clemenceau's spirit. Its goal was *not* to boycott German science and culture – by subtler means. It had been founded to contribute toward mutual understanding between nations, particularly in the scholarly world – as well as to undermine the boycott of German science. So the indignation about the step Einstein had taken was justified. His impetuous response to the occupation of the Ruhr had hurt international cooperation among intellectuals.

Thus from its inception until Einstein's readmittance in July 1924, the committee was forced to operate without his participation, hence without a representative of Germany. As long as Einstein had been a member in the past (until March 1923), he had rather posed more of a problem than a help to the committee. His retraction was regretted nonetheless, and a retraction of his retraction desired.

There was no sense in looking for another *German* scholar to replace Einstein. There were plenty more of the same but none who were willing to take on the burden of an internationalist and pacifist. So they were delighted to find a solution to the dilemma in the Dutch professor Hendrik Antoon Lorentz – who was, moreover, even a friend of Einstein. Bergson wrote a jubilant telegram to Nitobe on 23 April 1923:

> `Nomination of major Dutch physicist Lorentz, world celebrity like Einstein, seems to me absolutely necessary for presige and future of our committee Will explain by letter = Bergson`[603]

Lorentz was elected on the same day – at the thirteenth meeting of the twenty-fourth session of the Council of the League of Nations.[604] Einstein himself was consoled by the thought "that one of the purest and most excellent men had been elected on this committee in my stead, who with his acknowledged integrity and justness can exert an exceedingly salubrious influence."[605]

It took a few months for Einstein to change his mind entirely. Nitobe sent a confidential note about this to the secretary general of the League of Nations on 23 April 1924.[606] Luchaire – inspector general of the French school board – had informed Nitobe about a long conversation he had had with Hellmut von Gerlach, "a well-known German publicist, an intimate friend of Einstein's."

Hellmut von Gerlach – a co-founder of the New Fatherland League and since 1918 on the controlling bodies of the German League for the League of Nations (*Deutsche Liga für den Völkerbund*) – reported to Luchaire that Einstein regretted the ill-considered gesture of his resignation and had asked whether a return in a more or less official form was possible.[607] "Excellent [...]!" was the recipient's note.

On 16 May Gilbert Murray wrote to Einstein that, if he should have an eye toward renewing his membership, the committee "would unanimously welcome his membership." Einstein answered on 30 May in a confidential letter,[608] that he was willing to become a part of the committee again. He pointed out that a year ago he had withdrawn not out of a lack of trust in the committee but in the League of Nations as a whole. His best friends had censured this step. He was prepared to reenter, because he hoped thus to be able to contribute toward an improvement in Franco-German relations. In the event that he was not elected (which, after all that had happened, was understandable), he intended to work for the committee.

The question of his readmission was discussed on 16 June at the fourth meeting of the twenty-ninth session of the Assembly of the League of Nations. Henri Bergson could "see nothing but advantages in having Professor Einstein occupy a seat in the committee." The Assembly agreed. "*It was resolved that Professor Einstein should sit on the committee as the representative of German science.*"[609]

So Einstein was invited by the secretary general of the League of Nations, Sir Eric Drummond, to reenter as a member. Einstein sent his formal acceptance on 25 June 1924.

Einstein to the secretary general of the League of Nations, 25 June 1924:[610]

*25 June 24.*

*To the Secretary General of the League of Nations in Geneva.*

*Esteemed Sir,*

*I accept herewith reelection into the Committee on Intellectual Cooperation with cordial thanks. After my past performance, this election is an act of especially generous high-mindedness and therefore fills me with special joy. I will constantly try diligently to do my utmost in the service of the good cause.*

*In great respect,*

*A. Einstein.*

Einstein's letter of 26 June 1924 – classified by Nitobe's office as "private" and for that reason not published in the *Official Journal*[611] – was nicely formulated but also an expression of weakness. Einstein came back as a penitent sinner. Even if he did want to represent the interests of German science with energy (which would have meant energetic criticism of the French-led boycott), he could not. His earlier conduct and his current request had at least partially tied his hands.

His willingness to renew his collaboration received much attention in the press (as did, in fact, everything that Einstein did). In response to a question by an agent for the *Gazette de Lausanne,* Einstein explained that he had accepted reelection because the overall situation had changed and the prospects for the future of the League of Nations had become more favorable. He was convinced that the League of Nations with all its organizations, including the international tribunal in The Hague, was the only instrument that could be relied upon to avoid the worst catastrophes. Einstein argued that mankind, particularly the European situation, was under serious threat and it was therefore the preeminent duty of every individual in a position to do so to collaborate in the work of the League of Nations. He had just chosen to work in the Committee on Intellectual Cooperation, for self-evident reasons. He emphasized the necessity of complete objectivity and free expression of the committee's view, whose members were nominated by the Council of the League of Nations and were completely independent with regard to the statements it issued.[612]

At the fourth meeting of the committee (from 25 to 29 July 1924) Einstein was welcomed among its ranks again.

The address that President Henri Bergson delivered on this occasion was published the following day – on 26 July 1924. In it Bergson said, among other things, "that the committee was pleased and proud that Einstein, who is world famous, was a part of it. His scientific feat was one of the most powerful that humans had ever accomplished in extending the limits of human thought. Einstein had worked the wonder that his theories, whose complexity could discourage professional scholars, passionately attracted and moved the whole of mankind. Bergson recalled that by applauding interrelations among nations, Einstein advocated attitudes resembling the *ideal of the League of Nations* and expressed the hope that Einstein's participation on a committee of the League of Nations might win over large proportions of humanity to this ideal already converted to his scientific speculations."[613]

The German consulate in Geneva sent a full report on this meeting to the Foreign Office on 12 August 1924.[614]

According to this report the (three) subcommittees on intellectual property, intercommunication among universities, and bibliography convened before this committee meeting. The resolutions by the subcommittees were then approved by the Committee on International Intellectual Cooperation. It reports, the meeting "acquired added special interest by the presence in the Committee on Intellectual Cooperation of the readmitted member, Professor *Einstein*," to whom the president, Professor *Bergson*, devoted "amicable words of welcome in his opening speech." The Foreign Office was also informed that, as "has been learned," *Einstein* had allowed himself "to be completely snared by the French and the members of the Committee on Intellectual Cooperation, most of whom are under French influence."

## 2.7.4 Einstein's collaboration

The German consulate also informed the Foreign Office on 12 August 1924 that during the meeting an announcement was made "by the French minister of instruction that the French government was ready to explore its interest in involving the committee in the founding of an international institution for intellectual cooperation in Paris. The French government would make available the necessary accommodations and funds through the intermediary of the League of Nations." The following commentary was added: "It quite evidently constitutes an attempt to anchor French influence in the committee and to influence their work permanently along French lines. – The committee accepted the offer, conditional on approval by the Council of the League of Nations, with enthusiastic thanks."

What the consul forgot to mention was that the committee had been short of money *from the very beginning*. This had motivated the committee to launch an appeal to *all* member states at the preceding third meeting in November 1923. The *French* government had been the only one to respond, and did so generously. During the fourth meeting of the committee (25 to 29 July 1924), "a letter by the then serving French minister of culture, François Albert, was read out in which he announced the following: The French government had gained the impression from perusal of the minutes of the meetings and the resolutions passed by the committee that it was absolutely essential to come to the assistance of the committee. [...] it therefore was offering the funding and housing for an institute of intellectual cooperation to be founded in Paris."[615]

The gesture by the French government was not altruistic; but that was no reason for the committee members to decline the offer – Einstein among them. Why they subsequently let the French government largely dictate the rules of engagement is an entirely different question. Despite the protest by the committee's president, two officials from the French Ministry of Culture "drew up the agreement to be closed between the French government and the League of Nations as well as the draft bill to be laid before the French parliament" without consulting the committee.[616]

Einstein himself reported about the committee meeting for the *Frankfurter Zeitung* of 29 August 1924. He pointed out that every issue was treated as if Germany "were already a member of the League of Nations," and said that his discussion partners were without exception of the view "that Germany would obtain a permanent seat in the Council of the League of Nations, like the other major states. [...]. All were also of the opinion that after successful completion of the conference in London, Germany should join the League of Nations" (which is what Foreign Minister Stresemann wanted and what eventually happened in 1926).

A short time later Einstein commented on other aspects of the meeting – in the *Berliner Tageblatt* of 17 September 1924. French newspapers had alleged that, unlike Professor Gilbert Murray from England, who had demanded a guarantee that the Parisian institute of intellectual cooperation offered by the French government be truly *international,* Einstein had "approved the French offer without qualification."

Einstein's response was: "The fact is that a lively debate followed [...] about the position that the Committee on Intellectual Cooperation should take regarding this offer, during which I pointed out the great threat this project posed for the committee's situation, in full agreement with Messrs. Murray and others, who were not all Englishmen. My point was the following: The committee can only be affiliated to an institute if it is truly international in character. Inherent in this is, in my opinion, that *its seat be in Geneva,* and that it be *maintained jointly by many states.* But this does not prevent the French government from creating and maintaining a French institute in Paris that is commissioned to undertake, on a case by case basis, activities at the committee's instigation. For assistance of this form our committee would be grateful to the French government just as it would be toward the governments of any other state supporting our work in this way. After lengthy, sometimes heated discussion, the committee arrived at *unanimous approval of this position,* and the reply to the letter we had received was formulated in this sense. This debate in particular demonstrated to me most finely the will for objectivity prevailing in this committee of the League of Nations."

Einstein may have been inconsistent in many instances, but the allegation that he was a mouthpiece for the French was simply a mean-spirited misinterpretation. He certainly did his best to act in the interest of Germany and German science. He was intent on improving relations between nations and having Germany regain admission in international organizations. His article in the *Frankfurter Zeitung* is evidence of this.

Bergson's welcoming address at Einstein's return shows the high expectations the committee had placed on his membership – it confirms once again that Einstein's fame was more their object than his personal participation. The surviving

17. IX. 25

Sehr geehrter Herr Prof. Krüss!

Koppel bekam eine dicke Gänsehaut, als ich ihm den Vorschlag unterbreitete. Er klagte über geschäftliche Schwierigkeiten. Dann fasste er sich und ging zum Gegenangriff über. Das Plan mit den Büchern erschien ihm nun nicht wirkungsvoll. Dagegen gefiel ihm das Projekt der gegenseitigen Einladung verdienstvoller Gelehrter. Die Bücher könne man von den Verlegern gratis erhalten. Er sei bereit, mit etwa 4 anderen zusammen (kluges Dezimalsystem!), die noch aufzutreiben wären, die Verpflichtung zu übernehmen, die aus den Einladungen erwachsenden Kosten zu übernehmen, wobei er in Aussicht nahm, dass die auf den Mann entfallenden Auslagen von der Grössenordnung 5000 M seien. Er meinte, man solle unverzüglich eine Kommission bilden, die über die zu unternehmenden Schritte beriete.

So Koppel. Ich füge nichts hinzu als die Hoffnung, dass etwas geschehen möge, was von den Franzosen als liebenswürdige Geste empfunden wird. Sie haben in diesem Falle keine „Prestige-Politik" getrieben, und es wäre sehr hübsch, wenn dieser Gesichtspunkt auch bei unserer Reaktion auf ihren Schritt nicht weit in den Hintergrund gestellt würde.

Es grüsst Sie freundlich

Ihr ergebener

A. Einstein.

**Figure 31: Albert Einstein's letter to Hugo Andres Krüss from 17 September 1925 – regarding Franco-German relations.**

files of the Committee of the League of Nations document that during the next two or three years Einstein certainly did "try diligently" to do his "utmost in the service of the good cause." But it was not easy for him. Objective hurdles stood in the way. But he certainly *tried*.

His involvement in the following are worth special mention:

1. Distribution of aid by the Red Cross to needy students and scholars.
2. The Bibliographic Subcommittee, hence he also worked on the international exchange of books.
3. International coordination of research on phototelegraphy.
4. The creation of an International Institute for Mining.
5. A panel of the International Labor Organization.

Perhaps the most important outcome of Einstein's participation in the committee of the League of Nations was his correspondence with Sigmund Freud, which was published under the title 'Why War?'[617] This topic will be discussed later because it dates to the end of his period of membership.

Einstein to Krüss, 13 September 1925:[618]

*13 IX 25.*

*Esteemed Dr. Krüss,*

*The idea you mentioned of a train-wagon-load of books seems to me to be a very good one. I immediately telephoned Mr. Koppel to schedule an appointment. On the way home it occurred to me how, in my opinion, one could make this planned gift much more valuable. In my opinion, one should have a rubber stamp made and stamp each book along the following lines*

*''To young French students from German ones in remembrance of the memorable visit of ... and ... in Berlin on 13 September 1925.''*

*Thus this visit and the gift would leave a lasting trace in France as well, even if part of the press did not give it enough notice. If it would not be possible to find an agent competent or willing enough to be treated as ''young German students,'' one could, of course, say a little less colorfully and effectively ''from German academia'' or ''from the German University Association'' or ''from German libraries,'' depending on which academic body could be enlisted for it. This would have a much greater effect on the French than if the gift came from a public authority; I know that they are very sensitive about such nuances.*

*In great respect*
*A. Einstein*
*Haberlandstr. 5.*

## 2.7.5 Einstein's position – an object of desire. Einstein's deputy

Despite his good intentions, after two years of membership Einstein's willingness to spend time on tedious and time-consuming details faded. It was also partly the fault of the committee's president and secretary as well as the director of the Parisian institute. To make the most out of Einstein's fame, they overwhelmed him with work and appointed him to subcommittees he was not qualified for. Einstein obviously had other things to do as well. Politics was only his "minor subject." He was still going on lecture tours besides. He had little opportunity to concentrate on the founding of the International Institute of Intellectual Cooperation in Paris, for instance, because he was away in Argentina, Uruguay and Brazil between March and June 1925. But he was back home in time to attend the fifth committee meeting (27–30 July 1925).

One of the most important reasons for Einstein's waning interest was that the committee began increasingly to lose its proclaimed independence and became the plaything of national interests.

The nascent International Institute of Intellectual Cooperation was supposed to support the committee and relieve its members of time-consuming details. But it became an instrument of French cultural policy. Einstein had welcomed the founding of the institute but at the same time criticized the way it was being politicized. At its official inauguration on 16 January 1926 he expressed his reservations in a dinner speech.

Einstein's dinner speech, 16 January 1926:[619]

This year leading politicians of Europe have for the first time drawn the consequences of the realization that our part of the Earth can only flourish again if the latent battles among traditional organized states stop. [. . . ].

This great goal cannot be reached solely on the basis of treaties. Intellectual preparations are also of primary importance. [. . . ]. In consideration of this the League of Nations called into existence the ''Commission de coopération intellectuelle.'' This committee is supposed to be an absolutely international agency detached from all politics, aimed at establishing a link in all areas of intellectual life between national cultural communities isolated by the war. [. . . ].

This committee has convened twice a year. In order make its work more effective, the French government decided to create and maintain a permanently operative Institute for Intellectual Cooperation which is now being inaugurated. [. . . ].

It is an easy and gratifying business to cheer and applaud and to remain silent about what one regrets or does not approve of. [. . . ]. So I will not shy away from adding some criticism to this birthday greeting:

Every day I have occasion to notice that the greatest obstacle to the work of our committee is a lack of trust in its political objectivity. Everything must be done to strengthen this trust; and everything must be avoided that could damage this trust.

When the French government establishes and maintains as a permanent organ of the committee an institute in Paris on state funds and with a French citizen as its director, this must necessarily lend the impression to outsiders that *French influence must predominate in the committee.* This is intensified by the fact that the president of the committee itself has hitherto been a Frenchman. Although the men in question are highly esteemed and appreciated by everyone everywhere - the impression still remains.

After Einstein expressed doubts about too strongly dominating French interests he "received the assurance that no considerations were being entertained to relocate the seat of the committee to Paris."[620] He advocated locating the institute in Geneva for the same reason, which a few member states of the League of Nations also wanted. He was generally against such a powerful position by the French in the new institute. Einstein tried unsuccessfully to convince Lorentz. There were heated debates also in the Assembly of the League of Nations over locating the institute in Paris. But ultimately, the League of Nations accepted the gift by the French government. It was decided at the same time that the institute should be accountable to the committee, which would form its governing board.

For all its skepticism, even the Foreign Office had to acknowledge that there was opposition to use of the Parisian institute as a tool of French cultural politics by "the other side (the English, neutral countries and *Professor Einstein as well*)." Soehring recorded the specific elements of French dominance at the institute: "thus the chairman of the governing board, the chairman of the institute's board of directors, and the director of the institute are French. It furthermore turns out now from facts recently made known that virtually half of the staff employed at the institute, including the academic auxiliary staff, are all French."[621]

At the committee meeting on 17 January 1926, Einstein and Marie Curie protested against the obviously politically motivated appointment of the fascist minister of education, Alfredo Rocco, to the committee of the League of Nations and as Italy's representative at the Parisian institute – as successor to Ruffini, an opponent of Mussolini.[622] To no avail. The Italian government supposedly threatened to withdraw completely from the League of Nations. So much for the political independence of the committee and its proclaimed selection of scholars "*with a view to their academic stature irrespective of their citizenship*." Einstein's reasons for joining the committee had thus actually become defunct.

Einstein would suffer another disappointment at the eighth meeting of the committee (26 to 29 July 1926). At issue was questionable passages in history textbooks initially discussed at the sixth meeting in July 1925 at the proposal of Committee Member Casares.[623] Einstein had argued at earlier meetings that depictions liable to cause misunderstandings and hate between nations ought to be eliminated from history class. At the eighth meeting it was resolved, in the contrary sense, that the national committees did not have to set forth their reasons for refusing to implement externally suggested changes to their history textbooks. At this and coming conferences Einstein was put in the frustrating position of seeing his suggestions ignored or voted down. He became a spanner in the works of international politics. Einstein fought back, but he was the weaker, inferior party.

In a sense Einstein was also a victim of the structural faults of the committee of the League of Nations. It occasionally expected far too much of its members.

Einstein's time schedule in July 1929, for example, would have been:

| | |
|---|---|
| 13–16 July | bibliographic subcommittee, |
| 22–26 July | plenary meeting of the committee, |
| 27 July | board of directors, |
| 29 July | administrative council, |
| 31 July | consultative committee of intellectual workers. |

If he had conscientiously followed this schedule, he would have had to abandon altogether doing serious scientific research for an entire month. Consequently, Einstein was glad to arrange to have someone else attend in his stead.[624] Given these circumstances, his departure from the committee or a permanent substitution was positively received by many, Einstein himself among them.

As early as 1922 it was suggested that he name a substitute. He tried to find one but with little energy and even less success. On 1 October 1922 he informed Comert that he was planning to have his friend, the psychologist Wertheimer – professor at the University of Berlin – named his proxy.[625] Wertheimer was away on vacation in Prague, however, so Einstein could not reach him to find out what he thought of the idea. He worried that Wertheimer would hesitate to accept the offer. His second candidate, Prof. Troeltsch, was also out of town.

Wertheimer was either not asked in the end or else he declined (likewise Troeltsch). In any event, the German consul, Dr. Nasse, reported to the Foreign Office on 4 January 1923 about a conversation with the secretary of the committee, de Halecki.

De Halecki had inquired whether Einstein was expected to return home from Japan at all.

> I could not give a definite answer, of course, to the next question about whether he would be back in time for the planned convention in the summer. In connection with this Halecki mentioned that he wanted to inform me confidentially about an incident that had left a strange impression within the committee: In response to a suggestion that he name a substitute, Einstein had first singled out someone (Halecki did not want to reveal his name to me), who had then declined and Einstein finally said he could not find anyone willing to substitute for him. One would have to conclude from this, Halecki thought, that Germany was still quite negatively disposed toward anything having to do with the League of Nations. I pointed out to him that in view of the political activities by the League thus far, this should actually not be so surprising - whereupon we both thought of Upper Silesia and out of a mutual sense of delicacy moved on to another topic.[626]

The difficulty of finding a suitable substitute is also reflected in Einstein's letter from 9 January 1925 to H.A. Lorentz. Einstein commented with resignation: "I fear we cannot find any person (for the local committee) of sufficient standing at the present moment, who is regarded by the people here as trustworthy and one of their own. We shall probably have to wait until the political barometer indicates a maximum again."[627]

Einstein wanted Prof. Langevin to represent him at the meeting of the board of directors of the Parisian institute on 25 March 1926 (who had already agreed to do so at Einstein's request).[628] The fact that Langevin was French did not seem to bother him. But his petition was rejected on 9 March 1926, because it did not satisfy existing regulations.[629] After further futile attempts Einstein had to admit to Oprescu on 16 April 1926: "I have still not been able to find a substitute for the board of directors. It is supposed to be a man of influence, you know, but they will hesitate until the issue with the International Research Council has been set right."[630] On 14 May 1926 he reported once again, empty-handed.

Einstein to the secretariat of the committee of the League of Nations, 14 May 1926:[631]

> 14 V 26.
>
> To the Secretariat of the Commission d. Coopération Intellectuelle
> Esteemed Sirs,
>
> It is unfortunately not possible for me to come to Paris for the meeting on 20 V of the Board of Directors. I ask you please to present these apologies of mine there. Unfortunately it has still not been possible for me to designate a substitute, because the man I have in

view would still like to await developments in international relations with Germany in the area of the sciences.

In great respect,

*A. Einstein.*

The files do not reveal *whom* specifically he had chosen as his substitute. Presumably it was Fritz Haber, whom he later wrote, on 16 June 1931: "I had long sought to have you and your great organizational skills enlisted for Geneva." But satisfaction of Einstein's wish as early as 1926 would have required Haber to go against the wishes of corporate academia. Despite his great respect for Einstein, Haber was too ambitious and patriotic to do that.

Einstein did not have the support of the academic institutions. His attempts to win over Max Planck, in particular, were in vain.[632]

Einstein's opinion of the committee and the Parisian institute was totally different from that of his colleagues at the Berlin academy. The Prussian Academy of Sciences informed the Prussian Ministry of Science, Arts and Culture on 30 January 1926 that the Cartel of German Academies was of the unanimous opinion that "no use would arise for the Academies' scientific endeavors" from Germany's entrance in the organizations founded by the League of Nations. That is, the Committee of Intellectual Cooperation and its subordinate International Institute of Intellectual Cooperation. So the academy directed "the request to the Reich government, not to assign any representatives for Germany in the mentioned international organizations and in the planned affiliated German 'national' Committee of Intellectual Cooperation without first contacting the authorized academic organizations." It was pointed out "that the seat of the International Institute of Intellectual Cooperation is in Paris, that the Institute's operating costs are approved by the *French* government, and that the Institute's annual and administrative reports are laid before *French* chambers." The signatories were the academy secretaries Lüders, Planck, Roethe and Rubens.[633]

Presumably Einstein's favorite candidate, Fritz Haber, was involved in the final selection of *Krüss* as his proxy. A strictly confidential discussion that took place on 25 February 1926 indicates this, to which Fritz Haber had issued invitations on 17 February 1926.[634]

Fritz Haber's letter from 17 February 1926 to Secretary of State Zweigert, Minister of State Schmidt-Ott, Privy Councillor von Harnack, Managing Director Krüss and Professor Schubotz (excerpt):

The ministerial officials, Head of Department Heilbron, Head of Department Richter and Senior Civil Servant Donnevert, are meeting Professor Haber informally on Thursday the 25th inst., 8 o'clock in the evening at the Club von Berlin, Jägerstr. 8, to discuss a topic that I know is of vital interest to you. I therefore ask you please to attend this discussion, which will involve a simple dinner.

Respectfully,

Haber

The participants of this consultation were people officially concerned with matters related to the League of Nations and foreign cultural policy: they were representatives of the Reich Ministry of the Interior (Donnevert), the Foreign Office (Heilbron), the Prussian Ministry of Science, Arts and Culture (Richter), the Emergency Association of German Science (Schmidt-Ott), the Kaiser Wilhelm Society (Harnack) – and *Hugo Andres Krüss* (managing director of the Prussian State Library). The invitation was issued by the person most closely connected to Einstein from among them all and best suited to exert an influence on him.

It *appears* as if the name *Krüss* was even Einstein's salvation from his ordeals. In 1926 Einstein proposed H.A. Krüss as his proxy on the administrative council of the Parisian institute, adding, however, that his acquiescence had yet to be obtained. Hugo Andres Krüss was then *elected* pursuant to this proposal. This seems to have "escaped" Einstein, as he later commented. But Einstein may very well also have been caught unawares. Maybe when Einstein asked Mr. Krüss to represent him in the future, he was only informing him of what sly Krüss already knew about.

Einstein to Oprescu, 5 March 1927:[635]

Schöneberg, the 5th of March 1927

To
Dr. Oprescu
Commission de Coopération intellectuelle

Dear Mr. Oprescu,

I recently spoke with Prof. Krüss and told him that I would like to suggest him as my proxy

a. for the Committee
b. for the Board of Directors of the Parisian Institute
c. for the Administrative Council of the above-mentioned Institute.

Mr. Krüss accepted unconditionally, so I can hereby propose him. As far as I can recall, the Committee can make this resolution only at its plenary session.

With kind regards,
yours,
A. Einstein.

Thus Hugo Andres Krüss became his heir for the second time. The first time had been the year before – with his membership in the Bibliographic Subcommittee.

Einstein was glad to be relieved of a burdensome responsibility.

This was not such a harmonious arrangement as it might seem. Krüss acceded to this office, or rather, elbowed his way into it. *He* realized that Germany's membership in the League of Nations signaled the beginning of a new era for the German Reich. The postwar period was over. Like the Foreign Office, Krüss, too, was much more flexible and far-sighted than the academic institutions about the position to take regarding the various organs of the League of Nations. He was also very well informed, as far as the Genevan committee was concerned. He had *taken the trouble to inform himself.* His words and actions in the past years point to the conclusion that Einstein's seat had long been an object of his desire. If Einstein could not be thrown out of the committee, he could at least be pushed aside and reduced to a walk-on extra.

The Foreign Office had also realized that in the changed international situation membership in the committee of the League of Nations was not problematic anymore.

The consulate in Geneva informed the Foreign Office on 31 July 1925:

> During the last meeting of the Committee, the American member Vernon Kellog made an interesting announcement that particularly impressed the Committee itself as well as the public at large. For Kellog stated that the Americans had hitherto not shown particular interest in the Committee on Intellectual Cooperation, because they had not drawn sufficiently into account the committee's plans and practical work. But now that various projects have assumed definite form and were approaching materialization with the founding of the International Institute of Intellectual Cooperation, Kellog vouched for his citizens that the Americans could in future be won over to a greater degree to the idea of intellectual cooperation among nations and that the enterprise could soon expect to benefit to a greater degree and more effectively from support by America. [. . . ]. In the wake of Kellog's announcement the delegate for Uruguay, Buero, also felt obliged to make an analogous statement on behalf of Latin-American countries.[636]

Hugo Andres Krüss sensed the new direction that the winds were blowing – his connections at the Foreign Office were very good indeed. Already in 1924, when the question of readmitting Einstein to the Genevan committee was under discussion, hence before the decision to have him rejoin had been made, Krüss took action. (At that time it was still in his capacity as manager at the Ministry of Culture, not even before he had become managing director of the Prussian State Library!) In a private letter from 7 June 1924, Krüss informed the Foreign Office: "I am considering spending some time in Lausanne from the 15th of June and would like to take this opportunity to visit the German consulate in Geneva one time and get briefed about the details on local impressions of the activities of the League of Nations in the area of culture and, in particular, academia. I am especially interested in the Committee on Intellectual Cooperation."[637] Although, he said, "not much has come out of this area so far," he was keen "on staying up to date and one generally learns more verbally than from reports." Dr. Soehring accordingly informed the German consulate in Geneva on 10 June 1924 of these intentions and requested that "everything be done to smooth the way for Mr. Krüss's plan.[638] At the same time – on 12 June – Soehring asked Dr. Krüss to uncover "underground ties between the Committee of Intellectual Cooperation and the International Research Council in Brussels." That is, the people at the Foreign Office suspected that the committee was an instrument of anti-German activity. So Krüss set off for Geneva not just on his own initiative but as a spy for the Foreign Office.

We may presume that Soehring's position on Einstein's readmission to the committee on 16 June was negative. Neither Soehring nor Krüss considered Einstein a valid "representative of *German* science," to use the exact formulation by the League of Nations. While *Krüss's* activities were treated as in conformance with German interests, the German consulate in Geneva viewed Einstein's anticipated return to the committee at the same point in time as "Einstein's private step concerning the League of Nations."[639]

Through the intermediary of the German consulate, Krüss spoke with Undersecretary General Nitobe of the Secretariat of the League of Nations, and with his rapporteur for the committee, Oprescu. "The private character of Mr. Krüss's visit in Geneva was emphasized to Messrs. Nitobe and Oprescu [...] in order to prevent that the presence of Mr. Krüss be correlated with the readmission of Professor Einstein to the Committee of Intellectual Cooperation and perhaps be taken as a sign of an official rapprochement."[640] In other words, Krüss's trip was disguised as a *private* matter, while he was, in fact, checking out the situation at the instruction of the Foreign Office. By not contacting Einstein but visiting Nitobe, Oprescu and the German consulate in order to obtain information about the committee, his message was clear. "Official" Germany did not regard Einstein as its representative and did not want to be represented by Einstein.

A short time later Krüss became a member of the Bibliographic Subcommittee. He had his foot in the door. Krüss corresponded with Oprescu about other matters as well but did not even consider contacting Einstein directly. On the other hand, Oprescu's conduct leaves the impression that *he,* too, was acting behind Einstein's back.

A remark in Krüss's letter from 22 October 1924 to Oprescu sounds dubious: "information has come to us from academic circles to the effect that the members of the Committee on Intellectual Cooperation are considering accepting another German member on the Committee besides Professor *Einstein,* in order to establish a live link with German universities and the German intellectual world. [...]. If a plan along such lines does exist, the choice of the German scholar in question would be of decisive importance to the relationship between German science and the Committee. The selection should therefore be made with the greatest care and with a thorough knowledge of existing circumstances. A person-to-person discussion of this point on the occasion of your presence here could perhaps be useful as well."[641] If Krüss himself was not the source of this "information," it would certainly have been a highly welcome development. Krüss, no doubt, passed this "information," along with his own commentary on it, to Oprescu with the object of guiding his thoughts in a direction to his own liking. Krüss *wanted* the members of the committee to consider accepting another German member on the committee besides Professor Einstein. Oprescu was *supposed* to know that German officialdom (namely, Krüss on behalf of the Ministry of Culture) wanted to have a decisive say in the filling of such a post. Krüss could only be more specific in private *dialog.*

In mid-November Krüss hoped to be able to welcome Oprescu in Berlin.[642] On 15 November 1924 Oprescu wrote to Einstein about another matter, but he did not mention his imminent visit to Berlin. Oprescu and Krüss evidently wanted to negotiate without Einstein's knowledge.

At the meeting arranged by the Foreign Office for 6 February 1925 re. the "attitude of the German scholarly world toward other countries," already mentioned and quoted in detail above, Krüss finally had an opportunity to say what he thought of Einstein's membership on the committee of the League of Nations. His audience were people of influence in foreign cultural policy at the Foreign Office, at the Ministry of Culture, at the Ministry of the Interior, the Emergency Association of German Science and from the Technical University at Dresden as well as from the Universities of Berlin, Bonn, Frankfurt am Main, Kiel and Munich.

From the minutes of the discussion on 6 February 1925 at the Foreign Office:

> *Dept. Dir. Krüss* emphasized that Einstein, who himself constantly emphasizes his supranationality, could not be a valid representative of the German nation at the League of Nations Committee on Intellectual Cooperation. Considering that, under the circumstances, rejecting an invitation to Germany to participate at the Parisian Institute is out of the question, we should now focus on choosing *really very good competency* for election to this post with the *backing of the whole of German science.* Abstention on this issue would be worse, because the other side would never have difficulty finding *willing outsiders* as representatives of Germany, who would then lack the crucial characteristics they ought to have in guarding the interests of German science.[643]

In August 1925 Oprescu traveled again to Berlin. Dr. Soehring informed him via the German consulate that he was pleased to receive him "in the period from the 14th–16th of this month."[644] Again there was no mention of Einstein. Krüss, for his part, wrote Oprescu a private letter while vacationing in Madonna di Campiglio to ask him "to delay his visit to Berlin by a few days, because I shall only have returned there on the 27th of August."[645] He wanted to answer Oprescu's questions *orally,* not in writing. One issue was "the question you asked about a suitable person for the Institute" (i.e., the International Institute of Intellectual Cooperation in Paris).[646] Einstein's official authority was not even addressed; he was simply ignored. Krüss acted as if he, not Einstein, was member of the League of Nations committee. And Oprescu, the secretary of the committee, accepted it.

But Oprescu's travel plans went ahead as scheduled. Krüss returned from his vacation only to find a letter by Oprescu dated 17 August. From Privy Councillor Soehring Oprescu had also found out that Krüss had meanwhile "been appointed managing director of the Prussian State Library and therefore had left his former office." Krüss assured him that he would continue to devote special attention to the issues of international cooperation and was sure that he would "henceforth perhaps even be better placed to do so [...]."[647]

From a letter by Krüss we gather that Oprescu was again "planning to take a trip to Berlin in September 1925.[648] Again there is no mention of Einstein, even though Einstein's letters to Krüss document intense contacts between them at that time. Oprescu must have *known* that Einstein would be back from his South American tour in September: Einstein had sent a telegram to Geneva from Montevideo on 28 April 1925 to inform him that he would be returning to Europe that June."[649]

Judging from a memorandum for the Reich Minister dated 26 February 1926, Soehring was counting on Einstein's post in the Genevan committee becoming available in the foreseeable future (along with that of Schulze-Gaevernitz as head of the division for scientific cooperation at the Parisian institute). Soehring was confident that the new representatives would find positive resonance in academic Germany, "which is currently unfortunately lacking with these two gentlemen."[650] He would consequently not be pleased to have to report shortly afterwards that Einstein had been elected as a sixth member on the formerly five-member board of directors of the International Institute in Paris.[651]

Soon afterwards, on 8 March 1926, a meeting took place at the Reich Ministry of the Interior "regarding the formation and composition of a National Committee on Intellectual Cooperation."[652] Einstein was not invited. It was decided there to ask numerous associations and organizations in the arts and sciences to submit proposals by a given deadline. Einstein was not mentioned. To assure direct government influence on the National Committee, it was determined that the position of secretary be entrusted to *an official appointed by the Reich Ministry of the Interior.*

There was anyway no obligation to consult Einstein on appointments concerning *state representatives* to the institute of the League of Nations. Following the initiative of the Greek government,[653] many states (thirty-three states in 1928) took it upon themselves to delegate *state representatives,* in order to be able to exert a direct influence on the institute's work. In July 1927 the head of department VI (Culture Department) at the Foreign Office, Envoy Freytag, was appointed delegate of the German Reich at the institute of the League of Nations.[654]

Germany's admittance to the League of Nations on 10 September 1926 marked another stage in Hugo Andres Krüss's efforts gradually to oust Einstein. A German, Alfred von Dufour-Feronce, was thereupon appointed to replace Nitobe as one of the three undersecretary generals of the League of Nations – with authority over the International Committee of Intellectual Cooperation. Shortly before he took office, at the end of December 1926, Dufour-Feronce expressed in unmistakable terms what he thought of the committee and Einstein's membership: "a committee composed of the Oxford scholar Professor Gilbert Murray, the German Einstein, the Frenchman Bergson, etc., and with an institute in Paris that is mainly financed by France, Poland and Czechoslovakia. Commentary superfluous! But my new colleagues [...] tell me that this intellectual cooperation will only take up about a quarter of my time."[655]

A few months later, as we saw above, Albert Einstein himself suggested on 5 March 1927 that Hugo Andres Krüss become his substitute on the committee as well as at the Parisian institute! That Krüss "accepted unconditionally" is not in the least surprising. On 10 March 1927 Einstein even expressed pleasure that through Krüss "an effective link is established between the committee and local authorities and scholars." Einstein was glad to be rid of yet another obviously

**Figure 32: Albert Einstein and Hugo Andres Krüss, managing director of the Prussian State Library (from H.A. Krüss's diary)**

burdensome office. Not through his own doing, though, but through pressure applied by others, he had now found his *substitute*.

He cannot have been happy about it. Why else would he have written to Oprescu on 3 May 1928: "I thank you for your kind lines. My doctor unfortunately categorically forbids my coming to the meetings in July. May I in this case send a substitute? *If yes, must it be Mr. Krüss?*"[656] So Krüss was certainly not his choice candidate. But what else could he have done? None of his mind-mates and friends, including Fritz Haber, were willing to take this "odious burden" upon themselves.

Einstein to Krüss, 10 March 1927:[657]

Schöneberg, the 10th of March 1927

Esteemed Dr. Krüss,

I had already nominated you at the meeting of the Institute's administrative council as my proxy, adding however that your acquiescence was still outstanding. Mr. Oprescu informs me now that the administrative council had thereupon elected you last year already in case you approve, which had escaped me. The election is renewable every summer, which is just a formality, though. I am very pleased that through you an effective link is established between the committee and local authorities and scholars.

I would now like to ask you right off whether you would like to represent me at the next meeting of the administrative council. (I enclose the invitation.)

With amicable regards,

Yours,

A. Einstein.

*P.S.*[658] *I cannot find the invitation to the meeting of 26 III at the Parisian institute. But I would like to check the exact time again. As a rule, the meetings start at 10 o'clock.*

On 16 March Einstein finally *did* find the program for the meeting and sent it to Krüss.[659] But Krüss evidently could not wait. Promptly on 11 March Krüss informed the secretary of the committee, Oprescu, of Einstein's request and asked that the invitation and agenda be sent to him because "Mr. Einstein apparently has lost or misplaced" them.[660] He also requested the by-laws, the rules of procedure, earlier minutes and the committee's membership list. Oprescu replied on 14 March with the announcement that he would ask his friend, the director of the Parisian institute, to send Krüss what he wanted and also informed him who else was on the board of directors of the Paris Institute (besides Einstein or Krüss):

- Gilbert Murray, professor at the University of Oxford;
- de Reynold, professor at the University of Berne;
- Destrée, formerly Belgian minister of sciences and arts;
- Rocco, Italian minister of justice;
- Vernon Kellog (as Calkin's substitute), member of the National Research Council of the USA and secretary of the American National Committee on Intellectual Cooperation.[661]

On 15 March the requested documents were sent out to Krüss.

Einstein was presumably not very forthcoming in such matters. He had evidently either lost or misplaced more than just the invitation for the *coming* conference. This incident illustrates how casually Einstein treated official affairs and how thoroughly and purposefully Krüss took up his new responsibilities. *He* found the business enjoyable. *He* was in his element.

That is how Krüss's future involvement on the committees of the League of Nations became, as he wrote Oprescu, "a continuation of our relations started a year ago in Geneva."[662] Einstein, "probably referable as oriented politically to the left," gave his approval "even though from the political point of view, Mr. Krüss is denotable as positioned on the right."[663]

Einstein must have been doubly pleased to have found a substitute when he could not attend the meetings of the committee and the Parisian institute in 1928/29 by reason of serious illness. The *politically* orchestrated exclusion of Einstein thus appeared as though it were *medically* dictated and a solution of Einstein's own preference.

### 2.7.6 Einstein's confession

Einstein's illness offered him occasion to express more definitely than ever before his position among the intellectual elite and his relations with Germany's political class. He had already once said that he "did not feel like [...] representing people who would surely not have elected me as their representative and with whom I [...] do not share the same views." Only now did he become fully aware of his situation.

On 16 September 1928 Einstein wrote to Krüss, still marked by serious illness and uncertain what the future held in store for him. He confessed that he had just filled a "gap" during hard times when no one else had the courage to take upon themselves "*the odious burden of internationality*." But now, he implied between the lines, he was no longer needed. His "*roots among the German intelligentia*" *were far too weak* to be a strong link between this "intelligentia" and the committee of the League of Nations. Physically at the end of his strength, Einstein acutely realized the potentials and the limits of his political effectiveness.

Einstein to Krüss, 16 September 1928:[664]

*Scharbeutz 16 IX 28.*
*near Lübeck*

*Esteemed Mr. Krüss,*

*Thank you very much for your friendly news and the reports you drew up. I am very glad that you have familiarized yourself with the Genevan business with such a positive attitude and energy. At the beginning of October I am returning to Berlin and very much look forward to receiving the promised oral report. My health situation is still quite rotten so it is not likely that I shall be able to attend the out-of-town meetings in future. I am very glad that you are inclined to substitute for me on the panel of the labor organization.*

*I do not regret at all not being able to participate personally at the meetings of the League of Nations anymore. It was clear to me from the very start that I was as unsuited as could be for such kind of work. It was solely the circumstance that, under the prevailing mentality of our ''intellectuals,'' no other person with an international name could be found willing to assume the odious burden of internationality that compelled me to jump in and fill this gap. I did so even though I was fully aware that my roots among the German intelligentia were far too weak to form an effective link. Through your generous participation this deficiency has now been lifted in the best possible way. I am very curious to hear about how you liked the committee and its mentality.*

*With best regards,*
*Yours,*
*A. Einstein.*

This same tone and argumentation reappears in a letter to the secretary of the Genevan committee a few days later.[665]

Einstein to Oprescu, 26 Sep. 1928:[666]

**Figure 33: Convalescent Einstein in Scharbeutz by the Baltic Sea, 1928.**

*Scharbeutz. 26 IX 28.*

*Dear Mr. Oprescu,*

*Cordial thanks for your letter. I am very happy that Mr. Krüss has familiarized himself so thoroughly with the committee's affairs. This is particularly important to me because I cannot count on coming to the meetings myself again. I look forward to seeing you in Berlin and being able to be present at the envisioned consultation. I think it important that Mr. Krüss also be there; for, first of all, his favorable opinion on the matter arouses less distrust than mine, because he does not have the reputation of a bias for internationality like I do, second of all, he has valuable connections in government as well, through which he can further the cause indirectly. Finally, it is also good to draw him into the foreground to strengthen his interest further in the matter.*

*With best regards, yours,*

*A. Einstein.*

Besides conceding having been "unsuited" as a member of the committee, Einstein knew he had been representing interests of the intellectual elite and the German Reich as a whole at a time when no one else was willing to take on the risk of earning the reputation of a "traitor" or an "appeasement politician." Walther Rathenau had become a victim of such nationalist sentiments. Einstein knew he was isolated and alone on the open field, without any backing by German "intellectuals." He resigned, allowing the impression even to arise that he was glad that a person was taking his place who did not have a "*reputation of a bias for internationality.*"

Einstein's clairvoyancy about his social position may have come from a consciousness of being on the brink of his grave. This opinion is nevertheless not solely explicable by his health condition. Years later, on 20 April 1932, he gave his reasons for joining the committee and his social position in Germany in virtually the same words.

Einstein to the secretary of the International Committee on Intellectual Cooperation, 20 April 1932 (excerpt):[667]

> [. . . ]. I became firmly convinced at an early stage that I was not suited to do fruitful work within this organization. [. . . ]. I would also like to take this opportunity to comment that I had accepted the election as member only because under the political conditions of the time it would have been difficult to find another suitable person of standing among German academics with an interest in the international problems.

Any hopes Einstein attached to his membership in the committee proved to be illusory. He wanted to stay above the quibbles among the parties but still got caught up in the wheels of politics. Politically, he fell between two stools. Any effect he may have had in expressing his opinion on the bigger political issues was solely due to the fame arising from his extraordinary *scientific* achievements. Often the only thing sought after was Einstein's *name,* certainly not his personal opinions.

Einstein's scientific merit brought him perhaps unique liberties. He was not bound to any party or organization. His time was truly his own. He could decide whether or not to give lectures as he pleased. He could say what he thought – without regard for others. He could afford to do all this, because he had accomplished so much. His materially secure position was another significant factor of his independence. The price he had to pay for it was, of course, very high. Einstein was basically a lonely man. He had done infinitely much for the reputation of German science and for the German "fatherland." So it was tragic that his "roots among the German intelligentia" were so weak. (But then again, why should he have been firmly rooted in a largely conservatively minded social class that was even to a considerable degree anti-Semitic?!) Another price of his independence was his frequently vacillating political judgment, even though he remained, despite everything, on the *left* end of the political spectrum. Einstein was a kind of intellectual rarely seen in his century. He was prototypical of the intellectual per se.

Einstein's political naïveté and gullibility was also tragic. In his private life he he could be incredibly severe (his attitude toward his first wife is one instance) while otherwise being tolerant and obliging. Never maliciously intentioned himself, he could not believe that others could be any different. His knowledge of human nature was not very good.

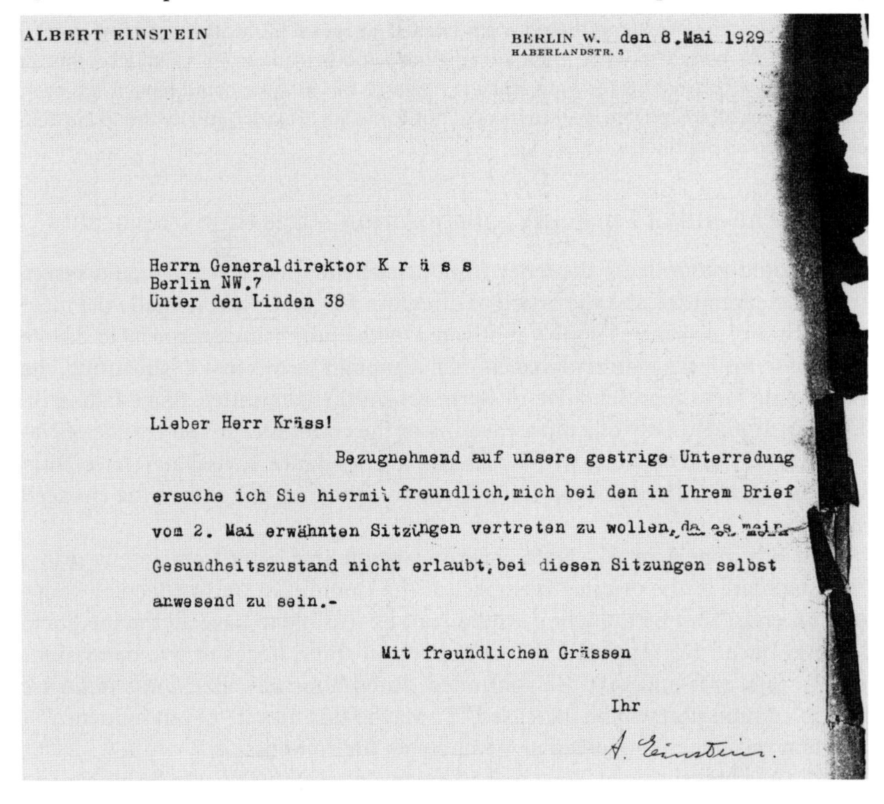

ALBERT EINSTEIN

BERLIN W. den 8.Mai 1929
HABERLANDSTR. 5

Herrn Generaldirektor K r ü s s
Berlin NW.7
Unter den Linden 38

Lieber Herr Krüss!

Bezugnehmend auf unsere gestrige Unterredung ersuche ich Sie hiermit freundlich, mich bei den in Ihrem Brief vom 2. Mai erwähnten Sitzungen vertreten zu wollen, da es mein Gesundheitszustand nicht erlaubt, bei diesen Sitzungen selbst anwesend zu sein.-

Mit freundlichen Grüssen

Ihr

A. Einstein.

**Figure 34: Albert Einstein's letter to Hugo Andres Krüss from 8 May 1929.**[668] **Later singed by the war but still salvageable from the burning State Library.**

So his letter to Hugo Andres Krüss from 16 September 1928 was an incomprehensible act of trust. He had allowed the recipient to see into his inner soul and let him know where he was vulnerable. Krüss never let down his guard like that with Einstein.

Another letter of Einstein's from 20 September 1929 (hence a year after the letter from Scharbeutz) demonstrates how downright naïve he could be. Krüss had previously expressed the wish to report to Einstein about the committee meeting in Geneva. In his letter from 20 September Einstein thanked him, telling him it would be too troublesome for him to travel all the way "into the city" to meet him and suggesting instead that they meet halfway in Wannsee at Friedrich Karlstr. 18 (tel. 5259) "at friends [...] where there is also a garden with a wonderful view."[669] We find out who these friends were from the regional directory 'Adreßbuch für Berlin und seine Vororte 1919–1932' as well as from the city directory 'Berliner Adreßbuch' from 1931. The owners of the buildings at Karlstrasse 13, 17 and 18 are listed there as: Toni Mendel, P. Mendel and the physiologist Bruno Mendel. The telephone number 5259 was connected to Karlstr. 13, which according to the city directory for 1931 (page 1513), belonged to *Toni Mendel*. This Toni Mendel was Einstein's lover at that time. Hugo Andres Krüss was supposed to go to this house owned by Jews. We cannot assume that Einstein wanted to reveal his most intimate secrets to him. Even so, added to what Krüss already knew (particularly from his involvement in settling the affairs connected with Einstein's divorce in 1919), he could have spied even closer into Einstein's private life. Krüss could not come, incidentally, and informed Einstein as much on 23 September 1929.

### 2.7.7 The end of Einstein's collaboration – this time irrevocably

Krüss could gather from the letter from Scharbeutz that Einstein would resign from the committee and the board of directors *for good*. Having made the most of his chance, Krüss was doubly motivated by the promising prospects to devote himself to his tasks with renewed energy. Although he was just a "substitute," he was the de facto representative of Germany on the committee. Since falling ill, Einstein attended only one more meeting of the committee of the League of Nations (1930) – and he did so with Krüss also in attendance. Krüss's new friendship with the fascist Alfred Rocco, whom Einstein abhorred, epitomizes the changed situation.

"In July of next year" – Dufour-Feronce wrote in a letter from 29 July 1930 – "the mandate of the original members of the Committee on Intellectual Cooperation ends." But he thought it would "not be feasible to have all the members leave at once." He suggested that it happen in three stages of five demissions each – 1931, 1932 and 1933. He continued, "then hopefully the *Nordic influence* on the commission will be increased."[670] Maybe this meant "Aryan influence" – Einstein was one of the "original members on the commission."

As already mentioned, Einstein had long wanted *Fritz Haber* to be his successor.

Fritz Haber to Einstein, 16 June 1931 (excerpt):[671]

I have no idea in which parts of the world you are exercizing your represention of German culture right now. I write you, because at the moment I am about to travel to Paris for 5 days and I am sending a carbon copy of this letter to Mr. Krüss. The Committee on Intellectual Cooperation has appointed a temporary preliminary committee and enlisted a professor from each of the major countries. They have also selected me as one of the 5 participants and we are supposed to have initial consultations on the 17th and 18th of next month in Geneva about what should be presented afterwards to the plenary committee for resolution, namely,

''In what way could international intellectual cooperation have an effective impact on the coordination or efficiency of scientific research.''

The most important documentation in connection with this is the report on what the International Institute has done thus far in preparation. I accepted the post because I assumed that you and Mr. Krüss orchestrated my election and now I ask you please to let me know, as far as the state of affairs allow, what function I shall be assigned for this preliminary meeting.

Einstein's letter to Dufour-Feronce from 20 June 1931 followed in this vein. Einstein also added there – in due diplomatic form – his preference that Krüss *not* become his successor. He had nothing against Krüss's membership on the executive committee but he explicitly wished that Fritz Haber – not Krüss – become his successor on the International Committee "as a *particularly suitable and influential personality*."

Einstein to Dufour-Feronce, 20 June 1931:[672]

Prof. Albert Einstein Caputh near Potsdam,[673] the 20th of June 1931
To Mr. Albert Dufour-Feronce
Société des Nations
Geneva, Switzerland

Esteemed Mr. Dufour-Feronce,

I have meanwhile heard, to my delight, from Mr. Haber that he has accepted his election on the Study Committee for the Exact Sciences. As far as Dr. Krüss is concerned, I was of the opinion that by his election on the Executive Committee he would be permanently connected to the Committee. It was only under this precondition that I expressed the hope that Mr. Haber would be my successor as member on the [International] Committee, as a particularly suitable and influential personality. I would otherwise not have permitted myself to suggest this.

It will probably not be possible for me to come to Geneva in July and I may surely presume that, in this case, Mr. Krüss can assume the Committee meetings. In my view it would even be desirable if for the coming election period he would assume the position on the Committee that I had invested this past election period.

In great respect,
A. Einstein.

Haber had forwarded to Krüss a copy of his letter to Einstein from 16 June 1931.[674]

Understandably, Haber did not forward Einstein's *reply* to "silly and conceited Mr. Krüss, who is, however, hand in glove with Mr. Dufour-Feronce, the German representative on the Council of the League of Nations."

Einstein to Fritz Haber, 20 June 1931:[675]

Caputh, the 20th of June 1931

Dear Haber,

I had long sought to have you and your great organizational skills enlisted for Geneva. I would also like to have you on the main committee as my successor instead of silly and conceited Mr. Krüss, who is, however, hand in glove with Mr. Dufour-Feronce, the German representative on the Council of the League of Nations.

The Study Committee for the Exact Sciences, onto which you have been elected was formed at the instigation of Mrs. Curie and has as its main focus the international organs of scientific reporting, which has thus far foundered against German resistance and even more so English resistance. Your committee is well composed and will be receptive to your suggestions.

Cordial greetings,

Yours, Albert Einstein

Krüss thanked Haber in a letter from 22 June and requested meeting with him after the next academy meeting: He wanted "to repeat his request, then taking the usual refreshment for once not at the Schlosskonditorei café but at my home. Mr. Einstein might perhaps also come along then."[676] This closing conjecture probably implies that Einstein was not inclined to come along but would not turn down an invitation by his friend Haber.

There is no documentation on whether Einstein actually did come along. But either way, it was clear that the apparently good relations between Krüss and Einstein were history. In 1932 Einstein withdrew his membership on the committee together with others. Krüss enjoyed the company of a growing number of people on the political right, and at least *one* of them being a close friend of his – the minister of justice for fascist Italy (and later president of the University of Rome), Alfredo Rocco.

As early as 1930 – at least, according to 'Einstein: On Peace' – Einstein already informed Dufour-Feronce that he would "not be going to Geneva anymore." He had resigned.

Einstein to Dufour-Feronce (excerpt):[677]

[. . . ]. My decision not to go to Geneva is founded on experience unfortunately teaching me that the Committee, on average, does not embody a serious wish to guide its efforts toward significant progress in repairing international relations. [. . . ]. Precisely because I would like to apply my energy to the creation of an international, tribunal and regulatory authority superior to the individual states and because this goal is very close to my heart, I believe I must leave the Committee.

The Committee has given its blessing to the suppression of cultural minorities [. . . ].

The Committee has furthermore advocated such a lukewarm position regarding the problems of combating chauvinistic and militaristic tendencies in education in the individual countries, that serious efforts in this fundamentally important area cannot be hoped of it.

The Committee always abstained from lending moral support to persons and associations who have committed themselves in a radical way to acting for an international legal system and against a military one. The Committee never made an attempt to resist the inclusion of members known to be advocates of entirely different tendencies than the ones it would be their duty to represent. [. . . ]. If I had any hope, I would act differently, you can be certain of that.

Einstein's collaboration was thus, in fact, at an end. So it is surprising how avidly Dufour-Feronce tried to coax Einstein to remain on the committee (even though he did not visit him while he was in Berlin). His letter dated 29 June 1931 is evidence. But from what we know about Dufour's views and attitude until that time, this cannot have been a change of heart. It was rather an expression of fear that Einstein's resignation could damage the committee's reputation and his own ambitions. Einstein evidently was not supposed to leave with the "first stage" of departing members.

A. Dufour-Feronce to Albert Einstein, 29 June 1931 (excerpt):[678]

Thank you very much for your amicable letter of the 20th of this mo., which I answer only today, because last week I had to travel to Berlin for a few days on business.

Can you really not arrange to come here for at least a few days? You know how extremely well liked you are, how pleased the other members on the committee are to see you and how much they value your collaboration. If it is really not possible for you to come here yourself, Dr. Krüss will then represent you, of course, but perhaps you can part with the Havel lakes, after all, for a short time and give us Genevan lakeside dwellers the pleasure of being able to welcome you here.

When he nevertheless received an invitation to the meeting of the committee and the board of directors in July 1932 in Geneva, Einstein responded on 20 April 1932 that he regarded his official term as having ended with the year 1931. He in any event believed he could no longer be of any use to the committee. The recipient of this letter was the new secretary of the committee of the League of Nations, J.D. de Montenach.

Einstein to the secretary of the International Committee on Intellectual Cooperation, 20 April 1932:[679]

Albert Einstein

Berlin, W., the 20th of April 1932.
Haberlandstr. 5

To the Secretary of the
C.I.C.I.
Société des Nations
Geneva.

Esteemed Mr. Secretary,

Upon my return from a longer trip abroad, I am only able to answer your letter from 5 February of this year today. To my knowledge my mandate as member of the C.I.C.I. expired with the year 1931. With your letter of 5 February, however, which contains the invitation to the meetings in July, I apparently seem not to be correctly informed about this.

I became firmly convinced at an early stage that I was not suited to do fruitful work within this organization. It would therefore be amenable to me, and it would appear justified, if another person took my place as member of the Committee. This would be the more desirable since it will be difficult for me to spare the time next summer to attend the Genevan conference.

I would also like to take this opportunity to comment that I had accepted the election as member only because under the political conditions of the time it would have been difficult to find another suitable person of standing among German academics with an interest in the international problems. This has luckily changed for the better in the meantime. It would certainly be possible now to find a personality for this function who is, besides, more vitally connected with public life here than is the case with me.

Please discuss this letter with Mr. President and let me know the state of affairs as soon as possible.

In great respect,

A. Einstein.

Montenach acknowledged receipt of the letter on 22 April 1932 and informed Einstein that he had forwarded the letter to the president of the committee, Professor Gilbert Murray and to Dufour-Feronce for a response to the questions broached in it.[680] Dufour-Feronce answered on 23 April. Again the same routine: Of all people Dufour-Feronce, Krüss's fellow conspirer, now was hoping Einstein would "accept a reelection"! Of all people Dufour-Feronce, who had once spoken so deprecatingly about the scholars' committee, was asking to see Einstein again in order to say good-bye!

Dufour-Feronce to Einstein, 23 April 1932 (excerpt):[681]

I gather with great regret from your letter [. . . ] to Mr. de Montenach that you have, sadly, finally decided that you do not want to be a member of the Committee on Intellectual Cooperation anymore. Also in view of the friendships you have made with a number of other members of the committee, I was certain that you would reconsider, after all, and accept a reelection. [. . . ] and thus we, who value your presence at the committee meetings so much, must bow to the inevitable.

If, however, you do find it possible to come to Geneva for the next meeting of the committee, which is starting on 18 July and will only take up 6 days, not just I but - as I am convinced - all your colleagues would be very especially pleased because they would otherwise - at least many of them - perhaps have no other opportunity to see you again and say good-bye. [. . . ].

At its meeting in January the League of Nations decided to reschedule the renewal of the committee to next September. Your membership accordingly runs until September of this yr., and your successor could consequently only be elected then. Thus you are, under all circumstances, still a member of the committee in July, and for this reason, too, I hope that you will be able to attend this unfortunately last meeting for you.

Envoy Freytag also received Einstein's departure with "honest regret" in a letter from 2 June 1932 to Dufour-Feronce and recommended "most warmly Mr. Krüss's candidacy in the name of the Foreign Office." He praised Mr. Krüss's personal abilities and achievements also as a science administrator. This latter point may have had its merit but it is remarkable how completely the original conception of the committee's composition was forgotten, ten years after its establishment. Nothing at all is mentioned about *scholars* of international renown or members *independent* of their national governments.

Letter by Envoy Dr. Freytag to Dufour-Feronce, 2 June 1932 (excerpt):[682]

Among the members of the Committee on Intellectual Cooperation who, pursuant to the statutes, are resigning this year is also Professor Einstein. In connection with our honest regret at having to see him leave as a German representative is the concern that a suitable successor be found. I have considered this question carefully here and also discussed the matter with interested domestic German competencies. We ultimately agreed that the managing director of the Prussian State Library, Dr. Krüss, is the suitable person. Dr. Krüss [. . . ] has been versed in the committee's work since a number of years and from his former employment at the Prussian Ministry of Culture, as head of the University Division, he has long been familiar with the problems attached to the organization of science [. . . ]:

I would therefore like to recommend most warmly Mr. Krüss's candidacy in the name of the Foreign Office.

Dufour was intent on conquering the opposition to Krüss's candidacy. So he asked Alfredo Rocco, the Italian minister of justice and culture, on 14 June 1932 to apply to the Italian foreign minister and his colleagues on the committee to back Krüss. His aim was to break the resistance based on the committee's founding creed that "the International Committee on Intellectual Cooperation may only be composed of men and women from the sciences," which posed the main obstacle to Krüss's candidacy. As a friend of Krüss and a nonscientist himself, Rocco was likely to receive his arguments readily. The strongest argument was that Max Planck and Schmidt-Ott also supported Krüss's candidacy. (This same Max Planck once opposed Einstein's membership on the committee, after asking Einstein in vain in 1924 "to found, that is, to bemother, a [German] National Committee on Intellectual Cooperation.")[683]

Dufour-Feronce to Minister Rocco, 14 June 1932 (excerpt):[684]

Following our brief conversation in Milan about the candidacy of Dr. Krüss as successor to Prof. Einstein in the International Committee on Intellectual Cooperation, it is my pleasure to enclose a copy of a letter I received from the director of the Sciences Division of the Ministry of Foreign Affairs in Berlin, Minister Dr. Freytag.

From this letter you will gather that the Minister of Foreign Affairs recommends Dr. Krüss most warmly as successor to Prof. Einstein. On the other hand, I also have the support of his Excellency Dr. Schmidt-Ott, president of the Emergency Association of German Science and through the intermediary of Dr. Schmidt-Ott also that of Prof. Planck, president of the Kaiser Wilhelm Society and privy councillor.

Very many persons are of the opinion that the International Committee on Intellectual Cooperation may only be composed of men and women from the sciences. Those who advocate this opinion will naturally be against the election of Dr. Krüss [. . . ].

If you are of one mind with me on this issue, I take the liberty of suggesting to you to ask your minister of foreign affairs to lend his support to Dr. Krüss's candidacy at the next session of the Council of the League of Nations when it deliberates on this matter.

Additionally, Mr. Minister, I would be very grateful if you would exert your influence in this direction on your colleagues at the meeting of the International Committee on Intellectual Cooperation next July.

It was most important to defuse the objections raised by the committee's president Murray.

Gilbert Murray had responded to a letter by Dufour-Feronce by writing on 21 June 1932 that he wished Krüss would continue his work for the committee as a consultant and that a *reputable scholar* would become Einstein's successor – supported by an administrator like Krüss. But Murray also admitted that the decision on the succession was actually a German affair and it would be difficult for him to contradict a nomination advanced by the German government (thus essentially pointing out how to undermine his own objections).[685] On the same matter the British member of the Executive Committee and secretary of the British National Committee on Intellectual Cooperation, Sir Frank Heath, assured Dufour in a letter from 24 June 1932 that he would back Krüss's candidacy. But he was also worried about Murray's reservations and about being blamed for acting like a narrow-minded philistine. To bolster his confidence about how to proceed, he wanted to consult with Murray.[686] Dufour replied on 27 June that it was better *not* to discuss Krüss's nomination with Murray and briefed him on Murray's position on the Einstein–Krüss matter. His arguments were that Krüss would not be willing to serve as substitute again and that his candidacy was being supported by the Foreign Office, other ministries, "& c. & c."[687]

Such intrigues were too much for Murray. Judging from Dufour's characterization from 31 July 1928, he was anyway not up to the task as committee president.

Dufour about Murray, 31 July 1928:[688]

For all his good intentions of being international, he does remain the insular Englishman, who is purposefully good-willed but cannot display a natural charm connected with much esprit, as was the case with Lorentz. Even so, Gilbert Murray is undoubtedly the best that could be elected on the committee and it could be that with time he will lose the somewhat strict Puritanism, albeit somewhat softened by his mild voice, but I don't really expect so.

Such preparatory work having been successfully accomplished, Dufour-Feronce was pleased to be able to inform Mr. Krüss on 29 June 1932 that "if I am not mistaken, you have the best chances of becoming Professor Einstein's successor on the Committee on Intellectual Cooperation," and that the decision would be made in September.[689] Dufour also asked Krüss to name a substitute after consulting with Freytag. On the same day he wrote to Freytag that the president of the committee still had some reservations about appointing Krüss as Einstein's successor but ultimately considered "the election of Einstein's successor more a German affair."[690] He asked Freytag to assist Krüss in selecting a substitute and offered his own suggestion: Professor Konen[691] from the University of Bonn.

Krüss replied on 4 July 1932 that in the event he was elected, he would nominate Prof. Konen (Bonn)[692] as his substitute.[693] In September 1932 the Council of the League of Nations finally elected Krüss as member of the committee. Krüss expressed his delight about it in a letter from 30 September 1932, thanking the secretary of the committee, de Montenach, "for your kind share in this matter."[694]

A few weeks later Hitler acceded to power. Krüss aligned himself promptly and ably with the Third Reich. There was not much of an alternative for a managing director of the Prussian State Library anyway: either to distance oneself and therefore leave the service *or* to remain in office and, on 25 August 1934, take the oath: "I shall be loyal and obedient to the Führer of the German Reich and nation, Adolf Hitler, abide by the laws and conscientiously fulfill my official duties, so help me God."[695] There are no indications that he – as opposed to Einstein – would ever have considered such an oath as a "gruesome ordeal" and "humiliation."[696] But others also had taken this oath of loyalty to Hitler, Max Planck, for instance. The only difference was that Krüss officiously did more than he needed to.

Einstein distanced himself immediately and definitively from the Third Reich. The price he had to be willing to pay was expulsion from the Berlin academy and exile. Krüss's reaction was different. *Krüss* immediately and definitively placed himself on the side of Einstein's opponents.

Not sure that he could attend the meeting of the executive committee of the Committee on Intellectual Cooperation on 11 and 12 April 1933 in Paris, Krüss wrote a long letter to the undersecretary general of the League of Nations and director of the Section on Intellectual Cooperation, Massimo Pilotti. The purpose of this letter from 31 March 1933 – written just hours before the "day of the Jew boycott" – was to place the new leaders of Germany in a good light. While there seemed to be no limit to the Nazis' bloodthirsty rampages, he was lauding "the unheard of discipline with which the national upheaval is following its course," and asking for forebearance during the German people's defensive struggle. He could have maintained dignified silence, but he chose not to. He asked that his letter be read out during the meeting.

Krüss to Pilotti, 31 March 1933 (excerpt):[697]

[. . . ]. Through unrestrained incitement, lies and blaspheme, the German people has been forced to defend itself in active battle. The unheard of discipline with which the national upheavel is following its course is not receiving the attention and acknowledgment it deserves. Instead an aggressive propaganda campaign is mounting against Germany that is not only harmful to German economic interests but also intends to slight the national reputation of the German people. [. . . ].

At the State Library ordinary business is continuing completely undisturbed as before. The existence of a national organization among civil servants and library employees has a decisive influence on this; the library is conscientiously following the strict orders issued by the supreme leadership and has been a model of order and discipline.

Thus the State Library is exemplary in revealing that the German nation wants to continue doing its work in peace. If it now sees itself compelled to fight for its survival, every nation and all self-respecting people should understand this is a state of emergency. The responsibility for the unforeseeable consequences rests on those who started the attack on Germany and forced the German people to defend themselves actively.

It is high time that common sense and good will throughout the world be mobilized everywhere to put a stop to this fatal action against Germany.

I trust that there will be an opportunity to discuss this matter at the meeting of the Executive Committee. [. . . ].

Otherwise I ask you please to make use of my account wherever you deem it possible for the purpose of enlightenment [. . . ].

Krüss's original fears were unjustified. He was able to attend the meeting in person. He used the opportunity to report on how he interpreted the developments in Germany and to comment on the German government's position on the "Jewish problem."[698]

"Jew boycott day" on 1 April 1933 offered him a good opportunity to do so.

When "the university was occupied by the S.A. and Jewish students as well as Jewish assistants and certainly also the lecturers were expelled," on that day "the users cards were taken away from Jewish visitors of the State Library." Managing Director *Krüss* saw to it "that this took place smoothly":[699] He did not prevent it, he did not protest; he assured a *smooth* "processing of the business." He let the Aryans in, the Jews had to stay outside.

During the meeting in Paris on 10, 11 and 12 April 1933, Krüss noted,[700] – he had spoken with many "persons of importance [...] about the circumstances in Germany [...]. The Jewish issue was of primary interest." Krüss noted having

pointed out that the Jewish issue in Germany would have to be evaluated not just according to racial aspects but also to economic aspects [. . . ] I used the procedures of the large institute under my authority, which employs 400 persons and is visited daily by 3,000 persons, to illustrate how disciplined and moderately the national uprising has been happening in Germany [. . . ].

Dr. Babcock and Dr. Kittredge will report in New York to the Carnegie Foundation and the Rockefeller Foundation on the basis of my descriptions.

No explanation was necessary for the two Italians Pilotti and Rocco. My close friend of many years, Minister Rocco, told me 6 weeks ago already in Rome that Mussolini had adjusted his European policy 5 years ago to what has been turned into fact in Germany.[701]

As revealing as Krüss's attitude is, apparently *none* of his discussion partners spoke out seriously in objection. The British member of the board of directors, Heath, explicitly stated "that Germany was in fact set before the choice between a nationalistic regime and a Bolshevist one."

The travel report does not reveal whether his predecessor on the committee was mentioned by name. Commentary is superfluous about what he thought of Einstein. Einstein was just one of the persons Krüss thought had evaluated the developments in Germany "untruthfully" and "unfairly."

Barely back from Paris, he applied to the Foreign Office on 13 April 1933 to have "another German personality [...] *who embodies the essence and mentality of modern nationalistic Germany* and hence could act as a balance against Thomas Mann," delegated to attend the meeting of the subcommittee on literature and arts of the committee of the League of Nations in Madrid from 3 to 7 May 1933, because "neither the League of Nations nor the Spanish government are inclined [...] to change anything about Thomas Mann's commitment to attend."[702] The Nobel laureate Thomas Mann was a *member* of the subcommittee.

The book burning on 10 May 1933 did not move the managing director of the Prussian State library particularly either. He waved it off as a mere "symbolic act" – as if the very *symbolism* of it was not even the problem. He said that the holdings of the State Library and other libraries had "not been affected by it at all." At the same time he deemed it appropriate to praise "the old national flag and banner of the National Socialist movement" as "a symbol of the new unified Germany, which is resolved to build its future upon the best traditions and youthful energies that have carried the banner of the national rebirth to victory."[703]

In the summer of 1933 Krüss prevented a protest from being read out before the committee of the League of Nations that the council of the American Association of University Professors had submitted against the infringement in Germany of the freedom of instruction and learning.[704] Krüss wrote to the secretary of the committee, de Montenach, on 20 June 1933: "I would like to think that in the interest of all parties this resolution or similar kinds of statements *not* be brought before the Commission de Coopération Intellectuelle. Its object is primarily of a political nature and any such reference would necessarily elicit an immediate corresponding statement on the part of Germany. The Committee has hitherto legitimately demonstrated the greatest restraint concerning matters that could have the slightest appearance of a political or critical evaluation of a country's internal affairs."[705] The consequences of such a "regrettable disturbance" to the committee's work, he continued ominously, could "not be predicted."

The fact that members of the committee and elsewhere abroad viewed the developments inside Germany with some sympathy made the job considerably easier for people like Krüss. The German embassy in Washington informed the Foreign Office regarding the petition that Krüss had successfully obstructed: "With reference to newspaper reports about the protest by the American Association of University Professors to the League of Nations," we have encountered understanding for the German government's Jew laws, "particularly among educated individuals."[706]

On 20 June 1933 Krüss seems to have already forgotten that just three months before, on 31 March 1933, he had requested that his own statement about the *political* developments in Germany be read out before the next committee meeting of the League of Nations. He evidently granted himself liberties that he would not allow others.

In September 1933 he wrote Prof. Gilbert Murray, president of the committee of the League of Nations: "I am confident that your better judgment will assure that you continue to separate the area of our Intellectual Cooperation from things currently the subject of political debate."[707]

A trip to Chicago for the annual conference of the American Library Association from 27 September to 4 November 1933 gave Krüss the "opportunity to explain the situation in Germany to the persons listed below, who have a special standing in science administration in the United States [...]." This is mentioned on the *first page of the 15-page travel report.*[708] Page 2 lists the names and occupations of eleven important persons.

Under "7. General impressions," Krüss remarked, among other things:

> Everywhere I encountered the urgent wish to get more detailed information about the situation in Germany independent of incidental and sensationally exaggerated news in the press. There was generally also a reasonable understanding for the basic ideas and successes of the nationalist German reconstruction [. . . ].
>
> A particularly noteworthy result of conversations with diverse people was the degree to which the issue of Jews is becoming increasingly recognized as a problem in America as well. [. . . ]. The majority of serious universities and colleges are against the Jews. The admissions procedures are treated correspondingly. Fraternities do not accept any Jews as members. [. . . ].
>
> The appointment of a Jew as American ambassador in Paris is incomprehensible, indeed it is frequently viewed as an affront; certain exclusive residential areas are closed to Jews. These are further examples of how pronounced emotional anti-Semitism is in the United States.''

A new side of Krüss is thus revealed. He was not only conservative ("from the political point of view, Mr. Krüss is denotable as positioned on the right"),[709] but also anti-Semitic. How else is one to interpret Krüss's opinion that "serious," that is, prestigeous, model American universities, were disposed "against the Jews." How else the fact that on "Jew boycott day" the managing director of the Prussian State Library had "the users cards taken away from Jewish visitors of the State Library" and saw to it "that this took place smoothly"?[710] He was certainly not a militant anti-Semite; others did the dirty work. Genteel discretion was his approach. As discreetly as possible keep the number of Jewish professors down, confiscate their user's cards, deprive them of ambassador posts and shut them out of exclusive residential areas. It is not contradictory that he should protect certain Jewish individuals. Prominent Nazis did so too (Hess, for instance, shielded Albrecht Haushofer and even Adolf Hitler kept Otto Warburg under his wing). Neither is Krüss's late entrance into the Nazi party (1940) contradictory: Reich Foreign Minister von Neurath decided to do so only in 1937. The tragic fact remains: Einstein's substitute was an anti-Semite.

So the National Socialist regime was not "naturally alien" to Mr. Krüss, as Werner Schochow contends.[711] *Krüss* had rather *immediately* and very willingly offered his exceptional abilities to the new regime. He did not struggle. The imperial civil servant and conservative culture policy-maker was urged by opportunism and attachment to his office to become a loyal servant of the Third Reich. What separated Einstein from Krüss was not peripheral political issues but core issues; not trivialities, but worlds. Their paths diverged clearly and finally in 1933.

He could not revel long in his accession to Einstein's seat on the International Committee of Intellectual Cooperation. In October 1933 Germany withdrew from the League of Nations, and therefore from all its organs. On 9 November 1933 Krüss dutifully resigned his offices in organizations of the League of Nations, thus his involvement in the committee and in the administrative bodies of the Parisian Institute of Intellectual Cooperation.[712]

### 2.7.8 Albert Einstein / Sigmund Freud: Why war?

Before he left office, Krüss still had to deal with a troublesome publication that could be regarded as the literary heritage of Committee Member Einstein: his correspondence with Sigmund Freund published in 1933.

On 9 May 1933 the Foreign Office wrote a disgruntled letter to Krüss about a pamphlet of correspondence between Einstein and Sigmund Freud on the topic 'Why war.' Krüss was asked not to allow the International Institute of Intellectual Cooperation to publish such things again.[713] Although only an oral response was expected, Krüss replied *in writing* on 11 May (unimpressed by the book burning that had taken placed in front of his door on the previous day).

Krüss's letter to Legation Councillor Zoelsch (Foreign Office), 11 May 1933:[714]

> I agree with you entirely that the brochure containing the correspondence between Einstein and Freud published by the Parisian Institute is completely unwelcome. As you will gather from the dating of the letters, Einstein's letter originates from July 1932 and Freud's letter is from September 1932.
>
> The whole affair stems from a resolution by the Committee on Intellectual Cooperation made as early as 1931, in particular, pursuant to a proposal submitted by the Permanent Committee on Literature and Art [. . . ]. The president of this committee was the Belgian socialist and former minister of the fine arts, Destreé.
>
> Destreé left the Committee on Intellectual Cooperation in 1932, likewise Einstein, so I believe the tendencies finding expression in the now post festum publication of the Einstein-Freund correspondence are finally dealt with. In any case, I for my part will continue to prevent such kinds of things, in which I may be sure of the backing of my Italian friend, the former minister of justice, Rocco. [. . . ].

The history of such letter exchanges goes back to the early days of the Committee on Intellectual Cooperation. Soon after its founding, the committee was asked "to stimulate correspondence between leading thinkers, in emulation of such as have emerged during the great epochs of European history; to select topics that are best suited to serve the general interests of the League of Nations and intellectual life of humanity; and to publish this correspondence from time to time."[715]

The first volume, bearing the title *A League of Minds,* contained letters by Gilbert Murray, Paul Valéry and others. In the fall of 1931 Steinig, an official of the League of Nations, traveled to Berlin to obtain Einstein's commitment to work on a second volume. No specific topic was specified. It was just important that it be a *long original letter by Einstein* intended for publication under the auspices of the League of Nations.

Einstein liked the idea very much. Steinig reported to Bonnet on 29 October 1931: "After Mr. Einstein underscored his interest in education as a means of securing peace, we decided together that he would accept, in principle, writing two letters to two different persons about this question: one letter would probably be addressed to Mr. Langevin, and Mr. Einstein suggested concerning himself here with an exchange of views between two representatives of French and German organizations, for example on the ways and means to influence the content of history textbooks in the two countries. One could, Mr. Einstein thought, continually improve historical accuracy by, on one hand, removing "tendential errors," and on the other, misrepresentations that incited and encouraged the feeling of national animosity. A second letter, written in the form of a questionnaire, should be addressed to Mr. Freud in Vienna [...]."[716]

They ultimately decided on the Einstein–Freud variant.

A communication by Montenach from 2 December 1933[717] to the secretary general of the League of Nations indicates how the planned publication was to be structured. Freud was supposed to send a short memorandum in reply to Einstein's letter containing questions about educational psychology of creating a peaceloving mentality ("la pacification des esprits"). The third step envisioned was a summarizing conclusion by Einstein.

After the first two stages of the work had been completed, Steinig wanted to return to Berlin between 3 and 6 December 1932 to discuss what remained to be done. But Einstein left Germany forever on 10 December 1932. This means that the issue "Why war?" was among the last that preoccupied Einstein inside Germany. It also means that a paper for the International Committee on Intellectual Cooperation was the final touch to his political activism in Germany.

Political developments in Germany determined that only a torso of the work would come into existence, composed of one letter by Einstein from Caputh dated 30 July 1932 and Sigmund Freud's reply from September 1932. It was printed in March 1933 in France.[718]

Einstein discussed with Freud an issue that "in the current state of affairs seems the most important to civilization: Is there a way to free humans from the bane of war?" Technological progress had made this "a question of the very existence of human civilization."

Einstein regarded himself as the person posing the questions and Freud as providing the answers. But the question already suggests Einstein's opinion of where the answer may be found: education as a means to break down psychological barriers. The "external, that is, the organizational side of the problem" seemed to him to be very simple: "the states create a legislative and judicial authority to mediate all conflicts arising amongst them. They commit themselves to subordinating themselves to the laws established by the legislative authority [...]."

In principle, this same idea had been occupying Einstein since the First World War: creating a world government and unconditionally relinquishing some freedom of action or sovereignty as a means to resolve international conflicts of interest and to preserve the peace.

Einstein's assessment of the "undoubtedly earnestly meant efforts of past decades" was discouraging. We are "at the moment far away" from creating such an organization, Einstein wrote; all past efforts had been in vain.

One basic reason for this failure was the ambitions of a small but determined group of people uninhibited by social considerations and restraints, "for whom war, armament and weapons trade are nothing but an opportunity to draw personal gain and extend their personal power." This is where Einstein discerned a clear social or societally in-built reason for war and warmongering.

But this, he wrote, was just a first step toward the answer. Delving deeper, he asked why the masses let themselves be manipulated by a minority for the purposes of war (disregarding those who "have made war into their profession"). His immediate answer is that the ruling minority, including people with a direct interest in war, has "schools, the press and mostly also religious organizations under its control" with which it transforms the "emotions of the general public" into a willing tool. Einstein was aware of the power of the media.

This answer did not completely satisfy him, though. "The question arises: How is it possible that the masses let themselves be incited to the point of self-sacrifice?" The answer "can only be: Humans have a drive to hate and destroy." Einstein explicitly pointed out (from bitter personal experience) that he was not just thinking of the uneducated but also specifically of the so-called "intelligentia," who were "the most susceptible to portentous mass suggestion."

Consequently, the question best answered by a psychologist like Sigmund Freund poses itself: "Is it possible to direct the psychological development of humans to become more resistant to the psychosis of hate and destruction?" Einstein looked to Sigmund Freund for the answer, but the question comprehends that humans are not contemned by fate to hating and murdering each other.

Freud admitted that Einstein had articulated most of the answers himself, actually "taking the wind out of his sails" so he could only "expand upon" what had already been said. Freud also concluded that war could only be prevented "if humans can agree to establish a central authority to which is conferred the final verdict on all conflicts of interest." He also had to concede resignedly that the existing League of Nations had remained toothless, because its members were not willing to relinquish powers to it. He questioned the conviction of many of his contemporaries that a "general penetration of the Bolshevist mentality [... could] put an end to war." The Bolshevist belief that satisfying material needs could "make human aggression vanish" was, he thought, illusory. It was possible to arouse the counterpart to the aggressive drive, Eros, however, and subject the passions to the "dictatorship of reason."

As Freud himself wrote, he was left to swim in the "wake" of Einstein's implied responses. Neither did he know how to impose a world government. He joined Einstein in thinking that – "perhaps"[!] – it was not a "utopian hope that the two impetuses of cultural insight and a legitimate fear of the effects of a future war will put an end to war in the foreseeable future."

It did not take long for the next great war to occur. After this war as well, much effort was put into creating the kind of durable peace Einstein and Freud were envisioning. How successful they have been is an open question. But one thing is sure. Even the threat of extinction of the human race has not got rid of war; it has not helped to victory the "counterpart to the aggressive drive, Eros." And in our day, seven decades after the exchange of letters between Einstein and Freud, we are confronted with the additional problem: What should and ought to be done if the eventual "*world*" government becomes a tool of a *single* state?

It is not the aim of this book to discuss whether Einstein's and Freud's argumentation is still relevant today. Some of the opinions may indeed have been "utopian" and "illusory." Maybe – as the above has suggested – more far-reaching questions have to be posed. If, ultimately, money *still* is reigning supreme, if the power of monopolies along with further monopolization of power still persists today, then the Einstein–Freud correspondence is no less current than it was seventy years ago.

## 2.8 Parting ways. Einstein and the end of the Weimar Republic

### 2.8.1 Apparently "more tranquil and undisturbed" – Einstein's summer villa

The year 1925 marks a turning point in the history of the Weimar Republic as well as in the biography of Albert Einstein. This fact may seem coincidental but Einstein's life-line had meanwhile become so closely intertwined with the history of Germany that trenchant changes in the political situation could not be without repercussions on this private individual as well.

Things seemed at first to have settled down again. In the fall of 1923 the inflation left more quickly than it had come. It was worth investing in Germany again; the influx of funds from abroad was soon correspondingly large (August 1924: the Dawes plan).[719] The more other countries invested in Germany, the lower the risk became that the debtor would go bankrupt. The clairvoyant support of German science now began to pay off: Germany was still a center of glamorous technological feats. The spectacular crossing of the Atlantic by airship in October 1924 is one example. Another is the breathtaking progress that modernization brought German industry.

And the Reich capital Berlin was one of its main centers. In 1926 the major power station in Rummelsburg was completed. In 1923 the developer, Berliner Messe-Aufbau GmbH, was established; soon afterwards the new trade fair center in the west of the city was completed. In 1926 the German General Electric Company (AEG) converted to assembly-line production. Berlin experienced an unprecedented construction boom. "Siemens City," a workers' community for employees of the Siemens electronics company, emerged west of the city. The streetcar system was electrified and the subway system made steady progress. Large new suburban residential areas appeared. The professional and social landscape changed in tandem with the sudden technological changes. The army of white-collar workers gathered strength in industry, trade, finance, etc.[720]

During this period, Germany's international reputation was restored. Germany used the extensive interests USA had in the country adeptly and, where necessary, would draw the Soviet card to bend the will of the Western powers, terrified of fraternization between Germany and Soviet Russia. Rathenau had already shown how successful this tactic was with the Germano-Soviet treaty of Rapallo. The treaty of Locarno, signed in October 1925 between seven (western!) European states, raised Germany's standing in Europe so much that it was able to come forward as a Great Power again. In the fall of 1924 already Germany applied for admission into the League of Nations; two years later the way was clear – on 10 September 1926 Germany became a member of the League of Nations.

In 1928 industrial production in Germany resumed its former level in 1913. A year later Germany was the second industrial power in the world behind USA. Joblessness fell from 25 percent at the end of 1923 to 11.5 percent in 1924 and to

8.5 percent in 1925. Social conditions among the masses improved perceptibly. It was at last possible to get something for one's money again and technological progress offered undreamt of products, like record players, radio receivers, motorcycles and cinema tickets. Once the fear of revolution had subsided, the German government opened the airwaves to the general public. Sports, fashion and film stars dominated the front pages of newspapers. The "roaring '20s" was an exciting era.

Einstein was inevitably much less in the public consciousness than at the beginning of that decade. The relativity rumpus was over and, despite everything, most people still could not understand his theory. There were other famous people for sensation-hungry Berlin and elsewhere: Max Reinhardt, Asta Nielsen, Charlie Chaplin, Marlene Dietrich, and the list goes on. Airship flight plans and trajectories had displaced announcements about Einstein and the theory of relativity in daily news reports. Even the Weimar presidents and ministers found much less use for him by the mid-1920s than at the beginning of that decade. The isolation of Germany dictated by the treaty of Versailles had been broken, the boycott of German science overcome. Germany was once again a major player in the world. What business did Einstein have there?

Einstein himself was glad that his "life has become more tranquil and undisturbed." *Work* was what he wanted to do and he also needed rest.

The attic at Haberlandstrasse 5, what they called the "tower room," was his place of refuge. But this did *not* remain a purely private matter that might have been ignored in accordance with the present book's conception. We owe references to the "tower room" to this circumstance.

Einstein had the attic directly under the roof remodeled into a study with two side rooms, one for use as a lounge and the other for storing his books. These renovations were made without an official building permit. In March 1927 the local police officer called attention to this fact and soon afterwards, on 23 April 1927, the rooms were officially inspected "in the presence of Prof. Einstein and his secretary." On 9 May 1927 the Municipal Building Inspectors forbade use of the rooms, under threat of a fee of 50 reichsmarks or three days imprisonment.[721] But Einstein was also provided the option of applying for a special dispensation by the police chief – which Einstein decided to pursue on 18 May 1927. On 11 June 1927 the dispensation was approved. The grounds read: "The petitioned dispensation [...] is granted for the duration of the use by the current inhabitant *in consideration of the public interest* in his research activities."[722]

Einstein to the chief of police of Berlin, 18 May 1927:[723]

Copy.

Professor
Albert Einstein.

Berlin W. 30, Haberlandstr. 5
the 18th of May 1927

Application for Dispensation

To the
Chief of Police
Berlin.

The Municipal Building Inspectors Office sent me the enclosed injunction, according to which I must vacate the study used by me alone and exclusively as such. I petition that this injunction be annulled.

Substantiation:

(1) The room is sunny - directed toward the east - and spacious, the window opening to the east. Area circa *20* sq[uare] m[eters].

(2) The official sent here by the Building Inspectors Office stated that according to the official listing the room was not supposed to be used as living space. The injunction fails to take into account that the window had been enlarged.

(3) The response by the official to my information that the parallel room in the abutting apartment was being used as a living room was that he was not interested in that. But I have a right to being treated the same as others.

(4) I had to have the space renovated and a separate access to it built from the stairway in order to be able to attend to my studies away from my apartment, in which I have been disturbed far too often. I disbursed a substantial sum for this, which is considerable for an averagely compensated civil servant/university professor.

(5) The room is supposed to be used only by me personally, not, however, by others. Any eventual hygenic defects could only affect me personally; I know, however, that most people in Berlin have to work in much less favorable accommodations.

(6) As a reputable scholar and teacher at the university, I have moral claim to special treatment with regard to my application concerning my study.

In great respect,
sig. A. Einstein.

Enclosure: Building Inspectors injunction

Soon it was a matter of *necessity,* not choice, that he slow his pace. The reckless way in which he and others had been abusing his health over the years took its toll. At the end of March 1928 Einstein suffered a heart attack on the way from the train station to the chalet owned by the industrialist Meinhardt at the upper edge of the town Zuoz (Switzerland), where he was vacationing. (He had additionally taken lodgings in nearby Davos because he had been invited to deliver a lecture there.) He had to wait until that summer before he was fit enough to travel to the Baltic to recuperate in Scharbeutz near Lübeck (which did not prevent him from pondering and calculating there "all day and half the night" as well). Even though he returned to Berlin in October, he still needed some time to regain his former condition.

His fame did not evaporate in the interim, as his fiftieth birthday on 14 March 1929 demonstrated. Reich Chancellor Müller sent him his congratulations with the statement: "Germany looks with pride on its great scholar, who has earned imperishible fame for German science."[724] Einstein replied "with pleasure and emotion," in "profound gratitude for the kind words of high esteem," adding, "It

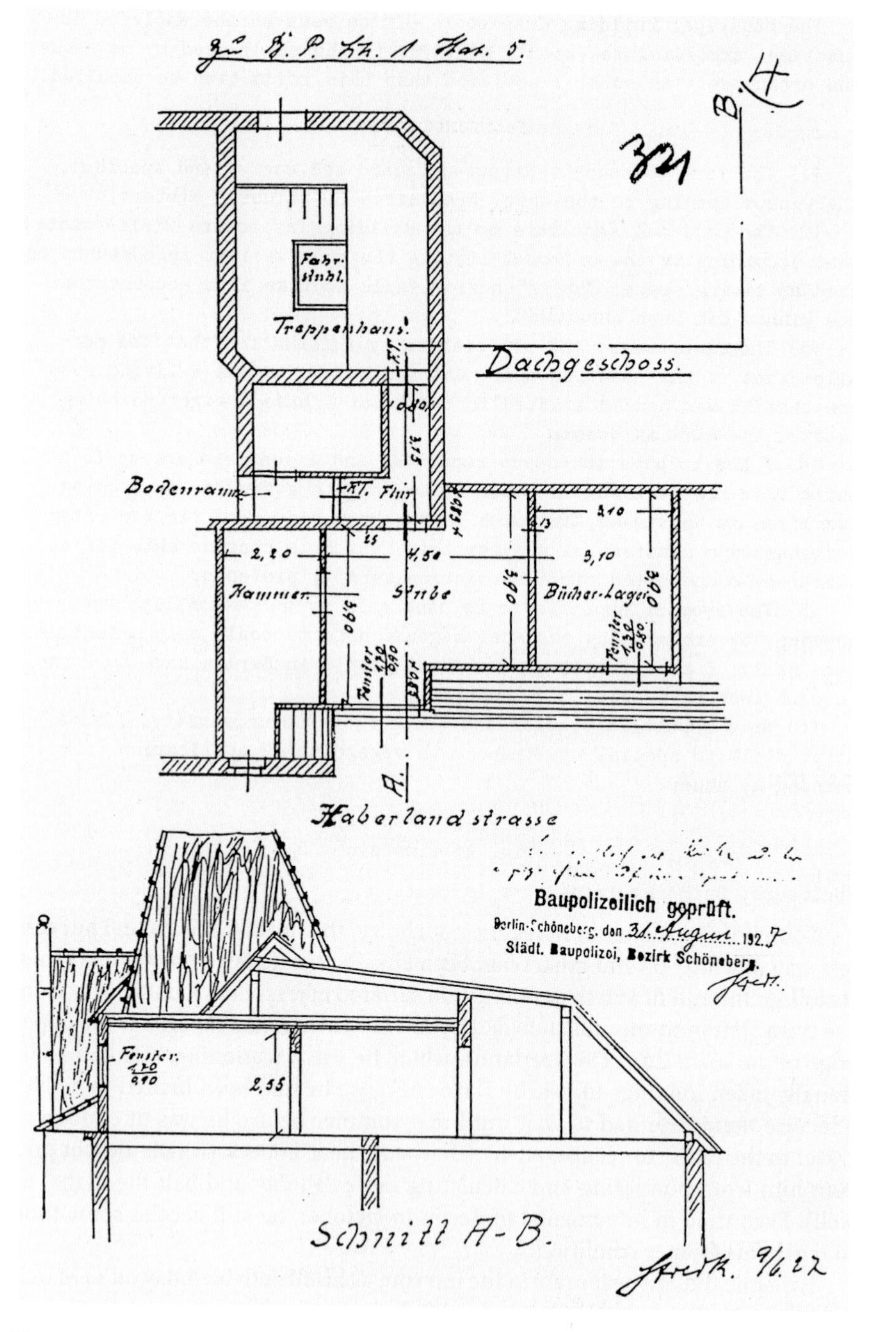


**Figure 35: Einstein's "tower room" – under the scrutiny of the Municipal Building Inspectors.**

is fine proof of the high regard intellectual pursuits enjoy in our country that its governors take such a live interest in someone devoted exclusively to intellectual ideals."

Einstein to Reich Chancellor Müller, 20 March 1929:[725]

Albert Einstein Berlin W., the 20th of March 1929
Haberlandstr. 5

To the Government of the German Reich
c/o Reich Chancellor Müller
Berlin W., Wilhelmstr.

Esteemed Reich Chancellor,

It is both with pleasure and emotion that I express my profound gratitude for the kind words of high esteem you addressed to me. It is fine proof of the high regard intellectual pursuits enjoy in our country that its governors take such a live interest in someone devoted exclusively to intellectual ideals. Just this awareness of the symbolic significance of the acknowledgment bestowed on me gives me the strength to accept such high recognition without a deep feeling of embarrassment.-

In great respect,
A. Einstein.

He spent this day in a villa in Gatow near Berlin as a guest of his doctor Janos Plesch. In the meantime "whole laundry-basket loads of post, telegrams and gifts" had arrived in his apartment at Haberlandstrasse 5.[726] The Prussian Ministry of Culture had purchased a bust of Einstein from the sculptor Isenstein and had it set up in the entrance hallway of the Einstein Tower in Potsdam. The weekly news reel, *UfA-Wochenschau*, covered the event as its lead story.

The City of Berlin was concerned to be among the congratulators. It was proud to count someone of Albert Einstein's calibre among its citizens. So the city decided to present a gift worthy of this honor. After inquiring about what fulfillable wishes Einstein might have, the city decided to present him with the gift of a plot of land. But the city representatives seem to have been pretty late in noticing the significant date. What happened next turned into a farce.

Long after Einstein's fiftieth birthday, on 10 May 1929, the Assembly of City Representatives consulted in closed session on a resolution submitted by the Municipal Council regarding an "honorary gift for Professor Einstein on the occasion of his 50th birthday."

425. Submission [. . . ] for resolution, in re honorary gift for Professor Einstein on the occasion of his 50th birthday.

The Municipal Council has decided to present to the greatest scholar of our century, our fellow citizen Professor *Einstein*, a parcel of land on the Havel on his 50th birthday.

The discussions conducted with Mrs. Einstein in this regard yielded a description of a parcel of land that meets all the great scholar's demands regarding tranquillity, pretty situation, access to the Havel for sailing as a sport, close connections with public means of transportation and convenience for deliverers. The land is located in Caputh, Waldstrasse 7/8, has a fine far view across the Havel and is described as excellently suited to Professor Einstein's purposes. We have obtained a notarized commitment from the seller and shall have a raised foundation added to the gently rising slope to improve the view and also lay gardens.

The building itself will be erected by Professor Einstein. The total costs for the acquisition of the land, landscaping and property sales tax come to about 20,000 reichsmarks.

We petition the resolution that:

The Assembly declare its approval that for the purpose of donating a piece of land in Caputh as an honorary gift for Professor Einstein on the occasion of his 50th birthday, c. 20,000 reichsmarks be made available from funds out of the real - estate purchasing budget.

Berlin, the 24th of April 1929.

Municipal Council
Böss. Busch.

But the Berlin authorities were not as openhanded as the Prussian Constituent Assembly had been ten years earlier, in 1919 (when it had been a matter of approving funds to promote Einstein's research). The municipal committee on the purchase and sale of real estate resolved on 30 April 1929 with reference to printed item 425: "The Committee does not consider itself authorized and asks the Assembly first to pass a resolution on the gift."

Albert Einstein decided to solve the dilemma and declined the city's present on the intention of purchasing the land privately instead. "Einstein refuses Berlin's gift," the *Berliner Lokalanzeiger* reported in its issue dated 15 May 1929. "As we hear, Professor Einstein has refused the birthday gift offered by the Municipal Council [...] with its associated obstacles. In a letter to the chief mayor[727] Einstein requests that it principally refrain from making the donation, because he could not reconcile the constant delays with the nature of a gift." According to the newspaper report, the chief mayor wanted to take measures to have the refusal retracted. The Municipal Council was even of the opinion that the Einstein family would "accept the gift in the end after all." There are no more articles in the *Berliner Lokalanzeiger* that could have informed us of the further developments in the affair.

One can only assume that the "honorary gift by the City of Berlin to Professor Einstein" was never presented. The files on the 'Municipal Council of Berlin, City Treasury/main finance administration' at the State Archive of Berlin[728] contain *no* information about any purchase by the City of Berlin of the parcel.

Herneck's reports may very well agree with the facts:[729] "Opposition to the gift to Einstein intended by the Municipal Council [arose ...] from right-wing, nationalistic and anti-Semitic representatives in the City Assembly." The true reason why the "honorary gift by the City of Berlin to Professor Einstein" remained a fine wish of Chief Mayor Böss was hence less the result of confusion within the Municipal Council of Berlin than *politics*. This was not the last time that the City of Berlin would embarrass itself.

Long before the intended recipient's official refusal, the Einstein couple already regarded as settled the construction of a summer house on that property – whether or not it be an "honorary gift." The notorized commitment by the seller was a sound enough basis.

The practical matters were in the hands of Elsa Einstein, and she certainly had a firm hand. *She* decided to have a picket fence installed at the beginning. When an employee of the Forestry Department pointed out that it was located on state property and not planted along the border of the property, she defiantly declared that it would remain where it was and that the matter could be settled with the minister.[730] And that is what happened.

On 17 April 1930 Albert Einstein applied for the *additional purchase* of a narrow strip of land.[731] Elsa Einstein is not revealed in this letter as the driving force behind the purchase. The true reason for the petition had apparently not been his wife's high-handedness but a sentimental protectiveness toward a stand of "a few fine trees."

Einstein's application was granted; on 24 April 1930 the head forester of the Prussian State approved the sale of a piece of land measuring 16 times 2.40 m.[732]

On the heels of this came a fresh desire to purchase another piece of property because, as the Einsteins thought, the site at the edge of the woods, which had hitherto been at the border of the property, would be the ideal place to build their summer house. The location is clearly indicated in Elsa Einstein's sketch with additions in Albert Einstein's hand. The owner of this property was the *Prussian State* – represented by the Kunersdorf Forestry Department. But the head forester was not amenable to approving Einstein's application of 29 April – for arguments that were not easy to refute: the "considerable augmentation of the fire hazard," the "straight course of the boundary would furthermore be extremely inconveniently disrupted," and it would mean "the start of private land purchases in Caputh that has long been sought from all quarters."[733]

The Prussian government had to make the decision and did so in Albert Einstein's favor. In a quarrel about a deviation from normal practice and yet another annoyance for the great scholar, the decision fell on the first alternative. On 1 May 1929 (hence even before the consultation on 10 May by the Assembly of City Representatives), the Prussian government informed the minister of agriculture, dominions and forestry that Einstein's application should be exceptionally granted.[734] The Department on Taxation and Dominions of the Prussian Government sent the head forester of Kunersdorf corresponding notice on 10 May 1929, along with the information that the construction could begin immediately.[735] A day later the minister informed the government of Potsdam that he approved of the sale of the 228 sq m of state property.[736] Einstein was informed of the purchase price on 22 May 1929 (3 reichsmarks per square meter – without tree felling).[737] He was notified at the same time that the construction could begin.[738] A later surveyal of the purchased property revealed that it was not 228 but 350 sq m, so its price was raised correspondingly.[739]

From this day forward the construction and use of the summer house became a purely *private* matter,[740] thus falling beyond the bounds of this book.

The intentions by the Municipal Council to give Einstein an honorary gift and the favorable treatment he received by the Prussian government document that there was still the political will to bind Einstein to house and property and therefore also to Germany. The time had not yet come when Albert Einstein would be politically undesirable and out of luck. Einstein was *still* a respected and influential personality in the German Reich. A few months later this would all change.

Einstein remained a sought-after man. The mere presence of Albert *Einstein* turned any public occasion into a special event. A petition gained particular importance if it was signed by *Einstein.* A book containing a foreword by *Einstein* was a special book. Not just Einstein was honored but anyone invited to opening ceremonies at a conference or other functions that *he* attended – such as the seventh German Radio Fair in August 1930 in Berlin.

Nevertheless, much had changed since the first half of the 1920s. His continued fame and good personal relations with notable representatives of the State, business and finance made it seem that everything was the same as before. But a new stage of Albert Einstein's life had in fact begun – not just because the social and political *environment* had changed. *Einstein himself* had changed. On one hand he was much more sedentary than earlier. Living like "travelling folk [...], chosen from among the swarm of philistines to dance on the tight-rope,"[741] was no longer his life style. Such phantasies belonged to a time long since past. For a while he lost all interest in travel. At Caputh he seemed finally to have settled down. Not Haberlandstrasse 5 in Berlin, but Waldstrasse 7 in Caputh had become the center of his life and work.

At the same time, his ties to Germany were beginning to loosen. For so long, Germany had wanted to be his fatherland and had, in fact, succeeded in doing so. But soon only the summertime was granted to Germany. Albert Einstein's plan was to spend the winters working in America (it was not his idea eventually to adopt America as his permanent home). But a process of alienation had already gradually begun to creep into his life.

### 2.8.2 Social milieu. Friends and acquaintance

Einstein was being courted by the worlds of science, politics and beyond. In science his opinion was of primary importance. Everywhere else, his *name* was what was of interest. The reputation he had made in science was supposed to be capitalized on otherwise elsewhere. The intended purpose obviously could collide with the man's convictions. But often a very rudimentary congruency was all that was needed. If his arguments were useful to the cause, all the better. Nevertheless Einstein's name came first and foremost, with inquiries about his opinion coming only as an afterthought, if at all.

**Figure 36: Fritz Haber (1930)**
**Fritz Haber for Albert Einstein's fiftieth birthday:[742] Dear Einstein!**
**From among the great things I have experienced in this world, the substance of your life and work grips me most deeply. In a few hundred years, the man on the street will know our times as the period of the World War, but the cultivated person will connect your name with the first quarter of this century, just as today the former remembers the wars of Louis XIV at the close of the 17th century as the latter remembers Isaac Newton. Then from each of us so much will remain as connects us with the larger events in time and, I should think, I should not remain unmentioned in your biography of any sufficient detail, as your partner in making more or less perspicatious comments about the Academy's business and in drinking more or less good coffee together. Thus I serve posterity and my personal conservation in history by requesting from my heart, on your fiftieth birthday, that you take care of yourself and thus stay healthy so that I can continue to joke around with you and drink coffee and silently be vain about being able to count among the group sharing life with you in a more intimate and personal sense.**
**Your friend,**
**Fritz Haber**
**14 III 29**

**Figure 37: Photos taken during Einstein's opening address at the 7th German Radio Fair in Berlin on 22 August 1930. (Three years later Josef Goebbels inaugurated the German Radio Fair at the same place in the presence of Adolf Hitler. Einstein's hope that technology would have a democratizing and pacifying effect proved to be a noble illusion.)** ➠

**Figure 38: A prominent foursome: A. Einstein, Prussian Minister of Culture Becker, the English author H.G. Wells, and President of the Reichstag, Paul Löbe, in Berlin 1929.**

Prof. J. Elbogen, letter from 10 March 1929:[743]

> You, esteemed Mr. Rosenthal, raise the question of Einstein's view of the world and so I would like to take the liberty of calling to mind a reminiscence. As a loyal colleague of the unforgettable Health Official Bradt, it is surely known to you that the Academy of the Science of Judaism only survived its most difficult period through its ability to send out publicity letters bearing Albert Einstein's name. At that time, nobody asked whether Einstein was a nationalistic or a religious Jew; in those days his name was very useful to us [. . . ].

Einstein knew this:

> So use me as you see fit [. . . ] but know that I am entirely aware that in this capacity I am more an object than a person.[744]
>
> I was not needed for my skills, of course, but only for my name. They anticipate from its promotional power considerable success among our rich fellow clansmen of Dollaria.[745]

Einstein's main purpose was to serve as "a famed bigwig and decoy-bird."[746]

Sometimes his own opinion was expressedly *not welcome*, particularly since one was never quite sure what he would say. That was why Blumenfeld warned the organizer of the trip to America in 1921, the Zionist Weizmann: "Please [...]

be very careful. Einstein [...] sometimes naïvely says things that we find unwelcome." So Weizmann preferred to do the talking himself. It probably suited him perfectly well that before an assembly of about eight thousand listeners on 12 April 1921, Einstein limited his words to: "Your leader Dr. Weizmann has spoken, and he has spoken very well for us all. Follow him, then you are doing the right thing. This is all I have to say."[747]

As long as knowledge about Einstein's significance was confined to professional circles, his public sphere remained within the "small world" of science. This was characteristic of his life during the Kaiserreich. What's more, a few months after his arrival in Berlin, war broke out and most of his friends and acquaintance were out on the battlefield. Secondly, Einstein was still preoccupied with settling his family affairs. Thirdly, between 1915/1916 he was gripped by a fit of creativity that left little time for anything else.

This changed as quickly as his reputation reached beyond scientific and national boundaries. That is, in a very short time. Einstein became a popular personality. Inviting him became the stamp of high society. Einstein the icon; Einstein the idol. He was even chosen to open the First International Congress on Sexual Research and play the violin part in Brahms's string quartet op. 18.[748]

Count Kessler noted on 4 February 1921 in his diary: "Première of 'Joseph' conducted by Strauss. Enormous, almost unheard-of success. The hall was a veritable 'tout Berlin' [everybody who was somebody was there]; the Reich chancellor, Simons, Seeckt, many ministers, all the leaders in society, art and literature, Albert Einstein, etc., etc. The applause at the end did not want to end."[749] Einstein, a fine show piece of the Reich capital.

The circles in which the Einsteins moved as private individuals give indications of their social milieu, particularly their invited guests to their apartment. One cannot conclude that the Einsteins' world was identical to other people's but certainly that there was no gaping chasm between them. Anti-Semites would not have invited them (nor vice-versa). Social interaction is, as a rule, conducted among persons wishing and able to talk to each other with a consequent mutual influence on each other.

So who were Einstein's guests during the early 1920s? Whom did the Einstein couple open their door to?

On 20 March 1922 Kessler noted: "Ate in the evening at Albert Einstein's. Quiet, pretty apartment in the west of Berlin (Haberlandstrasse 5), a slightly too copious and mass-production diner, lending this truly dear, almost childish couple a certain naïvité. The extremely wealthy Koppel, Mendelssohns, President Warburg, Bernhard Dernburg, shabbily dressed as always, etc. Radiant kindness and simplicity transported even this typical Berlin society from the conventional and illuminated them with an almost patriarchical, fabulous aura."[750]

So at least *two* persons were among them who had initially supported Einstein: the banker and major industrialist *Koppel* and the president of the Bureau of Standards (PTR) *Warburg*. It is not clear which of the two *Mendelssohns* was present, either Erich the banker or Franz von Mendelssohn, president of the German Industrial and Commercial Convention.[751] Bernhard *Dernburg* was, at that time, secretary of state at the Reich Colonial Office, later Reich minister of finance and at the time of this visit member of the Reich parliament, on the parliamentarian Committee on Foreign Affairs and also member of the working committee of the German League for the League of Nations founded in 1918. "President Warburg" presumably refers to *Emil* Warburg (possibly *Max* Warburg – in 1919 member of the German delegation for the peace negotiations at Versailles and at the time of this visit on the central board of the Reichsbank). Thus all influential or very wealthy people greatly interested in the success of Einstein's imminent "Frenchmen's trip," who must have exerted some influence on Einstein as well (as they could legitimately expect, being among Einstein's closer acquaintance). Incidentally: all the persons Kessler named were Jews.

On 14 June 1926 Kessler noted: "Dinner at Sammy Fischer's with Gerhart Hauptmann, the Einsteins, Hugo Simon, Jessner, Kerr couple, Secretary of State Hirsch, etc. After the meal, conversation with Hauptmann and Einstein [...]."[752] Thus a writer (Hauptmann), a drama critic (Kerr), a theater director (Jessner), a banker and patron of the arts (Simon) as well as a politician of the Social Democratic Party. As before, the majority were Jews (Fischer, Kerr, Simon, Jessner, Hirsch and others).

At the beginning of the 1920s many people and institutions were courting Einstein, but his apartment could not yet be called a "salon." His many trips abroad made Haberlandstrasse 5 more a temporary residence than a permanent meeting place for social occasions. Before that could happen the political situation and Einstein's own personal situation had to settle down. By the middle of that decade things had developed favorably. Einstein became sedentary, and his living quarters became a place of vibrant intellectual life.

The *social class* Einstein evidently felt he belonged to was the middle class, with a stronger affinity for bankers than for the lower echelons of society. *Professionally,* he was very closely connected with the most renowned and influential scientists of Berlin. The republic of letters extended into the private sphere; no distinction was made between "private" and "public." Otherwise, the Einsteins generally mostly moved among *writers, artists and medical doctors*. They evidently made the pleasantest and most stimulating company. At the same time they were a relieving counterbalance to the sphere of physics, which only few people were really versed in. The Einsteins mingled among social groups that were positioned toward the *political center, more socialist left-of-center than middle-class conservative*. The gamut was broad: ranging from Planck, who was virtually a follower of the kaiser, to liberal Laue and the communist Ernst Toller. Despite this variety, an atmosphere of tolerance and cultivated debate dominated. It goes without saying that Jews assumed an important place among Einstein's friends and acquaintances. However, besides the Jew Moritz Katzenstein, the German Gentile Max von Laue could also call himself one of Einstein's very best friends. On the whole, Einstein's apartment and summer villa developed into a "salon" of the highest calibre, a political, cultural and scientific point of crystallization in Germany of the 1920s. It was far from being the refuge of a hermit.

**Figure 39: "Two people of world fame ..." – the first encounter between Einstein and Tagore (from: *Berliner Zeitung,* 26 Sep. 1926).**

But despite this closer interweaving into the social fabric of the Reich capital, despite the hectics of city life, the Einsteins embarked on a gradual process of *alienation*. The serious illness in 1928 did its bit, certainly also their summer retreat in Caputh. But the refuge in Caputh was less the cause than the *result* of a growing distancing from life in Berlin. This distancing was consciously intended. Einstein purposefully had no telephone installed in his villa, for instance. (He could still be contacted through his neighbors, but only when he deemed it necessary.) The spatial remoteness also reduced the numbers of visitors, guaranteeing a quieter life style.

After the tumultuous 1920s were over, Einstein's milieu became noticeably more sedate. His participation in the "great world" of Berlin relaxed; the intensity of his government contacts diminished. German foreign policy no longer had any need for the prominence his name could evoke. Einstein even became less involved in the scientific life of Berlin. From 1930 on, he also started to travel more again and thus during his strongest bondage to his beloved Caputh, he was already preparing the way for his later (originally unintended) exile from Germany.

Einstein's attendance at the meetings of the Berlin academy reveals the following pattern:

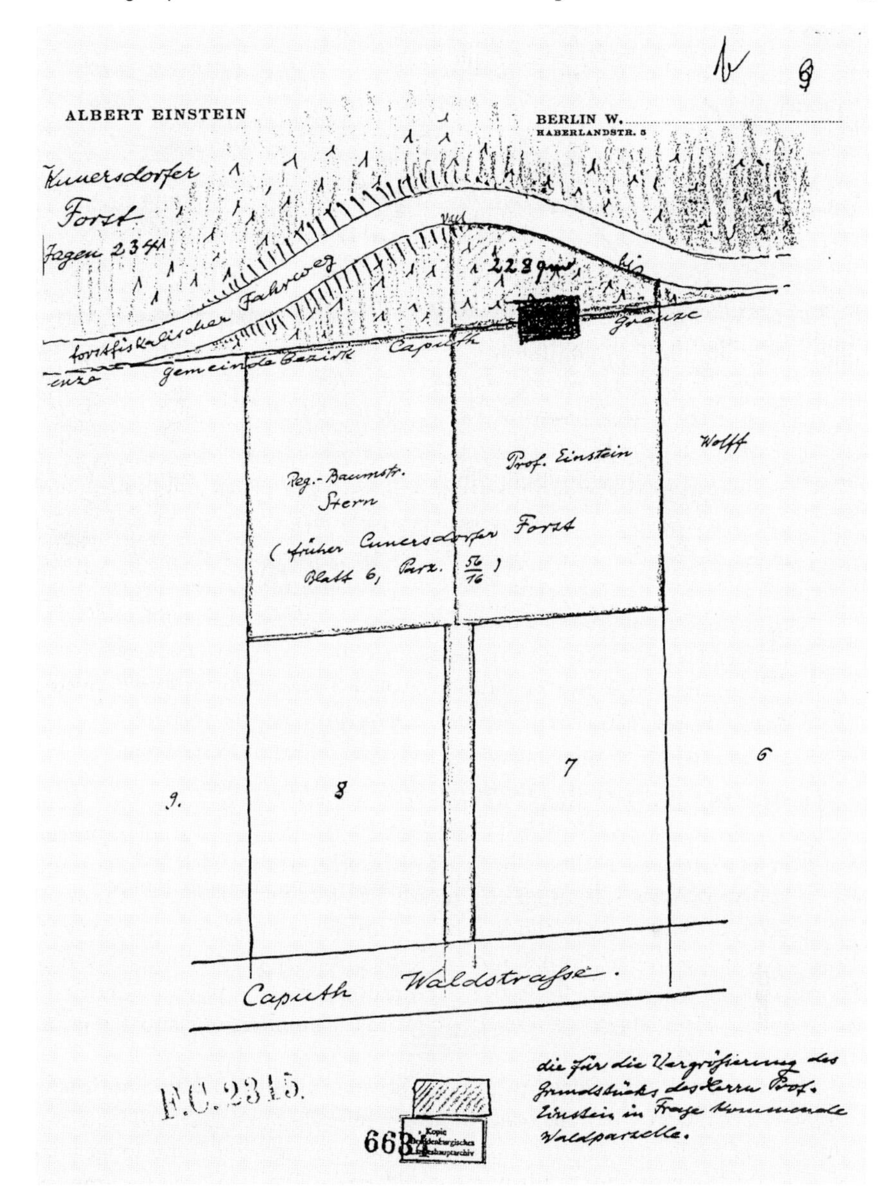


**Figure 40: A drawing of the parcel of land in Caputh with the planned additional land purchase.**

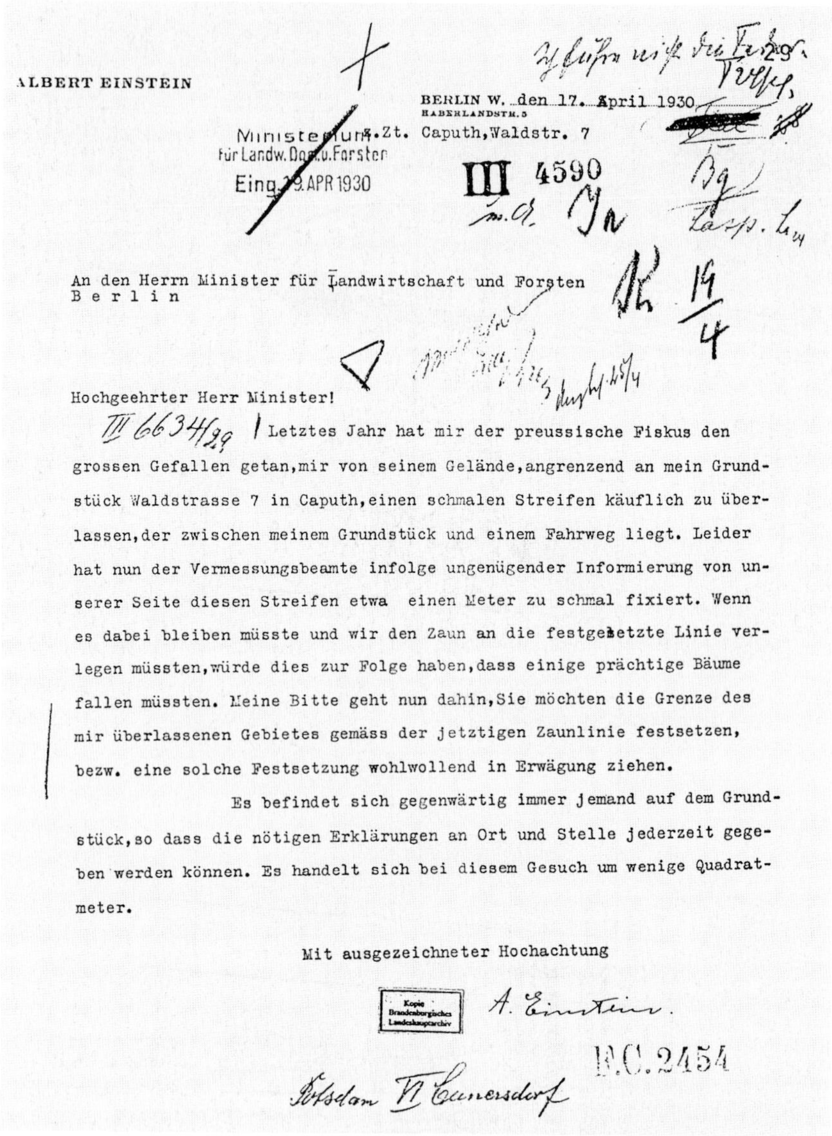

ALBERT EINSTEIN

BERLIN W. den 17. April 1930
HABERLANDSTR. 5
z. Zt. Caputh, Waldstr. 7

Ministerium für Landw. Dom. u. Forsten
Eing. 19. APR 1930

III 4590

An den Herrn Minister für Landwirtschaft und Forsten
B e r l i n

Hochgeehrter Herr Minister!

III 6634/29 Letztes Jahr hat mir der preussische Fiskus den grossen Gefallen getan, mir von seinem Gelände, angrenzend an mein Grundstück Waldstrasse 7 in Caputh, einen schmalen Streifen käuflich zu überlassen, der zwischen meinem Grundstück und einem Fahrweg liegt. Leider hat nun der Vermessungsbeamte infolge ungenügender Informierung von unserer Seite diesen Streifen etwa einen Meter zu schmal fixiert. Wenn es dabei bleiben müsste und wir den Zaun an die festgesetzte Linie verlegen müssten, würde dies zur Folge haben, dass einige prächtige Bäume fallen müssten. Meine Bitte geht nun dahin, Sie möchten die Grenze des mir überlassenen Gebietes gemäss der jetzigen Zaunlinie festsetzen, bezw. eine solche Festsetzung wohlwollend in Erwägung ziehen.

Es befindet sich gegenwärtig immer jemand auf dem Grundstück, so dass die nötigen Erklärungen an Ort und Stelle jederzeit gegeben werden können. Es handelt sich bei diesem Gesuch um wenige Quadratmeter.

Mit ausgezeichneter Hochachtung

A. Einstein.



F.C. 2454

Potsdam VI Cunersdorf

**Figure 41: Not quite voluntarily: Application for the additional purchase of a "narrow strip of land." Einstein to the Prussian minister of agriculture and forestry, 17 Apr. 1930.**[753]

**Figure 42: Elsa Einstein, Albert Einstein and *his beloved sailboat, "Tümmler"* [porpoise] – a gift from wealthy friends on Albert Einstein's fiftieth birthday.**

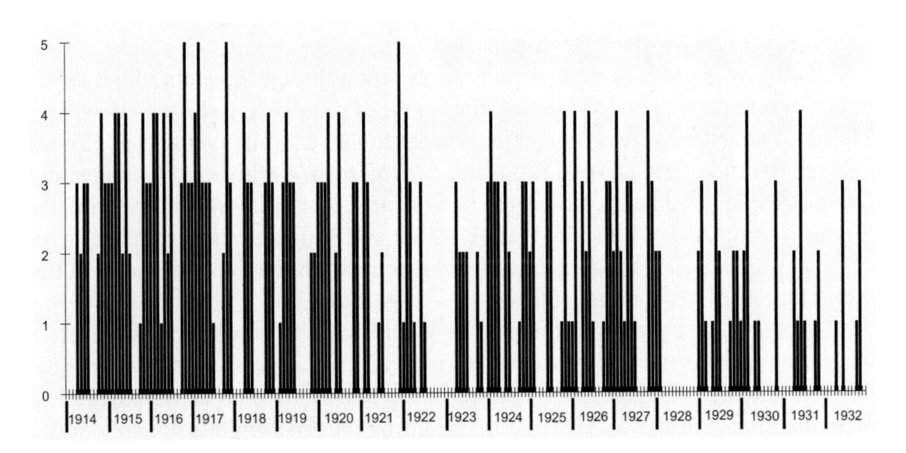


**Figure 43: Einstein's attendance record at the Berlin Academy of Sciences (no. of days each month on which he attended plenary sessions, meetings of the phys.-math. class and the Secretariat) set against the number of meetings per month each year (Statistical basis = *Einstein in Berlin 1913–1933*, Part II, Akademie-Verlag, Berlin, 1979).**

Einstein's attendance record at the Prussian Academy of Sciences

| year | 1914 | 1915 | 1916 | 1917 | 1918 | 1919 | 1920 | 1921 | 1922 | 1923 |
|---|---|---|---|---|---|---|---|---|---|---|
| days per yr. | 21 | 30 | 32 | 32 | 20 | 18 | 25 | 15 | 13 | 12 |
| year | 1924 | 1925 | 1926 | 1927 | 1928 | 1929 | 1930 | 1931 | 1932 | |
| days per yr. | 25 | 17 | 23 | 23 | 4 | 17 | 12 | 11 | 8 | |

Ignoring the fluctuations, the general trend is a flagging participation in the academy's activities.

The reasons are many. Einstein's serious illness in 1928 prevented him from attending the meetings of the academy for almost an entire year. His *need* to attend such meetings had also diminished, however. "Einstein loved life in the country so much that he began to skip the Physics Colloquium and even the weekly Academy meetings and only appeared at the Academy when Planck or he himself submitted a paper, and both happened infrequently."[754] Whether or not excusable, such frequent absences weakened his bonds there, causing a growing estrangement and eventual loss of interest.

One sign that he was reorienting himself intellectually away from Berlin was the resumption of his foreign travels since 1929 (he still had time *for that,* rarely so for academy meetings).

Einstein's growing isolation can also be explained by the fact that he lost more and more of his friends and closer acquaintances to old age, sickness and death. Formerly one of the youngest members, Einstein had become older. Only in rare cases did he see eye to eye with the younger generation. On 4 February 1928 his friend Hendrik Antoon Lorentz died. Einstein described himself as his "pupil and loving admirer." A few days after speaking these words at the grave of his fatherly friend,[755] Einstein fell seriously ill himself, coming close to death's door. In 1932 his friend Moritz Katzenstein died.

Finally we must take into consideration that although Einstein's high reputation in science was undiminished, the center of physics had shifted elsewhere. The latest novelties were no longer incarnated in Einstein but in quantum physics and scholars like Heisenberg, Bohr and Schrödinger. Berlin was no longer the German capital of physics. Göttingen had become more important. The "mighty fortress" composed of Einstein, Planck, Haber and von Laue also had its negative sides: new advances had to emerge *elsewhere*. Einstein fought a hard, and in the view of many of his younger colleagues, a futile battle.[756]

Einstein had a premonition that he would have to leave soon. In a letter drafted as early as 17 July 1931 to Max Planck, which was presumably never actually sent, he mentions his intention to give up his German citizenship acquired after the First World War. He was still planning on retaining his position at the Academy of Sciences. The reason he gave for this wish was "a concern for the numerous people financially dependent on me as well as [...] a need for personal independence."[757]

A year later it seemed that the schedule for the next stage in his life had been worked out and finalized. Einstein committed himself to working every winter semester at the research institute founded by Professor Abraham Flexner in 1930 at Princeton. He informed the ministry of this in writing on 13 September 1932. He inquired about "the extent to which my employment at the Academy of Sciences is reconcilable with this new situation." Senior Civil Servant Dr. Rottenburg responded to his letter on 17 September 1932 on behalf of Secretary of State Lammers, requesting a personal meeting to discuss the matter on 23 September.[758] Einstein's question about his membership in the academy was clearly *not* at issue. Von Rottenburg emphasized: "*Our discussion will go on the premise that you will retain your position as regular member of the Pruss. Academy of Sciences.*" We may presume that Secretary of State Lammers[759] thought likewise.

Albert Einstein to von Rottenburg, senior civil servant at the ministry, 13 September 1932:[760]

Albert Einstein Caputh near Potsdam, the 13th of September 32

To the Ministry of Science,
Art and Culture
c/o Senior Civil Servant von Rottenburg
Berlin NW.
Unter d[en] Linden

An inquiry by the Ministry of Science, Art and Culture of the Prussian Academy of Sciences regarding obligations of a contractual nature that I have assumed at the newly founded research institute of Professor Abraham Flexner in Princeton, gives me formal occasion for this letter. I already had the intention earlier, however, after consulting with Professor Plank,[761] and following his advice, to contact the Ministry.

The commitments I have entered into must prompt the question, of the extent to which my employment at the Academy of Sciences is reconcilable with this new situation.

It is clear that I shall have to publish a part of my scientific research in America as well. I have also obligated myself to be in Princeton every winter semester as of the winter of 1933. These obligations are naturally incommensurate with the conditions of my employment as member of the Pruss. Academy of Sciences, in which capacity I draw the salary of a full professor without teaching obligations. The question must therefore be raised whether maintaining my employment at the Academy is at all possible or desirable under the new conditions. Only in the event of a positive reply to this question would a rearrangement of the conditions of employment be necessary.

In great respect,
A. Einstein.

It is possible that Einstein wanted to demonstrate his *commitment* to the academy by attending *five* meetings of the academy between October and December – the last time on 1 December 1932. On 10 December he left the country to fill an appointment as guest professor at Princeton University in New Jersey, U.S.A. He could not know that this departure would be his last.

**Figure 44: Moritz Katzenstein (1872–1932) – he and Max von Laue: Einstein's two best friends.**

Two things still attached him very strongly to Germany: his colleagues and *his summer villa*. Elsa Einstein's letters document this: "Albert [...] has adjusted himself to Caputh, entirely! He lives here divinely as nowhere else. He tells me, too, no one can convince him to leave";[762] – "Caputh, where my husband was the happiest!"[763]

The Nazis made sure that Einstein left Germany *forever*.

### 2.8.3 Political developments: Republic moves to the right, Einstein to the left

During the first half of the 1920s the institutions of the German Reich, particularly the government and the Foreign Office, *welcomed and encouraged* Albert Einstein's travels outside the country. Einstein contributed importantly toward breaking Germany's political isolation specifically respecting the boycott of German science. They pressured Einstein into formally recognizing his Prussian citizenship because it was a matter of importance that he be *German*. When the Reich gained admission into the League of Nations and the boycott of German science fell apart, interest in Einstein's lecturing tours waned perceptibly, although the political benefits from the trips were still very welcome.

As we have seen, Einstein's membership in the *Committee on Intellectual Cooperation of the League of Nations* was *not particularly appreciated or supported,*

**Figure 45: Adolf von Harnack (7 May 1851–10 June 1930).**

*but it was at least tolerated* by the German institutions in the early 1920s. In this case they preferred not to consider Einstein a representative of the Reich. On the other hand, there was neither any interest nor any occasion to appoint another German scholar in Einstein's stead. When membership in the League of Nations became imminent, however, and participation on the committee of the League of Nations had prospects as a diplomatic tool, everything was done to elbow Einstein out and replace him with a politically acceptable substitute.

As far as Einstein's political activities *inside Germany* were concerned, the first signs that the institutions of the Reich were *distancing* themselves from Albert Einstein became visible much earlier and more directly. Einstein's deployment on the domestic front was but brief (as brief as the call for a "socialist Germany"). Just a few years after the November Revolution, Einstein was already being regarded at home as a political opponent and an alien – not just by members of the Free Corps, the Nazis and uninhibited anti-Semites. The escalation that took place during Einstein's emigration had its beginnings in the early Weimar period. All the same, the political and economic consolidation of the Weimar Republic during the *mid-1920s* became another turning point.

Einstein's membership and collaboration in the following organizations were of decisive importance:

**Figure 46: Five Nobel laureates still cheerfully together at table in November 1931. Nernst (1920), Einstein (1921), Planck (1918), Robert Andrews Millikan (1923) and Laue (1914). Einstein to Laue: "But I do often think that the small group of people, who were earlier so harmoniously attached to each other, really was unique."**

1. German League of Human Rights (formerly New Fatherland League)
2. Society of Friends of the New Russia
3. The board of trustees of International Workers Relief

Everything started with his membership in the *New Fatherland League* (*Bund Neues Vaterland*) dating back to 1915 and renewed membership in 1918. This is where Einstein's political roots lay. This is where he found his earliest confederates as a political activist. From among the many people who wanted something from him he was most receptive to requests for help coming from them. Einstein's membership in the other organizations came later: 1923 and 1926, resp. In 1926 Einstein was subscribed to all three organizations, but the emphasis of his political activities moved increasingly in the direction of Red Aid (*Rote Hilfe*).

A new trend in Einsteins political activism emerged toward the end of the 1920s. His attachment to the organizations listed above loosened somewhat. Instead of just signing articles drawn up by others, Einstein came forward increasingly frequently to put his name under political statements he had *drafted himself.* These declarations could no longer be simply attributed to or explained by his membership in one of these organizations. They particularly involved statements about the issue of disarmament and peace.

Another characteristic of Einstein's political involvement during the 1920s was the growing public attention his opinions attracted (in contrast to his polit-

**Figure 47: Soon, *in their name,* it would be proclaimed: *"The Prussian Academy of Sciences has* [...] *no reason to regret Einstein's resignation"*: The four secretaries of the Academy of Sciences; from left to right: Heinrich Lüders, Ernst Heymann, Max Planck, Max Rubner – 1932.**

ical activity during World War I). The media looked after him; Einstein became a media star. His opinion gained importance, especially in *mass organizations* with large memberships from the working class. The societal base of his influence grew broader. For this reason he was a risk factor, if not an outright threat to the political class, not solely from the intellectual content of his statements. Consequently, Einstein became the object of police observation.

Another phenomenon of political relevance in Albert Einstein's life that augmented over time was his activism on behalf of his "Jewish clansmen," particularly his advocacy of the World Zionist organization. In a certain way this activity moved *in tandem* with Einstein's membership and involvement in the German League of Human Rights (GLHR), the Society of Friends of the New Russia, and Red Aid. "*In tandem*" means it did not emerge directly from them but drew from another source. As a Jew in Germany during and after World War I, Einstein became increasingly aware of how severely Jews were being discriminated against as a minority in the German milieu. Hence Einstein, the alien, foreigner – and Jew.

*German League of Human Rights (formerly New Fatherland League)*

The *New Fatherland League* (NFL), banned by the High Command of the Official Seals on 7 February 1916, became active again at the end of September 1918, when "the labor pains of the approaching German revolution became perceptible."[764] In decrees dated 30 October and 4 November 1918, the High Command of the Official Seals finally lifted all existing decrees against the NFL.

The revived league sent the following demands to the Reich chancellor and the ministers in a number of telegrams to be presented to the assemblies: the immediate release of political prisoners, the introduction of the right of assembly, freedom of speech and of the press, countermeasures against the inculcation of militaristic attitudes in children's education, the appointment of a legislative national assembly with equal, secret and direct voting rights also for women and soldiers, the reassignment of key positions of the State as well as the "extinction of human poverty by socialization of the means of production." On 10 November 1918 Magnus Hirschfeld proclaimed the New Fatherland League's demand: to establish a socialist society. It was saluted by machine guns of the kaiser's officers.[765] Some of the earliest and most active members of the league were subsequently appointed to important posts: the banker Hugo Simon was made administrator of the Prussian Ministry of Finance, Hellmut von Gerlach was appointed to the Prussian Ministry of the Interior, Count Arco to the Ministry of Commerce, Kurt Eisner became prime minister of Bavaria. "The former secretary Ernst Reuter was arrested in Russia and organized 700,000 Volga Germans near Saratow as a socialist commune."[766]

The New Fatherland League's new agenda[767] called for the following:

- Implementation of socialism and centralization of the production of goods through systematic transferral of the means of production into the property of society as a whole.
- Elimination of military and class dominance, the infusion of democracy and socialism and securing a minimum standard of living. Engaged defense of all economically weak individuals.
- Political democracy. Sovereignty of the people, complete equality of all fellow comrades, without sexual, national, racial or religious discrimination. Protection of minorities. Complete freedom of the press, of speech and religion, and full rights of association and assembly. Democratization of security organs.
- Cultivation of individuality, marital and educational reforms. Selection and fostering of the specially gifted. Emancipation of the arts and sciences.
- Collaboration on international reconciliation. The triumph of justice over force in international relations.
- Elimination of militarism and of compulsory conscription.
- Supranational organization. Right of national self-determination. Establishment of a supranational league of nations with a world parliament, international tribunal and executive powers against rogue states.

This virtually "communist" agenda bore *Albert Einstein*'s signature as member of the main committee. (Other signers included: Magnus Hirschfeld, Max Wertheimer, Count Arco, Käthe Kollwitz, Walther Schücking and Heinrich Mann).

One of the first political actions Einstein was known to undertake after the end of the monarchy was on behalf of the arrested officials of the University of Berlin, including its president, at the end of 1918.

Max Born recalled:[768]

When the German Supreme Command suddenly capitulated towards the end of 1918 and the revolution broke out all over Germany, I was in bed with influenza and thus only witnessed the events in Berlin from a distance. Just after I had recovered, Einstein rang me up (the telephone functioned even during the wildest days) and reported that a student council had been formed at the university, modelled on the workers' and soldiers' councils (German soviets). One of its first actions had been to depose and lock up the Rector and some of the other dignitaries. Einstein, because of his left-wing political views, was believed to have some influence with the more radical of the students, and he was asked to negotiate with the 'council' in order to bring about the release of the prisoners and the restoration of reasonable order. Einstein had discovered that the student council met in the Reichstag building, and asked me whether I would accompany him. I accepted despite the weak state I was in after my bout of influenza. [. . . ] then three of us - Einstein had asked the psychologist Max Wertheimer to come too - went by tram to the Reichstag. I will not go into the difficulties we had in penetrating the dense crowds which surrounded the Reichstag building [. . . ]. Eventually someone recognised Einstein, and all doors were opened.

Once in the Reichstag building, we were escorted to a conference room where the student council was in session. The Chairman [. . . ] asked us to [. . . ] wait [. . . ]. So we patiently waited [. . . ]. Then our business was discussed; but the student council decided that it had no authority in the matter, and referred us to the new Government in the Wilhelmstrasse, issuing us with a pass for this purpose.

Accordingly we walked on to the Reich Chancellor's palace. [. . . ] Einstein was recognised at once, and we had no difficulty in getting through to the newly appointed President Ebert, who [. . . ] said that we would appreciate that he was unable to pay attention to minor matters that day, when the very existence of the Reich itself was in the balance. He wrote a few words on our behalf to the appropriate new minister, and in no time at all our business had been concluded.

We left the Chancellor's palace in high spirits, feeling that we had taken part in a historical event and hoping to have seen the last of Prussian arrogance, the Junkers, and the reign of the aristocracy, of cliques of civil servants and of the military, now that German democracy had won. [. . . ].

In those days we believed in the triumph of reason, of the 'brain'. We had yet to learn that it is not the brain which controls human beings but the spinal cord - seat of the instincts and of blind passions. Even scientists are no exception to this.

19 May 1919 Einstein signed a telegram together with Ernst Cassirer, Adolf Grabowsky, Hugo Haase and Maximilian Harden that asked the courts to delay the proceedings against Eugen Leviné until a time when the political atmosphere was less charged.[769]

The communist, Eugen Leviné, was the leader of the second Munich Soviet Republic, proclaimed on 14 April 1919. He abdicated on 27 April. On 1 May 1919 troops supporting the old government marched in and there was a blood bath among adherents of the Soviet Republic. On 13 May Leviné was arrested, sentenced to death on 3 June and shot on 6 June.

The minutes and publications of the New Fatherland League (*Bund Neues Vaterland*, since 1923 German League of Human Rights, GLHR) give information about Einstein's activities:

- "16 December 1919. For the first time since the war, an 'enemy' foreigner: Paul *Colin* from Paris speaks in Berlin before a large public assembly with Prof. Albert Einstein, who is one of the oldest members of the league."[770]
- 1921:[771] "14th/15th February. Trip by Prof. Albert Einstein, Harry Count Kessler and Lehmann-Russbüldt to the International Confederation of Unions to motivate it to take up the problem of reparations for Europe."[772]

Count Kessler wrote about this in his diary:

Berlin. 10 February 1921. Thursday. Lehmann-Russbüldt asked me early on whether I wanted to travel to Amsterdam with Albert Einstein by commission of a group of pacifists, to open contacts with the International Confederation of Unions about the Parisian resolutions. The people behind this mission include Eduard Bernstein, Gerlach, Walther Rathenau, Heinrich Ströbel, Hugo Simon etc. Therefore, 'New Fatherland League' [. . . ].

Amsterdam. 14 February 1921. Monday. Early morning in Bentheim border controls. Einstein, who, as it seemed, was traveling for the first time in a sleeping-car, observed everything with extreme interest. [. . . ] at the International Confederation of Unions, which we called on at three. [. . . ] Fimmern admitted to me that I theoretically was entirely right. But practically, the International Confederation of Unions lacked the power to implement these demands at the present time [. . . ]. Also, the moment for a test of strength by workers associations was very unfavorable, because unemployment was prevalent everywhere. [. . . ]. Otherwise, we could rest assured that he and Oudgaast would do everything to help us Germans and German workers. [. . . ].

15 Februar 1921. Tuesday. In the morning together with Einstein in the Reichsmuseum.[773]

Kessler's diary entry sounds as if *only* Kessler had done the talking and Einstein just sat there and listened. It may have been the count's self-embellishment but it was just as much a reflection of the real situation. Einstein was there for the show, once again being used as a "famed big-wig."

- In February 1922 Einstein was among the co-signers of the appeal issued by the League of Human Rights in France and Germany (the latter then still under the name New Fatherland League): 'To Democrats of Germany and France.'[774]

The appeal included the following demands:

Reparations for damages caused by Germany to its neighbor country, disarmament in Germany and France.

"In the interest of human civilization between both nations, mutual relations should be revived not just among the proletariat, industry and trade of both nations but also between representatives from the *sciences and arts*."

"Finally and above all, the German and French peoples must realize that the genuine foundation of a lasting peace is a *League of Nations* composed not of governments but of *nations* and France must declare its agreement that a democratic Germany will be accepted in this League of Nations."

Einstein actually practiced what he preached – for instance, just a few days later he departed on his "Frenchmen's trip."

- The GLHR counted Einstein's lecture tour to Paris in March 1922 as one of *her* activities within the framework of Franco-German dialog. ("Furthermore, end of March: Presence of Prof. Albert Einstein in Paris").[775]
- 1922: 15 July "Prof. A. Einstein, H. von Gerlach and Count Kessler raise protest against the social revolutionary process in Moscow."[776]
- 1923: "17 Jul. Major lecture in Berlin, 'Pro and contra the League of Nations' [...] Telegram with reference to letters by Fritj[of] Nansen, President of the Reichstag Löbe and Albert Einstein to the Reich Chancellor and Foreign Minister."[777]
- According to the minutes of the general assembly of the club of social sciences, *Sozialwissenschaftlichen Club e.V.*, dated 28 January 1924,[778] thirty-three ladies and gentlemen had themselves enrolled, *Albert Einstein* among them (along with Lehmann-Russbüldt, Harry Count Kessler, H. von Gerlach, Eduard Bernstein, Fritz Wolff, Dr. Kuczynski).[779]
- 1924: "30 Jun. Prof. Albert Einstein's personal audience with Reich Chancellor Dr. Marx about the fate of Erich Mühsam[780] and other political prisoners in Niederschönenfeld."[781]

Einstein to Reich Chancellor Marx, 25 June 1924:[782]

25 VI 24.

Esteemed Reich Chancellor,

I hereby take the liberty of requesting a personal audience with you about an urgent and important matter.

In great respect,

Prof. Dr. A. Einstein

Haberlandstr. 5

Noll[endorf]. 2807.

Einstein's intercession for Erich Mühsam does not mean that he shared the same convictions as this anarchist and communist. It documents his sense of injustice about his prison sentence (all the same, he did not seek the release of convicted right-wing extremists). Like the GLHR in general, Einstein also had no qualms about coming within arms reach of communists and Soviet Russians. Although he carefully maintained a critical distance from them, such actions *became more and more frequent.*

From membership in the GLHR it was but a small step to membership in the *Society of Friends of the New Russia* (especially since many other members of the GLHR among Einstein's *personal* acquaintance chose the same path: Eduard Fuchs, Lehmann-Rußbüldt, Hugo Simon, Count Arco and others).

### *Society of Friends of the New Russia*

The Society of Friends of the New Russia (*Gesellschaft der Freunde des neuen Rußlands*) was founded on 1 June 1923. On 27 June 1923 the society's first meeting took place behind closed doors. Einstein did not attend personally but had followed the invitation of "prominent personalities from various cultural fields" and had become a member (the familiar pattern: he did not have to seek allies himself, he *was sought after* and had the privilege of choice). The *manifesto,*[783] published soon after the founding of the society, already names Prof. Albert Einstein as a member of its central committee (the other members were: Theater Director Jessner, President of the Reich Parliament Löbe, the architect Poelzig, the writer Thomas Mann and others).

Excerpt from the "Manifesto" by the Society of Friends of the New Russia:

Since a number of years, a new economic and intellectual existence has been developing in Russia, drawing the attention of all civilized nations upon itself. It has hitherto not been possible to become sufficiently acquainted with the manifold forces at work there in molding new forms of the economy and society. Many preconceptions [. . . ] have obscured the picture.

And yet, Germany and Russia, in particular, who by virtue of their geographic proximity and somewhat related destinies are economically and intellectually dependent on each other, would have the greatest advantage from drawing closer together and supporting each other. [. . . ]

Today Germany is completely isolated from the world. [. . . ] worse still, [. . . ] the relentless militaristic and imperialistic policy of force followed by our French neighbors are eating away at the vitals of the German nation and the German economy and thus are also sapping away the energy of German culture.

One would think that this dire situation gives cause to welcome with joy any help kindly offered to the German people and German culture as an act of friendship worthy of due recognition. And more and more segments of the German nation are indeed beginning to grasp the importance of finding such a friend in *the new* Russia, after long years of complete isolation. [. . . ] in order to remove the disastrous effects of the World War, the cultural workers of both countries should unite in cultural communion. The 'Society of Friends of the New Russia' has been formed for this purpose; to organize our mutual acquaintance systematically and to work toward building a collaboration between the two countries upon this foundation.

The society's membership included well-to-do people from the middle class and its aim to break down the nation's diplomatic isolation through a Russo-German rapprochement certainly agreed with Germany's foreign policy. But influential political institutions, especially the Foreign Office and the Prussian Ministry of Culture, remained suspicious of the society, all the same. They favored the *Russian Institute* in Berlin over the Society of the Friends of the New Russia. This preference did not really have anything to do with the idea of seeking closer ties with Soviet Russia. Many others were trying to do so, even the leadership of the Army and armaments industrialists. Schmidt-Ott, Max Planck and General Seeckt (who certainly cannot be accused of sympathizing with the Bolsheviks), among many others, went on reconnoitering trips to Russia. The Russian Institute was favored over the society because of its *political* orientation.

The Russian Institute was founded on 17 February 1923 with the backing of the Prussian Ministry of Culture and the Foreign Office.[784] The Society for Eastern European Studies, headed by Schmidt-Ott, was represented by its vice-president, Prof. Hoetzsch, who was a member of the institute's senate. The Society for Eastern European Studies also acted as an official sponsor of the institute and in this capacity passed on funds made available by the Foreign Office (629,000 reichsmarks in 1923 until 1932).[785] The institute's staffing and program of courses clearly reflect its anti-Bolshevist orientation. The staff was composed of emigré professors from Soviet Russia. A note in the file dated 2 June 1923 indicates that the founding of the Society of Friends of the New Russia on the day before was interpreted as a direct "rival establishment" to the Russian Institute. The same note surmises that the "political stance of that association" had to be understood as "a positive stance toward communism."[786] A letter to the Foreign Office from 20 September 1924 states: "The initiative for founding the 'Society of Friends of the New Russia' came from groups with close ties to the Soviet embassy." In reply to an inquiry made by the president of the University of Berlin from 28 May 1924, the Ministry of Science, Arts and Culture pointed out that although nothing prevented German scholars from supporting the Russian relief organization through the university it still advised against it.[787] While generously supporting the "Russian Institute," the ministry shunned the Society of Friends of the New Russia.

Einstein made repeated public appearances as a member of the Society of Friends of the New Russia. For instance, he together with Count Arco sent out invitations in the name of the society to a talk by Prof. Fersmann, vice-president of the Academy of Sciences of the USSR, about 'Successes in science and technology in Soviet Russia." It was scheduled for 29 November 1926 in the plenary hall of the Berliner Herrenhaus on Leipziger Strasse. Einstein attended and also took part in the discussion. There was lengthy press coverage of the event.

Vortrag des Prof. Fersmann im Berliner Herrenhaus vor den Freunden des Neuen Rußland
Das Präsidium (von links nach rechts): Prof. Fersmann, Dr. Max Osborn, Dr. Graf Arco, Prof. Albert Einstein, Eduard Fuchs

**Figure 48: Einstein on the presiding committee at an event organized by the Society of Friends of the New Russia on 29 November 1926.**

### *Red Aid, International Workers Relief (IWR)*

Membership in the GLHR and in the Society of Friends of the New Russia had consequences for Einstein: his appointment on the board of trustees of *Red Aid* (*Rote Hilfe*). This meant a *further* step leftwards – into the periphery of the *Communist* Party. Red Aid was a relief organization of the Communist Party dating back to its foundation.

Excerpt from a report by the chief of police of Berlin dated 12 March 1924 about the International Workers Relief:[788]

I.W.R. has ties, according to Münzenberg [. . . ], not only to the Soviet government but also to the IIIrd International. [. . . ]. The I.W.R. implemented and carried out the International Workers Loan for the Russian government. [. . . ].

The I.W.R.'s ties to the IIIrd International (Comintern) are much closer, which is already explained by the fact that the former had originally been founded by the latter, as the 'International Committee for the Organization of Workers Relief for the Starving in Russia.' [. . . ]. The famous communist [Ms.] Kameneva has indicated the purpose of the I.W.R.: [. . . ]. With the further development, purposeful organization and continual perfection of international relief, the mission posed to the Russian and international proletariat, namely to consolidate and extend its dictatorship, will be successfully accomplished. [. . . ] The secretary general is the communist

Münzenberg. [. . . ]. The tools that the I.W.R. uses are the following: (1) establishing superficially seemingly apolitical 'Associations of Friends of Soviet Russia' [. . . ]. (2) Exerting influence by means of motion pictures [. . . ]. (3) charitable work, i.e., establishing an international relief drive for the populations of economically distressed countries [. . . ].

The I.W.R. is therefore not a neutral relief organization, like the ''Red Cross'' or the ''Salvation Army.'' Münzenberg himself admitted this without qualification [. . . ] during a meeting at my agency and stated that the I.W.R. was a class-based organization and primarily supported ''struggling workers.''

During the above-mentioned conversation with Münzenberg, he conceded that communist propaganda was one of the tasks of the I.W.R. [. . . ].

The I.W.R. has tried to conceal its purely communist mentality, of course, by also appointing to its extended committee personalities not belonging to any communist organizations.

A publication by the German League of Human Rights, *Die Menschenrechte, Organ der Liga für Menschenrechte,* issued a Christmas appeal in its issue no. 17, dated 1 December 1926, for assistance to the children of political prisoners. It also bears *Einstein*'s signature.

*Die Menschenrechte, Organ der Deutschen Liga für Menschenrechte,* 1 December 1926 (excerpt):

Christmas assistance for the political prisoners.

The German League of Human Rights would exceedingly appreciate giving a little joy this coming Christmas to the family members of political prisoners or to the political prisoners themselves.

We call upon our members to be most kind and assign some amount for this purpose with reference to the following circular letter by ''Red Aid'' [. . . ].

We need not waste any words about the extreme need that this coming winter will cause among the working class, above all for the family members of the political prisoners.

The undersigned persons, who have joined hands to form a board of trustees for the Children's Homes of Red Aid, consider it [. . . ] their duty to approach the business world as a whole with the request to support the social welfare organization, Red Aid, as best they can. *Christmas is at our doorstep!* Help us so that Red Aid can continue to carry out its works of solidarity. [. . . ].

In the hope that you will demonstrate your understanding for the difficult situation of the political prisoners and their families and not reject our appeal, we sign

most respectfully,

The members of the board of trustees:

The painter Hermann Abeking, [. . . ] Professor Albert Einstein, [. . . ] Health Official Dr. Magnus Hirschfeld, [. . . ] Käthe Kollwitz, [. . . ] Heinrich Mann, Dr. Thomas Mann, [. . . ] Professor

Max Reinhardt, the banker Hugo Simon, [. . . ] Professor Heinrich Zille [. . . ]

Einstein subscribed *publically* to an organization of the Communist Party of Germany (CPG)?! (It was founded on 12 April 1925; first chairman of the IWR section Germany: 1924 Wilhelm Pieck, 1925 Clara Zetkin.) From our point of view today, this is sensational, if not scandalous! Einstein was seeking solidarity with the political prisoners of the Weimar Republic – with "proletarian convicts." In those days it was superfluous to point out that many of them were *communists*. Everyone knew that. It was also sensational, if not scandalous that many other well-to-do people, who were anything but proletarian, were backing it as well, such as the writer Thomas Mann and the banker Simon!

A file from the Office of the Reich Commissioner for the Surveillance of Public Order containing a situation report notes pertinently: "The communists have known how to attract to some of their relief organizations, like Red Aid of Germany (RAG) and the International Workers Relief (IWR) famous intellectuals, who are not members and supporters of the CPG as collaborators and signatories to appeals. For instance, the board of trustees of the Children's Homes of Red Aid includes Professor Albert Einstein, the publisher S. Fischer and Thomas Mann and the board of trustees of the children's homes of the IWR includes Mrs. Einstein [...]."[789]

At the Fourth Congress of the IWR from 20 to 22 November 1927 in Berlin, Einstein was elected along with Klara Zetkin, Käthe Kollwitz, Maxim Gorki and others on the *Extended Central Committee of the IWR*.[790] The communist, Willi Münzenberg, was unanimously reelected secretary general of the IWR.

Einstein's commitment to Red Aid shines through in his reply to greetings that the children from the Mopr Children's Home in Elgersburg sent him for his fiftieth birthday. His tone was not just friendly, it also suggested his political leanings. The periodicals *Illustrierte Justiz-Zeitung der Roten Hilfe Deutschlands* and *Tribunal* published the letter in May 1929 (5th ser., no. 5, p. 6).

Einstein to the children of the Mopr Home in Elgersburg, 27 March 1929:

27 III 29

Dear children,

Your greetings and magnificent drawings delighted me. I would love to praise some of them especially, but that would not be nice to the others and not good for the spirit of brotherhood, which is the most important thing. Let yourselves be guided by the best. Read the letters by Rosa Luxemburg and never lose sight of the fact that people differ from each other more by their external destinies than by their feelings and actions.

Hearty thanks to you,
yours,
A. Einstein

In the same year and prompted by the same occasion Einstein wrote about Lenin, full of acknowledgment and admiration:[791]

In Lenin I admire a man, who has thrown all his energy into making social justice real, at the sacrifice of his own person. I do not consider his method practicable. But one thing is sure: Men like him are the guardians and reformers of the conscience of mankind.

On 8 August 1929 he wrote his friend Paul Levi – a former fellow combatant with Lenin and Rosa Luxemburg.

Einstein to Paul Levi, 8 August 1929:[792]

*Dear Paul Levi [. . . ]. It is uplifting to see how you as an individual person have purefied the atmosphere without restraint through acuity and a love of justice, a wonderful pendant to Zola. Among the finest of us Jews, something of the social justice of the Old Testament still lives on.*

Paul Levi had met Lenin "through Radek in Switzerland in 1915 or 1916 [...]. Levi was a Bolshevik then already."[793] Levi (Hartstein) was one of the co-signers of the treaty between the German government and Russian emigrants in 1917 that made it possible for Lenin and other Bolsheviks to travel back home to Russia and prepare the (October) Revolution.[794] On 4 December 1920 he was elected together with Ernst Däumig to preside over the CPG. It is true that Levi was replaced on 24 February 1921 (like other presiding members: Clara Zetkin), and shortly afterward expelled from the CPG because of his criticism of the "March 1921 drive."[795] But he remained loyal to the leftist ideal. Membership in the CP was not necessary to follow communist ideals. For Levi no less than Einstein. Three days before his sudden death (9 February 1930), Elsa and Albert Einstein visited Paul Levi for the last time.

It is doubtful whether Einstein always knew how he was being used and for what reasons. He was used and sometimes even misused; but he *consented* to this use and occasional abuse.

There was no explicit agreement between him and the Communist Party or the Communist International. Nevertheless in 1930 Einstein acted in *conformance* with their interests and actions, also participating in the international relief drive for China. He tried to win over the former ambassador Solf to the cause as well.

Einstein to Wilhelm Solf, 18 October 1930:[796]

Albert Einstein

Berlin W. 18 X 30<br>Haberlandstr. 5

To Minister Solf, Berlin

Esteemed Mr. Minister,

Today Mr. Engels was here to see me, who is promoting a drive for the famine region Chensi in China. A drive of this type also seems to me to be really appropriate, provided a way can be found to guarantee conscientious use of the collected sums. In this case I would be very willing to sign a fund-raising appeal. But since I cannot assess these circumstances at all, I think it advisable to pass this matter on to you for your opinion. The question is:

Can anything effective be done from here at all, and if so, how should such a drive be organized, and how can the public be *legitimately* convinced to trust that the collected amounts will be used appropriately?

I take the liberty of bothering you with this matter, because I know that you have a heart for East Asia and have an unmatched expertise in the local conditions.

Cordial greetings from yours very truly,

A. Einstein.

Not long afterwards Albert Einstein joined other prominent intellectuals (Lion Feuchtwanger, Arnold Zweig, Bert Brecht, Johannes R. Becher, Jürgen Kuczynski, Erwin Piscator and others) to call for the release of the international union official, Paul *Ruegg* (Noulens Hilaire), who had been arrested in Shanghai in June 1931 and condemned to death by court martial that October in Nanking (Einstein could not know that he was simultaneously a Soviet spy).[797]

In blind delusion about the social conditions in the Soviet Union, Einstein even advocated the on-going political trials there in 1931! The fact that he explicitly corrected an earlier statement means that the political right surely took due note of his statement and used it publically.

Albert Einstein's statement about a sabotage case in the Soviet Union:[798]

Professor Einstein on the Soviet Union

Prof. Hermann Mueninz, who is one of Albert Einstein's closer scientific collaborators and is currently vested in a teaching position in higher mathematics at the University of Leningrad, has been authorized by Albert Einstein to issue a statement for publication regarding the campaign by a group of European intellectuals against the prosecution of the 48 trouble-makers:

''I gave my signature at the time after some hesitation because I trusted in the competency and honesty of the persons who had approached me about this signature, and also because I considered it psychologically impossible that people bearing the full responsibility for implementing technical tasks of utmost importance could purposefully harm the cause they are supposed to be serving. Today I regret most profoundly that I gave this signature, because I have since lost confidence in the correctness of my views at that time. I was not sufficiently aware then that under the special conditions of the Soviet Union things were possible that are totally unthinkable to me under conditions familiar to me.''

As Prof. Mueninz reports, Einstein (who is also known to be a member of the Society of Friends of the New Russia) is keeping close track of the successful advances made toward the socialist construction of the Soviet Union. ''Western Europe,'' Einstein declared, ''will soon be envying you.''

### *Pacifism*

It is the not purpose of this book to retrace Einstein's biography. Here I would just like to refer to the book edited by Otto Nathan and Heinz Norden, 'Albert Einstein on

Peace.'[799] It compiles a large number of Einstein's letters, speeches and publications dating from 1928 until 1933 (and later). Ronald Clark's biography also records the fluctuations in Einstein's pacifist stance.[800]

Despite all his reservations and vacillations, Einstein stood on the left side of the political spectrum. He consequently lent his support to E.J. Gumbel – a man prominent for exposing the machinations of the reactionary right and criticizing German armament. As a result he became the brunt of scathing abuse and was forced to submit to disciplinary measures and legal persecution. His revelatory book on 'Four years of political murder,[801]' providing a comprehensive list of all the political assassinations (both by rightists as well as leftists) within that period, had been fanning the emotions for a while already. His conclusion had been that the most threatening danger to the Republic came *from the right*. In 1925 Gumbel was officially censured by the Faculty of Philosophy at the University of Heidelberg, purportedly for insulting German veterans of the World War. In 1931 the Ministry of Culture in charge still had enough courage to grant him an overdue promotion; furious proclamations by National Socialist professors and students were the consequence. Einstein took Gumbel's part in solidarity – as he had done in many other instances.

Einstein's statement in 1931 on the Gumbel case:[802]

> [. . . ]. I value his personality even more highly. His political activity and his publications are guided by a high ethical standard. The conduct of the academic youth against him is one of the most tragic signs of the times, esteeming so little the ideals of justice, tolerance and truth. What will become of a nation that persecutes such contemporaries and whose leaders pose no resistance to the mean masses? - I recently reread parts of Gumbel's book, 'Traitors succumb to the vehme,' and was once again moved by this man's noble mentality and energy.

How little Einstein was interested in power politics and how very different his attitude was from notable conservative personalities is revealed in answers he gave to a questionnaire from 1927. Konrad Adenauer also responded to the same question: "Should Germany conduct a policy of colonization?" The one respondant championed the ideology of "a landless nation," the other advocated more efficient use of the existing territory.

Chief Mayor Dr. Adenauer, Cologne:[803]

> The German Reich definitely must seek to acquire colonies. Within the Reich itself there is too little space for its large population. Somewhat more daring, strongly ambitious elements, who cannot be deployed inside the country but find a sphere of action in the colonies, are constantly going to waste. We have to have more space for our people and therefore colonies. Colonial activity by the German Reich in the form of colonial trade is less desirable, of course, than possessing our own colonies. [. . . ].

Prof. Albert Einstein:[804]

> The experts seem to be in agreement that by exploiting *still* uncultivated land as well as intensifying ground use through parceling latifundium property, the number of people employed in agriculture as well as the total yield from the land within the territory of the German Reich could be increased substantially. I consider this domestic colonization much more useful, safer and more agreeable than the colonization envisioned by your question for reasons of state on land across the oceans.

The quintessence of his position on war could be reduced to Tucholsky's motto: "Soldiers are murderers." Einstein saw as the only viable way toward eliminating war, not armament but disarmament, the renunciation of obligatory military service and conscientious objection. War meant the killing of people and this was, to Einstein, "common murder" – as he called it during an international conference between 4 and 6 January 1929 in Frankfurt am Main.[805] He publically castigated the military as a "disgrace to civilization."[806] A manifesto he co-signed with Thomas Mann, Romain Rolland, Sigmund Freud, Stefan Zweig and others states: "Military training is the schooling of body and mind in the art of killing. Military training is education for war. It is the perpetuation of a warrior spirit. It prevents the development of the will for peace."[807] In answer to a question in a periodical: "What would you do, if a new war broke out?" Einstein responded on 23 February 1929: "I would immediately refuse to perform any direct or indirect military service and would try to convince my friends to act the same way, and this irrespective of the assessment on the causes of the war."[808]

He described himself as a "militant pacifist,"[809] and thus also became an adamant opponent of the armament policy Germany was pursuing. He could fit nowhere. Neither the German nor the Soviet Russian, French nor British governments could join him in this conviction. *No* government, either then or now, could ascribe to such a policy. Because Einstein was a "*German,*" however, and came from *Germany,* the escalating conflict with the *German* government gained particular importance. Einstein was thus at odds with *every* political party – with the fascists, the Social Democrats and the communists. But primarily with the fascists. The others – the Social Democrats and communists – advocated disarmament, but certainly did not condone conscientious objection. They, too, had their para-military formations: the republican *Reichsbanner Schwarz-Rot-Gold* and the *Roter Frontkämpferbund.* The leader of the Communist Party, Ernst Thälmann, loved to wear a uniform and march in rank and file. The head of the International Workers Relief, the communist Wilhelm Münzenberg, scorned this "pacifist nonsense"[810] (but that did not stop him from making use of Einstein's support anyway).

Despite his politically leftist persuasion and activism favorable to the CPG and the USSR, Einstein continued to maintain his distance. It was "out of the question" for him to tolerate it "when people get killed as a result of an uncontrolled operation." He was "an opponent of any system of terror," and it "never occurred to [him] to approve of those methods employed in Russia" – he wrote in a letter dated 30 September 1931.[811] He could not be convinced to sign an appeal initiated by Henri Barbusse, because, as he explained to Barbusse in a letter on 6 June 1932, it "contains a glorification of Soviet Russia," adding: "I put much effort recently into making an informed opinion on the developments there and have arrived at quite grim results. At the top, personal battles by power-hungry persons using the most despicable means out of purely egoistic motives. Toward the bottom, complete suppression of the individual and freedom of expression. What value is there left in life under such conditions?"[812] A few days later, on 17 June 1932, he said with conviction about Barbusse, if he "happened to be a Russian, he would be somewhere in prison or in exile, if they had let him stay alive at all."[813] It was precisely because of a predictable biased belligerency with the Soviet Union that he changed his plans and decided not to attend the "World Congress in Amsterdam against imperialistic war" (but they elected him member of the German Committee against Imperialistic War anyway). He declined membership in this committee because the "Amsterdam Congress [...] was completely under the Russo-Communist Regime" and prominent Social Democrats were not admitted onto the Congress Committee.[814] According to a report by the News Collection Agency at the Ministry of the Interior (20 September 1932) he withdrew membership from the "League against Imperialism" with the reason: "Either you stop your Arab propaganda and take the Zionist movement into consideration or I can no longer be a member of your presiding committee."[815]

The communists and others did not respect this attitude. Criticism of cherished Soviet-land was sacrilege to them. When Hellmut von Gerlach petitioned to establish a committee "to make known all the Soviet government's measures of oppression," at the League of Human Rights and his petition was granted, the league was immediately labeled a "propaganda hive of the counterrevolution" and disparaged as the "League of Exploitation Rights."[816] The political right, for its part, continued to slight it as the "League against German Rights to Life" (the concrete case involved a "manifesto against anti-Semitism").[817]

The CPG gladly accepted statements of solidarity to celebrate the bond between the proletariat and the intelligentia. But if these very same intellectuals had the impudence to criticize the summary execution of intellectuals in the Soviet Union, the response was of a severity and ugliness that could hardly be outdone. On 4 May 1931 Münzenberg wrote an article in *Die Welt am Abend* on "The dormant conscience of intellectuals" and ended it directly addressing Heinrich Mann and Arnold Zweig with the words: "But where is the indignation by the 84 signatories of the demagogic manifesto against the USSR? [...] where is your voice? It stays silent as it stayed silent during the war and stayed silent throughout all the mass crimes against the working class! You are making yourselves accomplices to the crimes and will have to stand judgment, just as the murderers and criminals against the working class are now standing judgment!" (It would not be long before he wrote: "The traitor, Stalin, is you" – and Münzenberg himself became a victim of Stalin.)

Einstein's judgment maintained an independence and far-sightedness that can only be admired.

*Zionism*

During this same period Einstein made more frequent public appearances as a member of Jewish organizations and an advocate of Zionism. Inside Germany as well as abroad.

In 1929 Einstein joined the presiding panel of the Jewish League for Peace (*Jüdischen Friedensbundes*).[818] His wife, Elsa Einstein, was a member of the league's women's committee. On 12 January 1930, on the occasion of a public proclamation of "The peace program of the Jewry," Albert Einstein called upon the Jewish community, particularly the German Jews, to establish a "voluntary Jewish peace tax" among all Jews as an effective means for Jews to become involved in the work for peace and to secure it financially. The police searched the league's offices for proof of connections to the CPG. Nothing was found but the investigative authorities still considered the league "as leftist, nonetheless, judging from the credo of its members."[819]

Einstein's voyage to America in 1931 was, with the exception of his lectures in Pasadena (like the trip in 1921) another promotional campaign for the founding of a Jewish state.

The German Consulate General in New York submitted a thorough report on 21 March 1931 on Einstein's America tour in March 1931. The consul found it very agreeable that Professor Einstein's visit was a "boon to the reputation of Germanism" and that Einstein was "mainly described and celebrated specifically as a German scholar." He was less happy about Einstein's soliciting for the Zionist cause: "It does not seem out of the question that this effect would have been even stronger, had Einstein let himself be exploited less during his sojourn here for Zionist purposes and by pacifist organizations."

Report by the German Consulate General in New York to the Foreign Office in Berlin, 21 March 1931 (excerpt):[820]

[. . . ]. Professor Albert Einstein, who arrived here from Chicago on the 4th inst. in the morning and embarked again that same evening for his voyage homeward on the steamship ''Deutschland,'' was - as was true of his entire sojourn in America - at the focus of public interest.

It is a characteristic of the New Yorker psyche that Einstein's personality would set off outbursts of a kind of mass hysteria, without there being any clearly identifiable reason for it; this was true not only among groups with a particular propensity for this, such as the ''friends of peace'' and raving phantasts of newly formed mystical religious communities, but also in relatively tempered circles, such as, e.g., among the American sponsors of the Palestine Agency. To what extent the circumstance that approximately 2 million among New York's 7 million residents are Jews played a role is hard to judge; likewise the interplay between press and public, whether the former published its countless special features on Einstein because the readers wanted them or whether the latter took an interest in it, because the newspapers had awakened this interest even before Einstein's arrival and then kept it alive. [. . . ].

It was only at the Zionist dinner that Professor Einstein relented and let himself be recorded on a sound-track film, after the organizers had pointed out that refusal of this permission would have a negative impact on the outcome for the ''Palestine fund'' drive.

Numerous representatives of American pacifist organizations [. . . ] did not get their just desserts this time. Einstein only received them as a group on board the ''Deutschland'' [. . . ]. Even though Einstein's statements were very succinct and unimaginative, they set those present into an enthusiastic fit that expressed itself among other things in many people kissing Einstein's hands and pieces of clothing [. . . ]. Shortly before his departure, about 1,000 friends of peace had gathered on the pier with banners, etc., and breaking out into a stormy chant: ''No war forever!'' they tried to overrun the ship. [. . . ].

The high point of the events in honor of Professor Einstein's presence was a giant banquet at Hotel Astor, that the ''American Palestine Campaign, New York'' had organized as an inauguration of a promotional campaign [. . . ]. It was remarkable that a substantial proportion of those participating were non-Jews and that a number of leading Christian members of New York society were present, who usually tend not to present themselves publically in the society of Jews. In accordance with the purpose of the banquet, the ceremonial address covered exclusively Zionist or Jewish issues. Einstein's arguments reflected his sympathy with the Palestine problem and he pointed out that the Jews must go about resolving the Arab problem themselves. [. . . ].

The fundraiser organized along with the gala dinner yielded - including the profits from the banquet - a total of almost a quarter of a million dollars. The local public way of urging individuals to make contributions does not leave a very edifying impression. The largest pledge - in the amount of $ 50,000.- came from Felix Warburg. Individual pledges by attending Christians in the amount of thousands of dollars prompted stormy applause.

Very spirited proclamations followed even when reading out a heartily phrased welcome telegram to Einstein by President Hoover. [. . . ].

On the whole, Professor Einstein's visit can be regarded as a boon to the reputation of Germanism, because he has been mainly described and celebrated specifically as a German scholar. It does not seem out of the question that this effect would have been even stronger, had Einstein let himself be exploited less during his sojourn here for Zionist purposes and by pacifist organizations.

sig. Heuser

By the early 1930s little had changed about Einstein's reputation: The father of the theory of relativity was a *leftist and a Jew*. What better foe for the fascists? He fit the picture ideally. This is what the Nazis had been looking for from the very beginning: Jews and Reds to persecute and eliminate.

*Caution against the fascist threat. The year 1932*

1932, Einstein's last year in Germany, was filled with the most political activity. Einstein had a premonition of what could happen and did what was within his powers to prevent the worst.

- In February 1932: solidarity with the chief editor of the *Weltbühne*, Carl von Ossietzky, convicted in 1931 by the Reich Supreme Court for his article against Germany's secret policy of rearmament.
- In May, trip to Geneva to attend the "disarmament conference" hosted by the League of Nations that had been ongoing since February.
- On 12 June, testimony at court together with Willy Münzenberg on behalf of prosecuted members of the International Workers Relief.
- Einstein's letter to the lawyer, A. Apfel in re. the Ossietzky case.
- On 17 June, co-signing of a letter with Heinrich Mann and Käthe Kollwitz to the politicians Thälmann, Wels and Leipart.
- On 30 July, completion of his side of the exchange of letters with Sigmund Freund, published in 1933 under the title 'Why War?'

At the press conference on 22 May about his appearance in Geneva, Einstein had negative things to say about the conference – according to Clark.

Einstein supposedly said:[821]

Wars do not become less likely by drawing up rules of engagement for conducting war. [. . . ]. War cannot be made humaine, it has to be eliminated.

This is not a comedy, but a tragedy [. . . ] despite the fool's caps and buffoonery. We should stand up, every one of us, and decry this conference as a travesty! [. . . ]. If the workers of this world, men and women, decide neither to produce nor transport ammunition, war would be done away with once and for all. [. . . ]. The worst about most of the delegates is that they are stupid, dishonest and nothing but puppets [. . . ]. Each declaration of war ought to trigger global revolutions.

Einstein's pacifism may have been utopian, but not more utopian and certainly more sincere than the attempt to fix new rules of engagement for warfare. It was utopian because there wasn't a chance that workers would join such a radical pacifist movement (not to mention governments). Clark's contention that Einstein's performance was one of "his most unfortunate incursions into public affairs" is not so convincing. But it certainly cut the ties between him and the League of Nations permanently.

One answer to the Genevan "disarmament conference" came in the form of a call for another international conference on the prevention of a new world war. It was issued by Einstein, Romain Rolland, Henri Barbusse, Maxim Gorki, Upton Sinclair, Heinrich Mann and Theodore Dreiser at the end of May 1932. It was also supposed to take place in Geneva – on 1 August but had to be relocated to The Hague because the Swiss government refused to grant the necessary permission.

On 11 June 1932 appeal proceedings took place in Moabit, Berlin, against eight members of the International Workers Relief, who had illegally demonstrated in protest against the emergency decree. Einstein gave testimony in support of the defendants. A communist paper proudly reported: Einstein took part and "argued with great warmth on behalf of the IWR at yesterday's hearing."[822] It added, though: "The arguments presented by the co-founder and secretary general of the IWR, Willi Münzenberg, were even more convincing."

The defendants were acquitted. Einstein's efforts had been worthwhile. This was a very public appearance in the company of the arch-communist Willi Münzenberg. The fact that the CPG had scheduled an "assembly of international solidarity" for the following day only heigthened the political relevance of Einstein's appearance – it was a protest meeting by the International Workers Relief, which was dominated by the Communist International.

Just a week after these proceedings, on 17 June 1932, Einstein co-signed with Heinrich Mann and Käthe Kollwitz an open letter to Theodor Leipert, Ernst Thälmann and Otto Wels, calling for the SDP and CPG to join forces for the parliamentary elections scheduled to take place on 31 July 1932.

Appeal to Thälmann, Wels and Leipart, 17 June 1932:[823]

Berlin, the 17th of June 1932.

To
Theodor Leipart, Ernst Thälmann Otto Wels.

We, the undersigned, follow the developments in politics with the perception that we are nearing a horrendous fascist threat. In our view, the threat can be eliminated by a union of the two major workers' parties for the election. This works best by advancing common candidate lists.

The responsibility lies with the leaders; we stress this most emphatically. Only the obvious will of the workers to unite should decide. Such a decision is, at the same time, a matter of survival for the whole nation.

Heinrich Mann
Käthe Kollwitz
Albert Einstein
Adresses:
Heinrich Mann, Berlin-Wilmersdorf, Trautenaustr. 12
Käthe Kollwitz, Berlin N 58, Weisenburgerstr. 25
Albert Einstein, Kaputh, Waldstr.

Einstein gave his signature even though he was aware "that it would probably be easier to reconcile Cain and Abel."[824] And as things turned out, he was right.

Einstein signed at about the same time and with the same purpose an "urgent appeal," to which other names could be added to the original three signatories. They included Kurt Grossmann, Erich Kästner, Otto Lehmann-Russbüldt and Arnold Zweig. They demanded "a union of the SDP and the CPG for this election," with the recommendation, "best in the form of a common list of candidates, at the very least in the form of linked lists."[825]

**Figure 49: At an event of the League of Human Rights, 1932: Prof. Hohbohm at the lectern; from left to right: Prof. Rosenberg, Einstein, Gumbel.**

Käthe Kollwitz was supposedly the appeal's initiator. She had been "encouraged by a leading functionary of the CPG to write this appeal and sign it jointly with prominent partners from the left-of-center bourgeois camp."[826] Being inspired by a "leading functionary of the CPG" would mean, though, that this top official was departing from the "party line."

It is more probable that this initiative came from the German League of Human Rights. In any event this appeal by Heinrich Mann, Käthe Kollwitz and Albert Einstein conforms with an appeal by the league dated 8 June 1932, that read:

> All forces of the reaction have joined together to topple the Republic in a common assault. Not a single vote in favor of the rightist bloc will be wasted. This rightist bloc must be countered with a leftist bloc of republican parties. At the moment it is not a matter of establishing what errors the Republic has committed but saving the Republic's very foundations. [. . . ]. It is essential not to let any votes for the left go to waste either.
>
> That is why we urgently warn all bourgeois/republican parties, and on the other hand, all proletarian/republican parties to save every vote for the Republic by means of a linked candidate list.[827]

This corresponds with various newspaper articles, including the one published in the morning edition of the *Berliner Volkszeitung* on 8 June 1932. It reports that the GLHR would issue an appeal *in the coming days* to the bourgeois/republican parties and socialist parties "to agree to linked listings at the coming parliamentary election, in order to counter the aggressive bloc by the right with a united left."

None of the addressees responded to the appeal by Mann, Kollwitz and Einstein, at least not directly. Between 17 June and 31 July 1932, many speeches were held and election conventions organized. Thälmann, in particular, demanded countless times that an antifascist united front be formed; on 10 July 1932 the CPG held an "antifascist congress of unity." But never a word about the appeal from 17 June. Neither the social democratic *Vorwärts* nor the communist *Rote Fahne* (nor any other newspaper) published the appeal. There was the occasional more or less vague allusion, but no personally addressed reply to the signers. The CPG and SDP both categorically rejected any linking between their candidate lists.

In issue 6 (1932) of the *Internationale*, reprinted in a shortened version in the *Rote Fahne* on 5 July 1932, Thälmann explained "our strategy and tactics in the fight against fascism." The idea of linked listings was dismissed in unmistakable terms. Not Einstein, Heinrich Mann or Käthe Kollwitz were the brunt of his attacks, but the hated Trotsky: "Mr. Trotsky preaches about forming a 'bloc' between CPG and the SDP 'against fascism' [...]." But such suggestions were "counterrevolutionary suggestions." Thälmann continued, "One does not negotiate with people, who have defected permanently to the enemy camp, *one does not form any blocs with them.* [...]. One cannot land a blow on Hitler fascism *without* fighting the bitterest of battles *against* social democracy and for its isolation [...] and *without* the strategy of dealing the decisive blow against social democracy." Any official response to Albert Einstein, Heinrich Mann and Käthe Kollwitz was thus moot. One doesn't negotiate with, let alone write letters to counterrevolutionaries.

So the CPG showed the SDP one broadside after the next. The Social Democrats did not need to do more than point out that the CPG did not want to join hands with the SDP against the fascists. The leaders of the SDP retorted with the warning "Social Democrats! Don't let yourselves be misused as canon shot by the communist strategems!" and forbade any lower-level pacts with the CPG.[828] "A united front among the working class against fascism could only happen outside the CPG."[829] In these circumstances, it was only consistent that the leaders of the General German Federation of Trade Unions (*Allgemeiner Deutschen Gewerkschaftsbund*) would declare they saw "no chance of success for attempts to unite."[830] Thälmann was not the only one to consider it beneath himself to reply to the appeal of 17 June 1932. Otto Wels and Theodor Leipart acted likewise.

On 18 July 1932 the Leftist Cartel of Intellectual Workers and the Liberal Professions organized a "major public meeting" to discuss the appeal by Albert Einstein, Käthe Kollwitz and Heinrich Mann (which had since attracted other less well-known signers as well). The *Welt am Abend* from 18 July carried the announcement of the meeting but refrained from publishing the appeal from 17 June 1932 itself. It merely referred to the demand that the SDP and CPG "should form a united bloc on the basis of a linking of lists" and adds the question: "*Is this the right way to a united anti-fascist front?*"[831] The answer was preprogrammed.

The last deadline for submitting election proposals for the new parliament – the *15th* of July 1932 – had already passed. It was final. So the "major public meeting" could have only *one* purpose: to tell people that the way suggested by Albert Einstein, Käthe Kollwitz and Heinrich Mann could *not* be the right way. The "debate" could have only *one* purpose: to win votes *for the CPG* and hence also for the "antifascist united front" of the type sought by the CPG – to the exclusion of the "leaders" of the SDP and the Federation of Trade Unions (who themselves could think of nothing they wanted less than to follow a *common* path with the SDP and the CPG in the battle against the "horrendous fascist threat"). The meeting had the corresponding outcome. The signers of the appeal of 17 June 1932 (later joined by Theodor Plivier, Lehmann-Russbüldt and others) who attended as well as the ones who did not (perhaps not even having been invited) were set before a tribunal. As the *Weltbühne* reported on 26 July 1932, they went "on a roller-coaster ride" with them. Plivier and Lehmann-Russbüldt were "summoned," who were present, to "come forward" only to have them "sit down again." It was a far cry from "a meeting aiming at free debate."

There was no point: Cain and Abel would not be reconciled. The fraternal strife was much larger than their abhorrence of the common enemy. Einstein's dream of a union of SDP and CPG was a fine dream, but no more than that.

A few months later the CPG and the Communist International about-faced again and Einstein was once again a sought-after man. Soon the equally humiliated Heinrich Mann also advanced to become the figurehead of the communists' popular front policy (now also in attempts to establish a 'bloc' with the SDP leadership).

A possible date for a new attempt to call for a united front of CPG and SDP soon presented itself: the parliamentary elections on 6 November 1932 – after the communist chairwoman by seniority, Klara Zetkin, had just opened the new parliament (on 30 August 1932), the Reich president had promptly dissolved it again by decree of 4 September 1932. Einstein apparently was disillusioned by the failure of his appeal of 17 June 1932 and took its lesson to heart. Another final deadline for parliamentary elections was scheduled for 5 March 1933. In February 1933 Heinrich Mann, Käthe Kollwitz and fourteen other personalities issued yet another call for a joint listing between the CPG and the SDP. This time Einstein's signature was not among them.

Nevertheless, his efforts to banish the fascist threat continued.

A final, desperate attempt to save the civil liberties guaranteed by the Weimar constitution was made in February 1932. Einstein joined many others to announce preparations for a conference under the banner, "free speech."

While still on travels in the USA, Einstein issued together with Heinrich Mann and the lawyer Rudolf Olden an appeal to "all progressive thinkers, in the broadest sense," to advocate "regaining and safeguarding" freedom of the press, the right of assembly, free speech and instruction and to stand up for them publically.[832] Numerous notables then joined the board of this initiative. They included Lion Feuchtwanger, the statistician Robert Kuczynski, the architect Prof. Martin Gropius, Thomas Mann, Carl von Ossietzky, the publisher Ernst Rowohlt, the sociologist Prof. Tönnies, Harry Count Kessler, Municipal Building Officer Martin Wagner, as well as Dr. H. Wegscheider – the woman who had been involved in establishing the Einstein Donation Fund. Many organizations also wanted to participate in the conference. They included the Social Democratic Party of Germany and the Socialist Workers Party of Germany, the communist-directed organizations German Militant Committee against Imperialistic War, the International Workers Relief, the Leftist Cartel of Intellectual Workers and the World Youth League – but apparently not the CPG itself.[833] On Saturday, 18 February 1933, the initiative board met; on 19 February the conference. Prof. Evert from the journalism institute at Leipzig (Zeitungswissenschaftlichen Institut) gave the first talk – on the topic 'Freedom of the press.' There followed opening addresses, including statements by Albert Einstein and Thomas Mann. "Afterwards the aged Professor Ferdinand Tönnies spoke and declared his support, on principle, of the freedoms of instruction and speech, which are not supposed to be limited." When, finally, the former Social Democratic Prussian Minister Wolfgang Heine spoke about "freedom of the arts" and protested against "officially encouraged brutality," the crime commissioner in charge of assemblies intervened and declared the conference dissolved.

A few days later, on 27 February 1933, the parliament building (Reichstag) was in flames. That was the beginning of what the Nazis regarded as the great "purge." Soon books were being burnt and synagogues and people.

### *Monitoring by the political police*

The police did not need to wait for the end of the Weimar Republic to take an interest in Albert Einstein. The surveillance agencies of the State (the political police, the Office of the Reich Commissioner for the Surveillance of Public Order or the News Collection Agency stemming from it, which was located at the Reich Ministry of the Interior) had long been recording what the political activist Einstein was thinking or doing. So the Gestapo later only continued what had already been begun during the Weimar Republic.[835]

As early as 27 July 1923, the Reich commissioner for the surveillance of public order reported to the secretary of state at the Reich Chancellery about the German League of Human Rights (recorded in the file 'New Fatherland League (now) League of Human Rights').[836] This report also contained information about the league's program, which essentially took over that of the New Fatherland League: "The elimination of armed force as a tool of political conflict" and "cooperation in materializing socialism." The New Fatherland League/League of Human Rights was criticized for having "called together very well-attended meetings after the war [...] that were *particularly well received by the working class.*" The New Fatherland League was "recently trying to win over the *broader public as members*, whereas it had hitherto purposefully accepted only a very limited number of politically or academically prominent individuals."[837]

The exact wording: "very well-attended meetings [...] that were particularly well received by the working class. [...] recently trying to win over the broader public as members, whereas it had hitherto purposefully accepted only a very limited number of politically or academically prominent individuals," was used as early as 24 August 1922 in a report by the Reich commissioner! The threat posed by the league was, consequently, not just its agenda but also primarily its former elitist nature. *Albert Einstein* himself had evidently condoned this elitist course.

# Dringender Appell!

## Die Vernichtung aller persönlichen und politischen Freiheit

in Deutschland steht unmittelbar bevor, wenn es nicht in letzter Minute gelingt, unbeschadet von Prinzipiengegensätzen alle Kräfte zusammenzufassen, die in der Ablehnung des Faschismus einig sind. Die nächste Gelegenheit dazu ist der 31. Juli. Es gilt, diese Gelegenheit zu nutzen und endlich einen Schritt zu tun zum

## Aufbau einer einheitlichen Arbeiterfront,

die nicht nur für die parlamentarische, sondern auch für die weitere Abwehr notwendig sein wird. Wir richten an jeden, der diese Überzeugung mit uns teilt, den dringenden Appell, zu helfen, daß

## ein Zusammengehen der SPD und KPD für diesen Wahlkampf

zustande kommt, am besten in der Form gemeinsamer Kandidatenlisten, mindestens jedoch in der Form von Listenverbindung. Insbesondere in den großen Arbeiterorganisationen, nicht nur in den Parteien, kommt es darauf an, hierzu allen erdenklichen Einfluß aufzubieten. Sorgen wir dafür, daß nicht Trägheit der Natur und Feigheit des Herzens uns in die Barbarei versinken lassen!

Chi-yin Chen / Willi Eichler / Albert Einstein / Karl Emonts / Anton Erkelenz / Hellmuth Falkenfeld / Kurt Großmann / E. J. Gumbel / Walter Hammer / Theodor Hartwig / Vitus Heller / Kurt Hiller / Maria Hodann Hanns-Erich Kaminski / Erich Kästner / Karl Kollwitz / Käthe Kollwitz Arthur Kronfeld / E. Lanti / Otto Lehmann-Rußbüldt / Heinrich Mann Pietro Nenni / Paul Oestreich / Franz Oppenheimer / Theodor Plivier Freiherr von Schoenaich / August Siemsen / Minna Specht / Helene Stöcker / Ernst Toller / Graf Emil Wedel / Erich Zeigner / Arnold Zweig

**Figure 50: Appeal to the CPG and the SDP to form a united front against fascism. 1932. Poster.**[834]

It is not surprising that the *Christmas appeal for political prisoners* would feature in the files of the Reich commissioner for the surveillance of public order.[838] Frequent reference there to the name *Einstein* otherwise as well is not surprising either, for instance, in the "strictly confidential!" Special Report by the police headquarters in Fürth near Nuremberg, dated 6 November 1924.[839] They also record Einstein's foreign trips for the purpose of "exploring initial contacts with foreign friendly organizations" as well as his signature under the appeal issued by the League of Human Rights 'For an agreement with France' from February 1922 (that also bore the signatures of E. Bernstein, Käthe Kollwitz, Heinrich Mann, R. Kuczynski, F. Nicolai and others).

A note in the files of the Reich commissioner for the surveillance of public order dated 22 January 1926 reads: "The League is one of the mainstays of pacifist propaganda in Germany. Based on their political inclinations, its members probably mostly belong to the Democratic, Social Democratic and Communist Parties. [...]. The current president is the famous statistician Dr. Kuczynski [...] communist influence on the league [is] growing"[840] (Kuczynski, a member of the presiding board of the German League of Human Rights, was simultaneously chairman on the Reich Committee on Implementing a Plebiscite on the Dispossession of Former Royal Dynasties). The League once again drew suspiciously close to the Communist Party when it demanded the acquittal and release of the condemned American trade unionists Sacco and Vanzetti in 1927.

A report drawn up on 27 July 1926 by a member of the Royal Hungarian legation about members of the German League of Human Rights includes a 28-page list of names. "Prof. Albert Einstein, Berlin W 30, Haberlandstrasse 5" appears there as the fourth on the list of nonalphabetically sorted entries.[841]

Thanks to his membership on the board of trustees of the children's homes supported by Red Aid,[842] Einstein's name was added in the catalog: "Suspicious persons, who have made themselves politically conspicuous," compiled by the Office of the Reich Commissioner for the Surveillance of Public Order as early as September 1926.[843] The Gestapo later took over these records by personal order of Himmler's deputy Dr. Best.[844]

A document dated February 1927 in the files of the Office of the Reich Commissioner for Surveillance of Public Order notes that Einstein was one of five members of the presiding committee of the Society of Friends of the New Russia and describes the society's purpose as follows: "The society's aim is to penetrate the German scholarly world in order to inject Russian ideas into it. Einstein is a member of the Russian scientific association, International Scientific Center, which deals with scientific research."[845]

The previously cited 'situation report' by the Office of the Reich Commissioner for the Surveillance of Public Order, dated 23 November 1927, reads: "The communists have known how to attract to some of their relief organizations, like Red Aid of Germany (RAG) and International Workers Relief (IWR), famous intellectuals, who are not members and supporters of the CPG as collaborators and signatories to appeals. For instance, the board of trustees of the Children's Homes of Red Aid includes Professor Albert Einstein, the publisher S. Fischer and Thomas Mann and the board of trustees of the children's homes of the IWR includes Mrs. Einstein, Lehmann-Russbüldt, secretary of the League of Human Rights, and the city physician, Dr. Max Hodann."[846]

Einstein is not specifically named in the Reich commissioner's report to the Reich minister of the interior dated 18 March 1927.[847] But a position he had eventually adopted: Conscientious objection, "not just [...] in the case of an offensive war but also in the case of a defensive war," it characterizes as politically questionable. It leads to a repudiation of the Defensive Land Forces (*Reichswehr*) of the Reich.

The police was particularly diligent in its investigations of Red Aid of Germany or International Workers Relief of Germany. Efforts intensified as the end of the Weimar Republic drew near. During the morning of 1 September 1922 a large contingent of policemen and crime investigators raided the offices of the IWR on Wilhelmstrasse 48. About ten document files were confiscated. Rudolf Olden, attorney at the Superior Court of Justice, sent a protest to the state commissioner for Prussia on the same day.[848] He stressed that the IWR "was not a division of the Communist Party" and pointed out that its activities were being supported by such notable scientists and artists as Albert Einstein, Maxim Gorki, Romain Rolland, Heinrich Mann and Käthe Kollwitz. This did not either impress or persuade the State authorities. The response from the Ministry of the Interior on 27 September 1932 emphasized that the IWR "was operating along communist lines and in this way was actively supporting highly treasonable goals."[849] On 7 February 1933, the IWH's offices were searched again. Rudolf Olden protested again, yet again in vain: meanwhile the Nazis seized power.

On 8 October 1932, hence just a few weeks before the end of the Weimar Republic, the News Collection Agency of the Reich Ministry of the Interior informed intelligence agencies of the individual German states about the "CPG antiwar effort – German Militant Committee against War." In naming the "personalities belonging on the committee" they placed "Professor *Einstein*" at the top of the list. Helene *Stöcker,* Heinrich *Mann,* General von *Schönaich* and Otto *Lehmann-Russbüldt* (author of the book 'The bloody International' [Die blutige Internationale] on the armaments industry) were also named. *Again* Einstein appears as *first* on their list of German members of the Permanent World Committee on Combating Imperialistic War along with the communist Willi Münzenberg.[850]

At one time celebrated and officially promoted, Albert Einstein had thus been reduced to a political alien and enemy of the State, even before the demise of the Weimar Republic.

Chapter 3

# The Third Reich

## 3.1 Shouts of triumph by a band of murderers

The fateful day came on 30 January 1933: The president of the German Reich, Paul von Hindenburg, appointed Adolf Hitler as chancellor. It was all perfectly legal. At the last two national elections the Nazi party, officially called the National Socialist German Workers Party (NSGWP), established itself as the strongest party in Germany: on 31 July and 6 November 1932 they reaped 37 percent and 33 percent of the votes, respectively. People whom Max Planck had once described as a "band of murderers" rose to power (although Max Planck himself did not quite realize it yet).

The Weimar Republic was quickly choked to death. New parliamentary elections were scheduled for 5 March. On 2 February all demonstrations by communists in Prussia and other northern German states were banned. On 17 February Göring, as Prussian minister of the interior, issued an order authorizing the police to use firearms; a week later he formed an auxiliary Prussian police force, the forerunner of the notorious Secret State Police, the Gestapo. In the evening of 27 February the parliamentary building, the Reichstag, was in flames. Thus the hounding of communists, Social Democrats and leftist intellectuals began. The SDP and CPG were persecuted and plundered, later banned. On 28 February there followed an emergency decree that invalidated essential basic rights guaranteed by the Weimar constitution. Good-bye to freedom of the press, the rights of association and assembly, secrecy of the mail, and the inviolability of home and property. The parliamentary elections on 5 March yielded the desired result: 44 percent for the Nazi party; the eighty-one mandates for the CPG were annulled on the spot. On 21 March the newly elected parliament convened in the garrison church in Potsdam: this "day of Potdam" was celebrated with due pomp and ceremony. On 22 March the first concentration camp was set up in Dachau near Munich. On 23 March parliament passed an enabling bill, thereby condemning itself to insignificance. The first of April was turned into "Jew boycott day."

It was to be expected that Einstein's opponents would cheer in triumph. But the person they wanted alive had slipped away. They so much would have liked to get their hands on this Einstein, their declared enemy for such a long time.

Einstein was in Pasadena (California) when Hitler became Reich chancellor. On 10 March 1933 while on his return trip to Europe, Einstein sent word that he was not going to return to Germany. On 28 March, on board the "Belgenland," he sent his official withdrawal from the Academy of Sciences in Berlin. On 4 April, finally, from the Belgian seaport Ostende, he applied for release from Prussian (i.e., German) citizenship. But Einstein did not stop there. He took every opportunity he could to lambaste the new regime.

It only enraged the fascists more.

On 1 April, on "Jew boycott day," the *Deutsche Tageszeitung* published a caricature on the theme: "A pathetic loony" with the accompanying text: "The boots of the German legation in Brussels was ordered to cure a loitering Asian from the delusion he was a Prussian." This cartoon anticipated what later took place: Einstein was kicked out. That is how he was dismissed from a country that had

**Figure 51: 1 April 1933, "Jew boycott day."**

wanted to be his home and had once been so proud of having an Einstein among its citizens. This was its gratitude.

When Einstein finally abandoned his former pacifist stance in July 1933, this hate knew no bounds. Einstein's argument now was: Germany "is obviously working toward a war using every means"; consequently especially France and Belgium were "in serious danger and definitely dependent on their armies." He continued: "Under the present circumstances, as a Belgian, I would not object to military service, but would gladly take it upon myself with the feeling that I was serving the salvation of European civilization."[851]

He might have imagined before that such a situation could arise. But Einstein obstinately stood by his "militant pacifism" until there was no other way, all the while encouraging young people to refuse military service. He knew perfectly well that nations could be overrun and rejected any right of self-defense even then; his appeal for conscientious objection applied "irrespective of the assessment on the causes of the war."[852]

His pacifist mind-mates were understandably shocked and dismayed. Einstein's justification could not undo his betrayal.

Rolland, who had once upon a time effusively admired Einstein, wrote to Stefan Zweig on 15 September 1933: "Einstein is more dangerous to a cause as a friend than as a foe. His genius lies only in his science. In other areas he is a fool. The explanations he gave 2 years ago in America in favor of conscientious objection were absurd and unfounded. [...]. Believing and leading young people to believe that their objection could stop war was criminally naïve: because it is far too obvious that war will exist anyway [...]. Now he is making an about-face and betraying conscientious objectors with the same flippancy that yesterday he had given them with his support. He should stay out of any activism! He is just made for his equations."[853]

Irrespective of how one judges Einstein's pacifism and his change of heart, it is indisputable that his about-face was directed against the Nazis and had to infuriate them.

The file "Einstein's Theory of Relativity" is profuse testimony of what happened during this time.

On 20 April 1933 (surely not coincidentally on Adolf Hitler's birthday) the German Society for Intuitive Physics [Deutsche Gesellschaft für anschauliche Physik] allowed "Mr. Minister of Science, Art and Culture to publish the enclosed appeal with the request that he lend his support and promote the battle to liberate German science from Jewish Einsteinian theories."[855] Minister Rust replied with thanks and wishing good success. The next day a certain Mr. Kienitz sent him an essay on 'Einstein,' that a year before had not even been accepted by National Socialist newspapers. In his letter he glibly confessed: "Of course this essay is somewhat biased to the extent that I don't make any effort whatsoever to point out any scientific achievements by Einstein; the Jewish press has taken care of that most abundantly. Heil Hitler!"[856] The minister acknowledged receipt. On 8 July 1933 a Mr. Sandgathe petitioned for financial support for printing his book on 'The End of Einstein's Theory of Time' [Das Ende der Einsteinschen Zeittheorie]. The ministry sent the manuscript to Prof. Gehrcke for a referee report. His verdict was that it did not cite other publications sufficiently; in substance, the report was restrained but, on the whole, still positive.[857] But Sandgathe was

Der Hausknecht der Deutschen Gesandtschaft in Brüssel wurde beauftragt, einen dort herumlungernden Asiaten von der Wahnvorstellung, er sei ein Preuße, zu heilen.

**Figure 52: "A pathetic loony," *Deutsche Tageszeitung,* 1 Apr. 1933 Johann von Leers:[854] Juden sehen Dich an [Jews are looking at you]. NS press and publishers, Schöneberg, Berlin, 1933, p. 28: "*Einstein.* Concocted a heftily disputed 'Theory of Relativity.' Was highly celebrated by the Jewish press and the unsuspecting German public; his thanks were an atrocious foreign campaign of lies against Adolf Hitler. (Not yet hanged.)"**

not satisfied. He protested, saying "the opinion is useless."[858] The new reviewer (Prof. Konen, Bonn) recommended it but requested not being mentioned in the publication.[859] (Perhaps his positive report was just a product of fear; Konen was forced into retirement anyway in 1933 already). Sandgathe got his money and sent his thanks.

The Einstein opponents became bolder and more insolent as the days went by; they had nothing to fear. The new state was bolder and more insolent as well.

On 28 May 1933 the *Sonntag Morgen* published an article headlined 'Astronomy now teaches: The universe is getting bigger and bigger. A dizzying topic.' It contains the statement: "The physicist Einstein, who rightfully has become impossible for us politically speaking and as a teacher, contributed fundamental building blocks of these new ideas."[860] A Mr. Golle from Düsseldorf sent in a

**Figure 53: Book-burning on 10 May 1933.**

complaint and demanded the following: (1) relativity theory be removed from the school curriculum; (2) Einstein's name should not be named anymore or at best ridiculed; (3) newspapers be ordered to have articles on this issue reviewed beforehand.

Base abuse and derision of Einstein of the kind Mr. Golle had suggested became one of the most popular forms of Nazi propaganda: The nationwide press service referred on 4 February 1934 to the "most prominent Zionist hawker," and "charlatan."

"Shortly afterwards the professor thought he had to defend himself against public attacks that placed him in the same camp with the communists thus putting him where his behavior and mentality make him belong."

"In 1932 Dr. Einstein was forced to concede in a Berlin law suit that he was member of communist associations."

"The agitator and rabble rouser."

The German legation in Venezuela joyfully reported to Goebbels's ministry on 1 December 1933 about a derogatory remark Einstein had purportedly made about a Chilean scientist: "The audacity of this statement and its lack of respect for Chilean (and implicitly also Latin American) science leaves a very bad impression, considering the sensitivity and vanity of Venezuelans here, and is more useful to us than most of what has been said in German quarters about the case Einstein." The advice is given "to arrange for wide circulation of this statement by Einstein in Latin America."[861]

Letter by the German envoy in Caracas (Venezuela), Tattenbach, dated 1 December 1933 (excerpt).

Content: ''The case Einstein.'' For the Reich Ministry of Public Enlightenment and Propaganda -

As known from my reports, the propaganda directed against the German government's Jew policy has capitalized on the ''case Einstein'' to a particularly high degree. Professor Einstein was constantly being presented as a victim of German barbarity, daring to chase one of the greatest scholars out of Germany, or even putting a price on his head, just because of his Jewish origins. That is how Einstein attracted much sympathy here. This sympathy ought to have experienced a considerable cooling now, after the ''United Press'' repeated a statement by Einstein in a report from Santiago de Chile of today's date that he is supposed to have made in a debate with a Chilean scholar, Julio Bustos, and reads as follows: ''I do not believe that there is any scholar in the world who could refute my theory (of relativity), least of all a Chilean scholar.'' The audacity of this statement and its lack of respect for Chilean (and implicitly also Latin American) science leaves a very bad impression, considering the sensitivity and vanity of Venezuelans here, and is more useful to us than most of what has been said in German quarters about the case Einstein. In my view it would be advisable to arrange for wide circulation of this statement by Einstein in Latin America. [. . . ].

The recipient of this letter was of one mind with the sender. He underlined the words "advisable to arrange for wide circulation of this statement by Einstein in Latin America" on 13 January 1934 and a copy of the letter was promptly forwarded (1) to the Iberian-American Institute and (2) to the ministry's press office "for utilization." Denunciations had become official form. It was sufficient that Einstein was "*supposed to have made*" a statement to justify a systematic campaign to discredit him. Suspicions and rumors were good enough; no verification was deemed necessary.

From the papers we gather that the journalistic servants of the Nazis immediately set about "utilizing" the envoy's report from Caracas. Their deviating allegation that Einstein was living in Chile was either outright misinformation or just a, now, merely venial sin.

On 17 February 1934 the *Deutsche Zeitung* carried a report under the headline: 'Professor Einstein insults a scholar' (and quoted verbatim from the report by the envoy in Caracas):

Professor Einstein, whose scholarly activities and human qualities have already given frequent occasion for criticism, has now once again done something that exposes him in his full *Jewish coarseness*. Einstein lives in Chile, hence as a guest of the Chilean people. A Chilean scholar recently had a scientific dispute with Einstein that was also conducted in public. During the course of this debate Einstein is supposed to have said: 'I do not believe that there is any scholar in the world who could refute my theory (of relativity), *least of all* a Chilean scholar.' As we have learned, Chilean science in particular has taken offense at the audacity of his statement. Over there they are beginning to understand why the new Germany is gladly doing without this questionable eminence.

*Der Tag* from 19 February 1934 adds the following commentary:

This impudence has now taught the very hospitable and chivalrous Chileans something about this guest's 'character,' which we have known about for some years now already. Einstein will have to take up his walking stick again. Maybe this queer old man will finally accept the appointment at the Hebrew University in Jerusalem instead of letting his curly shock of hair be photographed every day.

Professor Lenard was not to be missed among these informers. On 21 March 1933, a matter of a few weeks after Adolf Hitler took over the government, Lenard circulated a 'memorandum' with suggestions on state interference in the affairs of universities and colleges because, as he thought, it was "impossible that universities will cure themselves."

Philipp Lenard's 'Memorandum regarding staffing reform at German universities in the scientific and mathematical subjects':[862]

Lenard offered his assistance ''on what the Education Ministers of the individual States can achieve under existing conditions from energetically influencing the choice of individuals in professorial and other appointments.''

On matters related to mathematical and scientific subjects, he declared, he was ''willing to be of assistance to Ministries of Education in their task of checking candidates on matters of university staffing; evaluating, influencing, or as the case may be, rejecting them and substituting others instead.'' He proposed that Adolf Hitler ''instruct the Education Ministers of the German States to obtain his advice before making a decision in all university staffing issues affecting science and mathematics, which I would provide in brief form according to the highest criterion of German renewal.''

His reason was:

''If the new spirit of the Third Reich is to enter Germany and gain secure foothold, universities must be considered in the first place as very much in need of renewal. [. . . ].

The alien spirit is so firmly lodged in the majority of university teachers that it is impossible that universities will cure themselves; they have to undergo treatment by the Ministries of Education. [. . . ].

The system of professorial and other appointments (full and associate professors and private lecturers), which in the last 15 years had been left solely to universities, is in a badly rotten condition. The selection of candidates is usually justified to the Ministries on an emphasis on special scientific accomplishments. Besides the fact that these accomplishments alone should not be decisive, most are just artificial covers for preferences conducive to the professorial spirit described above. As many truly gifted scientists as are needed as university teachers do not exist at one time alive; but there certainly are enough people of German stock with the necessary scientific training and teaching ability who could serve as models for our German youth in the lecture halls and selfless administrators in university institutes, who have certainly not been sought after for some time now.''

The Reich minister of the interior supported the basic ideas of this memorandum on 22 April 1933 and "humbly proposed that Professor Lenard please be consulted, as a matter of principle, for his opinion on all university staffing matters concerning the sciences and mathematics."[863] Such a categorical decision suggests that it had come from Adolf Hitler personally.

In response to a request to evaluate a newspaper announcement, in October 1933 Lenard recommended staffing consequences, in order to get a handle on the "Einstein imbecility" inside the country, including reappointing professorial chairs. On this occasion Max Planck also came under fire. At that time he was still secretary of the Academy of Sciences and president of the Kaiser Wilhelm Society for the Advancement of the Sciences. He was criticized for having supported the "sly Jew Einstein" so much. It was no longer an issue of Einstein having any great influence in the Third Reich anymore, Lenard thought, especially considering that this "is politically so damaging on top of it all." Having fought against Einstein's theory in vain during the 1920s, Lenard now employed denunciation and *political* arguments in order to move the undesirable competition aside. He regretted being too old; otherwise he would have come personally "to discuss the matter with Minister Goebbels."

Lenard to the Reich Ministry of Public Enlightenment and Propaganda, 8 October 1933 (excerpt):[864]

[. . . ]. Regarding the inquiry of 2 Oct. re. propaganda against the Einstein imbecility, I inform you of the following:

1. The publications by Prof. Carvallo in Paris do not offer anything thorough enough to enter into. Better papers in the same direction are those by D.C. Miller in America [. . . ]; but even these do not provide any secure reference for fighting the Einstein imbecility abroad *and* domestically. [. . . ].

2. Proof of the complete invalidity of Einstein's theories does not depend entirely on these issues of light velocity, however; proof has been *supplied sometime ago already* by the complete failure of verifications that the sly Jew Einstein himself had pompously published at the Berlin Academy (*via Planck*); they do not hold.

3. The fact that despite this proof Einstein is still being upheld among scholarly circles abroad *and* at home comes from ignorance. This support is of 2 kinds: The *leaders* of the Einstein clique (von Laue, Heisenberg, etc.) are masters in mathematics (just as Einstein) but know too little about thinking along with nature and about scientific knowledge, to which mathematics still is not very applicable. The *followers,* the great majority of whom want to advance in academia by the help of the leaders, do not even know the Jew's ''theories'' properly,

because the thing really is very convoluted; they simply comfortably follow along.

Taken altogether, this Einstein clique is a strong force at all ''*scientific conventions*'' and in the academic system of appointments (and promotions). Whatever Einstein does not choose to agree with is brazenly pushed aside.

4. The issue is not that Einstein still has such an influence ''scientifically'' in the 3rd Reich, which is based on *untruths* and is politically so damaging on top of it all. But the *Ministries of Education in Germany should be mobilized* against it; they very much need to do so. [. . . ].

5. If I were younger, I would come in order to discuss the matter with Minister Goebbels, to which very much could be added. Mr. Minister might send a confidential spokesman to see me, who would, however, have to *understand science*. If desirable, I would also know whom to name among my pupils as an intermediary.

A communication to the Ministry of the Interior dated 17 March 1934 from Gestapo headquarters at Prinz-Albrecht-Strasse proves that Einstein's foreign activities were being closely monitored. It points out that Einstein was a regular visitor of the physician Dr. Bucky in New York.[865] (The gentlemen failed to realize, though, that this Dr. Bucky had been one of Albert Einstein's physicians in the Potsdam area near Berlin.)

Einstein heard about some of this back-biting and denunciation but much of it never reached him – it remained concealed in the files. What he did hear about could not have surprised him. Since the early 1920s such reporting had become a part of Albert Einstein's everyday life.

What good acquaintances and mind-mates from former days were capable of did come as a bitter surprise for the Einsteins, though. Some evidently thought they might hold Einstein responsible for all that had happened.

Dr. Marx to Elsa and Albert Einstein, 30 March 1933:[866]

My Dears,

Dear Elsa, I just read that your first husband wants to move away from Prussia. I understand his irritation, but I am convinced that this step will bring other consequences with it than he is seeking. [. . . ].

I ask your dear husband from the heart to reconsider his intention again and ask you please also to support my plea. Namely, I ask him to consider which would sound better [. . . ] whether word went round that: Famous Einstein said: now I will really prove that I am a good German. - Or whether it must go down in history as: Famous Einstein let himself be carried away by his anger and denied his Germanness - Native German though he is. [. . . ].

A copy of this "direct mail" letter somehow arrived into the possession of the German legation in Brussels already on 1 April 1933. The file provides no indications on how this letter from "a loyal German Jew" might have got there, from where it was promptly forwarded to the Foreign Office. It is also remarkable that the legation already knew on

1 April 1933 what was not contained in the letter: that the author of the letter was a *chemist* by profession. Did Julius Marx want to demonstrate his "loyalty" in both directions: toward Albert Einstein *and* the authorities of the Third Reich?

Albert Einstein was, as the envoy, Count Lerchenfeld, correctly suspected, "related to the letter writer." Julius Marx, born on 17 August 1958 (on 30 March 1933 he was thus already sixty-nine years old), was a son of Elsa Einstein's uncle Leopold Marx. Because Elsa was Albert Einstein's cousin, Julius Marx was also *related* to Albert Einstein. "The Ulm directories indicated Julius Marx's profession as that of a chemist. The editions for 1929 and 1931 list him at Frauenstrasse 34, the address of his stepmother Clementine, and 1933 at König Wilhelm Strasse 13. Afterwards he seems to have moved away from Ulm, because he no longer appears on the list of Jews residing in the municipal district [...], compiled at the end of January 1934, and no death notices appeared up to that time either (source: Kreisarchiv Alb-Donau-Kreis, Oberamt Ulm, no. 26). Owing to the destruction of the resident registration records toward the end of World War II, where he moved away to at that time is no longer reconstructable."[867]

So many people used to call themselves friends of Einstein and thought Germany would go under if he left the country that one would expect some energetic resistance or at least some solidarity. But in most cases there was no question of that. Even though many feared worse was to come, many stayed silent. The ranks of loyal friends remained paralyzed also by fear. It was thought that resistance could only make matters worse and it was also hoped that they would thus somehow be spared the bitter cup of fate. Instead came the accusation that Einstein's criticism of the Nazi government had only made everything much worse than it would otherwise have been. If not even *one* had remained true and steadfast, the biblical word of wisdom would have also applied to Einstein: "Not one was there, not even one..."

It was even harder to swallow severest criticism from *Jews*.

Elsa Einstein to Antonina Luchaire, 11 April 1933:[868]

The tragedy about my husband's fate is that German Jews all make him responsible for the so terrible things happening to them there. They believe that his conduct brought on reprisals and in their narrow-mindedness have issued the watchword to shun and hate him. So we get more hate mail from Jews than from Nazis. But the truth is, he had sacrificed himself for the Jews. He was fearless and did not let them down. Isn't it tragic that the same people who had taken him as an icon are now slinging mud at him? They are so intimidated and scared there that they issue one statement after the next with the finest assurances about how well they are doing there and that they all really had nothing to do with Einstein nor wanted anything to do with him. [. . . ].

My husband is constantly getting nasty letters. The poor, blind, stupid people there. German Jews regard him as their bane. Just read those highly undignified declarations dictated by fear and desperation by the Central Association of the Jewish community [. . . ]. The Jews [. . . ] there are so intimidated and fearful and assess their situation so wrongly that my husband cannot even reach them. They have all removed or even burnt his pictures and proclaim their hate in the most drastic ways. [. . . ].[869]

Some of these critics had been frequenting Caputh just a few months before. Musical Director *Kleiber* was one. He "was one of the people wanting to make Einstein responsible for the persecution of Jews in Germany."[870] Max Planck also thought similarly. He wrote Einstein on 19 March 1933 that the effect of Einstein's statements abroad was that "your racial and religious brethren will not get relief from their situation, which is already difficult enough, but rather they will be pressed the more."[871]

The Confederation of Jews of German Nationality [*Verband nationaldeutscher Juden*] "already now lodge a protest against this behavior by *cowardly refugees,* who escape from all personal responsibility and evidently don't care about the other consequences of their actions." The editors of the weekly *Freie Presse* listed the following persons has having "systematically eroded Germanism" during their sojourns in Germany: Georg Bernhardt, Alfred Döblin, Egon Erwin Kisch, Lion Feuchtwanger and Max Brod.[872] *Einstein* ought to have been added to this list of prominent "cowardly refugees."

The Reich League of Jewish Soldiers at the Front went even further. In a letter dated 4 April 1933 to Reich Chancellor Adolf Hitler, its members not only distanced themselves from the foreign "atrocity propaganda" campaign and "the *vile persons among their own ranks,*" but even offered suggestions about "a professional reshuffling of German Jews" along the lines of the Nazi party platform.[873] On 6 May 1933 the Reich league sent to Adolf Hitler a detailed program on Jewish "youth education in military fitness and labor service" as well as on a "professional reshuffling" of the Jews.

This same league had already written a letter to the American embassy dated 24 March 1933, decrying the "*irresponsible incitement* against Germany being conducted abroad by *so-called Jewish intellectuals.* [...]. These men, who in the overwhelming majority *never gave themselves off as Germans,* still posed as their champions and then abandoned their religious brethren at home at a critical moment to flee abroad, are now claiming the right to have a say in German-Jewish affairs." The Reich League of Jewish Soldiers at the Front requested that its letter be made known to the American public as soon as possible.[874]

## 3.2 Resignation from the Academy of Sciences

The arrival of fascism in Germany was unexpectedly quick but not unforeseeable to Einstein. He had been keeping abreast of the political developments as an active participant. The fascist threat had compelled him to change his mind and accept Abraham Flexner's offer in summer 1932 to work in the United States in the newly established Institute for Advanced Study at Princeton. Einstein did not want to leave Germany and decided to spend only the wintertime each year at Princeton. He could not know that his lecture on 20 October 1932 at the University of Berlin would be his last. On 29 November he attended a reception by the Soviet ambassador in Berlin during the Russo-German medical week. A few days later Einstein left for the USA. This was his final departure from the country he had worked in so well for so long. While still in America he heard about the fascist ascension to power in Germany.

The first news of terrorization by the fascists reached him on his voyage homewards. Einstein made the decision then, with a heavy heart, not to return to Germany. He spent a short time in Belgium and England. After that he immigrated permanently into the USA.

At the end of March 1933 the tension escalated within a matter of days, indeed of hours, to the point that open conflict became inevitable.

On 28 March Einstein submitted his resignation to spare the Prussian Academy of Sciences and the Bavarian Academy of Sciences the humiliation of throwing him out.[875]

Red Star Line S.S. Belgenland
28 III 33
To the Prussian Academy of Sciences, Berlin.

The current state of affairs in Germany compels me to resign herewith from my position at the Prussian Academy of Sciences.

For 19 years the Academy has given me the opportnunity to devote my time to scientific research, free from all professional obligations. I know how very much I am obliged to her. It is with reluctance that I withdraw from this circle, also because of the intellectual stimulation and the fine human relationships I enjoyed throughout this long period as a member and have always valued highly. But under the present circumstances I consider my position's inherent dependence upon the Prussian government intolerable.

Respectfully yours,
Albert Einstein

To remove any doubt about his motivations, Einstein simultaneously published a statement that appeared in French on 28 March 1933 in the *Journal des Nations*, shortly afterwards in other papers as well, both in French and German. One even appeared in the *Kölnische Zeitung* on 30 March 1933, albeit with commentary and under the headline: 'How Einstein foments!' [Wie Einstein hetzt]!

Einstein's public statement (*Kölnische Zeitung*, 30 March 1933):

The acts of brutal force and oppression directed against all people of liberal mind and against the Jews, these acts which have taken place in Germany and are still taking place, have fortunately jolted the consciences of all countries that have remained true to the humanitarian ideal and to political freedoms. The International League against Anti-Semitism has won the merit of defending justice by uniting together nations not contaminated by this venom.

We can hope that this reaction against it will suffice to keep Europe safe from a relapse into barbarity of epochs long since past. May all friends of our so direly threatened civilization concentrate all their efforts toward eliminating this world malady.

Signed: Albert Einstein.

The *Kölnische Zeitung* also reported that Einstein had already withdrawn his membership in the Academy of Sciences and had inquired at the German embassy in Brussels about what needed to be done to relinquish his citizenship of the Prussian state as quickly as possible.

The newspaper's commentary read:

After so coarsely denying any feeling of inner attachment, Mr. Einstein really ought to be given the opportunity to break the superficial tie of citizenship with the German nation without any lengthy formalities. He certainly was able to conduct research and work in Germany without impediment, as virtually no other scholar in our country. But not even that prevents him from talking now about German 'barbarity' and hoping that a kind of union be formed in Europe against us. May he also move away to where he believes himself to be safe from the 'venom' of nationalist mentality. It is only good when such minds leave.

Max Planck and others also thought that in the given situation it would be the best thing for everyone if he withdrew from the Academy of Sciences. It would be the most dignified form of separation for both parties.[876] Planck told Albert Einstein so in a letter dated 31 March 1933 from Munich (evidently still unaware that Einstein's resignation had been received by the academy the day before).

A plenary session of the academy took note of Einstein's resignation on 30 March and was "of the opinion that Mr. Einstein's withdrawal obviated further measures on their part."[877] Formally speaking, there was another option. The Ministry of Culture explicitly mentioned the option (because that is what it wanted) of conducting a disciplinary procedure that would have resulted in official expulsion from the academy. But the academy never seriously considered doing so.

The academy nevertheless let itself be swayed into issuing a statement to the press that plunged it into disgrace. Its author, Heymann, was the academy secretary present in Berlin at that time. The text declared that the academy had "no reason to regret Einstein's resignation." After avidly courting Einstein's Prussian citizenship at the beginning of the 1920s, the academy now preferred to allege that Einstein had acquired his "Prussian citizenship" in 1913 "by virtue of his acceptance in the Academy," to imply that it had merely suffered Einstein's Prussian citizenship against its will.

Press release by the Academy of Sciences on Albert Einstein's withdrawal of membership in the academy:[878]

Berlin, 1 April 1933

The Prussian Academy of Sciences was shocked to learn from newspaper reports about Albert Einstein's participation in the slander campaign in America and France. It has demanded an immediate explanation from him. Einstein has since given notice of his withdrawal from the Prussian Academy of Sciences on the grounds that under the current government he could no longer serve the Prussian State. Because he is a Swiss citizen, he apparently also intends to give up his Prussian citizenship, which he had acquired in 1913 by virtue of his acceptance in the Academy as a regular full-time member.

Einstein's agitatorial behavior abroad is particularly offensive to the Prussian Academy of Sciences, because it and its members have felt intimately attached to the Prussian State since times past; and for all its strict restraint in political matters, it has always emphasized and preserved the national idea. For this reason it has no cause to regret Einstein's resignation.

For the Prussian Academy of Sciences
Heymann
Permanent Secretary

There was an epilogue to this that goes to the credit of the members who raised objections.

One member of the academy, Haberland, protested on 2 April 1933 that such a statement would be published without due resolution by the academy; he also objected to the brusqueness of the statement.[879]

Max von Laue protested immediately by telephone and sent his complaint in writing on 3 April. He applied for a special plenary session of the academy on 6 April to discuss Heymann's press release. Laue was of the view that the academy had to distance itself publically from Heymann's statement.[880]

At the meeting on 6 April, Academy Secretary Ficker naturally backed Heymann; the majority of the membership then approved Heymann's actions. Laue did manage, however, to push through his motion that the minutes contain the addition: "that no member of the phys. and math. class had had an opportunity to have a say on the Academy's statement of 1 April 1933 on the Einstein case."[881]

Heinrich von Ficker informed Albert Einstein on 7 April about the academy's resolution. While expressing his esteem for Einstein's scientific merit he also faulted him for his political attitude.[882] He rebuked him for "spreading false judgments and unfounded suspicions to the detriment of our German nation." He would have expected that, "without regard for his own political attitude, he would have placed himself on the side of those who must defend our nation at this time against a flood of defamation."[883] On the following day he informed Max Planck, who was vacationing in Taormina (Sicily) at the time.

Planck also gave his sanction to the officiating secretary, Ficker, that the academy's official communication to Einstein had been written "in an entirely honorable and dignified manner. [...]. It is another question, though, how the Einstein case will go down in the annals of the Academy, and there I do have a certain worry that it will not count among the Academy's pages of glory. His scientific eminence is far too great."[884] But Planck also thought it "not yet out of the question" that Einstein "had not been involved and sharply rejected the infamous hate and propagandistic slander campaign that has spewed forth against Germany since the nomination of Hitler as Reich Chancellor."[885] This naïve statement reveals that Planck evidently was unaware of how much Einstein had developed politically by the beginning of the 1930s.

Einstein responded to the published announcements on 5 April 1933. He rejected the accusation that he was involved in any "slanderous campaign" against the German nation, adding that he had "seen no sign whatsoever of such incitement anywhere," and repeated his criticism of the political situation in Germany. He demanded that his resignation statement of 28 March be published.[886]

It is amazing that Einstein's wish was granted, even by the Ministry of Culture – facilitated by a simultaneous press release by the academy. It shows that the political class did not yet dare to ignore the academy's formal independence completely. The academy had committed the indignity of declaring that it had no reason to regret Einstein's withdrawal but not all elements of decency and honor had yet been lost. Einstein's retraction of his membership was published within the context of the academy's press release on 11 April.[887]

At the request of Nernst and Laue, the last meeting of the academy in re. Einstein took place after Planck had returned from his vacation – on 11 May 1933.

At the plenary meeting on 11 May Planck pointed out officially before the academy how very aware he was of Einstein's significance to science (he was probably also motivated by an urge to rectify himself at least partially for posterity): "I believe I am speaking on behalf of my fellow professional colleagues within the Academy as well as on behalf of the overwhelming majority of all German physicists when I say: Mr. Einstein is not just one among many prominent physicists, but Mr. Einstein is the physicist whose papers, published in our Academy, gave physical knowledge in our century a profundity whose significance can only be measured against the achievements of Johannes Kepler and Isaac Newton. It is important to me to express this so that posterity does not one day get the idea that Mr. Einstein's professional colleagues in the Academy were not yet capable of fully grasping his importance to science."[888] But Planck also commented that, by his political actions, Einstein had himself eliminated the possibility of staying in Germany. This statement was followed by the academy's resolution "to declare this matter now closed."[889]

Thus one of the most productive and finest stages in the history of the academy – indeed in the history of science, in general – also "was closed." The quartet Planck/Nernst/Haber/Einstein had weathered many a storm in its day; but fascism smashed it asunder.

These painful days confirmed that *Planck*, a profoundly decent person on the personal plane, was a devoted and loyal servant to whichever government happened to be in power. He was driven by a strong sense of patriotism no less than by fear. Nobody had worked harder than Planck to promote Einstein. He used every opportunity to further his reputation. He had also frequently shielded him from much younger, occasionally inconsiderate colleagues. But now he had neither the energy nor the courage to do more for Einstein. He was not able to realize that no compromise was possible between Einstein and the Nazis. As Planck noted, "an abysmal chasm in political respects" separated him from Einstein.[890] Einstein's description of Planck's comprehension of politics – at that time an exaggeration – later proved terrible reality: Planck approached public affairs as naïvely as a child and understood as much about politics "as a cat about the Our Father."[891]

All the same, for the rest of his life Planck was a broken man. It is hard to imagine how a person could bear so much suffering. His eldest son Karl died in 1917 after being wounded during the war. His daughter Grete died in 1917 a week after giving birth to a child. In 1919 his second daughter Emma also died in childbirth. "Planck's misfortune wrings my heart," Einstein wrote to Born on 9 December 1919. "I could not hold back my tears when I saw him [...]. He was wonderfully courageous and erect, but you could see the grief eating away at him." From his children, only the youngest, Erwin was left. The Nazis' seizure of power threw Max Planck into a deep emotional conflict. He retained his official posts because he believed he could thus save what there was left to save. The end result was grim. His attempt to intercede for Fritz Haber and other Jewish colleagues in 1933 (probably in May 1933) during an audience with Hitler resulted only in a tantrum by Hitler, whereupon Planck recalled, "I was left with no other choice than to remain silent and to take my leave."[892] *He* had not used the government, rather it had been the Nazi government who used *Planck*. He was someone they could put forward at home and abroad as a presentable representative of the Third Reich (the public gathered very little about Planck's mental dilemma). Even so, as a loyal defender of the theory of relativity, Planck became the target of internal and public attacks by Philipp Lenard, Johannes Stark and other luminaries of "Aryan physics." After his eightieth birthday – on 22 December 1938 – Planck "at last" resigned his office as secretary of the Academy of Sciences. On 15 February 1944 his family villa in the Berlin suburb of Grunewald was hit in an air raid. Virtually all his belongings, including his diaries, records and correspondence, were consumed in the raging fire. Soon afterwards the eighty-six-year old suffered the hardest blow of all: his son, Erwin, his "sunshine, [his] pride [and] hope," was found guilty of collaborating in the assassination attempt on Hitler at the end of 1944 and condemned to death. He was executed on 23 February 1945. Since Erwin's death Max Planck could find no more pleasure in life. In mental anguish and suffering a terrible physical ailment Max Planck died in Göttingen on 4 October 1947.[893]

Fritz Haber, Einstein's enthusiastic friend in better times was there, too. He did not contradict when the plenum of the academy thought on 6 April that there was nothing to regret about Einstein's withdrawal,[894] even though he personally was of a different opinion. On 11 May he wrote Einstein: "The Academy resolved yesterday, after endless debate, [...] to let your withdrawal be and exempt itself with a comment in your honor in its minutes, where future historians of the Academy will find it [...]. Your friend Haber."[895] But friend Haber did not have the courage to speak up and protest. He surely thought if he ducked under and kept quiet he would be spared. Academy Secretary Ficker later recalled: "During and after the meeting many members, Mr. Haber among them, criticized Einstein's actions."[896]

Haber could not know that very soon he, too, would be forced to emigrate and his friend Einstein would then write him[897]:

*I was very happy to receive such a long and substantial letter from your hand and especially to see that your former love for the blond beast has cooled a little. Who would have thought that my dear Haber would appear as my advocate on the Jewish, indeed the Palestine issue! [. . . ]. I hope you don't go back to Germany. It really isn't a good deal to work for a level of the intelligentia composed of men prostrate before common criminals, who even to a certain degree sympathize with these criminals. They could never disappoint me because I never had any respect or liking for them, apart from a few individuals (Planck 60% noble and Laue 100%). There is nothing I wish more dearly for you than a truly humain atmosphere in which you can be happy again (France or England). Best is (for me), though, always to be in contact with a couple of fine Jews; those couple of thousand years of civilized past do mean something, after all! Now that you won't be returning to Teutonia, I hope to meet you again sometime under calmer skies. Good times I wish you and your Hermann until then, and warm greetings.*

Haber did not live long after this collapse. On 30 April 1933 he informed Minister Rust that he would not take advantage of the exceptional clause allowing Jews who had served in the Great War to remain in the civil service. He said he could not continue to direct the Kaiser Wilhelm Institute of Physical Chemistry under the existing new conditions. He emigrated to England. The fall for proud Haber was too precipitous. In the summer of 1933 he wrote to Weizmann: "[...] I was one of the mightiest men in Germany. I was more than a great commander, more than an industrial captain. I was a founder of industries, my work was essential for the economic and military expansion of Germany."[898] On 29 January 1934 Fritz Haber died in Basel. On the occasion of the anniversary of his death, on 29 January 1935, the Kaiser Wilhelm Society organized together with the German Chemical Society and the German Physical Society a memorial ceremony in his honor.[899] The president of the University of Berlin took this as an "affront against the National Socialist state."[900]

Nernst, like the others, also gave his sanction to the academy's hard-hearted statement. Neither did he want anything more to do with Einstein, who not long ago had been so much sought after and the object of such generous patronage.

From among the academicians, only one remained truly loyal to Einstein: his friend and deputy Max von Laue. Laue refused to let himself be intimidated, not even when the Nazis withheld from him the priviledged academy position that Albert Einstein had vacated. At the opening address before the nineth convention of physicists and mathematicians in September 1933, he boldly defended Einstein's stance. He "drew a parallel between Galileo's inquisition about his advocacy of the Copernican solar system and the treatment of Einstein's theory of relativity – meaning Einstein personally."[901] It not only underscores how courageous Max von Laue was personally but also that in 1933 it was still possible for a scientist to take Einstein's part without risking his life or his professional position. Laue, apolitical Laue, assumed the purest ethical and political position.

This is a good place to review the attitudes of former Einstein enthusiasts, colleagues and collaborators:

- For people like Friedrich Schmidt-Ott and Carl Bosch "Einstein" was a nonissue; at least there are no known relevant public statements after the arrival of the fascists in government.
- Erwin Freundlich, who could otherwise be so persistent, if not downright headstrong, willingly accepted the renaming of the "Einstein Institute" into Institute for Solar Physics.
- Max Planck defended Einstein's oeuvre (thus his own as well) but also confessed that "an abysmal chasm in political respects" separated him from Einstein.
- Fritz Haber and Walther Nernst said nothing when the plenum of the academy thought there was nothing to regret about Einstein's departure.
- Musical Director Kleiber – a former regular guest in Caputh – was among the Jews who held Einstein responsible for the extermination of the Jews in Germany.
- Very few dared to raise objections. Von Laue and Haberland defended the scientist and individual and thus also scientific freedom.
- The number of resignations among those who remained in Germany was also negligible. (They include the ambassador to the USA, von Prittwitz, and Max Liebermann – at that time honorary president of the Prussian Academy of the Arts.)
- The majority of those who stood by Einstein were themselves already vulnerable to the same fate: persecution, escape and emigration.

As soon as courage was called for, the ranks of loyal friends rapidly thinned. Love can be as fleeting as a person's utility value.

Einstein's withdrawal from the academy and his public declarations had defined the line of battle. Afterwards the Nazis were not able to continue on as they pleased. Hitler's initial policy was to keep a watchful eye on the foreign reaction to the German government's anti-Jewish measures. Germany could not simply say "we shall do as we please." The Jews had to be "hit hard," but bystanders should not be given the opportunity to misinterpret the situation and "decry the Germans as barbarians." A Jew who had accomplished something of genuine importance for humanity could not simply be eliminated. The world would not understand.[902] But this initial restraint was soon forgotten.

When Planck came to Fritz Haber's defense in May 1933, Hitler spluttered furiously: "Jews are all communists, and it is they who are my enemies; it is against them that my fight is directed. [...]. A Jew is a Jew; all Jews stick together like burrs. Where there is one Jew, all kinds of other Jews gather right away. It would have been the duty of the Jews themselves to draw a dividing line between the various types. They did not do this; and that is why I must act against all Jews equally."[903]

Einstein also helped force the Nazis to show their true colors. But the world let itself be fooled anyway. It did not want to believe that the German government's anti-Semitic policy would end in the extermination of the European Jews.

To spare himself further formalities Einstein authorized his friend Max von Laue on 7 June 1933 to cancel his membership in other institutions.

Einstein to Laue, 7 June 1933:[904]

*Oxford, 7 VI 33.*

*Dear Laue,*

*I heard that my unsettled relationship with German corporate bodies in which my name still appears on the membership lists could cause difficulties for friends of mine in Germany. That is why I ask you please to make sure sometime that my name be struck from the lists of these bodies. They include, e.g., the German Physical Society and the Society of the Order pour le mérite. I explicitly empower you to arrange this for me. This is probably the right way, because this way new theatrical effects are avoided.*

*Amicable greetings, yours,*

*A. Einstein.*

*P.S. Your little book on wave mechanics appeals to me exceedingly.*

The *Deutsche Reich* certainly did not want the conciliatory image that the *academy* had acquired in dealing with Einstein's membership. In this case, his simple *cancellation* of citizenship was not good enough. He had to be pushed out. But that was not so quickly achieved.

### 3.2.1 Expatriation

On 28 March 1933, on his return voyage from the USA on board the "Belgenland," Einstein wrote not only to the Prussian Academy of Sciences but also to the German authorities about his intention to drop his Prussian citizenship. He inquired of the German legation in Brussels about the requisite formalities.

Albert Einstein to the German legation in Brussels, dated 28 March 1933:[905]

Copy

Red Star Line S.S. Belgenland

To the German Consulate General in Brussels.

I permit myself herewith to address the following inquiry to you. I am a Swiss citizen but through my employment at the Prussian Academy of Sciences at the same time a Prussian citizen. I have resigned the mentioned position by letter. What steps must I take in order to give up Prussian citizenship?

Please send your reply, if possible by return post, to me at the address: Mr. César Koch, Rue Gaucet 50, Liège.

Very respectfully,

sig. Albert Einstein.

The legation complied and answered *by return post*. Under the above indicated address the legation informed Einstein on 30 March 1933: "to initiate the loss of your Prussian citizenship, you must submit an application to the president of the authorized administrative district for dismissal from affiliation with the Prussian State."[906]

Pursuant to this information the Einsteins applied to the senior president of the Province of Brandenburg on 4 April 1933 for release from Prussian citizenship (which was identical to "German" nationality). The addressee, "Oberpräsident des Landes Brandenburg," was not the right recipient so the letter was forwarded to the president of the District of Potsdam, "because Einstein had last resided in Caputh."[907]

The release almost took place as requested. In response to an inquiry by telephone on 16 June 1933 from the Foreign Office, both the district administrator of Belzig[908] and the responsible Revenue Office had no objections to the release, "because Einstein has satisfied his tax and other obligations."[909]

Why it did not happen had two reasons.

The first was delays owing to the Einsteins' negligence. Valuable time was lost because Elsa Einstein forgot to enclose the necessary documentation with the application, in particular the marriage certificate. "As soon as these certificates have been received by the government," a note in the file from 16 June 1933 reads, "the release will be issued." According to a letter by the president of the administrative district dated 15 July 1933 to the Prussian minister of the interior, the district president had instructed Einstein "to present his marriage certificate, a certified copy of his employment diploma and written approval by his wife. Mrs. Einstein submitted her statement of approval and informed us that the personal documents were still packed away owing to the move of their household and would be submitted as soon as possible." The documents had still not been received by 15 July 1933, however.[910] The Prussian minister of the interior forwarded a copy of the letter on 24 July 1933 to the Foreign Office.

The second reason was that the Reich authorities had meanwhile erected barriers against the release from state citizenship.

On the basis of a circular order from 15 June 1922, "regarding the nonissuance of certificates upon dismissal from citizenship for tax reasons," the Beelitz Revenue Office had changed its mind and raised objections to issuance of the certificate. With reference to this, the district president informed the Prussian minister of the interior on 15 July 1933, "Einstein's dismissal could not yet be declared according to §22 of the Reich and State Citizenship Law of 22 July 1913. I shall next determine what the objections by the Revenue Office are based on."[911] In consultation with the Reich authorities, the relevant Revenue Office in Beelitz had "discovered" a new source of income, a tax loop-hole, as it seemed – in reality a lucrative chance to tap the expatriates. Suddenly Einstein had *not* satisfied his "tax and other obligations," after all.

Another barrier to the release from German nationality was the Law on the Revocation of Naturalizations and Release from German Citizenship, passed on 14 July 1933.

The Reich Ministry of the Interior was firmly resolved to apply this law to Einstein's case as well. Either no one at the ministry knew about Einstein's application or it had not been given much attention. On 22 July 1933 the Reich minister of the interior (Frick represented by Secretary of State Pfundtner) told the Prussian minister of the interior, "Professor Einstein is reportedly supposed to have applied to the president of the administrative district in Potsdam for his release from Prussian citizenship." Evidently, nothing more definite was known. The more resolute was the prejudicial outcome of his application. "Because Einstein falls among those persons to whom §2 applies of the Law on the Revocation of Naturalizations and Release from German Citizenship from 14 July 1933 [...], may I request that the district president be advised to forego any decision on the release application until further notice."[912] The Prussian Ministry of the Interior immediately contacted the president of the administrative district "by telephone [...] accordingly."

From then on, the district president in Potsdam no longer had a say in the matter. His message from 14 November 1933 to the Prussian minister of the interior arguing that he had no objection to approval of Einstein's application for release ("after the Revenue Office (Mark) had abandoned the originally raised objections")[913] was as good as immaterial. He bowed to directives from above and "took no further steps in this matter."

In knowledge of the "now closed" investigations by the district president in Potsdam, the Prussian minister of the interior realized that the matter had become urgent. On 30 October 1933 he wrote to the Reich minister of the interior[914] (with a copy to the Foreign Office)[915] that the matter should be resolved promptly, "so as not to give Einstein the opportunity to lodge a complaint about alleged delays by the relevant Prussian authority in deciding on his application."

Further negotiations since July 1933 took place between the Reich Ministry of the Interior, the Foreign Office and the Prussian Ministry of the Interior. The president of the administrative district of Potsdam was henceforth a marginal figure.

The *Deutsche Tageszeitung* had already anticipated the desired outcome in vulgar and unmistakable terms in its statement about 'Jew boycott day' on 1 April 1933. Einstein was not supposed to be released but kicked out.

The only question for the Reich authorities was *how* to go about it.

On 22 July 1933 the Reich minister of the interior had commissioned the Prussian minister of the interior with "speedily instructing the president of the district of Potsdam to forego any decision on the release application until further notice."[916] They hastened to prevent Einstein from being relieved of his Prussian citizenship without first getting a hefty kick from behind.

A short time later, on 14 August 1933, the Reich Ministry of the Interior submitted by express to the Foreign Office and the Prussian minister of the interior a list of Reich subjects. Pursuant to §2 par. 1 of the Law of the Revocation of Naturalizations and Release from German Citizenship from 14 July 1933, their German citizenships were to be officially revoked.[917] This list of nineteen names included: Rudolf Breitscheid, Albert Einstein, Lion Feuchtwanger, Friedrich Wilhelm Foerster, Helmuth von Gerlach, Friedrich Heckert, Max Hölz, Heinrich Mann, Wilhelm Münzenberg and Ernst Toller. After consultations on 16 August 1933, "the list was approved [...] except in the case of Einstein."[918]

What had happened?

The Reich minister of foreign affairs, Konstantin Baron von Neurath, expressed reservations; certainly not out of gratitude for Einstein's appearances abroad in 1920 as having been conducive to the "German cause." No, even he had no shame or decency. Neurath just thought the application ought to be granted and the case settled without much fanfare. At the consultations on 16 August, Mr. von Neurath – represented by Mr. von Bülow – took the "position that, particularly among friendlily disposed foreigners, the danger of a damaging effect of the law applied to Einstein would be so serious that he regards it as urgently desirable to refrain from doing so, even though the Foreign Office as well as the interior agencies are convinced that Einstein's conduct alone justifies application of the law."

Secretary of State von Bülow (Foreign Office) to Secretary of State Pfundtner (Reich Ministry of the Interior), 17 August 1933 (excerpt):[919]

As you know, during yesterday's discussion about the first application of the Law on the Revocation of Naturalizations and Release from German Citizenship, the case of Professor Albert Einstein has been left open, because there are reservations about its effect abroad. Because pursuant to §2 paragraph 3 of the law the decision is made by the Reich Minister of the Interior in agreement with the Reich Minister of Foreign Affairs, I did not fail to seek the opinion of Mr. von Neurath, who is currently on vacation, about the Einstein case. Mr. von Neurath upholds the position that, particularly among friendlily disposed foreigners, the danger of a damaging effect of the law applied to Einstein would be so serious that he regards it as urgently desirable to refrain from doing so, even though the Foreign Office, as well as the interior authorities are convinced that Einstein's conduct alone justifies application of the law. According to his view, a solution to the problem could perhaps be found in that, at the same time and in connection with the promulgation of the decision by the Reich Minister of the Interior about the first persons coming into consideration, [. . . ] an official statement be issued of roughly the following content:

Formal application of this law to Einstein has been dispensed with because he forestalled this procedure by applying for release from Prussian citizenship acquired by his appointment as member of the Prussian Academy of Sciences. However, Einstein belongs to a number of communist organizations, including the ''Association of Friends of the Soviet Union'' [Verein der Freunde der Sowjetunion] and the communist ''International Workers Relief,'' who are belligerently working against nationalistic Germany and in numerous proclamations of their own have upheld a position that goes against the duty of loyalty to the Reich and the nation and damages German interests.

In consultation with Mr. von Neurath I ask you, esteemed Mr. Pfundtner, please to grant your approval to the treatment of the Einstein case proposed herewith and to effectuate the approval of Reich Minister Frick.

This was presumably von Bülow's *personal* moderating influence, hiding behind his superior's authority. His biography certainly supports this interpretation. But in the end von Bülow, in his capacity as secretary of state, could only do what the more powerful demanded of him.

**Bülow, Bernhard Wilhelm Otto Viktor** (born 19 Aug. 1885, deceased 21 Jun. 1936). Father: personal adjutant of Kaiser Wilhelm II. Jurisprudence studies. 1909–1911 private foreign travels. 1911 entry into the diplomatic service and attaché in Washington and in the Foreign Office. 1914/15 fought in the World War. After being wounded, returned to the foreign service. 1915/16 legation secretary in Constantinopel and Athens. 1917 again at the Foreign Office. Participated in the peace negotiations of Brest Litovsk and Bucharest. 1919 resignation from the diplomatic service in protest against the treaty of Versailles. Since 1923 again at the Foreign Office as head of the department on the League of Nations. 1930 appointed by Foreign Minister Curtius, upon nomination by Reich Chancellor Brüning, as secretary of state at the Foreign Office. "Molded by a sense of duty toward the old Pruss. aristocracy," "always polite and kind," "inspired with an almost religious sense of patriotism." "A sense of duty prevented him [...] after the upheaval of 1933, from resigning his office."[920]

This "sense of duty toward the old Pruss. aristocracy" had not prevented him from taking his hat in 1919, however, in protest of the Versailles treaty. His constant endeavor to revise the Versailles treaty was incomparably stronger than his reserve toward Hitler, Göring, Frick and others. Bülow had less concern for the victims of the new *domestic policy* than for the threat against German *foreign policy*. He accepted the persecution of democrats; in 1933 he was neither willing nor capable of resigning. Although his moderating efforts "intended to prevent worse," in fact they helped cloak the true intentions of the Third Reich from foreign countries. Like others, he provided the ways and means for the Third Reich "to implement its foreign policy gradually and without emphatic countermeasures from abroad."[921]

The only high official in the German foreign service to quit his job was the ambassador to Germany in Washington, von Prittwitz: "Prittwitz placed his loyalty to a liberal republican Germany over indiscriminate service to the State and the nation."[922]

It was not on matters of principle but *tactical issues* that caused the difference of opinion to persist for so long between the Reich Ministry of the Interior, responsible for the final decision, and the consulting Foreign Office.

Initially, or so it appears in notes from 22 September 1933, Reich Interior Minister Frick seemed to be amenable to the procedure supported by the Foreign Office.[923] But he first wanted to talk to Prime Minister Göring. *Whether* he spoke to him about this cannot be ascertained from the files at the Foreign Office. But we may suppose that Göring's opinion ultimately determined the further course taken. Very suddenly, latest since 26 September, Frick pursued a hard line and demanded that Einstein be *officially expatriated*.[924]

Meanwhile, the check on the citizenship issue instigated by the Prussian minister of the interior had yielded the unquestionable result that Einstein was a citizen of the Reich.[925] In other words: the whole process of expatriating Einstein had been begun without absolute certainty that he was even a Reich German. In resolving the issue of his citizenship, the Prussian minister of the interior presented copies of the relevant documentation from the academy from 1923 as well as the correspondence from that time between the Foreign Office and the Prussian Ministry of Science, Arts and Culture.[926] In a way, Einstein had been turned into a Prussian again just in order to be able to throw him out properly. This resembles the punitive procedure of running the gauntlet in former times: when required, the delinquent was first nursed back to health in order to be able to beat him to death afterwards.

In the fall of 1933 there was a hardening of the front between the Reich Ministry of the Interior and the Foreign Office. The foreign minister insisted on his standpoint. On 18 November 1933 the Reich minister of the interior was again informed: "From the point of view of the Foreign Office, [...] it must still be regarded as desirable that Professor *Einstein* not be expatriated from German citizenship but rather be released in accordance with his application."[927] Frick, in turn, responded to the foreign minister on 7 December 1933: "Precisely because of the reputation Einstein enjoys as a scholar abroad, I consider his fomenting expiations as particularly reprehensible and am therefore against granting his application with a simple release. Einstein's anti-German activities would presumably not stop in any way if he was released from his citizenship. He would rather book this to his account as a personal victory over the German government. But this leniency toward this particularly prominent and antagonistic traitor would only appear to the government as incomprehensible weakness. That is why I ask you please to reconsider the standpoint you have been defending up to now again and, as the case may be, am gladly at your disposal for a top-level discussion."

On February 1934, finally, – ten months after the Einsteins' submission – Frick carried his point. At the meeting on expatriations that took place at the Reich Ministry of the Interior on 8 February 1934, the Foreign Office dropped its objection to expatriating Einstein. Revenge was apparently more important than foreign policy considerations.

Einstein's expatriation on 24 March 1934 was made public in issue no. 75 of the *Deutsche Reichsanzeiger und Preußische Staatsanzeiger*, dated 29 March 1934.[928] As in other cases of expatriation, the publication took place a few days *later*, to allow the Gestapo sufficient time to intervene "in the interest of securing the private capital."[929]

With indirect reference to Einstein's expatriation, Frick informed the district governments and the Prussian Ministry of the Interior on the same day – on 24 March 1934: "Revocation of German citizenship pursuant to §2 of the Law of 14 July 1933 must be considered as a severe, dishonorable punishment [...]. It should therefore only be imposed on those who have incriminated themselves particularily seriously against the national community. Also for the sake of the public image to be made from revocation of national citizenship, it is advisable not to blunt the potency of this weapon by too frequent use." It was "supposed to hit primarily those [...] who have emerged as political troublemakers, whereby the punishment is particularly suited for individuals who have misused their influence or professions in order to incite the public by word and in writing against the New Germany [...]." "Primarily those incorrigible persons are to be punished, who have not desisted from denigrating the German image abroad since 30 January 1933."[930]

In the period between 23 August 1933 (first expatriation) and 1 February 1937 (eighth expatriation), "just" 375 persons were deprived of their citizenship.[931] Compared to the number of political convictions, this figure is relatively low (but such expatriations were increasingly primarily motivated by the attendant confiscations of wealth).

The Prussian minister saw no reason not to grant *Elsa* Einstein's application for release from German citizenship, even after her husband had already been expatriated (see the letter dated 10 August 1934). The Reich minister of the interior likewise had "no objections."[932] The Prussian minister of the interior informed the president of the district government of Potsdam of this on 29 August 1934. He wrote that "in agreement with Mr. Reich Minister of the Interior," he "would approve the application" – "in the event that Albert Einstein should have upheld the application for his wife's release – with her acquiescence."

If the president of the district government thought he might have another opportunity to decide on the case of Elsa Einstein, he was mistaken.

On 27 April 1937 the expatriation was extended to a total of thirty-six family members. This time Einstein's wife *Elsa* was among them.[933] But there was no more chance of "immediately arresting her at the border." A deceased person had been expatriated: Elsa Einstein had died on 20 December 1936. Although she was expatriated "without capital confiscation," Elsa Einstein's personal wealth had already been confiscated in 1933/1934 – as we shall see. "Without capital confiscation" also means: the expropriations from Elsa Einstein had been *illegal* even according to the Nazi's own interpretation of the law.

*Formally,* the matter at the Foreign Office was closed.

Einstein's retrospective commentary, written on 24 August 1948 was: "After I had already stated my abstention, I was still thrown out with due pomp by the Hitler government. It is somewhat similar to the case of Mussolini, who is known to have been executed by hanging, even though he was already dead."[934]

But the Foreign Office diligently continued to keep abreast of the developments on the issue of Einstein's citizenship.

Report by the German embassy, Washington, D.C., from 6 April 1934:[935]

At the House of Representatives, a draft bill [. . . ] has just been submitted that intends to confer American citizenship on Professor Albert Einstein. I have the honor of submitting 5 copies of the draft bill in attachment. Whether the draft will become law appears doubtful.

Luther.

Report by the German embassy, Washington, D.C., dated 27 June 1934:[936]

The draft bill regarding the naturalization of Albert Einstein was not enacted at the 73rd congress, which has just now closed session. The draft has thus become invalid.

Luther.

Der Reichsführer-SS
Der Chef des Sicherheitshauptamtes

№ 036

Geheim!

# Erfassung führender Männer der Systemzeit

Juni 1939

I.

Zusammenstellung und Statistik aller erfassten Personen.

Erfasst wurden insgesamt 553 Personen

Davon entfallen auf:

| | | | |
|---|---|---|---|
| Gruppe 1:: | Marxisten - Kommunisten | 192 | " |
| Gruppe 2:: | Liberalisten - Pazifisten | 82 | " |
| Gruppe 3:: | Konfessionelle Parteien | 76 | " |
| Gruppe 4:: | Rechtsopposition | 73 | " |
| Gruppe 5:: | Oesterr.Systemgrössen | 48 | " |
| Gruppe 6:: | Wissenschaftler | 17 | " |
| Gruppe 7:: | Künstler | 18 | " |
| Gruppe 8:: | Schriftsteller-Journalisten | 47 | " |

Wissenschaftler.

1. Aster, Ernst von
2. Barth, Karl
3. Bülck, Walter
4. Dessauer, Friedrich
5. D'Ester, Karl
6. Einstein, Albert
7. Freud, Sigmund
8. Hodann, Max
9. Hoetzsch, Otto
10. Kantorowicz, Hermann
11. Kelsen, Hans
12. Nawiasky, Hans
13. Nussbaum, Arthur
14. Radbruch, Gustav
15. Siemsen, Anna
16. Sinzheimer, Hugo
17. Strupp, Karl

Übersicht

**Figure 54: 'Record of leading men...' – that means: leading men and women of the opposition. It could not have been expected otherwise than that Einstein would appear on this list compiled by the Reich Leader of the SS. But that so few scientists would be among them is troublesome (even if the distinction scientist/pacifist is problematic from a methodological point of view). (Here: page 1 with two excerpts from subsequent pages – Source: Bundesarchiv – external branch in Zehlendorf, Berlin, formerly Berlin Document Center)**

Newspaper clippings forwarded to the Foreign Office, filed in the matter of "Expatriation, case of Einstein, Professor":[937]

From *The Daily Telegraph,* London, 2 October 1940:

New York, Tuesday (1 Oct.) Prof. Einstein, the father of the theory of relativity, who was expelled from Germany in 1933 because of his Jewish origins and later went to America, was today granted citizenship of the United States. He lives in Princeton, New Jersey.

Press announcement from *Weltwoche Zürich* dated 8 August 1941:[938]

Of famous emigrés in the U.S.A. The great physicist Albert Einstein, former expert at the [Swiss] Federal Patent Office in Berne and erstwhile professor at the University of Zurich, who was dismissed from his offices and expatriated from Germany in 1933, acquired American citizenship roughly two years ago. Today the demand is being made by the World's Christian Fundamental Association in Chattanooga, Tennessee, that Einstein's citizenship be taken away again, because he is an atheist.

World War II had already begun. Einstein was still being persecuted as an important enemy. And that, indeed, is what he was.

## 3.3 Confiscations

### 3.3.1 Bank account

By chasing Einstein out of Germany the job was not yet finished. His possessions still remained to be confiscated.

Initially the Foreign Office was not only against expatriation but also against expropriation in Einstein's case. Mr. von Bülow urgently advised against confiscation of his personal wealth, especially considering that "it probably only involves modest amounts, but foreign propaganda would accuse us of having destroyed or taken away Einstein's scientific material, etc., in order to prevent the great scholar from pursuing his research."[939]

But Einstein was wealthier than von Bülow thought. If it is true that he used up his last savings to build the house in Caputh in 1929, his subsequent income must have been considerable.[940] (The alternative is that he did *not* spend his entire savings on building the house.) Einstein was a comparatively rich man.

It did not make a difference that *Elsa* Einstein was cosignatory to Albert Einstein's accounts. It was justification enough that "the account holder [...] is the wife of Einstein, who has been engaged in communist activities."[941] The grounds against *Albert* Einstein consequently read: "E. was engaged in communist activities."[942] Thus the whole family became penally liable for the political activities of a single member.

On 10 May 1933 – as a sort of accompaniment to the book-burning on the same day – this letter by the Secret State Police addressed "To Professor Albert Einstein and spouse, last residing in Berlin, Haberlandstr. 5":[943]

C e r t i f i e d   c o p y .

Office of the Secret State Police Berlin, the 10th of May 1933.
$I^5$ $70^{23}$ A.-18.

To
Professor Albert Einstein and spouse,

last residing in
B e r l i n ,
Haberlandstr. 5.

In order to maintain public security and order and also to prevent future anticipated subversive communist activities, I hereby expropriate, in accordance with § 14 of the Police Administration Law of 1 June 1931 (S[tatute] B[ooks] p. 77) in conjunction with § 1 of the Decree by the President of the Reich for the Protection of the Nation and the State from 28 Feb. 1933 (R[eich] L[aw] G[azette], I, p. 83),[944] your deposits and securities at the Dresdner Bank, namely:
at deposit branch 19:

Professor Albert Einstein
deposits: RM 5 088.- plus interest as of 31 Dec. 1932
securities: RM 10 000.- 7 % Dt. Reichsbahn preferred stocks
RM 15 000.- 8/6 % 10th Mitteldt. Bodenkred. Anstalt gold mortgage bond
Mrs. Else[945] Einstein.
deposits: RM 773.- plus interest as of 31 Dec. 1932
securities: Fl. 500.- 4 % Hungarian Gold-Rente C.C. shares

Deposit branch 58:
Professor Albert Einstein
deposits: RM 291.70 plus interest as of 31 Dec. 1932
deposits: RM 2 000.- plus interest as of 23 Mar. 33

Mrs. Else Einstein
deposits: RM 522.71 plus interest as of 31 Dec. 1932
deposits: RM 4 000.- fixed-term deposits
plus interest as of 23 Mar. 33
deposits: RM 151.- sep. account ''travel money''
plus interest as of 31 Dec. 32
securities: Kr. 4 000.- 4 % rated Ungar Kronenrente J/D
RM 50.-4½ % 5½ % 43th Preuss. Pfandbriefbank Liqu.
gold mortgage bond J/J
RM 30.- ditto Certif. / Cps. issue 26
RM 5 000.- redemption value - RM 100.- face value I
German communal redemption loan + 1/5 drawing
RM 250.- redemption value - RM 50.- face value
Chemnitzer municipal redemption loan + drawing
RM 2 000.- 8/6 % I. Berliner Hyp. Bk.
Gold-Comm.-Obl. A/O.
RM 200.- A.E.G. stocks
RM 140 Harburg Gummi Phönix stocks
RM 960.- conv. Schultheiss-Patzendorfer Brauerei stocks
RM 1 800.- Reichsbank shares
RM 500.- Commerz- und Privatbank stocks
securities: Separate account ''travel money''
RM 5 000.- 6 % Dt. Reichspost treasury bills
from 31.33 A/O

for the benefit of the Prussian State.

Furthermore, I expropriate the contents of safe deposit box no. 31 rented by Mrs. Else Einstein at deposit branch 58 of the Dresdner Bank.

Contraventions are punishable pursuant to §4 of the decree of 28 Feb. 33.

sig. Diels.
For correct copy:
Hasler
Office clerk

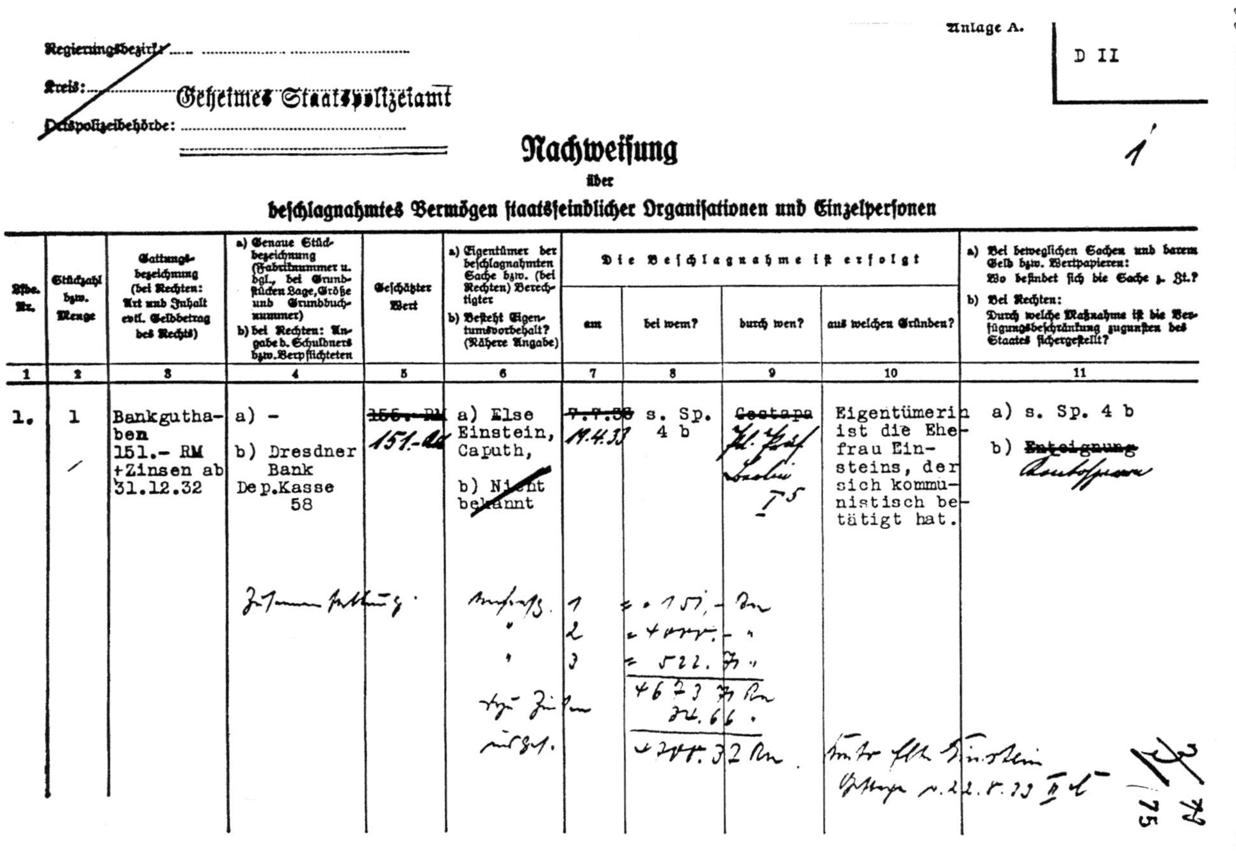

Anlage A.

D II

Regierungsbezirk: ..................................

Kreis: Geheimes Staatspolizeiamt

Ortspolizeibehörde: ..................................

**Nachweisung**

über

**beschlagnahmtes Vermögen staatsfeindlicher Organisationen und Einzelpersonen**

| Lfde. Nr. | Stückzahl bzw. Menge | Gattungsbezeichnung (bei Rechten: Art und Inhalt evtl. Geldbetrag des Rechts) | a) Genaue Stückbezeichnung (Fabriknummer u. dgl., bei Grundstücken Lage, Größe und Grundbuchnummer) b) bei Rechten: Angabe d. Schuldners bzw. Verpflichteten | Geschätzter Wert | a) Eigentümer der beschlagnahmten Sache bzw. (bei Rechten) Berechtigter b) Besteht Eigentumsvorbehalt? (Nähere Angabe) | Die Beschlagnahme ist erfolgt: am | bei wem? | durch wen? | aus welchen Gründen? | a) Bei beweglichen Sachen und barem Geld bzw. Wertpapieren: Wo befindet sich die Sache z. Zt.? b) Bei Rechten: Durch welche Maßnahme ist die Verfügungsbeschränkung zugunsten des Staates sichergestellt? |
|---|---|---|---|---|---|---|---|---|---|---|
| 1 | 2 | 3 | 4 | 5 | 6 | 7 | 8 | 9 | 10 | 11 |
| 1. | 1 | Bankguthaben 151.– RM +Zinsen ab 31.12.32 | a) – b) Dresdner Bank Dep.Kasse 58 | ~~156.– RM~~ 151.– RM | a) Else Einstein, Caputh, b) ~~Nicht bekannt~~ | ~~7.7.33~~ 19.4.33 | s. Sp. 4 b | ~~Gestapa~~ | Eigentümerin ist die Ehefrau Einsteins, der sich kommunistisch betätigt hat. | a) s. Sp. 4 b b) ~~Enteignung~~ |

75

Figure 55: "Notification of confiscated assets . . ." of Elsa Einstein, *"wife of Einstein, who has engaged in communist activities."*

Elsa Einstein's safe deposit box was opened on 23 November 1933.[946] It contained: 30.60 Spanish pesetas, 561.– Argentinian pesetas, 4,210.– French francs, 1,000.– Belgian francs, 1.70 "Swedish coins," 0.97 "Norwegian coins" and German Goldmark gold pieces worth 250.– marks.

The conversions into reichsmarks at the time amounted to:[947]

| | | | |
|---|---|---|---|
| 30.60 pesetas | = | 10.40 | RM |
| 561 Arg. pesetas | = | 541.40 | RM |
| 4,210 French francs | = | 691.30 | RM |
| 1,000 Belgas (Belgian francs) | = | 116.78 | RM |
| 1.70 Swedish kronor | = | 1.20 | RM |
| 0.97 Norwegian kroner | = | 0.70 | RM |

Thus the cash contained in the safe deposit box totaled 1,611.78 RM.

If we also take into account the summer villa and property (owner: Ilse Kayser – née Einstein – and Margot Marianoff – née Einstein) valued at roughly 16,200 reichsmarks and the sailboat, which was sold for 1,300 reichsmarks, the Nazis confiscated from the Einsteins a total of approximately 80,000 reichsmarks. We won't go into the details of how much this would have been worth today. Let it suffice to say that 1 square meter of building land in the best of locations (by the edge of woodland, raised situation with an unobstructed view of the Havel lakes) cannot be had nowadays for 3 marks.[948] The confiscated assets and securities had four times the value of the real estate.

Albert Einstein's "communist activities" served as the legal excuse. But no one bothered even to ask for any *proof* before helping themselves to the bank accounts and securities. (The attempt made to find any in the summer villa had failed miserably.) But even despite a raised tolerance for arbitrary applications of the law, considerable mental acrobatics were necessary to interpret the Decree by the President of the Reich for the Protection of the Nation and the State from 28 Feb. 1933, designed to "counteract subversive communist acts of violence," as grounds for confiscating bank accounts, not just of Albert but also of *Elsa* Einstein.

There was also a remarkable degree of confusion about the issue of "seizure" [Beschlagnahme] versus "expropriation" [Enteignung]. On 10 May 1933 the *Gestapo* ordered the "*expropriation*"; whereas, according to the Prussian *minister of finance* (April 1934), it was initially a "seizure." The assets purportedly "had been *confiscated* [eingezogen] pursuant to the decree of 7 July 1933 for the benefit of the Prussian State."[949] "Seizure" is the word used by the Reich minister of the interior in the 'Announcement about the second expatriation list' (second list dated 24 March 1934): "The personal wealth of all persons are hereby *seized*" (my emphasis, S.G.). Frick had overlooked that Albert and Elsa's assets and securities had already been *expropriated* by the Gestapo on 10 May 1933.[950] On 3 April 1937, on the other hand, it was not a matter of expropriating Elsa Einstein. Although Albert Einstein's expatriation was extended to his wife Elsa, it stipulated: "The confiscation of personal wealth is *not* connected with implementation of the revocation of German citizenship."[951]

The German Reich also demanded that Albert and Elsa Einstein pay what was called Reich evasion penalty tax (*Reichsfluchtsteuer*). It was imposed on whoever left the Reich without obtaining official permission from the Reich authorities.

The Reich evasion penalty tax was first imposed on 8 December 1931 more than a year *before* Hitler took over the government. Originally it was only imposed on Reich citizens intending to emigrate whose financial worth as of 1 January 1931 exceeded 200,000 reichsmarks or whose income during the calendar year 1931 exceeded 20,000 reichsmarks.

On 18 May 1934 the ordinance was extended to *all* emigrés whose financial worth on 1 January 1931 (or anytime afterwards) exceeded 50,000 reichsmarks or whose income in 1931 (or any year afterwards) exceeded 20,000 reichsmarks.[952]

"The extensions made in 1934 and the subsequently imposed implementing regulations did not just cause a changed result but also reflected a changed purpose. The original measure had the goal of *impeding* emigration – namely, emigration by wealthy Reich citizens who were intending to move their assets out of the country in the form of tangible property or money transfers. The later supplementary measure was aimed at *exploiting* emigration – this time primarily the emigration of Jews, who were leaving the country in order to start a new life abroad."[953]

So Einstein would have been obliged to pay this tax even before the Nazi take-over. The despicable thing was not so much the tax per se as the fact that the fascists still expected it to be paid after the Einsteins had been *expropriated.*

On an order dated 11 April 1933, the responsible revenue office fixed the preliminary Reich evasion penalty tax at 15,675 reichsmarks. According to the tax notice of 10 March 1933, it was based on the assumption that the Einsteins' total wealth came to at least 62,700 reichsmarks.[954]

The Einsteins filed an appeal against the regular implementation of the Reich evasion penalty tax. "They argued that the Reich evasion penalty tax only applied to those who wanted to take their capital abroad with themselves and were thus obligated to relinquish a part of it. This did not apply to them because the wealth they owned in Germany had been confiscated."[955]

In its session on 3 November 1933 the Fiscal Court at the Revenue Office of the Province of Brandenburg rejected the Einsteins' appeal and insisted that they pay a Reich evasion penalty tax in the amount of 15,675 reichsmarks plus the costs of the legal proceedings.[956] The court explicitly rejected any further appeal in this matter. Thus the Einsteins were expected to pay 15,675 reichsmarks to the German Reich in addition to their estimated total worth of 62,700 reichsmarks (plus legal fees for the appeal).

The decision was sent to the Einsteins on 12 November 1933 by regular mail.

Because the Einsteins evidently did not consider it necessary to lodge another (clearly futile) appeal against the decision and both the Secret State Police as well as the Prussian minister of finance refused to deduct the outstanding Reich evasion penalty tax from the confiscated funds, the Brandenburg Revenue Office was instructed to issue a tax warrant.[957]

Nothing was forthcoming from abroad and the useless quibbling between the various Reich authorities had to be put a stop to. On 21 April 1934 the Reich

Ministry of the Interior ordered that outstanding taxes generally (hence not just in the case of Albert and Elsa Einstein) had to be paid out from the confiscated funds. But this decision brought Einstein, and others in his position, no closer to his confiscated property.

### 3.3.2 Summer villa

"On the basis of § 1 of the Law on the Confiscation of Communist Wealth from 26 May 1933 (Reich Law Gazette, I, p. 293) and the Law on the Confiscation of Anti-National and Anti-State Wealth from 15 July 1933 (Reich Law Gaz. I, p. 479), in connection with the Implementing Order by the Prussian Minister of the Interior from 31 May 1933 (Statute Books, no. 39)," the president of the administrative district of Potsdam announced on 28 January 1935 that "the parcels of land entered in the Title Register of Caputh, Administrative District of Zauch-Belzig, Volume 40, Sheet No. 1155, run. nos. 1 and 2, for the daughters of the former Reich German Professor Albert Einstein, the deceased Mrs. Ilse Kayser, née Einstein – sole heir to the editor Rudolf Kayser, cur. Leyden (Holland) – and Mrs. Margot Marianoff, née Einstein, are hereby seized and confiscated for the benefit of the Prussian State."[958]

In legal terms, not "Albert Einstein's summer villa" was confiscated but *the house and piece of land belonging to his daughters* in Caputh. But such a distinction was negligible for the Nazis. The confiscation took place regardless.

A note in the file records that the State confiscated the property on 19 November 1935.[959]

Formal justification for the confiscation posed some problems for the authorities, however.[960] The end result was predetermined, but unlike the matter concerning the bank accounts, the authorities at least *tried* to find a legal basis for their confiscation of the summer villa.

In a document by the State Police for the Administrative District of Potsdam to the president of the District of Potsdam regarding 'Confiscation of the property of former Professor Albert Einstein in Caputh' dated 26 May 1934, the undersigned Count Helldorf discussed the legal grounds and factual preconditions for the confiscation.

**Count Wolf Heinrich von Helldorf** (born 1896) was a hussar officer in World War I and involved in the Kapp Putsch as a member of the 'Rossbach' Free Corps. Afterward he lived in exile in Italy until 1924. He became a member of the Nazi party in 1926 and was SA leader in Berlin from 1931. In 1932 he became a member of the Prussian parliament and on 12 Nov. 1933 of the Reichstag. As a leader of the SA Group Berlin-Brandenburg he was presumably complicitous in the politically motivated arson of the Reichstag. The arsonist who was subsequently convicted to death, van der Luppe, had contacted Count Helldorf's circle of acquaintance in February 1933. From March 1933 Helldorf was police chief in Potsdam. Helldorf was notorious, even among Nazis, as a morally decrepit drinker, impostor and bilk – as someone who only "was heard of as seducing and eloping with other people's wives."[961] The on-going proceedings at the highest party court about such behavior were aborted on 13 Aug. 1935. Helldorf had been appointed police chief in July 1935. When Goebbels wanted information about the "Jew evacuation, communist and other political affairs in Berlin," Count Helldorf was the man to ask. On 5 Aug. 1944 Count Helldorf was expelled from the Nazi party and executed on 15 Aug. for his cognizance of the conspiracy of 20 July 1944.

In his document from 26 May 1934, Helldorf referred to Einstein's official expatriation, conducted at the same time as the confiscation of his wealth per § 2 of the Law on the Revocation of Naturalizations and Release from German Citizenship of 14 July 1933 (Reich Law Gazette, I, p. 480), published in the *Reichsanzeiger* of 29 March 1934. He explained that the property had served "subversive purposes and endeavors to the extent that Einstein had worked there on pacifist and communist ideas (e.g., for Red Aid) and had met there with CPG leaders. [...] According to the decree by the Reich Minister of the Interior of 28 Aug. 1933 [...] and of 18 Nov. 1933, [...] private property can be confiscated from individuals for the benefit of the local state treasury without restitution if it has been employed to further communist or other subversive causes against the nation and the State. It suffices for the objects to have been used at any given time for the mentioned purposes." In closing, the signer requested that the addressee inform the district administrator of Belzig to confiscate the property for the benefit of the Prussian State and to issue notice, because he had to report the results to the Secret State Police.

As the president of the District of Potsdam pointed out in his letter of 5 July 1934 to the district administrator of Belzig, however, no proof had yet been found about meetings between Einstein and CPG leaders. He asked the district administrator to forward a report to him directly about the known facts of the matter. In his reply of 13 July 1934, the district administrator reported with reference to a briefing by the head official in Caputh that "much comings and goings in the Einstein house had been observed as well as negotiations behind closed doors." Who the persons involved had been could not be indicated, however. "This ought to be ascertainable, though, by skilled interrogation of the maid at that time, Schiefelbein. According to the police listings, Herta Schiefelbein was registered in Berlin, Haberlandstrasse 5."

Thus Herta Schiefelbein, born on 29 December 1906 in Salzhof near Spandau, residing in Berlin, Alt-Moabit 105, was summoned on 5 September 1934 for interrogation by Dr. Ostrowski.

Already a year ago, on 30 August 1933, Einstein's next-door neighbor, Wolff, had been questioned by the Gestapo in Berlin. This, unfortunately, is the only thing revealed in the files.[962] Undoubtedly the questioning involved Einstein but evidently did not produce the desired results.

The interrogation of Herta Schiefelbein yielded nothing. Ms. Schiefelbein could not name any suspicious persons.

Statement by the housemaid Herta Schiefelbein to State Police Inspector II, I b, Berlin, on 5 September 1934:[963]

I was working as a housemaid from 15 June 1927 until 1 June 1933 for Professor Albert Einstein, Berlin W 30, Haberlandstr. 5. During my employment very many personalities visited there like Gerhart Hauptmann, Professor Planck, Professor Ehrmann, Professor Lichtwitz, Professor Maier, Musical Director Kleiber and many others. There was always much comings and goings in the household. Many pacifists and Zionists also visited there. I often had an opportunity to hear their conversations but could not gather anything suspicious from them. At that time it would not have struck me as unusual if anything of a political nature was being discussed. The conversations with the visitors were carried out in public, that is, not behind closed doors.

Einstein left for America with his wife as early as December 1932 and did not return as a consequence of the political upheaval in Germany. To my knowledge, Einstein is supposedly staying in California and working there.

The summer house in Caputh was closed down each year in the months of October or November and the apartment on Haberlandstr. was reinhabited. That is what happened in December 32 as well, when E. went to America and never returned.

I cannot give any kind of information of a political nature about E. and his daughters. To my knowledge, the property in Caputh belongs to the children of E. The daughters are children of Mrs. Einstein from her first marriage.

Referring to the interrogation, the president of the district of Potsdam informed the Prussian minister of finance on 26 September 1934: "According to the findings of the State Police in Potsdam, the property had supposedly served subversive purposes and causes insofar as Einstein had meetings there with leaders of the CPG and with pacifists. Proof that the property served subversive causes could not be procured, however – even after interrogating the housemaid – [...] although the suspicion remains that subversive conversations took place on the property."[964] At the same time, he requested "a decision on whether to confiscate the property for the benefit of the Prussian State."

Thus no *proof* of Einstein's "guilt" could be produced. Yet *insinuations and suspicions* were enough grounds for confiscation. In a letter dated 2 January 1935,[965] the Prussian minister of finance, in agreement with the Prussian Secret State Police – by order of the chief and inspector dated 13 December 1934 – applied for a change in the title register of Caputh, district of Belzig, Volume 40, sheet no. 1155, run. nos. 1 and 2, to the benefit of the Prussian State. The reason was: "Einstein, as is generally known, promoted Marxist causes, in particular, and to be precise, has been actively promoting these causes on the mentioned property."

Dr. Rudolf Kayser executed a power of attorney on 19 January 1935 authorizing his lawyer Giese[966] to attempt to have the president of the District of Potsdam nullify the seizure. Attorney Giese accordingly filed a complaint on 24 January 1935. It pointed out that the country dwelling was *the property of Ilse Einstein* (and thus upon her death, of Rudolf Kayser) *and Margot Marianoff (née Einstein).*

The Prussian minister of finance rejected the claim on 2 May 1935.

The justification was: "The fact that the country house in Caputh inhabited by Einstein was legally the property of his stepdaughters does not pose an obstacle to the confiscation, because §1 of the Law of 26 May 1933 attaches the preconditions for confiscation to establishment of subversive use of the object to be confiscated, without regard for ownership issues."[967] The minister of finance had already argued this point on 4 April 1935 to the deputy chief and inspector of the Prussian Secret State Police.

On 28 January 1935 the president of the District of Potsdam seized the property, without waiting for a decision on the appeal by Attorney Giese.[968]

Notice.

On the basis of § 1 of the Law on the Confiscation of Communist Wealth from 26 May 1933 (Reich Law Gaz. I, p. 293) and the Law on the Confiscation of Anti-National and Anti-State Wealth from 15 July 1933 (Reich Law Gaz. I, p. 479), in connection with the implementing order by the Prussian Minister of the Interior from 31 May 1933 (Statute Bks., no. 39), the parcels of land entered in the Title Register of Caputh, Administrative District of Zauch-Belzig, Volume 40, Sheet No. 1155, run. nos. 1 and 2, for the daughters of the former Reich German Professor Albert Einstein, the deceased Mrs. Ilse Kayser, née Einstein - sole heir is the editor Rudolf Kayser, cur. Leyden (Holland) - and Mrs. Margot Marianoff, née Einstein, are hereby seized and confiscated for the benefit of the Prussian State.

Potsdam, the 28th of January 1935
The President of the District of Potsdam
by proxy: Zwicker

This did not change the fact that the district president had already ordered the confiscation of the property on *10 January* on identical legal grounds ("On the basis of § 1 of the Law [...] are hereby seized and confiscated for the benefit of the Prussian State," in a letter to the district administrator of Belzig).[969] The very same man had thus seized the summer villa and grounds *twice*. Things must have been pretty chaotic there in their haste to revenge the Jew Einstein.

The confiscation conflicted with the law signed by Reich Chancellor Hitler and Interior Minister Frick on 26 May 1933. It was illegal. Even if proof could have been found that "Einstein, as is generally known, promoted Marxist causes" (*actually on the property!*), it would not have been sufficient grounds. According to § 4 of the Law on the Confiscation of Communist Wealth from 26 May 1933 (Reich Law Gaz., Part I, no. 55 of 27 May 1933), it would have been necessary to prove that "promotion of communist causes *was intended,*" when Ilse Kayser and Margot Marianoff allowed their stepfather to use the summer villa. Nor was the Law on the Confiscation of Anti-National and Anti-State Wealth from 15 July 1933 (Reich Law Gaz. no. 81 of 15 July 1933) sufficient grounds, unless one accepted that it was left to the discretion of the Reich minister of the interior to decide what was to be deemed as subversive behavior against the nation and the State.

The district president had not taken into account in his order of 28 January, however, that Einstein's stepdaughter Ilse Kayser owned another parcel of land measuring 568 square meters.

Ilse Kayser had purchased this property on 9 November 1932 from the neighboring property owner Wolff.[970]

The acting head official in Caputh (in his capacity "as the local police agency") noticed that the property purchased on 9 November 1932 had not yet been confiscated and informed the district administrator of Belzig on 21 January 1935. He pointed out that because both parcels – the already confiscated parcel – and the neighboring parcel belonging to Ilse Kayser "form a closed unit, it would seem appropriate, under the given circumstances, to confiscate this parcel" as well.

The district administrator replied, somewhat crossly, on 26 January 1935: "Why have I not been informed earlier about the existence of this parcel?" He requested that it be "ascertained whether this parcel had been used or whether it had been intended to be used to promote communist or subversive activities against the nation or the State."[971]

The head official (Mayor Krüger) excused himself on 1 February 1934 for the oversight with the argument that the title numbers had never played a part before in the correspondence with the police chief in Potsdam. He asserted furthermore "that this parcel had been used in the same way as the seized property, to promote communist or subversive activities against the nation and the State."[972] But Krüger (member of the Nazi party, later also of the SS) was no better able to produce any evidence for this contention.

Finally, the acting administrator of the District of Zauch-Belzig asked the district president on 18 March 1935 "to seize [this] other parcel recorded in the Title Register of Caputh, Volume 43, sheet 1238 [...], which Einstein had purchased from the master potter, Robert Wolff, in Caputh on 9 November 1932 [...] and to confiscate it for the benefit of the Prussian State," because it "likewise" had served "to promote communist and subversive activities."[973]

In a letter from 27 March 1935 to the Prussian minister of finance in this matter, the president of the District of Potsdam requested approval to confiscate the property.[974]

The Prussian minister of finance consulted with the Gestapo on 4 April 1935. In its reply from 15 April 1935, the Gestapo criticized the "lack of details" in the report by the president of the District of Potsdam from 27 March 1935 about how the property "had served to promote subversive activities" – but did not demand another report. It simply "presumed that in this case it can be proved that Einstein has used the property [...] for subversive activities, particularly in connection with leaders of subversive organizations."[975]

There must have still been a vestige of a sense of propriety and legality at the Prussian Ministry of Finance. Otherwise the ministry would not have answered the district president on 2 May 1935: "The Office of the Secret State Police, to whom I sent the report dated 27 March 1935 [...] for comment, had no objections to the confiscation of the parcel in the Title Register of Caputh, Volume 43, sheet no. 1238, insofar as it can be proved that *Einstein* has used the property with the knowledge of the owner for subversive activities, particularly for meetings with leaders of subversive organizations. As the report only briefly brushes on this question, I humbly request more details."[976] The phrase "*with the knowledge of the owner*" was nowhere to be found in its exchange with the Gestapo, however. This was simply an assumption by the Ministry of Finance, indicating what it was (still) thinking. Earlier – on 4 April – the ministry had taken a quite different stance: *toward the Gestapo* it had left no doubt about the culpability of Einstein and his stepdaughters. The Gestapo's position was stronger. It simply obviated the necessity for proof of Einstein's "guilt."

The district administrator seems to have requested proof again. In any event, on 7 June 1935, Mayor Krüger wrote him: "It could be established that the owner of the property repeatedly spent longer periods of time at this location with the former Reich citizen Professor Einstein as well as meeting with leaders of subversive organizations. [...]. The purchase of the property unquestionably took place only in order to have the family living on the property later move away so that they could then continue their subversive activities with less disturbance. Shortly after the purchase of the property, notice was already given. There is thus no doubt that Einstein also used or intended to use the property in question for subversive activities with the owner's knowledge."[977]

The "proof" given for Einstein's activities on this property were thus: first, the *assumption* that Einstein occasionally stood on this piece of land ("with leaders of subversive organizations") and, second, the contention that he *wanted* to use the property for subversive purposes.

Pretty flimsy grounds. Krüger was apparently aware of this when he wrote: "Evidence in writing on this naturally cannot be produced."

Mayor Krüger wanted the Prussian State to confiscate, in addition, a mortgage that Ilse Kayser had issued to the master potter, Robert Wolff, on 1 July 1930 in the amount of 10,000.– marks. In search of leverage, "Wolff's political attitudes" were investigated (district administrator's inquiry dated 26 February 1925 to the official head of Caputh). The response was: "About W.'s political attitudes we can report that before the changeover of power he was close to the Wirtschaftspartei. He joined the NSGWP as a member on 1 March 1933; his withdrawal of membership again is dated 1 May 1934."[978]

"Withdrawal from the NSGWP after just 1 month of membership" – suspicious, but not enough to go on! The president of the District of Potsdam apparently found the intended confiscation of the mortgage went too far. He informed the administrator of the District of Belzig on 28 June 1935 (who informed the mayor of Caputh on 17 July) that confiscating the mortgage was out of the question, "because there were no legal grounds for confiscation. Nowhere has subversive behavior by Mrs. Kayser been alleged yet. [...] nor is it proven that by issuing the mortgage Mrs. Kayser was intending to further subversive causes."[979] Penal liability for the political activities by a family member did not extend quite that far. In the case of the *mortgage,* they had to admit that Albert Einstein's stepdaughter was an independent legal person.

The district president pressed his case again in his letter to the Prussian minister of finance dated 18 July 1935: "In view of the political activities of the Jew Einstein (e.g., Red Aid) the property served, with the owner's knowledge, subversive purposes to the extent that Einstein worked there. Regarding the liquidation, I report that the mayor of Caputh is willing to purchase both parcels at the estimated value for the Municipality of Caputh."[980]

Accordingly all that was necessary to justify confiscation was for "*Einstein to have worked there.*" All doors were thrown wide open to arbitrary license.

On 6 August 1935 the Prussian State allowed itself to be registered as title owner at the District Court of Potsdam.[981]

More time was again needed to complete the liquidation of the property.

The parties interested in occupying the buildings were:

1. The Municipality of Caputh – with the explicit backing of the Nazi party District Office for the Region of Kurmark and the Potsdam Office of the State Police. The intended purpose was accommodations for youth associations and for a permanent children's home.[982]
2. The 'German Student Association – Group: German Political College' intended "to use this building as a training camp with political – agricultural – sports activities."[983]

The district president supported the application by the Municipality of Caputh. In his letter to the Prussian minister of finance of 25 May 1935 he emphasized: "The two buildings do not fit the purposes of the NS Student Association in any way. [...]. I therefore request that the Einstein properties in Caputh be permitted as a children's home."[984]

The property was eventually granted to the Municipality of Caputh. The assessed value at that time was 16,200 reichsmarks. But according to an application by the municipality and approval by the district president, the municipality acquired the building and grounds for just 5,000 reichsmarks. On 27 August 1936 the building and grounds were sold to the Municipality of Caputh.[985]

A contract on the sale of the Einstein real estate was signed as early as 28 February 1936. But it was rejected by the judicial officer at the Potsdam District Court because of inaccuracies on the forms.[986]

There is evidence that intentions to have a "Jew-purged" neighborhood existed even *before* Einstein's summer villa was seized.

Then Jewish children were "cleared out" of Einstein's summer villa. On 1 May 1931 the Country Children's Home Caputh had been opened – later renamed Jewish Children's Country School and Home Caputh. Hitler's takeover of power had caused an influx of Jewish children from all over Germany so the head mistress, Gertrud Feiertag, had rented Einstein's villa, among other buildings (including private quarters) as of 1 May 1933. On 20 April 1934 seven children, one schoolgirl and one adult were accommodated there (from among a total of eighty-three children, sixteen schoolgirls and twenty-three adults); on 24 April 1934 another eighteen new names were registered.[987]

The children were moved out again on 23 April 1935. But that did not satisfy the mayor. His wish and goal from the outset had been to close down the home and create a "Jew-free" Municipality of Caputh.

On 12 March 1935 the district administrator of Zauch-Belzig applied to the district president of Potsdam "not to recognize the school but to have it dissolved immediately so that the Jewish children would be removed from Caputh."[988] Initially without success. Following the promulgation of the Nuremberg Laws (15 September 1935) the State deemed the official permit for the Jewish private school in Caputh useful. Nevertheless, the president and administrator of the district and the mayor continued to review ways to limit the school's sphere of activity.

When a complaint was filed that the window panes of the home's main building (Potsdamer Strasse 18) had been shattered on 19 February 1935, the district president of Potsdam did not look for the perpetrators. He preferred to turn the victims into culprits.

He wrote to the Office of the State Police in Potsdam on 8 July 1935:

> There is a Jewish Country and Children's Home in Caputh that, according to information lately sent to me, is becoming a gathering place for young Jews originating from regions all over Germany to satisfy their obligatory schooling. According to the report by the administrator of the district of Belzig, there is considerable discontent and annoyance about this among the inhabitants. It has already been the case that the window panes of the home have been broken. I request kind comment about whether any measures are being planned, in particular the closing down or dissolution of the home.[989]

Two days after the children had been moved out of Einstein's summer villa, on 25 April 135 Krüger wrote to the district administrator of Belzig:[990]

> It is a consternating fact [. . . ] that not just the home accommodates Jewish children but also a whole series of private lodgings.
>
> The result of this is that our German children must live in the same buildings as Jewish children and thus the naturally developing sensibilities of our German children must necessarily get obstructed, just like good plants are naturally always stifled by weeds.
>
> If one would make do with an appeal to the landlords, some of whom have doubtlessly been offering lodgings out of necessity, one would be faced with an educational campaign that would cost some time, because even the leader of the women's association [. . . ] in Caputh has been rooming Jewish children.
>
> Thus my request is, if possible to influence the district administrator toward sending the Jewish children back to the existing home and forbidding such lodgings in private households.
>
> Heil Hitler!

On 6 September 1935 the head mistress of the children's home, Ms. Feiertag, complained to the local police authorities that her Jewish children had been insulted, threatened and attacked by local ones.[991] The statement by the father of these children that "such cases are based on an interest in each other"[992] did not appeal to Mr. Krüger. On 20 September 1935 *he* complained to Ms. Feiertag about "provocative behavior" on the part of the Jewish children and scolded her: "It ought to appear appropriate that you teach the children in your charge that they make an effort to assume a different attitude on the street!"[993] On the verso of the complaint filed by Ms. Feiertag he noted, the father ought to see to it that his children "stay as far away from these weeds as possible." It was "just typically Jewish to first be cheeky and then cowardly file a complaint. In a similar case, Meier should file a complaint with the police."

But the children's home could not be closed down, and the "weeds" uprooted as quickly as Krüger wanted. He had to bow to bigger interests of the Reich. It was Adolf Hitler's wish "to host the Olympic Games 1936 [...] under all circumstances in Berlin and thus counter attempts to move them elsewhere."[994] So the Jewish children were granted a brief respite.

After *Reichskristallnacht,* the Night of Broken Glass, the dearest wish of the district president of Potsdam, the administrator of the District of Zauch-Belzig and the mayor of Caputh finally was satisfied. On the morning of 10 November 1938 – a few hours after the *Reichskristallnacht* – Nazis forced their way into the Children's Country School and Home of the Municipality of Caputh and began to demolish and pillage the place. The children and adults were chased out. Only a few adults dared to return in the following days to retrieve what little they could from the ransacked buildings.[995] On 26 February 1939 the home was shut down.

In May 1939 the chief mayor of Berlin "approved" the "acquisition of the property in Caputh situated at Potsdamer Str. 18 [...] from the Jewess Gertrud Feiertag [...] as well as the neighboring isolated property registered [...] by the Jewess Hildegart Littmann."[996] Hildegart Sara Littmann was expatriated on 20 December 1939 and thus also *dispossessed.* According to the Revenue Office of Beelitz, her private wealth totaled 209,503 reichmarks.[997] All of Gertrud Sara Feiertag's wealth was confiscated on 1 February 1943 on the basis of the Law on the Confiscation of Communist Wealth for the benefit of the German Reich.[998] In the declaration of personal wealth dated 10 May 1943, she crossed out as inapplicable all the lines regarding private capital. The only things left to her name were a few pieces of kitchen furniture, a Gasag share for 30 reichsmarks and a Bewag share for the same amount. On 5 July 1943 she was "evacuated" together with her subletter, Sara Rosskamm – that is, transported to a concentration camp.

Once Ms. Feiertag and the children had been run out of the home into the gas chambers of Auschwitz,[999] the Municipality of Caputh was "purged of Jews at last." Mayor Krüger had finally got rid of his detested "weeds." It was also thorough proof that Einstein's expulsion had just been the very beginning.

### 3.3.3 Sailboat

Einstein's sailboat was not forgotten in the confiscations.

His dearly loved "Tümmler" (porpoise) had been presented to him by wealthy friends on his fiftieth birthday. It was reportedly the gift Einstein had been most delighted with.[1000]

The research was conducted with true Prussian thoroughness. Once the *decision* had been reached, correctness was no longer appropriate. Ideology took precedence.

The police seized the boat on 12 June 1933.[1001] A report appeared already in the *morning* edition of the *Vossische Zeitung* of the same day nevertheless. So implementation of the order had clearly been fixed hours beforehand. The otherwise so precise Prussian bureaucracy would not worry itself about such subtleties.

Einstein's son-in-law, Rudolph Kayser, who was still living in Berlin at the time, wrote to his lawyer, Dr. Vogt, about this notice in the paper. Einstein had given his son-in-law the authority to act in his name if need be, but he had no illusions. He was not willing simply to accept what came without a fight.[1002]

Dr. Vogt corrected the factual errors in the press release and insisted on the owner's right of disposal.

Attorney and Notary Public Dr. Vogt to the administrator of the District of Zauch-Belzig (excerpt):[1003]

[. . . ]. Dr. Kayser is the son-in-law of Professor Einstein, who is sojourning outside the country, and his lawful agent.

He recently read in the morning edition of the Berliner Vossische Zeitung from 12 July 1933 the following notice:

''Einstein's racing motorboat seized.

Professor Einstein's racing motorboat, at anchor at a wharf in Caputh near Potsdam, was seized and secured for the Reich. Einstein purportedly had the intention of taking the boat, which is valued at 25,000.- RM, abroad.''

This was particularly astonishing to hear, since Professor Einstein does not own a racing motorboat, certainly not of the mentioned value, but just a sailboat, which had been entrusted to the Schuhmann shipping wharf in Caputh, Potsdamerstrasse for storage. A few days ago he [Kayser] went to see the wharf owner, Schuhmann, and asked him whether he thought it possible to have it moved.[1004] He gave professional advice of various types without he [Kayser] or his wife arriving at any specific instructions. He explicitly stated that any decision about any eventual transfer had not yet been reached. It was a matter of course that all necessary steps would have been taken at the authorized authorities before the transfer abroad. But no such move had been envisioned. Not the least cause exists for seizure of the boat, if it actually did take place. [. . . ].

It had. By order of the Gestapo (Office of the Secret State Police) dated 16 August 1933, the boat was officially confiscated for the benefit of the Prussian State.

Apparently unaware of this order, the district president of Potsdam "seized" the boat again on 7 November 1933 "and confiscated it for the benefit of the State of Prussia."[1005] Einstein's cherished sailboat had thus been seized *twice* (on 16 August and on 7 November 1933) and confiscated *twice*. The Gestapo set the district president right about the facts and pointed out to the district administrator of Belzig on 2 February 1934 that 16 August 1933 was to be regarded as the boat's confiscation date.[1006]

According to recollections by the (recently appointed) mayor on 2 October 1945, there was even a *third* attempt to seize it: "Police investigators from Berlin went to see Mr. Schuhmann and wanted to seize the sailboat (a light cruiser, *Jollenkreuzer*). Mr. Schuhmann said that the boat had already been sold. The police investigators went to the mayor. There they received the same confirmation that the boat had been sold and handed over to the hospital cashier from Nowawes."[1007]

The grounds given for the confiscation (in the order of 7 November 1933) were: the Decree by the Reich President for the Protection of the Nation and the State of 28 November 1933 (Reich Law Gaz. I, 83) and the Law on the Confiscation of Communist Wealth of 26 May 1933 (Reich Law Gaz. I, p. 293), in connection with the implementing order of the Prussian minister of the interior dated 31 May 1933 (Stat. Bks. no. 39) and [...] the Law on the Confiscation of Anti-National and Anti-State Wealth of 14 July 1933 (Reich Law Gaz. I, p. 479)."[1008] In short:

Einstein, the communist and enemy of the State, was no longer thought worthy of being the owner of a sailboat.

On 9 January 1934 the Gestapo instructed the district president in Potsdam to sell the boat – provided the police authorities saw no possibility of using it.[1009] At the same time it urged that the character of the purchaser be established beforehand, "in order to prevent that the boat be acquired by another subversively minded person."

The Prussian prime minsiter and chief of the Secret State Police to the district president of Potsdam, 9 January 1934 (excerpt):

> From evidence produced by the local police authorities in Caputh, the following object has been reported to me as seized and confiscated for the benefit of the State of Prussia:
>
> 1 sailboat, former owner Prof. Einstein, stored at the Schu[h]mann wharf in Caputh, estimated value 1,500.- to 2,000.- RM.
>
> I humbly request suitable suggestions on its possible utility. Should official use of the objects by the State police authorities not be deemed suitable or possible, I request liquidation by means of a sale on the open market. [. . . ]. In order to prevent that the boat be acquired by another subversively minded person, the agencies should ascertain the character of the potential buyer beforehand. [. . . ].

The boat was immediately put up for sale, even though the police authorities only responded much later to say that they had no use for it.

The following advertisement appeared in the *Potsdamer Tageszeitung* on 28 February 1934: "Jollenkreuzer with auxiliary motor, accessories. Immediately. Massive mahogany, good condition, 20 sq.m. sail, stored Caputh, Potsdamer Strasse 27. Offers to Municipality of Caputh until 8 March."

The advertisement does not reveal that the boat had previously belonged to the Jew Einstein. It appeared among many other offers for a "crystal mirror," an "upholstered suite" or a "small sofa," for example, and with the exception of "fur coats" none indicated any prices. The bidders never found out about the decisive selection criterium, of course: loyalty to the Third Reich.[1010]

The first offer (1,200 reichsmarks) – *without* the "Heil Hitler" closure before the signature – came from the dentist Dr. Fiebig from Nowawes and was dated 2 March 1934.[1011] Perhaps also to emphasize the generosity of his offer, Fiebig pointed out that the boat was "in need of a complete overhaul."

The second offer (600 reichsmarks) – with the "Heil Hitler" embellishment – came from a man from Potsdam and was dated 5 March 1934.[1012]

The third offer (1,000 to 1,200 reichsmarks) – again with a "Heil Hitler" – came from a doctor from Berlin and was dated 6 March 1934.[1013]

> There is a handwritten note on the offer submitted by Dr. Fiebig to the Municipality of Caputh: "Relatives of Fiebig gave Einstein the money to buy the boat (Fiebig's own admission here at the office) [...]."

In accordance with the Gestapo's order, the authorities wanted to make sure that in future no "subversively minded persons" talk, think, or love each other and, after all that, even relax on it. So the district administrator asked the police chief at Potsdam on 14 March 1934[1014] – shortly after the expiration of the offer deadline – to run a check on the bidders. The check yielded that none of the three bidders had comported themselves "in a subversive sense." It was explicitly pointed out in two cases (Dr. Fiebig among them) that the applicants were "of Aryan origin."[1015]

Nothing would have stood in the way of the sale, if the Berlin Sports Club of the Reich Railway had not decided to make the purchase on 6 April, long after the submission deadline set for bids. Their aim was "fitness of our younger generation in our water-sports department."[1016] The offer was submitted after "consultation with the head of the municipality" of Caputh. Not surprisingly after this "consultation," a higher offer was made, in the amount of 1,300 reichsmarks. The head of the municipality knew perfectly well, though, that the deadline had been fixed for 8 March. In other words, negotiations were continuing on behind the backs of the regular bidders. The administrator of the district of Zauch-Belzig was also involved in these backroom deals behind the scenes. On 17 April 1934 he asked the district president of Potsdam to make the award to the Berlin Sports Club of the Reich Railway,[1017] firstly, because it had made the highest offer and secondly, because "this club would also guarantee that the boat would not be transferred into the ownership of subversive organizations." The district president of Potsdam supported the view that the Sports Club get the boat in his letter to the Prussian minister of finance dated 24 April 1934.[1018]

The first bidder caught wind of the foul play, though. He complained to the Prussian minister of finance on 25 April 1934 "bids are still being accepted and taken into consideration, and because secrecy about the bid amounts has been dropped since termination of the bidding period and it is then an easy matter to outdo the offers and become the highest bidder."[1019] He assured him that "the boat would be used by me personally for weekend excursions and is not going to be immediately sold again." Dr. Fiebig also did not fail to mention, of course, that in anticipation of the award he had abandoned "other promising options." As an additional argument he pointed out that he was a war veteran, married and father of an eight-year old son, and that in 1930/1931 he had been the leader of the Potsdam branch of the Academic Veterans Association of Berlin.

He also defended his legal right as first bidder in a telephone call with the district administrator on 25 April 1934. The district administrator was convinced. Dr. Fiebig eventually presented himself at the Ministry of Finance and stated his willingness to match the bid submitted by the Sports Club of the Reich Railway (i.e., 1,300 reichsmarks).[1020] The minister of finance ultimately could not turn a deaf ear to the argument "that under otherwise identical conditions, bidders who had submitted their offers on time deserve preference in principle."[1021] He communicated this standpoint to the district president of Potsdam on 2 May 1934. Fiebig's objection was acknowledged.

In May 1934 Einstein's sailboat was sold to Dr. Fiebig from Nowawes for 1,300 reichsmarks.[1022]

If Herneck's information is correct, there is at least *one* consolation in this situation: "The findings reveal that the purchaser and direct successor in ownership of Einstein's sailboat belonged neither to the NSGWP nor to one of its 'formations.' On the contrary, he financially supported five orphans of an antifascist, who had been convicted and executed by the People's Court in Berlin in 1944. He forbade his son from joining the Hitler Youth, against the expressed wish of the head of his high school."[1023] To that extent, the sailboat did, in the end, go to a "subversively minded person."

## 3.3.4 Any help from Switzerland?

Einstein, his relatives and his attorneys all tried to shield Einstein's personal property from Nazi hands. In vain.

And what did Switzerland do when Einstein asked for help? The country Einstein always felt sentimentally attached to; whose citizen he was and wanted to remain?

Nothing. On the contrary. His call for assistance was not just rejected, Einstein was virtually reprimanded for having "*acted,* for years, as a German."

After Einstein found out from the papers that his bank account in Berlin had been frozen, he asked the Swiss embassy in Brussels "to apply to the Swiss government [... to] intervene with the German or Prussian government in the following matter. My and my wife's bank deposits in Germany have been seized without justification. I am a Swiss citizen (citizen of the Canton of Zurich) and apply as such to the Swiss government to demand the release of my bank deposits by the German authorities."[1024] Ambassador Barbey contacted his superior for instructions on how he ought to proceed. The head of the Department for Foreign Affairs, Maxime de Stoutz, replied: Protection of "dual citizens" (owners of Swiss as well as another citizenship) had to be declined "to the extent that they had hitherto" placed themselves "under the protection" of another state, "or had otherwise publically posed as their member." The chief diplomat continued: "As far as Professor Einstein is concerned, he has apparently found it important always to be in possession of a Swiss passport. But that did not prevent him in any way from letting himself be celebrated on various occasions abroad as a representative of German science and to place himself under exclusive German protection. [...]. Moreover, he was a member of the 'Commission internationale de collaboration intellectuelle' as a German. [...]. It is known to us that Professor Einstein has traveled on a German diplomatic passport."[1025]

Perfidious:

- It was known, also in Berne, from the outset that members of the committee of the League of Nations had been appointed on the basis of their scientific merit *as individuals* and not as representatives of any country.
- On the occasion of Einstein's Nobel prize award in 1922, Switzerland ceded place to the German Reich, simply accepting the statement "Einstein is a Reichsdeutscher," in full knowledge that the scholar was a citizen of Switzerland.

- When Einstein asked for a Swiss diplomatic passport in 1925, Switzerland rejected his application. He thus had no other alternative. He *had* to travel on a German diplomatic passport.

The bottom line: "In view of the above-mentioned facts, it is obvious that the preconditions are satisfied for denying Einstein Swiss diplomatic protection."[1026] The real reasons, one would think, were mentioned afterwards: "Added to this is, the measures that the German authorities have currently been taking against him specifically regard his Reich citizenship and an alleged breach of his duties as a citizen." Consequently, Einstein was deemed guilty of having taken part in the "atrocity propaganda" against Germany. His antifascism was the genuine reason why Switzerland declined the requested assistance.

"In somewhat less drastic words the Swiss ambassador in Brussels relayed to Einstein the negative report from Berne. Einstein then explained his position to the Swiss ambassador in a personal interview followed by a letter presenting his views in writing." Einstein wrote: "It is true that I have been described in many official speeches as a 'German scholar.' I myself have never employed such terminology in my own statements." He also set the facts straight about membership on the committee of the League of Nations, pointing out that the elections were not made according to national quotas but on the basis of their international reputation as scholars. He stressed that he had been very closely tied to Switzerland since early youth.[1027]

The Swiss side sought "the most discreet discussion possible of this case with the German authorities." Because Einstein was still, nationally speaking, a "dual citizen" (he was at that time still also a Prussian), they informed him that any official intervention was out of the question. Their assurance to "draw attention to Einstein's case along official channels" in Berlin was probably made more for appearances' sake than out of sincerity. Ambassador Dinichert reported about the outcome of these "efforts" from Berlin in mid-June 1933 to Federal Councillor Motta: The inquiry at the Foreign Office had yielded that it "was little inclined to concern itself in any way with this matter. Professor Einstein had just recently been involved in what is called the atrocity propaganda and the German Ministry of Propaganda, which had a decisive say in the whole matter, was therefore biased against Professor Einstein."[1028]

In plain English: the Swiss ambassador did not want to upset anyone of the likes of Goebbels. So he yielded. Set before the choice between Goebbels and Einstein, he opted for the former.

We can gather the quality of Ambassador Dinichert's "efforts" from a report by the Foreign Office about information he had confidentially conveyed to them: "Einstein was, by origin, a Swiss citizen but had regarded and conducted himself as a German for decades. For instance, he was on the League of Nations Committee on Intellectual Cooperation, alongside a Swiss, as its German member. [...]. Mr. Einstein had requested diplomatic intervention by the Swiss government, because, as the envoy confidentially suggested, in order to extract his personal wealth, which is still in Germany. The Swiss government initially replied by pointing out that they had hitherto regarded Mr. Einstein, just as he himself had, as a German. It was therefore not possible to intervene on his behalf with the German government. Mr. Einstein thereupon stated that he had given up his German citizenship and was now only a Swiss citizen. The envoy added that, should this information be confirmed, his government would conclude that it would be legitimate to intervene on his behalf. Whether and to what extent they would, in fact, act on this intention still remains to be seen."[1029]

It was clear that the Nazis had nothing to fear from Switzerland. In Einstein's case they could do what they pleased. When Einstein was in need of help from Switzerland, it shook hands with the Nazis.

The efforts to retrieve Einstein's savings continued on. On 15 October 1952, on the basis of the Law for the Indemnification of Victims of National Socialism, Einstein filed for damages at the Claims Office in Wilmersdorf, Berlin. On the same day he joined Margot Einstein in applying for restitution of Elsa Einstein's estate. The process consumed much time. When the verdict finally came down on 30 July 1957 (largely in favor of the claimants), Einstein had already died.

Einstein was not destined ever to see his summer villa and sailboat again, let alone the confiscated money.

## 3.4 Conclusion

Hatred against people who would not distance themselves from Einstein and the theory of relativity knew no bounds. Planck, von Laue, Heisenberg and others were eventually vilified as "white Jews." The Nobel laureate Johannes Stark, stood out in particular, alongside Lenard, as the most renowned Einstein opponent. On 3 February 1933, even before Hitler had taken over the government, Stark wrote jubilantly to Lenard: "Finally our time has come!" At last "we can give our conception of science and research due attention."[1030] He did not have to wait long for a reward for his efforts: On 1 May 1933 Interior Minister Frick appointed Stark as president of the Bureau of Standards (PTR) in Berlin; in the following year, Stark took over leadership of the Emergency Association of German Science (Notgemeinschaft).

Johannes Stark on "white Jews":[1031]

It is generally known that during National Socialism's time of struggle the great majority of professors at German universities and colleges failed *disgracefully*. They were unsympathetic toward Hitler and his movement; and some even repudiated him. At many universities serious conflicts arose between the National Socialist student body and the faculty linked to the black-red [i.e., conservative/socialist] political system. In 1933 Reich Minister Rust justifiably had bitter words in this regard for the professoriate in Berlin. The dominant Jewish influence at German universities was the decisive reason for the political failure of the majority of German professors in the National Socialist struggle for German liberty.

Its strength lay not only in the fact that in numerous departments 10 to 30 percent of all lecturers were Jewish or closely related to Jews, but it was also primarily because Jews had the support of their Aryan comrades and pupils.

The political influence of the Jewish mentality at universities was well known. Its influence scientifically speaking was less well known but equally damaging, in that it crippled Germanic research, which is oriented toward reality, by instilling Jewish intellectualism, dogmatic formalism, and propagandistic commercialism. Furthermore, this Jewish influence attempted to educate students in the Jewish mentality, including primarily the upcoming academic generation.

Though ethnically Jewish lecturers and teaching aids were forced to resign their positions in 1933 and though Aryan professors married to Jewesses are presently also being discharged, the majority of Aryan comrades and pupils of Jews, who had previously supported Jewish power in German science openly or covertly, have retained their positions and keep the influence of the Jewish spirit alive at German universities.

In their unworldliness they counted on a swift end to the National Socialist government right up to the election of the Führer, and therefore withheld public announcements in support of him. Two years ago, however, their tactics changed. Now they behave outwardly nationalistically; former pacifists throng to military service; pupils of Jews, who had published numerous scientific papers together with domestic and foreign Jews and who had participated besides in congresses held by Soviet Jews, now seek connections within party and government agencies.

Aside from their nationalistic or even National Socialist activities, they also try to exert their influence at high places with the following arguments: They say that as scientific experts they and their candidates are indispensable to the implementation of the Four-Year Plan. Moreover, they say, they are recognized abroad as great German scientists and must have a leading influence on German science in the interest of its reputation. With this bluff they believe they can count on the fact that the authorities are not informed that their 'fame' abroad is an over-exaggeration of the result of their collaboration with foreign Jews and comrades of Jews.

The following is indicative of continued Jewish influence in German academic circles: Not long ago an influential doctor said to me: ''I cannot imagine medical science at all without Jews.''

The department of natural sciences at a major university recently nominated three pupils of Jews for a professorship, two of whom had published numerous scientific papers *together with domestic and foreign Jews*. The scientific book market in Germany is lately being deluged, primarily in physics, with books penned by domestic and foreign Jews and pupils of Jews, particularly through the Julius Springer publishing company in Berlin and Vienna, which was formerly completely Jewish and is today supposedly 50 percent Aryan.

While the Jewish spirit's influence on the German press, literature and art, as well as on the German legal system has been eliminated, it has found its defenders and proliferators in German science among the Aryan comrades and pupils of Jews at universities. Behind the screen of scientific impartiality and referring to its international recognition, the Jewish spirit continues on unweakened and even seeks to dominate by securing and consolidating its tactical influence on important positions.

Under these circumstances 'Das Schwarze Korps' does a great service with its courageous and fundamentally important observations by turning public attention to the damage inflicted by 'White Jews,' which threatens a segment of German intellectual life as well as the education of the academic youth.

It is remarkable that Planck, von Laue and others did not let themselves be oppressed by the campaign of insults. Notably Max von Laue, Einstein's friend, would not be irritated in the least by it. The minister might feel obliged "to express his disapproval about your conduct damaging to the purposes of the Institute and the State,"[1032] or censure him for letting the word "relativity theory" slip out of his mouth again.[1033] It changed nothing. Not even the minister's rejection of the academy's application of 30 November 1933 to appoint Max von Laue to the position vacated by Einstein had any effect. Lenard and Stark had energetically dissuaded the minister from appointing this publically known "close friend and champion of Einstein," to use Stark's words.[1034]

Courageous though these people were, the Nazis, nevertheless, had the *power* to turn them out of their offices. The Nazis proved their unscrupulousness more than enough. Foreign considerations were equally insignificant to them, as the war they had machinated and the extermination of the Jews demonstrated.

There *had* to be other reasons for this tolerance towards relativists.

A *reminant* of doubt persisted. The fear remained that their rigorous rejection of Einstein's theory could have been a mistake only *detrimental to themselves*. The theory of the condemned and expelled Jew, Albert Einstein, might possibly be more useful than the 'Aryan Physics' of Messrs. Lenard and Stark. They had cheered too soon, after all. In 1933 already Stark was complaining that the Nazis were bad comrades in the struggle to implement Aryan Physics. As Andreas Kleinert has shown, the Aryan Physics movement was *not* Nazi physics. "This impression is misleading [...]. The influence of Lenard and Stark was rather in constant decline throughout the Third Reich."[1035]

# Einstein's FBI file – reports on Albert Einstein's Berlin period

German archives are not the only place where Einstein dossiers can be found. Leaving aside other countries, at least *one* personal dossier exists *in the USA:* the Einstein File of the Federal Bureau of Investigation (FBI).[1036]

This file holds 1,427 pages. In our context the numerous reports about Einstein's "Berlin period" are of particular interest. Taking a closer look at them does not lead us beyond the scope of this book. On the contrary, these reports give a complex picture of Einstein's political activities during his Berlin period – albeit from a very specific point of view: the view of the American CIC (Counter Intelligence Corps) and the FBI of the first half of the 1950s.

The core of these reports is the allegation that Einstein had cooperated with the communists and that his address (or "office") had been used from 1929 to 1932 as a relay point for messages by the CPG (Communist Party of Germany, KPD), the Communist International and the Soviet Secret Service. The ultimate aim of these investigations was, reportedly, to *revoke Einstein's United States citizenship and banish him.*

Space constraints prevent a complete review of the individual reports here. So under the given circumstances a survey of the contents of the two *most important* reports will have to suffice for our purposes along with some additional information. These reports are dated 13 March 1950 and 25 January 1951.

### *13 March 1950*

The first comprehensive report by the CIC (Hq. 66th CIC Detachment)[1037] about Einstein's complicity in activities by the CPG and the Soviet Secret Service between 1929 and 1932 is dated 13 March 1950.[1038] Army General Staff only submitted this letter to the FBI on 7 September 1950.

The essence of the accusations regarding Albert Einstein were:

Prior to 1933 the Comintern and other Soviet Apparate were very active in gathering intelligence information in the Far East. Many International Communist functionaries were stationed in Shanghai and Canton [. . . ]. One means of communication used by these persons to contact Central Headquarters in Moscow was through the use of telegrams. However, these telegrams, which were always in code, were never sent directly to Moscow, but were sent to Agents in other countries, [. . . ] where they were re-copied and forwarded to telegram addresses in Berlin. One of these addresses was the office of Einstein.

Einstein's personal secretary (her name cannot be recalled) turned over the telegrams to a special apparat man whose duty was to pick up such mail from several telegram addresses which included Einstein's office, a watchmaker's shop known as Uhrenelb, [. . . ] and a ''Kartonagenfabrik'' (box factory) operated by Walter Schauerhausen,[1039] Berlin [. . . ], Neue Jacobistr.

Since those telegrams were in code, it is assumed that Einstein did not know their contents. However, it is reasonable to believe that Einstein did know that his office was being used by the Soviets as a

FEDERAL BUREAU OF INVESTIGATION

ALBERT EINSTEIN

PART 1 OF 9

BUFILE NUMBER: 61-7099

**Figure 56: The Einstein file.**

telegram cover address. [. . . ]. Einstein's Berlin staff of typists and secretaries was made up of persons who were recommended to him (at his request) by people who were close to the ''Klub der Geistesarbeiter'' (Club of the Scientists), which was a Communist cover organization and which served as a source of personnel for various Soviet MD and other illegal Apparate. Einstein was also very friendly with several members of the Soviet Embassy in Berlin, some of whom were later executed in Moscow in 1935 and 1937.

Einstein's telegram address was for some time under the supervision and protection of Richard Grosskopf, who is presently the Chief of the Berlin Criminal Police in the Soviet Sector [of Berlin], and who at that time was in charge of the KPD's passport falsification apparat under the alias of Steinecke. Grosskopf had issued a fake passport to an alleged Swiss citizen, [. . . ] Ruegg, who was operating as an agent in the Far East and who was chief of the Pacific Labor Union, agent of the Comintern, and co-worker of the Soviet MD [intelligence services] in China. At one time Ruegg had a large amount of intelligence information to forward and, due to security reasons, he was forced to use both Einstein's address and the watchmaker's address, Uhrenelb, Berlin. Ruegg was later arrested by Chinese police, and Grosskopf was arrested in 1933 by German security police. Grosskopf was succeeded as head of the passport falsification apparat by Adolf Sauter, who in 1933 dropped Einstein's address since Einstein had already left Germany.

Persons who are known to have used Einstein's address or who were aware of the fact that Einstein's address was being used are Richard Grosskopf, Adolf Sauter, Friedrich Burde, Wilhelm Bahnik, Johannes Liebers, Wilhelm Zaisser, Karl Hans Kippenberger, Alfred Kattner, Wilhelm Wloch, Dr. Guenther Kromrey, and Herrmann Duenow. Of these the only persons known to be alive are Grosskopf, Sauter, Zaisser, Wloch, Kromrey and Duenow. The others were either executed, are missing, or died in Spain.

Einstein was closely associated with the ''Klub der Geistesarbeiter'' and was very friendly with Fritz Eichenwald, Dr. Bobeck,[1040] Dr. Caro, Dr. Hautwermann[1041] and Dr. Kromrey, who were all members of the Club and who later became agents of the Soviets. Also associated with this Club were the two Fuchs brothers, both of whom were students at that time. One of the Fuchs brothers worked for the Communist Party MD and in 1934 was associated with the Abwehr Apparat of the Landesleitung Berlin-Brandenburg.[1042] He left Germany in 1934 or 1935 for Switzerland. The other Fuchs brother was Klaus Fuchs, who was associated with Apparat Klara[1043] and worked with Fritz Burde, and later with Wilhelm Bahnick. Klaus Fuchs was recently jailed in England for giving the Soviets A-Bomb information.

Further verification of the matter was deemed necessary.

*25 January 1951*

The report by the CIC (Hq. 66th CIC Det.)[1044] from Munich provides details about Einstein's purported complicity in the activities of the CPG, the Communist International and the Soviet Secret Service between 1929 and 1933. It responds to questions posed by the head of the FBI dated 13 March 1950. This information is reiterated in numerous subsequent reports and memoranda.

SECRET

21267

13 March 1950

SUBJECT: EINSTEIN, Albert

X-714

1. REASON FOR INVESTIGATION:

The following information concerning Professor Albert EINSTEIN's affiliation with Communists and the use of his office in BERLIN (N53/Z75) until 1933 as a telegram address by Agents of the Comintern and other Soviet Apparate is deemed to be of sufficient interest to warrant forwarding to higher Headquarters.

2. SYNOPSIS OF PREVIOUS INVESTIGATION:

None.

3. PRESENT INVESTIGATION:

Prior to 1933 the Comintern and other Soviet Apparate were very active in gathering intelligence information in the Far East. Many International Communist functionaries were stationed in SHANGHAI and CANTON for that purpose, and among them were experienced German Apparate men such as Hans BARICH, Friedrich BURDE, Johannes LIEBERS, Wilhelm ZAISSER, and Edith [illegible]. One means of communication used by these persons to contact Central Headquarters in MOSCOW was through the use of telegrams. However, these telegrams, which were always in code, were never sent directly to MOSCOW, but were sent to Agents in other countries, such as Egypt or France, where they were re-copied and forwarded to telegram addresses in BERLIN. One of these addresses was the office of EINSTEIN, which proved to be very successful since EINSTEIN received a great quantity of mail, telegrams, cablegrams, etc. from all over the world.

4. EINSTEIN's personal secretary (her name cannot be recalled) turned over the telegrams to a special apparat man whose duty was to pick up such mail from several telegram addresses which included EINSTEIN's office, a watchmaker's shop known as UHRENKLB, BERLIN, and a "Kartonagenfabrik" (box factory) operated by Walter SCHAUERHAUSEN, BERLIN So., Neue Jacobistr. The pick up man then distributed the telegrams to the various chiefs of the various Soviet Apparate in BERLIN, who in turn sent the information to MOSCOW by courier, military attache pouches, illegal radio transmitters, and other methods.

**Figure 57: Page 1 of the FBI report of 13 March 1950.**

The accusations and allegations were based on the following:

Source has furnished the following details regarding the illegal use of subject's Berlin office as a letter drop:

a. In a technical sense, the actual address of subject's office was not used as a letter drop. What was used was the authorized international cable address of subject, correspondence for which was delivered to subject's office by the postal authorities. This cable address is believed to have been Einstein Berlin or AlbertEinstein Berlin. (AN: The exact address can be determined by consulting a German cable address book from 1929 to 1930, unavailable here.)

b. The exact location of subject's office, to which such cables were delivered, is not known, but it was in the vicinity of Nollendorfplatz.

c. The office concerned was subject's private office (Privatkanzlei), which had no connection with any organization or institution.

d. It is not known how many persons were employed in this office nor who those persons were, other than that Source has hearsay knowledge of at least two (2) female secretaries, both of whom were Communist-sympathizers.

e. The exact time of activation of the Communist use of subject's cable address as a letter drop is not known, but when Source was instructed to establish a legal cable address for use in the same ''line'' [. . . ] in 1929, subject's cable address was already in use.

f. It is not known exactly who decided to utilise subject's cable address for conspirative purposes, but the ''line'' to which it belonged operated under the West European Bureau (WEB) of the Comintern (headed at that time by Georgi Dimitroff, deceased) in conjunction with the International Liaison Department (OMS) (Otdel Meshdunarodnovo Sviazi).[1045]

g. The reason for using subject's address was that the extent of international cable traffic received by subject, from all corners of the globe, coupled with subject's established international reputation, would provide a relatively innocuous cover for conspirative communication.

h. The person in subject's office through whom the arrangements were made for the use of the cable address was subject's chief secretary at the time. [. . . ]. This secretary had close personal relationships, probably of an intimate nature, with an international Apparat functionary (whom Source can not identify) and, through these relationships together with her own Communist sympathies, was drawn into conspirative work. [. . . ].

i. The following procedure was used in transmission of cables through this channel:

(1) An intelligence message was encoded (not enciphered) from the ''clear'' into a prearranged Comintern or Soviet code. [. . . ].

(2) This message was then re-encoded in the approved international cable code used by subject. [. . . ].

(3) This double-encoded message was then dispatched to the Einstein cable address. [. . . ].

(4) At subject's office, it was the duty of the senior secretary, who was at the same time in the employ of the Apparat, to decode all messages and give them to subject for reading. She thus was in a position to intercept all messages which did not pertain to subject, which therefore were Apparat business, and transmit them to the Apparat courier.

j. The courier or contact man of the Apparat responsible for liaison with subject's office came directly to that office to pick up incoming cables for the Apparat. Covert contacts were considered both risky and unnecessary in this case. [. . . ]. Known liaison men working subject's office are as follows:

(1) Richard Grosskopf for about three (3) months in 1929, at which time Source first obtained knowledge of the use of subject's cable address.
[. . . ]. The ''line'' ran as follows: Einstein's office [. . . ] - an oriental rug shop on Potsdamerstrasse - a not further identified firm on Potsdamer platz - Katschalski, (or Kaczalski) watchmaker at the corner of Beuthstrasse and Kommandantenstrasse, ''money office'' (Geldstelle, an address used for financial transactions and transmittals) for the ''Klara'' Apparat (international Apparat of the Soviet General Staff, IVth Department) yet linked with the WEB line - ''Uhrenelb'', a watch shop at Alte Jakobstrasse 93 in which Adolf Sauter, functionary in the Berlin Security (Abwehr) Apparat was a partner - Gebrueder Schauerhammer carton factory on Neue Jakobstrasse - a tobacco store on Koepenickerstrasse. This line was concerned with Far Eastern operations and closely allied with the Paul Ruegg [. . . ] net in Shanghai. Grosskopf [who] was at that time head of the Pass Forging Apparat in Berlin and had very close connections with Abramov Mirov [. . . ], was probably chosen for that reason. [. . . ]

(2) Fritz Burde, head of the German industrial espionage Apparat (''BB'') with close affiliations with the Apparat of the Soviet General Staff, IVth Department (''Klara''), took over the contact briefly after Grosskopf was released from the mission. Burde is reported to have been executed in the Soviet Union after having been active for the international Apparat in the Far East, specifically Shanghai.

(3) Willi Wloch, leading functionary of the IVth Department (''Klara'') Apparat, who replaced Burde and his successor as contact man when they were indisposed or not available. Wloch's whereabouts are not known; his brother, Karl Wloch, is presently in Berlin and active with the German-Polish Cultural Relations and Friendship Society ''Helmut von Gerlach-Gesellschaft''.

(4) ''Fritz'' (Last Name Unknown), Berlin-Wedding, old-time international Apparat functionary most closely connected with OMS who in 1933 was operating a radio net out of Prague, Cse. Neither Source nor the German political police were ever successful in ascertaining the correct name of ''Fritz'', who was known throughout the Apparat by that name or as ''Fritz from Wedding''. Fritz is believed to be still alive and active. Fritz remained chief contact man with the Einstein office until the cessation of Source's knowledge, late in 1931, except during his frequent absences, when contact was made by Wloch or Burde or the following.

(5) Alfred Kattner, receptionist at the Communist Party Central Committee building in Berlin prior to 1933, who was contacted by unlisted telephone by the various letter drops when cables were not picked up promptly by the contact man and who occasionally

made contact himself. Source knows that Kattner made the contact at the Einstein office at least two (2) or three (3) times. Kattner was arrested and doubled by the German police after the seizure of power by the Nazis and was eventually liquidated by the Communist underground in about 1934.

k. As is the general rule with conspirative activities of this nature and on such a level, the only persons who were supposed to have knowledge of such matters were those directly concerned. Those who can be regarded as having direct knowledge of the conspirative use of subject's cable address are: Richard Grosskopf, Fritz Burde, Willi Wloch, Alfred Kattner, Fritz from Wedding, and Jakob Abramov Mirov of OMS. Knowledge is probable on the part of Wilhelm Bahnik (deceased, successor of Burde as head of industrial espionage, close confidante of Burde, ''Fred'' Liebers (deputy and right-hand man to Burde, who could usually be assumed knowledgable on matters known to Burde, sent in 1935 to Shanghai, Hermann Duennow (assistant to Grosskopf in Pass Forging Apparat, now active in Berlin), Albert Gromulat Sr (deputy head of Quarters Apparat, which was charged with providing cover and contact addresses and which would be informed at least to the extent that the Einstein address was ''tabu'' for other Apparate and purposes), Hans Kippenberger (as overall head of the German Apparate would be likely to know of the use of the Einstein address by Russian Apparate, probably had more detailed, though not necessarily direct knowledge), Leo Roth (long-time secretary and right-hand to Kippenberger), and possibly Wilhelm Zaisser (who would have no reason for knowing except through his activities in China, where he may have obtained knowledge of the use of the cable address from that end).

Source has stated that he does not believe subject was aware of the true nature of the correspondence which was channeled through his cable address from the Far East. It is even possible that arrangements were made for the use of the cable address with subject's secretary [. . . ] without subject's knowledge. If subject were aware of the use of his address from the beginning, Source points out that the most logical approach to subject in order to get his approval for such use would be to convince him that it was in the interests of ''human rights'' [. . . ].

a. One incident which took place during the summer of 1930 indicates that subject must have had some knowledge of the use of his cable address for purposes other than his own. At this time Fritz from Wedding was the contact man for the line to which the Einstein address was attached, and one day Fritz came to Source in a very excited state. According to the story he told Source at this time, subject's secretary had been planning to go on a three (3) or four (4) week vacation and had been instructed to notify Fritz in advance of the date of her departure so that arrangements could be made to intercept the comintern cables before they were given to subject and turn them over to Fritz. In making his routine contact on this day, Fritz discovered that the secretary had already left without informing him, and the other secretary disclaimed any knowledge of cables which were supposed to be turned over to anyone except subject. No attempt was made by Fritz to contact subject himself, and the only remaining alternative was to await the return of the secretary. When the girl did return several weeks later, all cables were turned over and there were no unpleasant repercussions, although Fritz did remark that his superiors were perturbed about the delay, since there were some important activities in the Far East at the time.

b. In connection with the above-described incident, Source comments that the normal reaction of a man receiving cables from various points in the Far East which made no sense to him and which had no connection with his activities, yet were clearly addressed to his own cable address, would be to make inquiries of his office personnel and probably check with the postal authorities or even make a complaint about unauthorized use of his cable address. Source states, however, that there were no unpleasant complications and that the use of the cable address continued as before when the secretary returned and even during her absence. Nothing is known about whether subject asked for explanations from the secretary or what explanation she gave, but the fact still remains that there was no hitch in the procedure, which indicates that subject must have at least continued to tolerate the situation.

c. Source has no further knowledge of details which would indicate knowledge or lack of it on the part of subject, other than that subject was frequently active in supporting so called ''front groups'',[1046] especially those concerned with human rights and anti-fascist activities.

The following information has been obtained from Source regarding the ''Klub der Geistesarbeiter'' (Club of Intellectual Workers):

a. The Klub der Geistesarbeiter (hereinafter referred to as KdG), although it was looked down upon and scorned by the functionaries of the international Apparate (who referred to it as the ''Klub of Mental Acrobats''), actually served as a fertile recruiting ground for high-caliber Apparat connections. The KdG was formed (date unknown, Source's first knowledge in 1931, last in 1933) in Berlin by a group of German scientists and intellectuals, all of whom were affiliated in some way with the international Apparate. The purpose of the KdG was to provide those persons with a chance to get together for intellectual discussion and exchange, to maintain contact for conspirative purposes, and to provide a basis for gradual recruitment of promising young intellectuals for conspirative work within or in connection with the Apparate.

b. Source is not aware of the identity of the founders or leading figures in the KdG. The KdG did, however, maintain offices and meeting rooms in the ''Hochhaus am Alexanderplatz'', in which building the very popular Cafe Braun was (and still is) located. It is believed the club rooms were on the 4th floor (American style).

c. Subject was not to Source's knowledge active in the KdG, although many of its members were acquaintances and/or associates and friends of his. It is believed that subject may have been an ''honorary'' member at one time. [. . . ] many of the younger staff members of the Berlin Technical College (Technische Hochschule) were connected with the KdG, as well as some theater [. . . ] and motion picture [. . . ] people.

d. Source definitely stated, in response to the direct query, that KdG was the correct name of the organization, and that it should be possible to locate and further identify it through the Hochhaus address, Dircksenstrasse, corner of Alexanderplatz.

[. . . ]. Agent's Notes:

a. [. . . ] So far as possible, pertinent background information has been included in the report.

b. Clarifying Comments:
Source's knowledge of the use of subject's cable address for conspirative purposes is based primarily upon his association with Richard Grosskopf, who mentioned the existence of the Einstein office in the ''line'' he was establishing in 1929 only in passing, and through his connection with Fritz from Wedding [. . . ]. Source was unable to identify any possible leads for further information aside from people like Grosskopf and Dueenow,[1047] who are still apparently firmly with the Communists. He was unable to give a single lead regarding the identity of the secretary through whom the arrangements were made for using subject's address other than to state that she was the chief or senior secretary at least from 1929 to 1931. Source's primary knowledge, although all actually secondhand, is based on his connection with one unit in the ''line'' to which subject's address was attached.

c. Agent's Opinion:
It is the opinion of the undersigned that the information given by Source, as far as it goes, is probably accurate. [. . . ].

d. Specific Recommendations:
It is recommended that no further exploitation be made of Source in this case. The identity of the secretary could be accomplished through inquiries in Berlin or of persons who had contact with subject's private office at that time. It is further recommended that former RSHA (German Central Security Agency) personnel now available be queried regarding the identity of those officials who, shortly after 1933, worked on the case built up around the cables received by subject's office prior to 1933. All cables were picked up by the RSHA from the Central Telegraph Office (Haupttelegrafenamt) in Berlin and studied in the light of knowledge received by the RSHA that subject's cable address had been used by the Soviets

```
and/or Communists. Source does not know the identity of the offi-
cials who worked the case but does have knowledge that the case was
being worked on in 1935.
```

*23 February 1955*

On 23 February 1955 Einstein's long-time secretary was questioned "Concerning possible connection with espionage activities on behalf of Russia 1928–1933 in Berlin, Germany."[1048] The interview took place at Albert Einstein's home: 112 Mercer Street, Princeton, N.Y.

Helen Dukas denied any knowledge about spies contacting Einstein. She also denied knowing anything about the Club of Intellectual Workers. She stated that she had been Einstein's first and only secretary since 1928 – disregarding the assistance by Einstein's wife and elder daughter-in-law, both of whom had meanwhile died. Before she had started working for Einstein he had employed students on a part-time basis. Einstein had no other office in Berlin and had no other employees in his household. Helen Dukas's own tasks had not been limited to typing and other office work. She had also been cook and maid for the whole family. She informed the interrogator that Einstein was currently in a bad state of health.

The report was preceded by a reiteration of the earlier allegations about Einstein's activities in Germany. Two points were emphasized: First, the use of his cable address (presumably with Einstein's knowledge) for communist activities. Second, Einstein's friendship with members of the Club of Intellectual Workers who later became Soviet agents. Klaus Fuchs, at that time still a student, was specifically named in this connection.

Helen Dukas denied being acquainted with or having heard of Georgi Dimitroff and Jacob Abramov (Mirov) – adding the qualification that Dimitroff was known throughout Germany as an official of the Comintern and in connection with the trial concerning the Reichstag arson.

She also denied knowing persons who were supposed to have acted as couriers to Einstein's office or to have been involved in any way in Soviet spy activities.

She denied having any knowledge about contacts Einstein may have had to the communist underground but also pointed out that she always only had a limited knowledge of Einstein's activities. Between 1929 and 1933 Einstein's health condition was good and he was actively involved in many different things. She knew nobody interested in communism. Her circle of friends were mostly Jews; she was primarily interested in Jewish issues.

Asked about how often Einstein was on the telephone, Helen Dukas stated that Einstein received few telephone calls and had mostly dealt with his affairs by regular mail.

Finally she was asked about Einstein's family (sons, first and second wife, etc.): Einstein had two sons (Hans-Albert and Eduard) and two step-daughters (Margot and the elder one, whose name Helen Dukas had forgotten). As regards

Albert Einstein's state of health, Helen Dukas said that he had contracted the flu that winter and also had a heart condition.

*9 March 1955*

The cover letter to the report[1049] informed the director of the FBI that Helen Dukas had been extremely friendly and sincere throughout her interview. She had not been evasive nor seemed in any way cautious. She gave no indication that she might have been feeling observed. At the end of the interview she was not at all antagonistic toward the FBI agents but offered to help anytime.

On the basis of the given information further investigations were not considered necessary.

Helen Dukas had stated that she had been Dr. Einstein's sole employee since 1928. It was very probable that Elsa Einstein – meanwhile deceased – or his stepdaughter (name unknown) had served as head secretary from 1926 until her marriage.

Further information, the report continued, could not be expected because the supposed Soviet agents were either dispersed in many different countries or had died in the meantime.

That was why, if nothing spoke against it, the case files on Dukas and Einstein at the FBI's Newark office would be closed.

*2 May 1955*

The Newark Office informed the director of the FBI that Albert Einstein had died on 18 April 1955 in Princeton, New Jersey.[1050]

Additional investigations were thus obviated: "This matter is being closed."

The whole business regarding Einstein and his secretary was apparently resolved on a friendly note. The inquisitional zeal and every last trace of distrust suddenly vanished. The FBI agents were even downright impressed with Helen Dukas, the subject of their researches. They concluded that they had been following a wrong lead. 'Source,' whom Counter Intelligence Corps had relied on so much, was thus also dealt with and discredited.

But this business does not end here. The allegations in Albert Einstein's FBI file must be examined to see exactly how much of the information was a figment of Source's imagination. So many of the details in otherwise unrelated matters were correct. We also need to find out why the attitutes of the CIC and the FBI in 1955 changed so radically in such a short time. How does the erstwhile so suspicious Helen Dukas become prime witness? Finally, we also need to ask *who* issued the reports and *who* exactly was Source? More generally, how could fascist-stlye tendencies emerge just a few years after the end of World War II in a fundamentally democratic country, indeed in a former member country of the anti-Hitler coalition? Why was the same thing attempted in USA, with the *very*

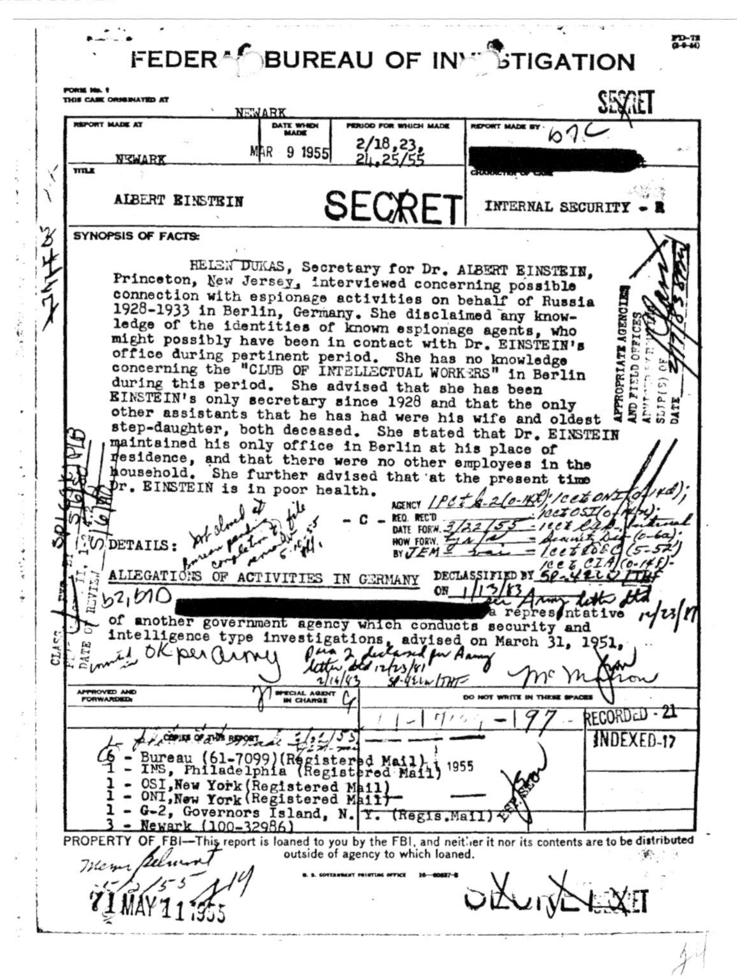

FEDERAL BUREAU OF INVESTIGATION

SECRET

| REPORT MADE AT | DATE WHEN MADE | PERIOD FOR WHICH MADE | REPORT MADE BY |
|---|---|---|---|
| NEWARK | MAR 9 1955 | 2/18,23, 24,25/55 | |

TITLE: ALBERT EINSTEIN — SECRET — CHARACTER OF CASE: INTERNAL SECURITY - R

SYNOPSIS OF FACTS:

HELEN DUKAS, Secretary for Dr. ALBERT EINSTEIN, Princeton, New Jersey, interviewed concerning possible connection with espionage activities on behalf of Russia 1928-1933 in Berlin, Germany. She disclaimed any knowledge of the identities of known espionage agents, who might possibly have been in contact with Dr. EINSTEIN's office during pertinent period. She has no knowledge concerning the "CLUB OF INTELLECTUAL WORKERS" in Berlin during this period. She advised that she has been EINSTEIN's only secretary since 1928 and that the only other assistants that he has had were his wife and oldest step-daughter, both deceased. She stated that Dr. EINSTEIN maintained his only office in Berlin at his place of residence, and that there were no other employees in the household. She further advised that at the present time Dr. EINSTEIN is in poor health.

\- C -

DETAILS:

ALLEGATIONS OF ACTIVITIES IN GERMANY

a representative of another government agency which conducts security and intelligence type investigations, advised on March 31, 1951,

RECORDED - 21
INDEXED-17

COPIES OF THIS REPORT

6 - Bureau (61-7099)(Registered Mail)
1 - INS, Philadelphia (Registered Mail)
1 - OSI, New York (Registered Mail)
1 - ONI, New York (Registered Mail)
1 - G-2, Governors Island, N.Y. (Regis. Mail)
3 - Newark (100-32986)

PROPERTY OF FBI—This report is loaned to you by the FBI, and neither it nor its contents are to be distributed outside of agency to which loaned.

SECRET

Figure 58: Report on the interrogation of Helen Dukas, 9 Mar. 1955.

*same* arguments that Germany had attempted hardly twenty years before: Einstein's expatriation (because of purported "*communist activities*")?

# Fact, fiction and lies

## Streets, places. Einstein's apartment

*All* the places, streets and addresses indicated in the reports about Einstein's Berlin period actually existed.

The noteworthy thing about this is less the informant's familiarity with the mentioned streets and places than their physical proximity. The geographical center of the described action was the region around Spittelmarkt in Berlin, even though the underground communist activities were dispersed throughout the *entire* municipal area.

Although Einstein's residence (along with his "office") is listed in the standard Berlin directory, it is not mentioned in the reports: Haberlandstrasse no. 5.

Consulting the German *Telegramm-Adressbuch* for 1929 and 1930 would have revealed whether Einstein had a "telegram address."[1051] They knew how to go about it. It would have been an obligatory thing to check out, considering the central role this address played in the intelligence reports. But it was not done. The argument that this information was "not accessible here" is no excuse. A book available to us today, fifty years later, was certainly obtainable then. Effort was all that was needed.

The result of such a check is: There is no listing of Einstein in the cable directory for 1929/1930. Einstein did *not* have any cable address!

Since Einstein's apartment was supposed to be a cable drop for messages, it is worthwhile to check and see whether it was suitable at all for such purposes.

The building at Haberlandstrasse no. 5 was built 1907/1908. A number of renovations were done subsequently to the façade as well as in 1919 when toilet plumbing, etc., was installed. An elevator is already documentable to 1911.[1052]

Einstein had resided at Haberlandstrasse no. 5 since the summer of 1917, and since 2 June 1919 in the apartment of Elsa Einstein, his lover and later second wife.

Elsa Einstein's apartment, that is, her father Rudolf Einstein's, was originally in the second story. Elsa and Albert Einstein must have moved upstairs to the fourth story sometime around 1920 (or earlier). *Two* newspaper articles from the time indicate this. The first appeared in various dailies (among others in the *Kieler Zeitung* dated 17 October 1920 and *Braunschweiger Landeszeitung* dated 21 January 1921). The first article reads: "There, where the high, old-fashioned styled building façades of the Bavarian quarter crowd into almost narrow alleys [...] is, *four flights upstairs,* Albert Einstein's Berlin apartment."[1053] The other article, likewise reprinted in various papers (*Kölner Tageblatt* of 2 September 1920 and *Leipziger Tageblatt* of 3 September 1920, among others) about 'A visit with Einstein' states: "And so there I was, standing in the corridor of the *four-story* building of Berlin Haberlandstrasse for the first time before Professor Einstein."[1054]

A sketch drawn by the stove fitter and district master chimney sweep, Georg Schwingel, before the gas water heater was installed in the bathroom in 1923, indicates the layout of the apartment. It shows that there were eight rooms in addition to a large hallway, kitchen, pantry, bathroom and water closet.[1055]

Besides providing us with the exact layout of the apartment, the master chimney sweep's sketch also shows us that it was *not* situated on the Haberlandstrasse side of the building but along Aschaffenburger Strasse. Mr. Schwingel's sketch does not agree with the structure of the building on Haberlandstrasse but does with the one along Aschaffenburger Strasse! So the address "Haberlandstrasse 5" was just the main entrance to their building. The photo published by Grüning and others[1056] is hence *not* correctly marked! The shapes of the windows visible on the sketch do not match the façade on Haberlandstrasse, but the one on Aschaffenburger Strasse. (Even if the cross mark had, in fact, been made by Albert Einstein, he had orientational problems in space in everyday life.) Banesh Hoffmann worked with the *same* marked photo and his reprint from 1976 reveals a handwritten note "Our apartment Haberlandstr. 5" in the same pen as the cross[1057] (in a handwriting that cannot be attributed to either Albert or Elsa Einstein!).

Other misleading statements about the apartment's location appear in Friedrich Herneck's book 'Einstein privat.' In answer to the question "What direction were the windows of the Biedermeier-room and the library positioned?" Einstein's former housemaid Mrs. Herta W. recalled, "I would think east or southeast, because there was always so much light in those rooms." Whereupon Herneck replied: "Southeast would fit nicely with the course of the street [. . . ]."[1058] These are very vague indications indeed, and Herneck was able to add "east or southeast" simply because it happened to agree so well with *his* conception of the "course of the street."[1059] On the other hand, other details Herta W. gave agree exactly with the building layout: The side entrance "ended upstairs in the hallway, directly next to the kitchen. At the bottom you did not come out onto Haberlandstrasse but onto Aschaffenburger Strasse."[1060] According to the official floor plan, the side entranceway leading "directly next to the kitchen" only matched the building with windows opening out onto Aschaffenburger Strasse! At the bottom of the stairs "you did not come out onto Haberlandstrasse but onto Aschaffenburger Strasse," provided "*you*" refers to deliverers or household employees. The floor plan also agrees with Hertha W.'s description of the rooms (which Herneck could not have known).[1061] Herneck's drawing according to former Miss Herta Schiefelbein's information agrees almost exactly with the true situation. The only significant discrepancy is the name of the street it looked out on: not Haberlandstrasse but *Aschaffenburger Strasse!* This is corroborated by a comparison against extant blueprints in the municipal building file.

For the uninitiated, hence also police officers, *this fact* would have complicated any surveillance of Einstein's apartment!

Until 1927 the purpose and therefore also the description of the rooms changed a few times. Konrad Wachsmann described the apartment as follows: "I think it was seven or eight rooms. If you entered the hallway, Einstein's bedroom was on the left" (called "Herrenzimmer" on the sketch) "behind that was the library"[1062] (1923 the "reception room")[1063] "and the salon" (1923 called "livingroom" (Wohnzimmer)), "in which the grand piano stood. From the salon you could pass through a sliding door on the right-hand side into the diningroom. Straight ahead was another door through which you came to a small hallway and from there to the bathroom. Elsa Einstein's and her daughter Margot's bedrooms also issued into this hallway. I do not know which room the daughter Ilse inhabited because she was already married. Behind the kitchen there were some more rooms for the staff. But I never saw them."[1064]

The apartment also included Albert Einstein's study rooms under the roof (the so-called "tower room" along with a small lounge and book storage room).[1065]

Unlike the apartment itself, Einstein's attic rooms looked out on *Haberlandstrasse*. They were only accessible by a flight of stairs because the elevator only reached the fourth story. Einstein had the tower room furnished a few months after his marriage. The room was, as Einstein wrote, "sunny – directed toward the east – and spacious, the window opening to the east. Area circa 20 sq m."[1066]

One very important aspect of this apartment for the present context was its *telephone* connection (tel. no. "Nollendorf 2807," from 1931 on: "Cornel. 2807"). Herta Schiefelbein remembered: "The main plug was in the small hallway in front of the guest room and the kitchen. But there were a number of secondary connections I could plug into [...] Herr Professor had two secondary telephones: one downstairs on his night-side table and one upstairs on his desk by the window [...]. Whenever you wanted to make a call, you first had to turn a crank. Then the [telephone exchange] office would call. In those days there wasn't any direct dialing yet, like now. All the Berlin districts were assigned to a telephone area, each with its own office with a special name. I cannot remember the name of our office anymore."[1067] The office that had slipped Herta Schiefelbein's mind was "Nollendorfplatz" – for short "Nollendorf."

All in all, – technically speaking – the Einsteins' apartment was ideally suited for conspirative purposes:

- two entrances/exits, with one (main stairway) leading to Haberlandstrasse and the other (servants' stairway) leading to Aschaffenburgerstrasse;
- the apartment's situation on the side away from the main entrance of the building (i.e., overlooking Aschaffenburger Strasse);
- Einstein's separate office rooms only accessible via the main entrance (not directly from the apartment);
- telephones in the apartment as well as in the study.

It was thus possible to enter the apartment as well as the building by one entrance and leave it by another. Konrad Wachsmann, the architect of Einstein's summer villa, explained how this could be done: "If someone appeared at Haberlandstrasse without notice and could not be turned away, [Einstein] sometimes escaped by means of the servants' exit. We once did that together. Some journalist came and absolutely insisted on interviewing Einstein. Because he had been turned away many times already, the journalist said he would wait at the apartment door until the professor came home. Einstein and I were sitting in the library and heard how desperately Mrs. Einstein was trying to get rid of the man. Professor Planck had made an appointment for that same afternoon. So her false excuse would have been exposed. "We have to help Elsa," Einstein finally whispered to me. He took me by the arm and led me through the salon into the diningroom and from there into the small hallway leading to the servants' stairway. Fortunately these stairs did not end on Haberlandstrasse but the adjoining street. We went down the stairs, walked to the entrance on Haberlandstrasse and took the elevator up to Einstein's apartment. The journalist was indeed still standing at the apartment door. But he had waited in vain. Einstein shook him off."[1068]

Outsiders could not know that the "tower room" also belonged to the apartment, which was likewise perfect for conspirative purposes. Two official searches through the apartment prove it, conducted by police inspectors and then by the SA in 1933. Herta Schiefelbein reported: "It must have been the beginning or the middle of April 1933, very early in the morning [...]. Three or four men in plain clothes stood at the door [...]. The men said "Kriminalpolizei" [...]. They did not ask for Herr Professor, though, they just wanted to know where Dr. Marianoff was. I said that he was out of town with his wife. They then asked which room he had been staying in. Dr. Marianoff had been staying in Herr Professor's room, because the Einsteins were away, of course, in America. Then they searched through everything there and asked me when the Marianoffs had left. One of the officers stayed with me in the kitchen, the others went away again, to Mrs. Kayser, presumably to check whether Dr. Marianoff and Margot weren't perhaps hiding there and whether my information that both had left town agreed with what Mrs. Kayser said. In order to prevent me from telephoning to warn anyone, one of the crime inspectors stayed behind with me. But he was only interested in Marianoff. He sat with me in the kitchen and since I was just having my breakfast I offered him a cup of coffee. But he always only asked me about Dr. Marianoff. [...]. They quickly glanced into the other rooms [...]. They did not go upstairs into Herr Professor's tower room. I suppose they didn't know about that room at all."[1069]

Einstein had evidently allowed *other* people to use his "tower room" as well (Einstein's secretary, Dukas, and his collaborator Mayer) – despite the restrictions imposed by the building inspectors."Of course he sat up there with Laue, Planck, Haber or Plesch. I believe I remember him entertaining students and other guests in the tower room as well. In addition, his study was also the working place of Miss Dukas and the calculator Doctor Mayer. So the restrictions on the tower room seem not to have been so rigorous."[1070]

None of this is *proof* that Einstein's apartment was used for subversive purposes, but certainly that it was very *suitable* for it.

## Institutions. The Club of Intellectual Workers

*All* the institutions mentioned in the secret service reports actually existed, even the suspicious *Club of Intellectual Workers!*

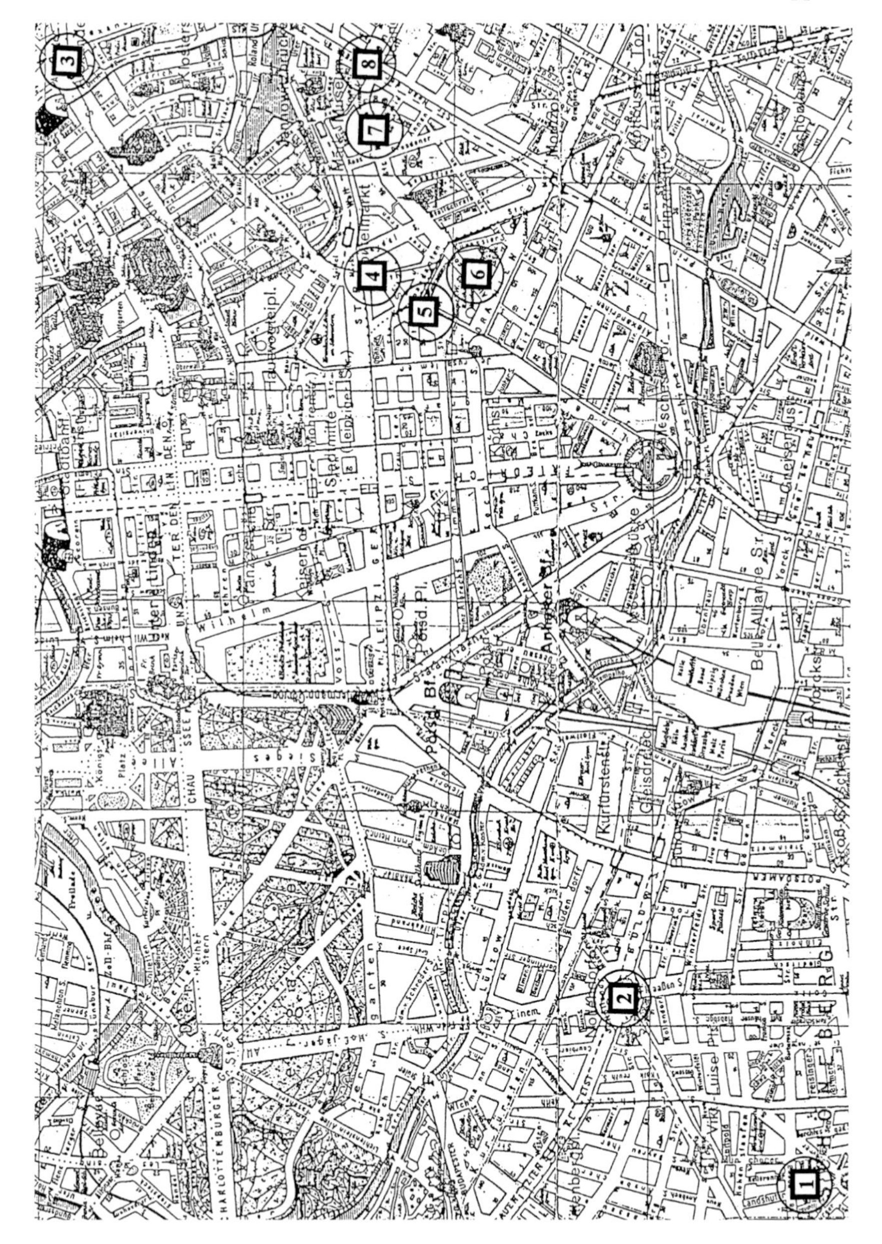

**Figure 59: Berlin streets and places mentioned in the CIC report.**

The Club of Intellectual Workers (CIW) was founded on 8 December 1931 at the Nationalhof during a meeting organized by the International Workers Relief (IWR). According to the police report, "numerous writers" and intellectuals took part. The club supposedly had roughly two hundred members – which had "developed out of the group of scholarly workers of the IWR comprising medical doctors, lawyers, writers, artists and other intellectuals." The police report dated 15 May 1932 described its purpose as "carrying communist ideas in among social circles that cannot be reached by usual propaganda tools."[1071] The club's founders were City Councillor Dr. Richard Schmincke, the member of the Medical Council, Dr. Fritz Weiss (first chairman), and the lawyer Dr. Hilde Benjamin (second chairman). On 27 February 1932 a youth group of the CIW was founded.[1072]

The discrepancy between the documentable leadership of the CIW and the information provided in the reports by the Counter Intelligence Corps (CIC) is striking. The persons named in the reports never played a central role in the CIW. Even the details provided by a former club member, Günther Kromrey decidedly after 1945, corroborate the membership.

Günther Kromrey was presumably describing the *youth group* of the CIW in his curriculum vitae dated 21 November 1949: I "received [...] in the KL-building[1073] the mission of forming a small-scale intellectual organization, in which I was only supposed to install specialists [...]. The core of this "club" were seven comrades [...]. The instructors included Albert Voigts, Eichenwald [...]. At that time the intellectual circle included: Felix Bobek [...]."[1074] The information about the CIW in the secret service reports actually applied to the CIW's *youth group!*

Kromrey described the club's working approach as follows:

> End of '29 Lala[1075] introduced me to two new comrades [. . . ]. With them [. . . ] an intellectual group was supposed to be cultivated. [. . . ].
>
> How did operations proceed?
>
> A) Personally following up addresses given to us or procured by us. Individual discussions.
>
> B) After a certain degree of maturity: being invited to the fortnight-long indoctrinations. There the guests spoke or we ourselves, whereby value was placed on *discussions*. Thereby gaining a deeper knowledge of these persons.
>
> C) Smaller, later larger missions according to the wishes of the supervisors, to whom I had to report respectively.
>
> D) Forwarding of information to ''Lala,'' about which he requested further explanations. In total, roughly 60 people went through this group, who did not come regularly every time but certainly with dedication. Visit of a few events of the League of the Intellectual Professions,[1076] in order to be able to fish intellectuals there.[1077]

A thorough scouring through the extant files and biographical sources on purported friends of Einstein among the CIW membership yielded absolutely nothing. Not a single mention can be found anywhere about Einstein's relations to the CIW or to its individual members, in particular Houtermans, Kromrey, Eichenwald and Bobek (let alone about any "friendly" personal ties with them).

## Persons: Richard Grosskopf / Helen Dukas

*All* the persons mentioned in the FBI file actually existed – with the exception, perhaps only of the alias "Fritz." This point is important, considering that the research on these unknown persons was incomparably more difficult than finding out about the streets, squares and institutions mentioned in the FBI file. CIC's and FBI's "source" must have been a very well informed person (and as they suspected, a person *from the communist underground,* someone who knew the material, an "insider").

The reports dated 13 March 1950 and 25 January 1951, mention the actors, some of them *concur* in both reports.

The list of persons named as purportedly having worked in the communist underground is long. Forty names appear.

The interesting persons among them were those who allegedly used Einstein's address or knew about it. According to the report of 13 March 1950, they were Grosskopf, Sauter, Burde, Bahnik, Liebers, Zaisser, Kippenberger, Kattner, Wloch, Kromrey, Dünow, Einstein's secretary, Abramov and "Fritz" from Wedding. The report dated 25 January 1951 *omits* the following as users/informed persons: Sauter, Bahnik, Liebers, Zaisser, Kippenberger, Kromrey, Dünow. Hence, what remains is: *Grosskopf, Burde, Kattner, Willi Wloch, Einstein's secretary and "Fritz."* From among these, only *Grosskopf* was still alive. The CIC knew nothing more about the fates of Einstein's secretary and "Fritz." Thus Grosskopf was not only the main person of interest to the CIC but also the only living witness in the matter. Because he was "currently chief of the Criminal Police in the Soviet sector of Berlin," however, he was *as good as dead* for the American secret service.

But the secretary, Helen Dukas, had also survived.

Consequently, investigations on the use of Einstein's address (and the so-called "office") had to concentrate on Richard Grosskopf and Helen Dukas.

Grosskopf's most important function during the period in question was correctly indicated in the FBI file: He was head of the communist passport forging organization. His alias – likewise correctly indicated in the FBI file and the only alias mentioned there was: "Steinke." His "successor as head of the passport forging organization" – likewise corresponding to the facts was: Adolf Sauter.

Grosskopf was directly responsible for the falsification of countless passports for functionaries of the Communist Party of Germany and other countries since the beginning of the 1920s. Without this organization the illegal communist machine could not have operated. His fake passports were so perfect that during his imprisonment in 1937/38 Grosskopf's expertise was put to use toward producing a fake-proof German passport![1078]

As head of passport forgery, but only in this capacity, Grosskopf had contact with Mirov (Abramov). Falsifying passports was his passion, he was busy enough with that. He was never subordinated to the CPG's intelligence service, Hans Kippenberger. He did not work either with him or with any other political apparatus. Grosskopf only provided services to them. For this reason alone, it is out of the question that he would have been in charge of "Einstein's cable address."

After a passport forging workshop was raided on 28 November 1932, the Gestapo was able to track down Richard Grosskopf. He was arrested on 3 May 1933.

During the Gestapo's interrogations Richard Grosskopf managed to withstand the abusive treatment. He lied, and when he was discovered he always promised henceforth to say the "whole truth." He could not remember any names and if he could name anyone, it was always just an alias. He even only had a vague memory of his own alias: "It is possible that, a longer time ago, I carried the alias *Steinke* within our organization."[1079]

The investigations made progress only toward the end of 1933, when the technical secretary of the Central Committee (CC) of the CPG, Alfred Kattner, became a traitor. As a result, on 15 November 1933 material originating from "Volta" (Grosskopf) were "discovered in secret hiding places of the former K.L.-building[1080] together with other C.C. material."[1081] The Gestapo agent Giering was *personally* "present at the disclosure of the hiding places."[1082] *He* succeeded "in finding other hiding places with the assistance of a well-informed confidante."[1083] This person was: Alfred *Kattner:* "The important finds in secret rooms of the Karl Liebknecht building were not least attributable to Kattner's finger-pointings."[1084]

Kattner revealed everything he knew. If there had been any information about Einstein, the Gestapo would have used it, if not immediately, then later on. Einstein is never mentioned in the relevant files.

Kattner's betrayal did not remain a secret to leading functionaries of the secret political machine of the CPG. He was murdered on 1 February 1934. The CIC was right in surmising that Kattner "had been liquidated by the communist underground."

On 11 January 1935 Richard Grosskopf was convicted by the People's Court (*Volksgerichtshof*) for anticipated high treason ("for preparing a highly treasonous enterprise in union with serious counterfeiting") to "a jail term [...] of nine years, less 20 months pretrial detention."[1085]

The files of the People's Court are comprehensive. *At that time* – 1933/34 – thorough legal research was still being conducted. The decrees to accelerate and simplify legal proceedings had yet to be issued.

The court's verdict alone filled fifty-five pages. The examinations and depositions counted many hundreds of pages (six binders, with the last sheet numbered 664). They include numerous transcripts of Grosskopf's interrogation (the first on 3 May 1933),[1086] but are mainly witness testimonies.

The twenty-month period of detention pending trial is indicative that the authorities' research was as thorough as could be. Even though the prime suspects did their best to deceive their interrogators and to reveal only what the officials already knew, it is likely that these proceedings exposed the activities of the "passport forging organization" completely. Witness testimonies and garrulous statements and confessions by codefendants in fear of punishment contributed substantially to this success.

After 1945 Grosskopf was subjected to numerous "personal checks," mainly by the East German Ministry of State Security, not despite his position as one of its officers but *because* of it. His personnel documentation is correspondingly copious, starting with his birth certificate and ending with the execution of his last will and testament. In the 1980s further information was compiled at the instruction of Minister Mielke for an "investigation of the activities of the intelligence and security organs of the GDR before 1945." The holdings of other GDR archives were consulted in addition to the ministry's own files, in particular the archive sponsored by the Institut für Marxismus-Leninismus,[1087] of the governing Socialist Unity Party (Sozialistische Einheitspartei or SED). Painstaking efforts were expended on combing through the existing archival material, making copies and reorganizing them. As a result, there are now comprehensive records on Richard Grosskopf.[1088]

Among these copious Grosskopf files, not a single reference to Einstein can be found; his name is not even mentioned.

The only "proximity" to Einstein was that Grosskopf maintained a forgery workshop in the cellar of the building at Aschaffenburger Strasse no. 14.[1089] That means, a few meters away from the service entrance to Haberlandstrasse no. 5, diagonally under Albert Einstein's apartment.

*It is pure invention that Grosskopf used Einstein's address.* There never was any connection to Einstein, either direct or indirect.

What remains is the question of what role *Helen Dukas* might have played.

She had no reason to refuse giving any true statements to the FBI.

Helen Dukas told the FBI agents, in conformance with the facts, that before she was hired, Einstein's wife and his elder stepdaughter had served as office assistants, besides a few part-time students. Her information that she had been Albert Einstein's *only* secretary since 1928 was also true. (This also means to say that he never owned an "office" establishment and that he had engaged many secretaries at one time was out of the question.)

The history preceding her employment is also sufficiently clearly documentable and credible. (Thus the CIC's contentions about the CIW's role in Helen Dukas's original hiring are pure nonsense.)

Owing to his failing health in March 1928 Einstein had to "engage a secretary for his office work at home. Placing a newspaper advertisement was not considered because it would have inevitably attracted masses of unwanted responses. Elsa mentioned the problem to Rosa Dukas, who headed the Jewish Orphan Organization. Miss Dukas recommended her sister. Helen Dukas appeared at Haberlandstrasse no. 5 on Friday, the 13th of April for an interview. At first she had rejected her sister's suggestion. She knew nothing about physics and had the feeling that it was all far above her capabilities. Finally she let herself be persuaded to give it a try."[1090]

"Helen Dukas's mother also originated from Hechingen, like Elsa Einstein."[1091] "Whoever came from Hechingen, even if only in the second generation," Helen Dukas wrote to Mr. Lemmerich on 6 July 1981, "needed no further recommendation." Elsa Einstein "naturally knew my mother and grandmother."[1092]

There was no room for "recommendation of Einstein's secretary by members of the Club of Intellectual Workers, a communist relief organization." Albert Einstein might perhaps have been duped, but certainly not resolute *Elsa* Einstein.

What Helen Dukas did not say was that since her hiring, other persons had also been employed in Einstein's household. But that might have been deemed a slight negligence. A lie it was not.

During the period in question (between 15 June 1927 and 1 June 1933)[1093] Herta Schiefelbein was employed as the Einsteins' housemaid (and cook), *not* Helen Dukas! Herta Schiefelbein had been employed temporarily after the former maid had suddenly left the Einsteins, taking a few valuables with herself.[1094] Contrary to Helen Dukas's statement, another person was also in the employ of the Einstein family household, a "cleaning woman from Schöneberg." She was responsible for the "dirty work [...]. She also washed the windows downstairs in the apartment [...]. But this woman did not come every day."[1095]

What weighed more heavily against Helen Dukas were her outright lies during her "interview." Smilingly, "helpfully," and "keen to inform," she adeptly hoodwinked the FBI agents.

She categorically denied any personal contact with communists. That would have to include a few members of her *own* family. Her brother-in-law Sigmund Wollenberger (husband of her sister Seline – the fourth Dukas daughter) was member of the CPG and likewise her nephew Albert Wollenberger since 1932! When Albert Wollenberger emigrated to the US during the 1930s, his Aunt Helen acted as his personal guarantor.[1096] "Einstein knew," Wollenberger told Michael Grüning, "that I was a communist and was a member of the CPG since 1932. He did not just tolerate it, he even respected it, although he was no Marxist, being rather more of a socialist by sentiment."[1097] "He knew that I had the *party mission* to qualify myself as a research assistant and member of the teaching faculty at Harvard University, to prepare myself professionally for the demands of the GDR."[1098]

His *aunt* could not have been ignorant of the fact that Albert Wollenberger was deported from the US in 1951, nor that his choice of residence in 1954 was *East Germany*. Wollenberger remembered: "When Senator McCarthy's witch-hunt started, I was arrested. After I had been sent the deportation order as a stateless person, Albert Einstein vouched for me. Thereupon I was released on bail."[1099]

Could his aunt have known nothing about all this? She was certainly not as naïve as she made herself out to be during her "interview." If anyone was naïve, it was the badly prepared FBI agents, who thought they could deceive Helen Dukas and ended up believing everything she deemed worth telling them. Helen Dukas was cleverer than the agents. She said much that was true but left out much as well.

Most importantly, she stayed silent about information that could have hopelessly incriminated her: that her apartment had been used for the purposes of the CPG (and probably also that she knew the name of the tenant).

We would have known nothing about all of this if *Luise Kraushaar* had not survived and left a record, including a "confidential" document dated 1986: 'Report on my conspiratory work between April 1931 until the end of 1937.'[1100] Luise Kraushaar was the former secretary of the nationwide head of the CPG's industrial reporting agency (*BB-Ressort*).[1101] That is – as the People's Court quite appropriately asserted – "the most dangerous apparat of the CPG there is [...]"[1102] (and, one would have to add, the most covert apparat).

She had not just been Wilhelm Bahnik's secretary but also his predecessor's, Fritz *Burde*. She knew personally (more or less closely, but certainly from direct personal contact) the following persons specifically named in the FBI file:

- Bahnik, Wilhelm ("Martin")
- Bobek, Dr. Felix
- Burde, Fritz ("Adolf," "Edgar")
- Dünow, Hermann
- Kippenberger, Hans ("Alex")
- Liebers, Johannes ("Fred")
- Roth, Leo ("Viktor")
- Welker, Helene

And she also knew *Albert Einstein's secretary!*

Luise Kraushaar reminisced:

The first illegal office I worked in from the spring of 1931 until about mid-1933, was located in Friedenau, Berlin, on a quiet, peaceful street that was easy to see down and on which any loitering observer would have been noticeable. I worked in one room of a larger apartment that Albert Einstein's secretary was living in with her sister. I unfortunately have forgotten the names, the street and the number of the building. The secretary was, like her sister, an older woman between 45 and 50. They were both very modestly dressed, very calm and friendly with me. Both of them left every day for work and I was mostly alone there. I think that they knew about the illegal nature of my work. But they did not know the substance and content of this work, of course. They never asked and our conversations consisted of the usual greetings and the payment of the rent. The degree of trust between us expressed itself in that I received the key to the apartment from them and could do what I pleased completely undisturbed.

Only Fritz Burde and Leo Roth knew about this apartment. Leo Roth probably discovered this quiet apartment and secured it. He came often to deliver material for me to transcribe or to pick it up again. I remember his visits so clearly, because he knew how to please people

> with insignificant gestures. He must have had a key to the apartment as well, because every once in a while he was there in my absence. After one such visit, a wonderful big apple lay on my typewriter table. With it a note: ''Bon appetit, Viktor.'' Viktor was his alias. Such gestures of kindness warm the heart of an illegally employed person and are deeply impressive.[1103]
>
> Leo Roth [. . . ] was very warm and kind. His girlfriend was the daughter of General Hammerstein-Equord,[1104] a pretty girl with long, blond locks of hair, at that time perhaps 20 years old. Because she brought us interesting notes about conversations by guests at her parents' apartment, I occasionally met her alone. Her given name has unfortunately slipped my mind. I was always pleased whenever I met them both, ''Viktor'' and his girlfriend.[1105]

During a conversation with an officer of the Minister of State Security sometime at the end of 1978/beginning of 1979, she made the following statement:

> I only knew the offices in which I was working. Some of them were surely apartments used as storage or for other kinds of work. I personally worked in various apartments (text processing). Two have remained in my memory. One belonged to Albert Einstein's secretary and her two sisters. They lived in Friedenau. The secretary was approx. 50 years old, a well-endowed, friendly lady. She probably knew nothing about the type of work we were doing. I do not know the pretext under which the apartment (one room) was rented. I unfortunately cannot remember anymore whether we stored unfinished business there. But it was probably mostly the case that I met with the responsible person in the evening hours in order to return to him the material and copies. When the passport office of the CPG machine (Karl Wiehn),[1106] which had been operating in our neighborhood, was exposed, we left the Friedenau apartment.[1107]

Luise Kraushaar's interviewer noted at that time: "The reports were written in two illegal apartments, 1931 until at least the end of 1932 at the home of one of Albert Einstein's secretaries. Comrade Kraushaar cannot recall the name anymore. She only still knows that it involved 3 sisters and that the office was situated in the apartment of the eldest of the three sisters."

With the exception of the comment about *three* Dukas sisters, the information in the two reports from 1978/1979 and 1986 are identical in substance.

Luise Kraushaar conceded that she could not remember everything precisely, despite her generally good memory ("My good memory was a great help").[1108] What she did think she was able to remember correctly was not always accurate either. Einstein's secretary was neither fifty years old nor "well-endowed," rather about thirty-six or thirty-seven and slim.[1109] We cannot exclude the possibility that Luise Kraushaar had seen *Elsa* Einstein there. *She* was around fifty years old and "well-endowed." The official tenant of the apartment was not Einstein's secretary but her sister Rosa. Even so, her "recollections" were otherwise remarkably precise. Matched against other independent sources, the basic allegation that espionage reports were written in the apartment of Einstein's secretary *had* to be right.

Hence there is no doubt *that* the apartment of Albert Einstein's secretary was used for the drafting of spy reports. (A perhaps important detail in this regard is the presence of a telephone in Rosa Dukas's apartments: Rosenheimer Strasse: "Steph. 5265" then "Cornel. 5265"; Hindenburgstrasse: "Cornel. 5265." She evidently kept her four-digit telephone number whenever she moved.) This section was indeed a "*quiet, peaceful street*" (as it still is today). I would think that there is no doubt about the rented room in the Dukas apartment having been abandoned upon the "exposure" of Karl Wiehn's passport establishment (because it was of existential importance for the industrial espionage reporting). This happened at the end of 1932. So: *until the end of 1932* the apartment of Albert Einstein's secretary was used for conspiratory purposes. It is also correct that the exposed passport workshop was located in the *immediate neighborhood* of Rosa Dukas's apartment – even though Luise Kraushaar could not remember the precise address anymore. The raided passport forging workshop was located at Kaiserallee no. 48a, that is, in the building on the corner of Hindenburgstrasse.

This block of residential buildings delimited by Hildegard, Livländische and Hindenburg Strasse (entrance no. 92a)[1110] was at that time newly built. The conveyance of the building had taken place on 24 September 1931.[1111] Therefore, the apartment could only have been occupied and used for the indicated purpose at the end of September/beginning of October 1931, at the very earliest.[1112] It is a mystery why the three Dukas sisters moved into a new building only to immediately sublet one of the rooms (which suggests that they were relatively well informed about the purpose of the rented room).

It is also credible that Helen Dukas knew nothing about the substance of Luise Kraushaar's work. That was normal. For reasons of security it was also normal that the communists informed themselves beforehand about the political attitudes of the residents of any sublet apartments. "Precisely because they were not party members, the apartments were perhaps also the most secure. But without exception, the residents were informed that they were making their rooms available to the CPG. The method of safeguarding them was so carefully conceived, that when many of these subletting residents were subsequently arrested, not a single one of them could be penalized. They could all prove that they had submitted an advertisement to the *Welt am Abend* that they had a room available for rent and that the people who had come over had identified themselves with a proper police registration." (Quoted from a speech by Hermann Dünow on 23 October 1967 before members of the Central Party Archive of the SED at the Institut für Marxismus-Leninismus.)[1113]

The fact that not just Luise Kraushaar but Leo Roth also had received a key to the apartment and that she could do what she pleased there "completely undisturbed" leads to the conclusion that all three Dukas sisters were either incredibly naïve or else they *knew* something, after all.

Moreover, Leo Roth wasn't just any communist undercover agent. He was one of their most important functionaries! He was responsible for "*special contacts*" at the CPG's intelligence service. After Rudi Schwarz was murdered on 1 February 1934, Roth also became director of the "security" (*Abwehr*) agency of the CPG political machine. In early 1934 he procured the murder weapon for Alfred Kattner's (the former technical secretary on the CPG's Central Committee) assassin and subsequently provided him with a false passport and helped him and his girlfriend Helga von Hammerstein leave the country.[1114]

Helen Dukas was simply not as "apolitical" and ignorant as she made herself out to be to the FBI.

If she managed to deceive the FBI agents, it was probably primarily because the FBI investigations into this matter were anything but professional.

Evidence of the superficiality of the FBI investigations is that despite constant reference being made to spies in the Soviet embassy in Berlin and that Einstein supposedly had many friends among Soviet diplomats, it never occurred to anyone to pay any attention to Einstein's son-in-law, the Russian Dimitri *Marianoff* – head of the film division of the Soviet Chamber of Commerce. It was known that he had married Einstein's stepdaughter. His birthdate was known to be 1 January 1889 (in "Weinitra or Venitza/Russia").[1115] His biography of Einstein, published in the USA was also cited.[1116] It was also known that he had immigrated to the United States. *Dimitri Marianoff* himself was never interrogated, however. Nor Margot Einstein. The possibility that Marianoff might have been one of the many Soviet agents active in Germany was never explored.

Einstein's stepdaughter Margot married the Russian Dimitri Marianoff on 29 November 1930.[1117] He was an important person among the many hundreds of employees of the Soviet Chamber of Commerce. A report by the Berlin chief of police, designated as "strictly confidential," entitled 'Report on the activities of the Berlin center of the O.G.P.U." and dated 18 August 1932[1118] states: *Dimitry Marianoff* and a woman *Asja (Susanne) Ari* are the "closest assistants" of Arthur Normann, the head of Soviet spies in Germany. Marianoff was, it continues, *constantly* socializing "in the consulate and embassy" and thus belonged among such people who for precisely that reason are "not conspicuous and arouse no suspicion."[1119] The Berlin police seem not to have realized that this Marianoff was Einstein's son-in-law.

*Einstein's son-in-law was a Soviet spy!* Dear son-in-law was more than just what Elsa Einstein described as "a gypsy, but a fine and interesting one"; "deep down in his soul [...] a decent, indeed a noble human being."[1120] There is good reason to believe that his many relationships, also with other women, did not just arise from his personal temperament (and even so, *not exclusively* for his own private gratification).

Since his marriage in December 1930, Marianoff lived in Einstein's city apartment, also during the summertime when Elsa and Albert Einstein spent longer periods of time in Caputh. He was still living there during Einstein's sojourn in the USA (since December 1932) – ultimately even staying in Albert Einstein's own bedroom (so right next to the telephone).[1121] Einstein's apartment was freely accessible to Marianoff. As Einstein later commented, his son-in-law had not inhabited his apartment for "eight years," as some people supposed, but "a couple of months long, without interruption."[1122] But he rarely came to Caputh[1123] – quite in contrast to his wife Margot. Between April and October of the years 1930, 1931 and 1932, Einstein lived in Caputh; from November/December until March of the following winters 1930/31, 1931/32, 1932/33 Einstein sojourned in the United States. His wife Elsa accompanied him every time, in the winter of 1930/31 Helen Dukas also went along. Consequently, since 1930 Einstein rarely used his city apartment. Marianoff was thus often alone there. We shall probably never know what he did then. But from what we do know, we may assume that Einstein's apartment was used for espionage purposes. Was the Schöneberg apartment in which Kippenberger first heard about the burning of the Reichstag[1124] perhaps *Einstein's apartment?*

The Soviet Secret Service had the keys not just to Einstein's secretary's apartment but also to his own!

## What about Einstein himself?

Einstein's signature under the 'Manifesto to the Europeans' in 1914 marked his entrance into political life. He remained true to the slant expressed in it, even though the manifesto was ineffective in its own time.

How Einstein acted in defeated Germany became an issue of international politics less during the Great War than afterwards: quite the contrary to the majority of Germans, above all, professors.

Ernst Reuter, secretary of the New Fatherland League, was one of the first communists he came in contact with. At the end of 1918 Reuter returned from Russia to Germany with Karl Radek[1127] – the man responsible for the revolution in Germany, from the Russian point of view, hence also in Lenin's eyes.

It is possible that Einstein's close friendship with Paul Levi stemmed from the beginning of the 1920s. On 8 August 1929 he would write him: "Dear Paul Levi [...]. It is uplifting to see how you as an individual person have purefied the atmosphere without restraint through acuity and a love of justice, a wonderful pendant to Zola. Among the finest of us Jews, something of the social justice of the Old Testament still lives on."[1128]

Paul Levi had met Lenin "through Radek in Switzerland in 1915 or 1916 [...]. Levi was a Bolshevik then already."[1129] Levi (Hartstein) was one of the co-signers of the treaty between the German government and Russian emigrants in 1917 that made it possible for Lenin and other Bolsheviks to travel back home to Russia and prepare the October Revolution.[1130] On 4 December 1920 Levi was elected together with Ernst Däumig to preside over the CPG. It is true that Levi was replaced on 24 February 1921 (like other presiding members: Clara Zetkin), and shortly afterward expelled from the CPG because of his criticism of the "March 1921 drive."[1131] But he remained loyal to the leftist ideal.

It was widely known that Einstein, who had described himself in the *Berliner Tageblatt* as "a Jew of liberal international bent,"[1132] assumed a leftist stance in conflicts during this period.

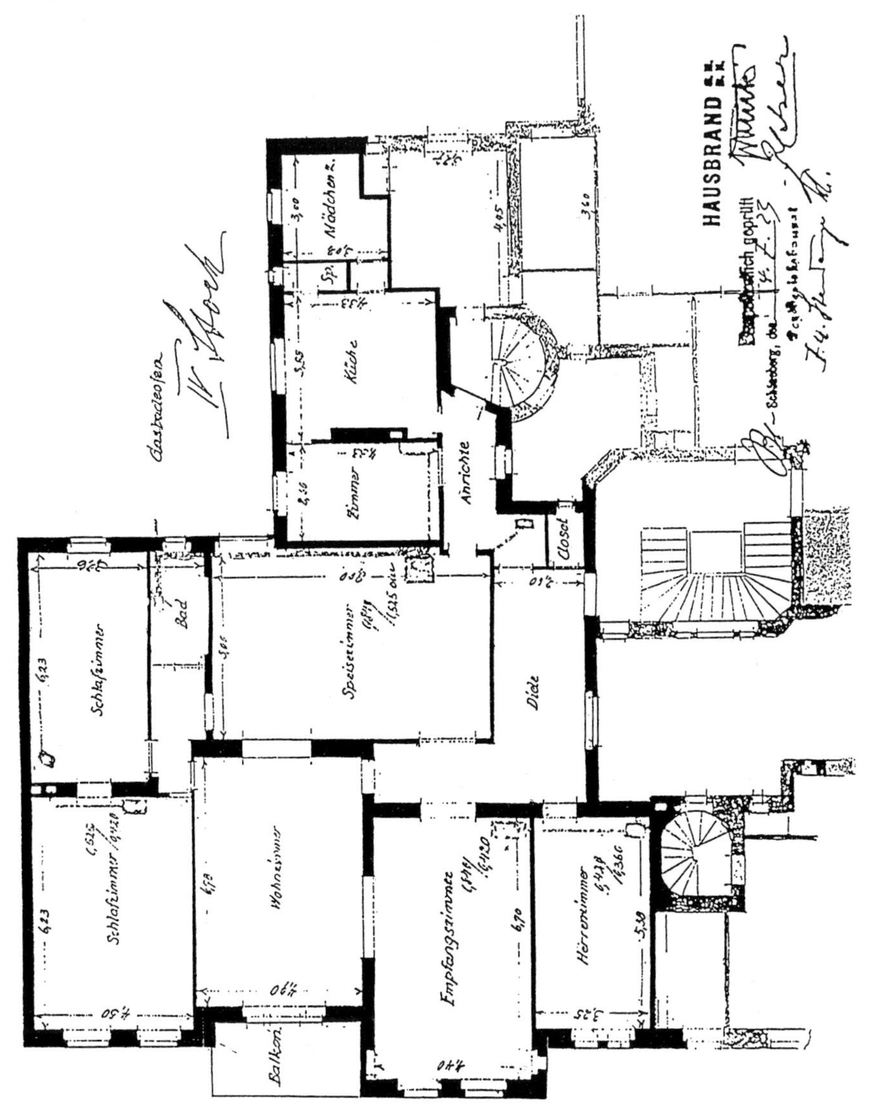


**Figure 60: Floor plan sketch of Einstein's apartment by the master oven fitter, Schwingel.**[1125]

From the controversies with the Nazi cohorts Weyland and Lenard, the whole world knew about it. The papers reported about it almost daily. His close friendship with Walther Rathenau was also public knowledge.

To the great displeasure of the Foreign Office and people advocating vengeance and a confrontational course, Einstein was appointed on the International Committee on Intellectual Cooperation of the League of Nations in 1922.

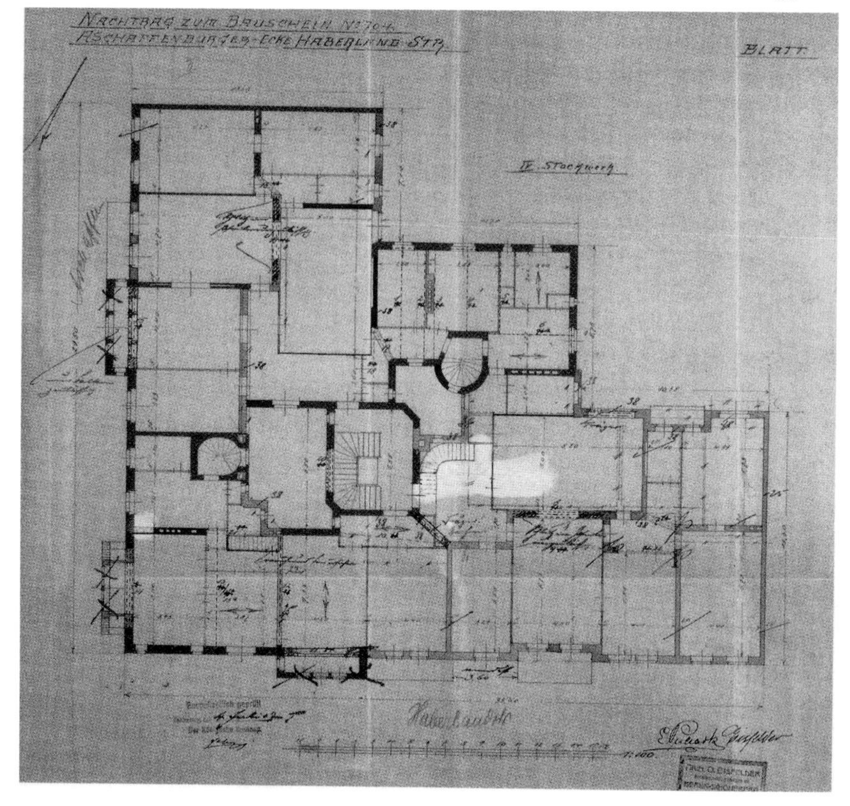


**Figure 61: Blueprint of the apartments in the building on the corner of Haberland Strasse and Aschaffenburger Strasse. Einstein's apartment is above.**[1126]

According to the minutes of the general assembly of the *Sozialwissenschaftlichen Club e.V.* dated 28 January 1924, Albert Einstein had enrolled himself as a member (others were Lehmann-Russbüldt, Harry Count Kessler, H. von Gerlach, Eduard Bernstein, Fritz Wolff, Dr. Kuczynski).[1133] On 14 June 1927 the Reich commissioner for the surveillance of public order had counted this particular club, among "radical pacifist, partly communist organizations" requiring observation.[1134]

The minutes of the German League of Human Rights in 1924 concern Einstein's political activities at the time: "30 Jun. Prof. Albert Einstein's personal audience with Reich Chancellor Dr. Marx about the fate of Erich Mühsam and other political prisoners in Niederschönenfeld."[1135] Einstein had applied for this audience on 25 June 1924.[1136] Erich Mühsam had been member of the Bavarian Soviet Republic. After his arrest, he was convicted to fifteen years fortress confinement. The files do not reveal how Marx responded to Einstein's pleas. What is remarkable is that Einstein only had to wait five days for his requested audience,[1137] a short delay. In any event, Mühsam was released after serving six years.

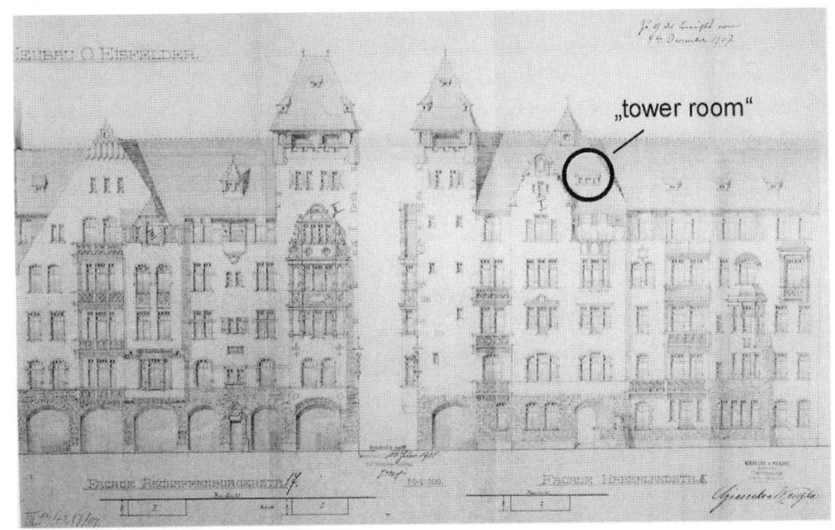


**Figure 62: Building façades toward Haberlandstrasse and Aschaffenburger Strasse (per architect's blueprint, 1907). The location of Albert Einstein's later study ("tower room") is indicated.**

No sooner was he free than Erich Mühsam appeared at the First Reich Convention of Red Aid in Berlin to give a talk together with Julius Gumbel.

Insofar as Einstein entertained close contacts not just with individual communists but also with the CPG *as a political party,* this was his membership in Red Aid. It was a *relief organization of the CPG,* widely referred to as such and perhaps its most successful one. Owing to his membership on the board of trustees of the children's homes supported by Red Aid,[1138] Einstein's name was added to the catalog: "Suspicious persons, who have made themselves politically conspicuous," compiled by the Office of the Reich Commissioner for the Surveillance of Public Order as early as September 1926.[1139] The Gestapo later took over these records by personal order of Himmler's deputy Dr. Best.[1140]

Einstein's signature also appeared under an appeal by the board of trustees of Children's Homes of Red Aid seeking "*Christmas assistance for the political prisoners*" dated 1 December 1926.[1141] Contributions were solicited also for "family members of political prisoners."[1142]

At the Fourth Congress of the IWR from 20 to 22 November 1927 in Berlin, Einstein was elected on the Extended Central Committee of the IWR. This too was carefully recorded in the files of the Office of the Reich Commissioner for the Surveillance of Public Order.[1143]

In Lenin he admired "a man, who has thrown all his energy into making social justice real, at the sacrifice of his own person. [...] a guardian and reformer of the conscience of mankind."[1144]

On 11 June 1932 the appeal proceedings took place in Moabit, Berlin, against

**Figure 63: Einstein in his study – the "tower room" – 1927.**

eight members of the International Workers Relief, who had illegally demonstrated in protest against the issuance of the emergency decree. Einstein gave testimony in support of the defendants. A communist paper proudly reported: Einstein took part and "argued with great warmth on behalf of the IWR at yesterday's hearing."[1145] It added, though: "The arguments presented by the co-founder and secretary general of the IWR, Willi Münzenberg, were even more convincing." The defendants were acquitted. Einstein's efforts on their behalf had been worthwhile. So Einstein made an appearance alongside the arch-communist Willi Münzenberg. Not secretly but very publically.

In the summer and fall of 1932 Einstein's name was repeatedly mentioned at the Ministry of the Interior in connection with communist activities. On 11 July 1932 the chief of the Berlin police informed the minister of the interior that, on the German front, Albert Einstein was member of a committee alongside three communists: Erich Mühsam, Willi Münzenberg and Klara Zetkin, whose purpose was to prepare an international anti-war congress at the initiative of the Soviet Union.[1146]

On 4 October 1932 Willi Münzenberg appeared in his capacity as member of parliament together with Attorney Rudolf Olden at the Prussian Ministry of

**Figure 64: Dimitri Marianoff and Margot Einstein on their wedding day, together with Albert Einstein, 29 Nov. 1930.**

the Interior to protest against the police imposed ban on anti-war events on the grounds of "highly treasonous goals for traitorous purposes."[1147] On the following day Attorney Olden wrote a letter to the ministry. With reference to the conversation on 4 October he emphasized the nonpartisan character of the German Committee against Imperialistic War underpinning his point, among other things with the argument: "Members of the committee include the physician Dr. Felix Boenheim, the writer Heinrich Mann, the former member of parliament Attorney Kurt Rosenfeld, Professor Albert Einstein [...]. The very composition of its members shows that the aim here is to draw together pacifists from every camp and that the committee's scope reaches well into the established bourgeoisie. [...]. The committee [...] is not a division of the Communist Party which, based on what has been said to me at the ministry, has been the suspicion."[1148]

The ministry would not be convinced. It stood by its view that it was a communist endeavor (hence Einstein, too, was a part of such an endeavor).

It is no coincidence that the Nazis reproached Einstein particularly for his links to Red Aid.[1149]

According to a progress report by the News Collection Agency of the Reich Ministry of the Interior (successor to the Office of the Reich Commissioner for the Surveillance of Public Order) dated 8 October 1932, Einstein was a member of the "German Militant Committee against War" (other committee members included Heinrich Mann, Otto Lehmann-Russbüldt and representatives from the weapons industry).[1150] The charge was that this campaign ("Kampfkomitee") intended "through its connections inside weapons factories [...] to mobilize the labor force there for the campaign against imperialistic war." The same report noted Einstein's membership in the World Committee on Combating Imperialistic War.

A news update by the Office of the Secret State Police dated 15 May 1933 notes: "Lately Münzenberg is attempting through intermediaries to approach all foreign newspapers that have been involved in any way in the atrocity campaign against Germany. He exploits the assistance of reputable journalists and scholars, such as Einstein, Tucholsky, etc."[1151]

Einstein's involvement in numerous organizations founded by the Communist Party was quite *well-known* inside Germany as well as abroad. These included his memberships on the board of trustees of Red Aid Children's Homes, in the Society of Friends of the New Russia, the International Workers Relief, and for a while also the World Committee against Imperialistic War.

However, not the least bit of evidence can be found about any of the covert activities that the American secret service later imputed to him. *Einstein was no communist,* not even a crypto-communist. He told the whole world what he thought about world developments. He had nothing to hide.

Einstein was no communist, but neither was he an anticommunist and he supported views that were often described as "communist" in a society steeped in hate and belligerency.

That he was occasionally misused and maliciously deceived is an entirely different matter. He cannot be held to account for what others thought and did – even people close to him, as it were "right under his window."

## Preliminary assessment

Most things were true, the most important things were not.

It is true that many of the communists named specifically in the FBI file worked for "Soviet apparate" in China.

It is also a fact that a radio link existed between China and Berlin (the seat of WEB, responsible for the Far East, likewise mentioned in the file; whether or not via Cairo is not documentable). It was only in 1942 that the Gestapo found out that: "Sometime around 1930 the greatest importance was laid on developing a separate Comintern radio network. [...]. From about 1932 on the Comintern was linked by radio with all its important sections."[1152] When Adolf Sauter was arrested by the police in Prague in 1938, he left recordings about his activities in the illegal apparatuses of the CPG before 1933: "At this time I was also joint owner of a watchmaker's store in Berlin, Alte Jacobstrasse no. 92 or 93. My partner was a

Jew and left Germany in 1933. [ ... ]. As long as I was joint owner of this company, I was employed by the Party for special tasks. I was incorporated into the existing international network of telegraph stations, received telegrams from China and other countries and delivered them to various persons who were known to me only by their aliases. In 1929 I joined the CPG officially."[1153]

The FBI file accurately identifies many collaborators in illegal CP agencies. Most importantly, the industrial reporting agency ("BB-Ressort," in fact an industrial espionage station for the CPG or the Fourth Division of the General Staff of the Red Army). The file identified the following persons as collaborators in this apparat before 1933 (and mostly afterwards as well): Fritz Burde (nationwide leader until 1932), Wilhelm Bahnik (nationwide leader 1932–1935), Johannes Liebers (photo man and a contact person for the Soviet Secret Service), Dr. Felix Bobek (photo man), Dr. Walter Caro (head of the "chemistry" station ("Chemie-Ressort") of the industrial espionage apparat), Dr. Fritz Houtermans (1929 to 1933 assistant/senior assistant/qualified lecturer at the polytechnic in Charlottenburg, Berlin) and Helene Welker (whose name first appears in 1955 during the "interview" with Helen Dukas).

It is also correct that numerous staff members of the polytechnic in Charlottenburg, Berlin, prior to 1933 (and later) worked for illegal apparatuses of the CPG: Dr. Fritz Houtermans, Dr. Günther Kromrey, Fritz Eichenwald, Bela Weinberger. The proportion of intellectuals in these agencies was generally very high. *One* of them was of the calibre to be able to discuss with Einstein the latest research in theoretical physics: Fritz Houtermans (who through his research with Atkinson on stellar luminescence might be called the "grandfather of the hydrogen bomb").[1154] Careful examination of the historical sources fails to yield any personal relations between Houtermans and Albert Einstein, however.

As far as the political machine, or apparate of the CPG are considered, the source CIC relied on was a very well informed person indeed.

The knowledge this "source" had about the main subject of this report, *Albert Einstein,* was less reliable (if at all) – details about his work, colleagues and relatives or living conditions.

Virtually every piece of information in this regard collided with the true facts. Fordian ideals must have been the spring of his assumption that a famous scientist must also possess a separate and very large office – with many stenotypists, a head and deputy secretary – and equipped with the latest technological achievements (including a "telegram address"). Much that could easily have been gathered from Einstein biographies or address directories, Source evidently did not know. Such details included Albert Einstein's place of residence and telephone number. He did not know who Einstein's secretary was; nor did he know whether or not she was still alive and where she lived.

Instead, vague suspicions were transformed into definite fact, or at least formulated as such. The informant must have learned about a few things that were approximately correct from hearsay. For instance, he seems to have heard a rumor that one room in Dukas's apartment was being used for purposes of the CPG. Out of this he painted a picture suited to his and his interrogator's needs.

**Figure 65: Adolf Sauter. Passport photo from the early 1950s.**

Nevertheless, the informant carefully installed "safety mechanisms" into his story. He knew that much was trumped up and sheer invention. Otherwise he would not have made *all but one* of the other people with purported knowledge about Einstein's complicity in activities of the communist underground vanish: "fallen in the Spanish Civil War," "executed in Moscow," "missing." This sole fellow witness was beyond the reach of the American secret service, because he was the inveterate communist and inspector of the *Volkspolizei* in the Soviet Occupation Zone/GDR from 1945 until 1951, when he became head of the East German secret service: Richard *Grosskopf.*

Albert Einstein's portrait was inserted at a suitable spot into a real, existing landscape – facilitated by the circumstance that Einstein's portrait was appropriately contoured and tinted to fit. Einstein was supposed to have been a dangerous communist, so he was fashioned into one.

## The Informant

We already have the identity of the source of the report dated 4 September 1953 about the controversy over relativity theory in 1919/1920, that is, it has been narrowed down as close to certainty as probability allows, by Klaus Hentschel in 1990.[1155] But this was – with all due respect – not an overly difficult task. The source could only have been *Paul Weyland:* Blacked out mentions of the informant's name measured *seven* letters in breadth; hence the name was *seven* letters long. Another hint: the report dealt with matters to which Weyland was most closely associated, which evidently still preoccupied him thirty years later. Third: there is proof that Weyland had worked for the intelligence authorities of the USA after 1945.[1156]

A much more difficult question to answer is who had provided the information to the Counter Intelligence Corps (CIC) reports dated 13 March 1950 and 25 January 1951. In this case also, we can narrow down the possibilities only to as close to certainty as probability allows (as long as the FBI continues to withhold part of its information).

Who was this informant? – Was this person whom the CIC did not want to identify male or female? And who passed the informant's information on to the CIC? Who, therefore, was the intermediary reporter (disregarding whether this intermediary added any "final touches" to the report to which he was willing to attach his signature)?

Our quest has to proceed by narrowing down the initial group of suspects by a systematic process of elimination.

The assumptions:

1. "Source" was *a natural person,* not an artificial figure composed of many different informants.
2. "Source" was – as the CIC itself suspected – someone *from the communist underground.* Only an "insider" could possess knowledge about the more subtle details.
3. "Source" had *survived* and passed his knowledge on only after the war. Otherwise he could not have been able to mention that Karl Wloch had become president of the Helmut von Gerlach Society in October 1949, nor that Bela Weinberger was employed in the Economic Planning Commission of the GDR. He was generally informed about who had survived the war, therefore also who had survived Nazi terrorism and Stalin's "Tshistka"[1157] of 1937/38.
4. The informant had, at a yet to be specified point in time, "*changed sides*" and stopped being a communist.
5. The informant was *one of the persons named in clear text in the CIC reports* because these persons had the most intimate knowledge of the events described.

The result of this long and laborious research is: the informant was *Adolf Sauter.* His name appears in the FBI reports and he was not just the informant but (what remained unknown to the CIC) he was *simultaneously* the intermediary reporter.

*Sauter* did not want to reveal the identity of the informant because he himself was the informant (or "source")!

1. The number *ninety-three* in the report dated 25 January 1951 is revealing: the street number of the watchmaker's store on Jacobstrasse, of which Sauter was demonstrably a partner. Who other than persons closely associated with a building would remember its exact street address decades later? With this precise little detail Sauter exposed himself. He even seems to have been filled with the insatiable wish to leave to posterity a coded trace of *his* authorship, *his fingerprint.*
2. At the prompting of the FBI, the CIC ran a check on the information contained in the report dated 25 January 1951. Although it yielded something about "Schauerhammer," nothing more was found about Uhrenelb, of all things. A proper search would even have been easy. The informant had gone too far with his information about "Alte Jacobstrasse 93" and then tried to cover up his traces. After first showing off about how detailed his knowledge was, he changed his mind and did not want the CIC to catch up with him. He did not *want* "Uhrenelb" to be found!
3. The informant knew that Grosskopf had headed the communist passport forging establishment. Hence he knew something that was known to very few people at the beginning of the 1950s: he knew *that* such an organization existed (even Grosskopf made no mention of it in internal documentation he drew up after 1945). The informant even knew Richard Grosskopf's *alias* ("Steinke" – as a rule, only aliases were known in the communist underground, very rarely the alias together with the person's identity). Finally, the informant also knew who had been Richard Grosskopf's *successor* – for just a few months: Adolf Sauter. Before 1951, Grosskopf's successor had been known to virtually no one aside from the Gestapo (after a long and hard search). Sauter could name the successor because he *himself* had been the successor.
4. While in police custody in Prague in 1938, Sauter put on record his career and his activities inside the CPG's political machine.[1158] The transcript reveals much unprompted information. Of particular interest within the present context are details that are substantially identical to information contained in the CIC reports (and appearing nowhere else):

   At this time I was also joint owner of a watchmaker's store in Berlin, Alte Jacobistrasse no. 92 or 93.

   I was incorporated into the existing international network of telegraph stations, received telegrams from China and other countries and delivered them to various persons who were known to me only by their aliases.''

   Approximately in July 1933, I was made successor to the head of the apparat concerned with the making of false documents, that is, as a result of that post becoming vacant.

5. The CPG's illegal activities were distributed over *many districts* of the Reich capital. Upper-class residential areas like Schöneberg were particularly pre-

ferred. All the streets and squares mentioned in the CIC reports are located in the vicinity of *Spittelmarkt*, however. To be precise: within the proximity of *Alte Jacobstrasse 93*. That was just where *Sauter* worked and lived. He projected *his own* residential area into the CIC reports.

6. Although not unimportant, the Club of Intellectual Workers (CIW) was all in all marginal. Yet it is given a central place in the CIC reports. This does not exclude the fact that, in *Einstein*'s case, it may indeed have been so important. The remarkable thing is, though, that *only* the information given by Günther *Kromrey* agrees with the CIC reports. The FBI reports as well as the archived transcripts by Kromrey leave the impression that Bobek, Caro, Swienty, Kromrey and others had been the real leaders of the CIW – with Kromrey its founder. In fact, Dr. Richard Schmincke, Dr. Fritz Benjamin and Dr. Hilde Benjamin[1159] had been the club's founders and leading officials. Kromrey names the following members (in agreement with the FBI report): *Bobek, Eichenwald, Kromrey and Caro.*[1160] Kromrey could not have knowingly passed on this information: in 1954 Sauter had described Kromrey as a "blind fanatic" who "is driven by an almost incredible hatred of anything noncommunist."[1161] It is that much more likely that Kromrey's acquaintance passed on to Adolf Sauter what was lying on Günther Kromrey's desk. This acquaintance was at the same time Adolf Sauter's long-time lover (since 1937). There was much to pass on, as well: Because Kromrey had joined the Nazi party in 1942, he had to testify repeatedly in writing before the Denazification Commission in 1946 as well as before the Central Party Control Commission of the Socialist Unity Party (SED) of East Germany. The CIW played an important role in his testimony (besides the CPG's industrial reporting agency).

And what leads us to the conclusion that Sauter was not just the informant but most probably also the *intermediary?*

1. As already mentioned, at the FBI's instigation the CIC ran a check on the information contained in the report from 25 January 1951. Unlike the "Schauerhammer" issue, the Uhrenelb tip remained a dead end. Purportedly neither the police in West Berlin nor the Municipal Administration had been able to find any information either on the telephone company, the watchmakers' guild or their suppliers. Perhaps *one* Berlin institution might have been so stupid or lazy, but not more than one. The CIC could be so easily cheated only because the person *responsible* for this research did not want any positive results. The person responsible for the Einstein matter inside the American secret service *did not want* the recipients of the report to find out anything of significance about "Uhrenelb, Alte Jacobstrasse 93" – this responsible official was *Adolf Sauter!*
2. According to the report dated 13 March 1950: Among those who had used or known about Einstein's address, "the only persons known to be alive are Grosskopf, Sauter, Zaisser, Wloch, Kromrey and Duenow. The others were either executed, are missing, or died in Spain." The report from 25 January

1951, however, only mentioned *Grosskopf, Burde, Kattner, Willi Wloch, Einstein's secretary* and "Fritz." Sauter is *no longer* among them. Sauter was not relegated among the dead but among the ignorant. This makes Grosskopf the sole survivor not counted among the missing. As an employee of the State Secretariat for State Security of the GDR he was out of reach for the CIC and not verifiable. The others in the know were "Source," whose identity the intermediary had not revealed.

3. When the CIC reports were drawn up, Sauter was working for the American secret service, of all places in West Berlin. "Springer, alias Sauter, "was working during that time [1950–1951] for the CIC in Berlin under the name Kramer" on a top secret "special mission."[1162] This mission may well have been to prepare reports about Albert Einstein's Berlin period.
4. Einstein's secretary still occupied "Springer" when he was already busy with quite different assignments. His lack of certainty about her seems to have been a cause for worry; perhaps he also found out that the FBI had been more successful. In 1955 Sauter ("Springer") finally found out the secretary's name *almost correctly:* "Ms. Lucas" – instead of "Ms. Dukas."

Excerpts from the correspondence between "Max Springer" (Adolf Sauter) and his coworker at that time, Kurt Rittwagen:

Kurt Rittwagen to Max Springer, 12 February 1955:

*Dear Max,* [. . . ] Lucas, *Bertha: This person is deceased per entry in the official registry on 31 May 1946 in the Wittenau sanatorium.*

"Max" to Kurt, 16 February 1955:

*Dear Kurt,* [. . . ] Lucas. *This time* I *fished up the wrong one. But maybe you can get further on this: The Reich Insurance Institution is at Fehrbelliner Platz. They still have very old files. Can you fish up the* Lukas *there who once worked in Einstein's secretarial office? (Until 1933). Thereafter abroad.*

Kurt to "Max," 19 February 1955:

*Dear Max,* [. . . ] Lukas: *I'll try to get further on Monday at the Reich Insurance Institution.*

"Max" to Kurt, 21 February 1955:

*Dear Kurt,* [. . . ] Lucas. *Wait a bit longer. Perhaps you'll be lucky after all. Otherwise you'll have to find a man who is a physicist, maybe one at the FU [Liberal University of Berlin]. Perhaps Student*[1163] *knows one. Would, of course, have to be at least a prof. and already have been a ''beast'' before 1933 and have been working in Berlin. Such people were most likely to have had contact with Einstein and his secretarial office. Then ask whether he might recall Ms.* Lukas *and whether any address was known. A for[eign] friend from South Africa had inquired, etc.*

Sauter did not *want* his reports to be thoroughly checked. That was why he wanted to be the person in charge of the check himself – to conceal his identity with that of the source he had used. The trick that the CIC and the FBI had fallen into was not supposed to come out into the open.

Another probable reason was that the CIC and the FBI were not allowed to learn what Adolf Sauter had been doing *before* he had submitted his CIC report. "Springer" did not want any delving into Adolf Sauter's past. It could have put him in trouble.

Why? What had he done?

A few stations of his life:

| | |
|---|---|
| 1929 | officially joined the CPG. |
| 7 February 1933 | together with the officiating head of the CPG's intelligence service, Hermann Dünow, and others: organized the security arrangements for the last convention of the CPG's Central Committee in Germany – in Ziegenhals near Berlin. |
| In July 1933 | after Richard Grosskopf was arrested, Sauter (at that time under the aliases "Ferry" and "Hugo") advanced to the leadership of the communist passport forging organization. |
| In June 1934 | left the passport forging organization, its administrative apparatus was tranferred to Saarbrücken at that time. |
| In fall of 1934 | serious differences of opinion arose between Sauter and Kippenberger, Roth, Ulbricht and other party functionaries. Two letters that Sauter (alias "Hugo") sent to "Adam" (Kippenberger) in October and November 1934 reveal the emotional state "Hugo" was in. |

These texts are as verbose as they are convoluted, making them as difficult to read as solving a "crossword puzzle," as Kippenberger commented. Apparently "Hugo" – the "greenhorn" – had been treated by his superior, particularly by Leo Roth (whom he later called "two Jews rolled into one")[1164] in a condescending way and had been thoroughly humiliated. At Roth's instigation Sauter was dismissed from his post at the Prague Emigré Committee and put out of action (his successor: Granzow, the organizer of Kattner's assassination).

Sauter's letter to Kippenberger is a mixture of self-criticism, defensiveness, minority complexes, reproachfulness and injured pride. His growing rage ended in threats. Sauter probably did not have anything specific in mind yet when he darkly forecasted a "Hugo affair" and threatened to land up "on the list of losses," to "resign" and "disappear." Nevertheless, the break with Kippenberger and his former accomplices seemed to have been clean. Hatred and revenge were the product of this humiliation.

In another fit of rage on 14 May 1935 in Prague, Sauter threatened to offer his services to the Gestapo. What the people involved did not want to believe, he actually set out to do on that very day. He left for Berlin and appeared at the Gestapo's Columbia concentration camp on 15 May 1935 to offer his services. He immediately passed their reliability test ("Gesellenprüfung"). On 15 May Dr. Felix Bobek and the designated nationwide leader of the "BB-Ressort," Ewald Jahnen, were arrested. The latter died a few months later as a consequence of a "more rigorous interrogation." The industrial espionage "BB-Ressort" was largely liquidated in the summer of 1935. Sauter's friend "Sem" and the prisoner-turned-Gestapo-aid Gerhard Diehl were among their number. We would have no written evidence of the part "Sem" and Diehl had played, if Diehl and his defense attorney had not pointed this out in his pardon plea. Diehl was condemned to death all the same. On 22 January 1935 he and Dr. Felix Bobek were executed in Plötzensee near Berlin.

During the war Sauter did not get enlisted into the Army. He supposedly was involved in special missions in the area of Kiev "under another name." What precisely he did is not known. Perhaps what his colleague Fromm proudly flaunted after the war: being an officer in command of a battery on the eastern front. As soon as he caught sight of a military hospital marked with a red cross on the Soviet side, his battery immediately opened fire. Fromm took pride in his hatred of Jews and bragged about extermination campaigns over there, when arms and legs were sent flying into the air.

Sauter was a confidant of the Gestapo until the end of the Nazi Reich and in this capacity took part in the persecution of antifascists, including members of the "Red Chapel" (*Rote Kapelle*).

Documentation of this is Ernst Rambow's statement during his interrogations by the National People's Commissariat for Internal Affairs of the USSR on 31 July and 1 August 1945. Rambow admitted to having worked as a Gestapo spy under Crime Inspector Schulz at the Reich Central Security Office and to having betrayed Anton Saefkow, Bernhard Bästlein and others. Rambow was sentenced to death on 25 September 1945 by the Soviet Military Tribunal of the Berlin Garrison. He was accordingly set before a firing squad on 12 November 1945.[1165]

In reply to the question "Did Schulz tell you which former CPG members were collaborating, like you, with the Gestapo?" Rambow said: "Answer: Yes, he did tell me. I knew about the following persons, who, like me, betrayed the illegal CPG to the Gestapo: [...] Sauter, first name unknown, party name "Ferry." With the Gestapo he carried the aliases "Schütz" and "Stein." He had been working voluntarily with the Gestapo since 1934 [...]. From the Gestapo he was assigned the mission of tracking down communists among workers in large factories and attempting to infiltrate the illegal Communist Party."[1166]

Rambow reiterated on 1 August 1945: "I have to say that the Gestapo agents Schelenberg and Sauter I mentioned in my first interrogation likewise infiltrated the illegal commun. organization and worked within it as communists."

Even though most of Hitler's active helpers' helpers remained relatively unscathed after the war, Sauter could not be certain that he would not end up before the courts of the Federal Republic of Germany. To be on the safe side he either told the CIC nothing about his activities before 1945 or only made vague allusions to them. He did everything to avoid raising any serious suspicion.

When the American secret service agent, Dallin, wanted to find out more about Rambow and thought Rambow had told on Saefkow, Sauter retorted on 22 October 1953: "Dear Mr. Dallin, [...]. You are posing a few more questions. [...]. Name of the alleg. traitor: Ernst Rambow (form. alias "Anton"). I consider this complete nonsense, [...]. He never had a political profile, he understood absolutely nothing about politics, just about the practical workings of the machine. A man of such a huge insignificance would never have been presented by

Saefkow-Jakobi-Bäuerlein to the liver group. [...]. The cobbler Rambow would never have aroused such an impression even among stupid people. [...] it really [would] not be a pity [...] if this fairytale about Rambow were sent to the fish and the man who really must have blown the whistle were found."[1167]

Dallin was not *supposed* to snoop around *Rambow* because he might chance upon *Sauter*'s own tracks. Rambow had been a comrade of his and – you never know – he might thus find out about "Springer's" Nazi past!

Sauter's irritation about having to conceal his past activities is also hinted at in documents drawn up in 1953/54 about the CPG's illegal political apparatus. Sauter wrote: With truly professional investigations at the end of the 1920s and into the 1930s, communist espionage could not have been nearly as successful as it was. "Even the police lacked detailed knowledge of a precise nature and lacked means of exposing the communist underground. In the subsequent period – 1930 and later – a whole new generation of true security specialists posed a severe threat to this apparatus [...]. A few of these experts are still around today and *it is quite incomprehensible that just because they had continued their security activities against this apparatus after 1933, which posed an equal danger to any country, they are today still being relegated onto the sidelines and are pursuing completely insignificant occupations.* According to the available information, they involve former security specialists who were simple crime inspectors and not politicians."[1168]

This was not written under Sauter's former alias (which could have revealed his obscure past) but under his new one, "*Springer*."

After the war, Sauter received a German identity card under the name of "Waitzer" (presumably on the basis of a false passport dating to his period in the communist underground). He worked for the western secret services under the name "Max Springer." In 1950/51 – when the CIC reports were drawn up, he had been ordered away to West Berlin. Because most people aware of his activities as an undercover agent of the Gestapo were no longer alive and even the Ministry of State Security looked for years in vain for the identity of the alias "Springer," he had a relatively easy time gathering information. He entertained relations with the cadre department of the SED, to the State Secretariat for State Security and ministries of the GDR. They were exposed only much later.

Luck did eventually run out on the versed secret service agent, though. "Springer" met his equal in the unscrupulous spy that the Ministry of State Security set on him: the plain-clothes man "Fritz" (real name Kurt Rittwagen). Together with an accomplice, this spy doped, tied up and – what he proudly called "properly delivered the so-called *Fricke* package [...] to the Ministry of State Security." After that the ground became too hot for Sauter in West Berlin.

Fritz could not pass up the opportunity of writing "Springer" a scornful letter:

*Dear Max, [. . . ]. I can understand that your organization will hold it against you, and justifiably so, that your three-year collaboration with me was completely senseless.*

*It's understandable, because I, in fact, already knew everything, whether it was through your information or your written papers for your organization, whether they were findings from the EMA [Residence Registrar's Office], or research on persons who were purportedly supposed to be operating in the East. [. . . ].*

*Your organization will hold it against you that they spent thousands of marks on a rotten egg. Because the most natural thing of course won't be left out, of presenting me as an agent sent from the East.*

*Dear Max, [. . . ] you were a communist yourself, once. And I am sure they won't let you ever forget that either, and will prop the chair against your door someday, too, despite your meritorious work.*

That is what eventually happened. There was no escape, presumably also because his Einstein story had been one big bluff. And besides, in 1955, at the end of the McCarthy era, Adolf Sauter was no longer needed.

# Conclusion

The Potsdam treaty with the victors had not even been signed when the accents were set for the East-West conflict: It was from Potsdam that American President Truman issued the order to drop the atomic bombs over Hiroshima and Nagasaki. Thus was dealt the final blow on the wartime opponent Japan. This first military deployment of the atomic bomb was simultaneously a demonstration of the USA's claim to global dominion and a veiled warning to the USSR. The Allied cause had not just defeated Hitler's Germany, it had also strengthened the Soviet Union into a global power, despite all its losses. The political and intellectual expansion of communism gained threatening proportions for the West. The Cold War, which was to characterize the second half of the twentieth century, had begun and more than once the terrifying spectre of another world war appeared. The polarity between East and West defined international relations and to a large extent domestic policy as well.

Churchill set the stage for the dramatic drop in the political barometer with his speech on 5 March 1946 in Fulton, Missouri: An "iron curtain" had been drawn along the line from Stettin to Triest to stop the advance of communism in Europe. In August 1946 Truman sent the largest aircraft carrier in the world, the "Franklin D. Roosevelt" to the Eastern Mediterranean to show to the Soviets that the USA would not tolerate a further advance into Greece and Turkey. On 12 March 1947, Truman proclaimed his policy of containment (the "Truman doctrine"): The USA had to stand by European nations under the communist threat. The "system of freedom in the West" had to counter the "totalitarian system of oppression in the East." The necessary basis for it was the economic potential of the United States – which thanks to the war was bigger than ever before. The USSR, on the other hand, although politically strengthened, lacked the material resources to be able to offer any substantial help to other countries. The Marshall Plan became the first and most important instrument of the American containment policy in Europe. The Soviet Union's answer came in the form of a rigorous manipulation of tools of political pressure and a constantly strengthening grip on the "People's Democracies" and particularly on the Soviet Occupation Zone within its sphere of influence. In February 1948 the communist coup d'état took place in Czechoslovakia. The Soviet Union's response to the currency reform in the Western Zones and in West Berlin in June 1948 was the blockade of West Berlin. The West's answer to the constant Soviet threat was the founding of the North Atlantic Treaty Organization (NATO). One month after the North Atlantic treaty was signed, the Basic Constitutional Law was also signed, on 23 May 1949, signifying the creation of the Federal Republic of Germany. The founding of the German Democratic Republic followed on 7 October. The division of Germany was complete. One month before, the Soviet Union had ignited its first atomic bomb. On 23 September President Truman announced that America's nuclear monopoly had been broken. Military might no longer lay clearly on the side of the USA. "A similarly severe shock for the USA was the developments in the Asiatic region. They had managed to settle with their former enemy *Japan* surprisingly quickly, but a new enemy emerged in the communist People's Republic of China, proclaimed in 1949. [...]. This new power proved no less militant in Asia than the USSR in Europe. Korea, divided since the end of the war, became the first point of attack when in 1950 troops advanced south from the communist north."[1169] On 25 June 1950 the Cold War escalated into heated battle: the Korean War began.

The mental and propagandistic side-effect of these developments was an almost hysterical fear of the communist threat. The USA and its allies were not just surrounded by enemies, they created them as well. Whoever was simply circumspect, even if not an outright enemy, but not willing to bow unquestioningly to the interests of the USA was quickly turned into an enemy. Equating critics with political enemy number one was the club that silenced any dissenters. "A law written in reaction to National Socialism during the war years, that made antirepublican propaganda punishable was now applied to communists, the cue word "un-American activities" appeared. This movement found its most famed, soon to become notorious champion in the Republican senator, Joseph McCarthy. His anticommunist witch hunt in government, the Army and cultural life defined the domestic political climate in the USA until 1954. But anticommunism persisted even afterwards for its own sake as the guiding principle of American politics."[1170] It necessarily followed that the vigilance against supporters of the Hitler regime was relaxed. It was no longer at the center of American interests or political calculations. Whoever could contribute to the fight against the communist threat was welcome, even if he had formerly been under the employ of the Nazis.

It also necessarily followed that the erstwhile so very welcome celebrity Albert Einstein would be transformed into the political enemy. On 2 August 1939 he had written a letter to President Roosevelt "emphasizing the necessity for largescale experiments to examine the possibility of producing an atomic bomb." Einstein justified himself on 20 September 1952 with the words: "the probability that the Germans would be working on the same problem with some prospect of success compelled me to take this step. I had no other choice."[1171] He was that much more emotionally affected when the bomb was actually deployed and the assessment of the true state of nuclear research in Germany proved to have been mistaken. But he did not stop at despair. Einstein became a passionate defender of peace. His opinion was published a few weeks after the dropping of the bombs that: "salvation of civilization and the human race was only possible by the creation of a world government whose laws would guarantee the security of the individual nations. New wars are unavoidable as long as sovereign states continue to arm themselves and keep their weaponry secret."[1172] The exercise of power over atomic weapons ought to be entrusted to an international organization. He was, as before, an intellectual stationing himself above belligerent parties. Thus Einstein was inconvenient to *both* sides. But he was a citizen of the United States of America, and the *American* media were the primary bearers of his message, and *American* citizens were his primary audience. Consequently his political activism collided primarily with the interests of the political class of the *USA*. On 31 January 1950 President Truman announced a program to accelerate the development of the hydrogen bomb. A few days later, on 12 February 1950 Einstein answered on American television. He warned that the development of the H-bomb and its inevitable triggering of an arms race could lead to the destruction of mankind. Thus Einstein dropped out of favor as an unwelcome alien. On 9 February 1950 Senator McCarthy celebrated his first notorious anticommunist speech. At the same time (HFP report dated 14 February 1950) John Rankin, member of the House of Representatives, accused "Professor Einstein of communist activities and declared that Einstein should have been deported from America a long time ago."

So credible reasons were needed for depriving Einstein of his American citizenship and deporting him as an undesirable alien (hence doing exactly what once upon a time the Nazis had done). As if on cue, on 10 February 1950 the Phoenix office of the FBI forwarded its first report about Einstein's activities at the close of the 1920s and into the 1930s to its director, J. Edgar Hoover, with the opinion that this information could be suitable for revoking Einstein's citizenship of the United States of America and expelling him for the country. On 13 February 1950 (the day after Einstein's appearance on television to express his opposition to the American H-bomb program), Hoover requested a detailed report on the basis of the FBI's files. He received the report on 15 February 1950.

Shortly before, on 13 January 1950 Klaus Fuchs confessed to security officials of the British Atomic Energy Research Establishment at Harwell that he had passed on top-secret information about the atomic bomb to the Russians. On 3 February he was charged. On 1 March 1950 he was given the maximum sentence (fourteen years imprisonment). The secret behind the Soviets' success in breaking the American nuclear monopoly seemed to have been revealed. Fears about the communist threat reached a new highpoint. Every intellectual positioned on the political left now became suspicious.

Albert Einstein fell within the range of communist hunters. The first detailed intelligence report about Albert Einstein's Berlin years was written on 13 March 1950, not many days after John Rankin's speech at the House of Representatives and Klaus Fuchs's conviction.

They got what they *wanted* – so careful verification of the reports seemed uncalled for. They seemed to have found what the FBI memorandum of 10 February 1950 was looking for: information apparently suitable for revoking Einstein's citizenship and expelling him as an undesirable alien. Therefore to repeat what had happened not twenty years before in fascist Germany.

# Abbreviations

| | |
|---|---|
| A-Caputh | Amtsarchiv der Gemeinde Caputh |
| ADGB | Allgemeiner Deutscher Gewerkschaftsbund (General German Federation of Trade Unions) |
| AdK-A | Stiftung Archiv der Akademie der Künste, Berlin |
| AEG | Allgemeine Elektrizitätsgesellschaft (electrical combine) |
| Am-Apparat | "Antimilitary" agency of the CPG |
| AoS | (Prussian) Academy of Sciences (Akademie der Wissenschaften) |
| BA-B | Bundesarchiv-Abteilungen Berlin |
| BA-K | Bundesarchiv Koblenz |
| BB | Betriebsberichterstattung (CPG's industrial reporting agency) |
| BBAdW | Berlin-Brandenburgische Akademie der Wissenschaften – Akademiearchiv |
| BDC | Berlin Document Center |
| BLHA | Brandenburgisches Landeshauptarchiv |
| BStU | Archiv beim Bundesbeauftragten für die Unterlagen des Staatssicherheitsdienstes der ehemaligen DDR (Archive at the Federal Commissioner for Documentation on the State Secret Service of the former GDR) |
| CIA | Central Intelligence Agency (USA) |
| CIC | Counter Intelligence Corps (US Armed Forces) |
| CICI | Commission Internationale de Coopération Intellectuelle (International Committee on Intellectual Cooperation of the League of Nations) |

| | |
|---|---|
| CIW | Club for Intellectual Workers (Klub der Geistesarbeiter) |
| Comintern | Communist International |
| CPG | Communist Party of Germany (KPD) |
| FBI | Federal Bureau of Investigation (USA) |
| FO | Foreign Office in Berlin (Auswärtiges Amt) |
| FRG | Federal Republic of Germany |
| GDP | German Democratic Party (DDP) |
| GDR | German Democratic Republic (former East Germany, DDR) |
| Gestapo | Geheime Staatspolizei (Nazi Secret State Police) |
| GLHR | German League of Human Rights (Deutsche Liga für Menschenrechte) |
| GNPP | German National People's Party (Deutsch-Nationale Volkspartei) |
| GStA | Geheimes Staatsarchiv Preußischer Kulturbesitz |
| HA | Hauptabteilung (main department) |
| HUB | Archiv der Humboldt-Universität zu Berlin |
| IOC | International Olympic Committee |
| IPSD | Independent Party of Social Democrats (Independent Socialists, USPD) |
| IWR | International Workers Relief (Internationale Arbeiter-Hilfe) |
| IRC | International Research Council (Conseil international de recherches) |
| KdG | see CIW |
| Klara | Acronym for the 4th department of the General Staff of the Red Army (center of military espionage in the USSR; "Grete" (OGPU) is an offshoot) |
| KPD | see CPG |
| KWI | Kaiser Wilhelm Institute |
| KWS | Kaiser Wilhelm Society (Kaiser-Wilhelm-Gesellschaft) |
| LA-B | Landesarchiv Berlin |
| MPG-A | Archiv zur Geschichte der Max-Planck-Gesellschaft Berlin-Dahlem |
| NATO | North Atlantic Treaty Organization |
| Nazi Party | see NSGWP |
| NFL | New Fatherland League (Bund Neues Vaterland) |

| | |
|---|---|
| NS | National Socialist |
| NSGWP | National Socialist German Workers Party (NSDAP) |
| OGPU | United State Political Administration (Secret Service of the USSR 1923–1934) |
| OMS | Otdel Meshdunarodnovo Sviazi (Comintern International Liaison Department) |
| PA-AA | Politisches Archiv des Auswärtigen Amts |
| Populists | Deutsche Volkspartei (DVP) |
| PTR | Physikalisch-Technische Reichsanstalt (Bureau of Standards in Berlin) |
| RAG | Red Aid of Germany (Rote Hilfe) |
| RSHA | Reichssicherheitshauptamt (Reich Central Security Office) |
| SA | Sturmabteilung (Nazi Storm Detachments) |
| SAPMO | Stiftung Archiv der Parteien und Massenorganisationen der DDR im Bundesarchiv |
| SBPK | Staatsbibliothek zu Berlin – Preussischer Kulturbesitz |
| SdN-Archives | Bibliothèque Archives de la Société des Nations à Genève |
| SDP | Social Democratic Party of Germany (SPD) |
| SED | Sozialistische Einheitspartei Deutschland (East German Socialist Unity Party) |
| SS | Schutz-Staffel (Nazi Defense Squadron) |
| USSR | Union of Soviet Socialist Republics |
| WEB | Western European Bureau (of the Comintern) |

# List of Figures with Sources

# Selected Bibliography

Adler, Friedrich: Vor dem Ausnahmegericht. Jena 1923

Aichelburg, Peter C. and Roman U. Sexl (eds.): Albert Einstein, sein Einfluß auf Physik, Philosophie und Politik. Braunschweig 1919

Alliata, Giulio: Verstand kontra Relativität. Leipzig 1922

Alter, Peter: Die Kaiser-Wilhelm-Gesellschaft in den deutsch-britischen Wissenschaftsbeziehungen. In: Rudolf Vierhaus and Bernhard vom Brocke (eds.): Forschung im Spannungsfeld von Politik und Gesellschaft. Aus Anlaß des 75jährigen Bestehens der Kaiser-Wilhelm-/Max-Planck-Gesellschaft. Stuttgart 1990, pp. 717–746

Arco, Graf von, Albert Einstein, Maximilian Harden, et al. (eds.): Beiträge zur Naturgeschichte des Krieges. Berlin 1919

Ardenne, Manfred von: Ich bin ihnen begegnet. Wegweiser der Wissenschaft, Pioniere der Technik, Köpfe der Politik. Düsseldorf 1991

Armstrong, Karen: Nah ist und so schwer zu fassen der Gott. 3000 Jahre Glaubensgeschichte von Abraham bis Albert Einstein. Munich 1993

Balfour, Michael: Der Kaiser Wilhelm II. und seine Zeit. Berlin 1996 (Ullstein-Buch no. 35651)

Barnett, Lincoln: Einstein und das Universum. Amsterdam 1950

Barnett, Lincoln: The Universe and Dr. Einstein. London 1949

Bartel, Hans-Georg: Walther Nernst. Leipzig 1989

Becker, C. H.: Kulturpolitische Aufgaben des Reiches. Leipzig 1919

Bernstein, Jeremy: Albert Einstein. Raum, Zeit und Kosmos. Munich 1988

Bernstein, Jeremy: Albert Einstein. New York 1978

Blohowiak, Donald W.: Querdenker an der langen Leine. Kühn wie Edison, klug wie Einstein, relativ wie da Vinci. Landsberg 1994

Blumenfeld, Kurt: Erlebte Judenfrage. Ein Vierteljahrhundert deutscher Zionismus. Stuttgart 1962

Bodanis, David: Bis Einstein kam. Frankfurt 2003

Born, Max: The Born – Einstein Letters 1916–1955, Friendship, Politics and Physics in Uncertain Times. London 2005

Born, Max: Mein Leben. Die Erinnerungen eines Nobelpreisträgers. Munich 1975

Born, Max: Einstein – Born Letters 1916–1955. London, New York 1971

Born, Max: My Life and My Views. New York 1968

Born, Max and Leopold Infeld: Erinnerungen an Einstein. Berlin 1967

Born, Max: Physik im Wandel meiner Zeit. Braunschweig 1959

Borries, Achim von (ed.): Selbstzeugnisse des deutschen Judentums. Fischer Bücherei 1952

Böttcher, Helmuth M.: Walther Rathenau – Persönlichkeit und Werk. Bonn 1958

Bracher, Karl-Dietrich: Das Gewissen steht auf. Lebensbilder aus dem deutschen Widerstand 1933–1946. Mainz 1984

Brian, Denis: Einstein. New York 1996

Brocke, Bernhard von and Hubert Laitko (eds.): Die Kaiser-Wilhelm-/ Max-Planck-Gesellschaft und ihre Institute. Studien zu ihrer Geschichte: Das Harnack-Prinzip. Berlin, New York 1996

Brocke, Bernhard von: Friedrich Althoff. In: Wolfgang Treue and Karlfried Gründer: Berliner Lebensbilder. Wissenschaftspolitik in Berlin. Berlin 1987

Brocke, Bernhard von: Wissenschaft und Militarismus. Der Aufruf der 93 'An die Kulturwelt' und der Zusammenbruch der internationalen Gelehrtenrepublik im Ersten Weltkrieg. In: Wilamowitz nach 50 Jahren. Darmstadt 1985, pp. 649–719

Brocke, Bernhard von: Wissenschaft versus Militarismus: Nicolai, Einstein und die "Biologie des Krieges." In: Jahrbuch des deutsch-italienischen Instituts in Trient. X 1984. Bologna 1985

Bucky, Peter A.: Der private Einstein. Gespräche über Gott, die Menschen und die Bombe. Düsseldorf, Vienna 1993

Bucky, Peter: The Private Albert Einstein. Kansas City 1992

Bülow, Bernhard Wilhelm von: Der Versailler Völkerbund. Eine vorläufige Bilanz. Berlin, Stuttgart, Leipzig 1923

Calaprice, Alice (ed.): Einstein sagt – Zitate, Einfälle, Gedanken. Munich, Zurich 1999

Calaprice, Alice (ed.): The Quotable Einstein. Princeton 1996

Carpa, Ulrich and Armin Grundwald: Albert Einstein. Frankfurt 1993

Cassidy, David C.: Werner Heisenberg – Leben und Werk. Aus dem Amerikanischen von Andreas und Gisela Kleinert. Heidelberg, Berlin, Oxford 1995

Cassidy, David: Einstein and Our World. Atlantic Highlands 1995

Cassidy, David: Uncertainty. The Life and Science of Werner Heisenberg. New York 1991

Castagnetti, Giuseppe, Hubert Goenner, Jürgen Renn, Tilman Sauer and Britta Schneideler: Foundations in Disarray: Essays on Einstein's Science and Politics in the Berlin Years. Max-Planck-Institut für Wissenschaftsgeschichte. Preprint 63, 1997

Castagnetti, Giuseppe and Hubert Goenner: Albert Einstein as Pacifist and Democrat during the World War I. Science in Context 1996

Clark, Ronald W.: Albert Einstein. Ein Leben zwischen Tragik und Genialität. Munich 1995

Clark, Ronald W.: Albert Einstein. Leben und Werk. Eine Biographie. Esslingen 1974

Clark, Ronald W.: Einstein, The Life and Times. New York 1971

Cuny, Hilaire: Albert Einstein – The Man and His Times. London 1963

Daecke, Sigurd (ed.): Einstein, Albert. Worte in Zeit und Raum. Freiburg 1996

Der Einsteinturm in Potsdam. Architektur und Astrophysik. Edited by Astrophysikalisches Institut Potsdam. ARS NICOLAI and Authors 1995

Die Berliner Akademie der Wissenschaften in der Zeit des Imperialismus. Part I (1900–1917). Berlin 1975; part II (1917–1933). Berlin 1975; part III (1933–1945). Berlin 1979

Die Juden in Deutschland. Edited by Institut zum Studium der Judenfrage. Munich 1935

Dokumentation über die Verfolgung jüdischer Bürger von Ulm/Donau. Compiled on commission of the City of Ulm 1961

Dukas, Helen and Banesh Hoffmann (eds.): Einstein, Albert – Briefe. Zurich 1997

Dukas, Helen and Banesh Hoffmann: Albert Einstein – The Human Side. Princeton 1979

Düwell, Kurt: Die deutsch-amerikanischen Wissenschaftsbeziehungen im Spiegel der Kaiser-Wilhelm- und der Max-Planck-Gesellschaft. In: Rudolf Vierhaus and Bernhard vom Brocke (eds.): Forschung im Spannungsfeld von Politik und Gesellschaft. Aus Anlaß des 75jährigen Bestehens der Kaiser-Wilhelm-/Max-Planck-Gesellschaft. 1990, pp. 747–777

Düwell, Kurt: Deutschlands auswärtige Kulturpolitik 1918–1932. Grundlinien und Dokumente. Cologne, and Vienna 1976

Eckart, Dietrich: Der Bolschewismus von Moses bis Lenin. Zwiegespräche zwischen Adolf Hitler und mir. Munich 1924

Einstein, Albert, Hedwig and Max Born. Briefwechsel 1916–1955. Commentary by Max Born. Munich 1969

Einstein, Albert and Max Born: Briefwechsel 1916–1955. Foreword: Heisenberg, Werner. Munich 1991

Einstein, Albert and Sigmund Freud: Warum Krieg? Ein Schriftwechsel. Zurich 1996

Einstein, Albert and Sigmund Freud: Warum Krieg? Internationales Institut für geistige Zusammenarbeit, Völkerbund 1933. Darantière, Dijon, March 1933

Einstein, Albert: Über den Frieden. Weltordnung oder Weltuntergang? Edited by Otto Nathan and Heinz Norden. Foreword by Bertrand Russell. Neu Isenburg, Cologne 2004

Einstein, Albert: Letters to Solovine. New York 1993

Einstein, Albert: Aus meinen späten Jahren. Stuttgart Reden, Aufsätze, Briefe. Berlin 1990

Einstein, Albert: Grundzüge der Relativitätstheorie. Wiesbaden 1990

Einstein, Albert: Mein Weltbild. Berlin 1988

Einstein, Albert: The Collected Papers of Albert Einstein. Princeton 1987 et seq.

Einstein, Albert: Akademie-Vorträge (Reprinting by Akademie der Wissenschaften der DDR). Berlin 1979

Einstein, Albert: Aus meinen späten Jahren. Stuttgart 1979

Einstein, Albert: Briefe an Maurice Solovine. Berlin 1960

Einstein, Albert: On Peace. New York 1960

Einstein, Albert: Mein Weltbild. Edited by Carl Seelig. Frankfurt/Main 1955

Einstein, Albert: The World As I See It. Zurich 1953

Einstein, Albert: Out of My Later Years. New York 1950

Einstein-Centarium 1979: Ansprachen und Vorträge auf den Festveranstaltungen des Einstein-Komitees der DDR bei der Akademie der Wissenschaften der DDR vom 28.2. bis 2.3.1979 in Berlin. Edited for the Einstein Committee by Hans-Jürgen Treder. Berlin 1979

Fischer, Ernst P.: Einstein & Co. Eine kleine Geschichte der Wissenschaft der letzten hundert Jahre in Portraits. Munich 1997

Fischer, Ernst P.: Einstein. Ein Genie und sein überfordertes Publikum. Berlin 1996

Flückinger, Max: Albert Einstein in Bern. Berne 1974

Foerster, Friedrich Wilhelm: Die jüdische Frage. Basel, Freiburg, Vienna 1959

Fölsing, Albrecht: Albert Einstein. New York 1997

Fölsing, Albrecht: Albert Einstein. Eine Biographie. Frankfurt am Main 1995

Fraenkel, Heinrich and Roger Manvell: Hermann Göring. Hannover 1964

Frank, Philipp: Einstein, Sein Leben und seine Zeit. Wiesbaden 1983

Frank, Philipp: Einstein. Sein Leben und seine Zeit. with a foreword by Albert Einstein. Wiesbaden 1979

Frank, Philipp: Wahrheit – relativ oder absolut? Zurich 1952

Frank, Philipp: Einstein – His Life and Times. London 1948

Frank, Philipp: Einstein. His Life and Times. New York 1947

Freeman, May Blaker: The Story of Albert Einstein. New York 1958

French, A.P. (ed.): Albert Einstein – Wirkung und Nachwirkung. Wiesbaden 1985

French, A.P.: Einstein – A Centenary Volume. Cambridge, Mass. 1979

Freundlich, Erwin: The Foundations of Einstein's Theory of Gravitation. Cambridge 1920

Freundlich, Erwin: Die Grundlagen der Einsteinschen Gravitationstheorie. Berlin 1920

Friedmann, Alan and Carol Donley: Einstein as Myth and Muse. Cambridge 1990

Friese, E.: Kontinuität und Wandel. Deutsch-japanische Kultur- und Wissenschaftsbeziehungen nach dem Ersten Weltkrieg. In: Rudolf Vierhaus and Bernhard vom Brocke (eds.): Forschung im Spannungsfeld von Politik und Gesellschaft. Aus Anlaß des 75jährigen Bestehens der Kaiser-Wilhelm-/Max-Planck-Gesellschaft. Stuttgart 1990

Galloni, Ernesto: Alberto Einstein, su visita a la Argentina. In: Anales de la Academia National de Ciencias Exaktas, Físicas a Naturales. Buenos Aires. Timo 32. Buenos Aires 1980

Gardner, Howard: So genial wie Einstein. Schlüssel zum kreativen Denken. Stuttgart 1996

Gehrcke, Ernst: Die Massensuggestion der Relativitätstheorie. Kulturhistorisch-psychologische Dokumente. Berlin 1924

Gelegentliches von Albert Einstein. Zum 50. Geburtstag 14.3.1929. Dargelegt von der Soncino Gesellschaft der Freunde des jüdischen Buches zu Berlin. (Berlin 1929)

Geruhn, Siegfried: Doch an Einstein führt kein Weg vorbei. Kampf und Untergang des größten Denkers der Menschheit. Hohenwestedt 1988

Glick, Thomas F.: Einstein in Spain. Relativity and the Recovery of Science. Princeton 1988

Glum, Friedrich: Zwischen Wissenschaft und Politik. Erlebtes und Erdachtes in vier Reichen. Bonn 1964

Goenner, Hubert and Giuseppe Castagnetti: Albert Einstein as a Pacifist and Democrat during the First World War. Max-Planck-Institut für Wissenschaftsgeschichte. Preprint 35, (1996)

Goenner, Hubert: Einstein in Berlin 1914–1933. Munich 2005

Goenner, Hubert: Einsteins Relativitätstheorien. Munich 2002

Goetz, Dorothea: Bild, Text, Material zu Leben und Werk Albert Einsteins. Potsdam 1979

Goldsmith, Marice, Alan Mackay and James Wourduysen: Einstein – The First Hundert Years. Oxford 1980

Gribbin, John and Michael White: Einstein – A Life in Science. London 1993

Grumbach, S.: Das annexionistische Deutschland. Lausanne 1917

Grüning, Michael: Ein Haus für Albert Einstein. Erinnerungen. Briefe. Dokumente. Berlin 1990

Grüning, Michael: Der Wachsmann-Report. Auskünfte eines Architekten. Berlin 1985

Grundmann, Siegfried: Einsteins Akte. Wissenschaft und Politik – Einsteins Berliner Zeit. Mit einem Anhang über die FBI-Akte Einsteins. Springer-Verlag Berlin, Heidelberg, New York. 2. Auflage 2004

Gumbel, E. J.: Laßt Köpfe rollen – Faschistische Morde 1924–1931. Flugschrift im Auftrage der Deutschen Liga für Menschenrechte. Publisher of the Germanl League of Human Rights.

Gumbel, E. J.: Vom Rußland der Gegenwart. 1927

Gumbel, E. J.: Verschwörer. Beiträge zur Geschichte und Soziologie der deutschen nationalistischen Geheimbünde seit 1918. 1924

Gumbel, E. J.: Vier Jahre politischer Mord. Berlin 1922

Guthnick, Paul: Der Ausbau der Sternwarte Berlin-Babelsberg in den Jahren 1921–1932. In: Sitzungsberichte Preußischen Akademie der Wissenschaften. Physikalisch-mathematische Klasse. Vol. 1932, pp. 491–499

Gutsche, Edith, Peter Hägele and Hermann Hafner (eds.): Im Vorfeld wissenschaftlicher Theorien. Am Beispiel Albert Einsteins. Marburg 1988 (Porträt-Studie no. 14, publ. by Studienleiter der Studentenmission in Deutschland–SMD)

Gutsche, Willibald: Ein Kaiser im Exil. Der letzte deutsche Kaiser Wilhelm II. in Holland. Eine kritische Biographie. Marburg 1991

Haber, Charlotte: Mein Leben mit Fritz Haber. Düsseldorf, Vienna 1970

Haber, Fritz: Aus Leben und Beruf. Aufsätze, Reden, Vorträge. Berlin 1927

Harig, Gerhart (ed.): Von Adam Ries bis Max Planck. Leipzig 1962

Harnack, Adolf von: Adolf von Harnack als Zeitgenosse. Reden und Schriften aus den Jahren des Kaiserreichs und der Weimarer Republik. Edited and introduced by Kurt Nowak. Parts I and II. Berlin, New York 1996

Harnack, Adolf von: Erforschtes und Erlebtes. Giessen 1923

Harris, Sidney: Wenn Einstein recht hat ... Basel 1997

Hartung, Fritz: Das persönliche Leibregiment Kaiser Wilhelms II. Berlin 1952 (Sitzungsberichte der Deutschen Akademie der Wissenschaften zu Berlin. Klasse für Gesellschaftswissenschaften. 1952 no. 3)

Hassel, Ulrich von: Die Hassel-Tagebücher 1938–1944. Aufzeichnungen vom Andern Deutschland. Edited by Friedrich Freiherr Hiller von Gaertringen. Berlin 1988

Heilbron, J. L.: Max Planck. Ein Leben für die Wissenschaft 1858–1947. Stuttgart 1988

Heilbron, J. L.: The Dilemma of an Upright Man. Max Planck as Spokesman for German Science. Berkeley, Los Angeles, London 1986

Heine, Jens Ulrich: Verstand & Schicksal. Die Männer der I. G. Farbenindustrie A.G. in 161 Kurzbiographien. New York, Basel, Cambridge 1990

Heisenberg, Elisabeth: Das politische Leben eines Unpolitischen. Erinnerungen an Werner Heisenberg. Munich, Zurich

Heisenberg, Werner: Gesammelte Werke. Vol. IV. Autobiographisches, Lautationes, Buchbesprechungen, Kernphysik, Münchener Festrede u.a.. Munich, Zurich 1986

Hentschel, Klaus: The Einstein Tower. An Intertexture of Dynamic Construction, Relativity Theory and Astronomy. Stanford 1997

Hentschel, Klaus (ed.): Physics and National Socialism. An Anthology of Primary Sources. Basel, Boston, Berlin 1996

Hentschel, Klaus: Der Einstein-Turm. Erwin Freundlich und die Relativitätstheorie – Ansätze zu einer "dichten Beschreibung" von institutionellen, biographischen und theoriegeschichtlichen Aspekten. Heidelberg, Berlin, New York 1992

Hentschel, Klaus: Interpretationen und Fehlinterpretationen der speziellen und der allgemeinen Relativitätstheorie durch Zeitgenossen Albert Einsteins. Basel, Boston, Berlin 1990

Hermann, Armin: Einstein. Der Weltweise und sein Jahrhundert. Eine Biographie. Munich 1996

Hermann, Armin (ed.): Albert Einstein/Arnold Sommerfeld. Briefwechsel. Sechzig Briefe aus dem goldenen Zeitalter der modernen Physik. Basel 1968

Herneck, Friedrich: Über das persönliche und wissenschaftliche Wirken von Albert Einstein und Max von Laue. Berlin 1980 (Berliner wissenschaftshistorische Kolloquien. Akademie der Wissenschaften der DDR. Institut für Theorie, Geschichte und Organisation der Wissenschaft)

Herneck, Friedrich: Einstein und sein Weltbild. Aufsätze und Vorträge von Friedrich Herneck. Berlin 1979

Herneck, Friedrich: Einstein privat. Herta W. erinnert sich an die Jahre 1927 bis 1933. Berlin 1978

Herneck, Friedrich: Albert Einstein. Ein Leben für Wahrheit, Menschlichkeit und Frieden. Berlin 1963

Highfield, Roger and Paul Carter: Die geheimen Leben des Albert Einstein. Eine Biographie. Berlin 1996

Highfield, Roger and Paul Carter: The Private Lives of Albert Einstein. London 1993

Hilberg, Raul: Die Vernichtung der europäischen Juden. Vols. 1– 3, Frankfurt am Main 1990

Hillgruber, Andreas: Kontinuität oder Diskontinuität in der deutschen Außenpolitik von Bismarck bis Hitler. Düsseldorf 1969

Hoffmann, Banesh: Albert Einstein. Schöpfer und Rebell. Unter Mitwirkung von Helen Dukas. Heidelberg 1997

Hoffmann, Banesh: Einstein und der Zionismus. In: Albert Einstein. Sein Einfluß auf Physik, Philosophie und Politik. Ed. by Peter C. Aichelburg and U. Roman. Braunschweig, Wiesbaden 1979

Hoffmann, Banesh: Albert Einstein. Schöpfer und Rebell. Unter Mitwirkung von Helen Dukas. Zurich 1976

Hoffmann, Banesh and Helen Dukas: Albert Einstein, Creator and Rebel. London 1973

Hoffmann, Banesh and Helen Dukas: Albert Einstein, Creator and Rebel. New York 1972

Holl, Karl: Pazifismus in Deutschland. Frankfurt am Main. 1988

Höxtermann, Ekkehard and Ulrich Sucker: Otto Warburg (Biographien hervorragender Naturwissenschaftler, Techniker und Mediziner, vol. 91). Leipzig 1989

Illy, József: Albert Einstein in Prague. In: ISIS no. 70, 1979, pp. 76–84

Infeld, Leopold: Der Mann neben Einstein, Ein Leben zwischen Raum und Zeit. Weymann Bauer Verlag 1999

Infeld, Leopold: Leben mit Einstein. Kontur einer Erinnerung. Wien, Frankfurt am Main, Zurich 1969

Infeld, Leopold: Albert Einstein: Sein Werk und sein Einfluss auf unsere Welt. Berlin 1956

Infeld, Leopold: Albert Einstein – His Work and Its Influence on Our World. New York 1950

Infeld, Leopold: The Quest: The Evolution of a Scientist. New York 1941

Jäger, Friedrich Wilhelm: Der Einsteinturm und die Relativitätstheorie. In: Der Einsteinturm in Potsdam. Architektur und Astrophysik. Edited by Astrophysikalisches Institut Potsdam. ARS NICOLAI and Authors 1995

Jammer, Max: Einstein und die Religion. Konstanz 1995

Jerome, Fred: The Einstein File, J. Edgar Hoover's Secret War Against The World's Most Famous Scientist. New York 2002

Joffe, A.F.: Begegnung mit Physikern. Leipzig 1967

Jordan, Pascual: Albert Einstein, Sein Lebenswerk und die Zukunft der Physik. Frauenfeld 1992

Jordan, Pascual: Physikalisches Denken in der neuen Zeit. Hamburg 1935

Kanitscheider, Bernulf: Das Weltbild Albert Einsteins. Munich 1988

Kant, Horst: Albert Einstein, Max von Laue, Peter Debye und das Kaiser-Wilhelm-Institut für Physik in Berlin. In: Bernhard vom Brocke and Hubert Laitko (eds.): Die Kaiser-Wilhelm-/ Max-Planck-Gesellschaft und ihre Institute. Studien zu ihrer Geschichte: Das Harnack-Prinzip. Berlin, New York 1996, pp. 227–243

Karamanolis, Stratis: Einstein und die Atombombe. Neubiberg 1988

Kessler, Harry Graf: Walther Rathenau – Sein Leben und sein Werk. Frankfurt am Main 1988 (Berlin 1928)

Kessler, H.W.: The Origins of the First World War. London 1984

Kessler, Harry Graf: Tagebücher 1918–1937. Edited by Wolfgang Pfeiffer-Belli. Frankfurt am Main 1961

Klatzkin, Jakob: Die Judenfrage der Gegenwart. Vevey 1936

Kleinert, Andreas: Paul Weyland, der Berliner Einstein-Töter. In: Naturwissenschaft und Technik in der Geschichte. 25 Jahre Lehrstuhl für Geschichte der Naturwissenschaft und Technik. Edited by Helmuth Albrecht, Stuttgart 1993, pp. 199–232

Kraus, Oskar: Offene Briefe an Albert Einstein und Max von Laue. Vienna, Leipzig 1925

Krause, Joachim, Dietmar Ropohl and Walter Scheiffele (eds.): Vom Großen Refraktor zum Einsteinturm. Zum 70. Jahrestag des Einsteinturms. Edited by Astrophysikalischen Observatorium Potsdam, dem Ministerium für Wissenschaft, Forschung und Kultur des Landes Brandenburg und dem Museumspädagogischen Dienst Berlin. Giessen 1996

Krüss, Hugo Andres: Deutschland und die internationale wissenschaftliche Zusammenarbeit. Budapest 1928

Kuhn, Thomas S.: Die Struktur wissenschaftlicher Revolutionen. Frankfurt am Main 1996

Kurz, Peter: Die berühmtesten Patentprüfer - drei biographische Skizzen. In: Mitteilungen der deutschen Patentanwälte. Vol. 85, 1994, pp. 112–122

Kuznecow, B. K.: Einstein. Leben - Tod - Unsterblichkeit. Berlin 1977

Laitko, Hubert et al.: Wissenschaft in Berlin. Von den Anfängen bis zum Neubeginn nach 1945. Deutscher Verlag der Wissenschaften, Berlin 1987

Lange, Annemarie: Berlin in der Weimarer Republik. Berlin 1987

Laue, Max von: Aufsätze und Vorträge. Braunschweig 1962

Laue, Max von: Gesammelte Schriften und Vorträge in 3 Bänden. Braunschweig 1961

Laun, Wilhelm Ludwig: Hier irrte Einstein: Fehler und Versäumnisse. Baden-Baden 1992

Leithäuser, Joachim: Werner Heisenberg. Berlin 1957

Lenard, Philipp: Erinnerungen. 1943

Lenard, Philipp: Große Naturforscher. Munich 1930

Levenson, Thomas: Albert Einstein, Die Berliner Jahre 1914–1932. Munich 2005

Levenson, Thomas: Einstein in Berlin. New York, Toronto, London 2004

Levinger, Elma: Albert Einstein. New York 1949

Lichthein, R.: Die Geschichte des deutschen Zionismus. Jerusalem 1954

Lohmeier, Dieter and Bernhardt Schell: Einstein, Anschütz und der Kieler Kreiselkompass. Heide 1992

Lorentz, Hendrik, Albert Einstein and Hermann Minkowski: Das Relativitätsprinzip. Eine Sammlung von Abhandlungen. With a contribution by Hermann Weyl. Stuttgart 1990

Lwow: Albert Einstein. Leben und Werk. Leipzig, Jena 1957

Marianioff, Dimitri and Wayne Palma: Einstein - An Intimate Study of a Great Man. New York 1944

Marsch, Ulrich: Notgemeinschaft der Deutschen Wissenschaft. Gründung und frühe Geschichte 1920–1925. Frankfurt am Main 1994

Max-Planck-Gesellschaft. Berichte und Mitteilungen issue no. 3, 1993: Dahlem - Domäne der Wissenschaft - Ein Spaziergang zu den Berliner Instituten der Kaiser-Wilhelm-/ Max-Planck-Gesellschaft im “deutschen Oxford.” Publ. by the Max-Planck-Gesellschaft

Mechle, A.: Albert Einsteins Jahre in Berlin 1914–1932. Albert Einstein-Gesellschaft Bern 1995

Melcher, Horst: Albert Einstein wider Vorurteile und Denkgewohnheiten. Wiesbaden 1982

Michelmore, P.: Albert Einstein – Genie des Jahrhunderts. Fackelträger, Hannover 1968

Michelmore, Peter: Einstein, Profile of the Man. London 1963

Michelmore, Peter: Einstein, Profile of the Man. New York 1962

Morsey, Rudolf: Georg Schreiber. In: Wolfgang Treue and Karlfried Gründer: Berliner Lebensbilder. Wissenschaftspolitik in Berlin. Berlin 1987

Moszkowski, Alexander: Das Panorama meines Lebens. Berlin 1925

Moszkowski, Alexander: Albert Einstein. Einblicke in seine Gedankenwelt. Hamburg 1921

Moszkowski, Alexander: Einstein the Searcher. London 1921

Müller-Markus, Siegfried: Einstein und die Sowjetphilosophie: Krisis einer Lehre. Dortrecht 1960

Münch, Hans: Die Gesellschaft der Freunde des neuen Rußland in der Weimarer Republik. Publ. by Gesellschaft für Deutsch-Sowjetische Freundschaft 1958

Nathan, Otto and Heinz Nordon: Einstein on Peace. London 1963

Neffe, Jürgen: Einstein. Eine Biographie. Rowohlt Verlag, Reinbeck near Hamburg 2005

Neumann, Thomas (ed.): Albert Einstein. Elefanten Press 1970

Nicolai, Georg Friedrich: Romain Rollands Manifest und die deutschen Antworten. Charlottenburg 1921

Nicolai, Georg Friedrich: Die Biologie des Krieges. Zurich 1919

Nipperdey, Thomas: Deutsche Geschichte 1866–1918 Munich 1991

Nivolle, Jacques: Paul Langevin. Berlin 1953

Nötzold, Jürgen: Die deutsch-sowjetischen Wissenschaftsbeziehungen. In: Rudolf Vierhaus and Bernhard vom Brocke (eds.): Forschung im Spannungsfeld von Politik und Gesellschaft. Aus Anlaß des 75jährigen Bestehens der Kaiser-Wilhelm-/ Max-Planck-Gesellschaft. Stuttgart 1990, pp. 778–801

Overbye, Dennis: Einstein in Love. A Scientific Romance. London 2001

Pais, Abraham: Ich vertraue auf Intuition. Heidelberg 2000

Pais, Abraham: Einstein Lived Here. Oxford 1994

Pais, Abraham: Raffiniert ist der Herrgott. Braunschweig 1986

Pauli, Wolfgang: Wissenschaftlicher Briefwechsel mit Bohr, Einstein, Heisenberg et al. Berlin 1996

Peare, Catherine Owens: Albert Einstein. Hamburg 1951

Pfizer, Theodor: Ansprache bei der Einstein-Feier in Ulm. Göttingen, Berlin, Frankfurt 1960

Pflug, Günther: Albert Einstein als Publizist 1919–1933. Frankfurt am Main 1981 (Kleine Schriften der Deutschen Bibliothek)

Plesch, Janos: Janos erzählt von Berlin. Munich 1958

Plesch, Janos: Ein Arzt erzählt sein Leben. Munich 1949

Plesch, Janos: The Story of a Doctor. Gollancz, New York 1947

Poliakov, Léon and Joseph Wulf: Das Dritte Reich und seine Denker. Dokumente und Berichte. Wiesbaden 1989

Popović, Milan (ed.): In Albert's Shadow. The Life and Letters of Mileva Marić, Einstein's First Wife. Balt 2003

Prittwitz und Gaffron, Friedrich von: Zwischen Petersburg und Washington, Ein Diplomatenleben. Munich 1952

Pyensen, Lewis: The Young Einstein. Bristol 1985

Rathenau, Walther: Tagebuch 1907–1922. Editing and commentary by Hartmut Pogge. Düsseldorf 1967

Rathenau, Walther: Briefe. Dresden 1930

Rathenau, Walther: Der Kaiser. Eine Betrachtung. Berlin 1920

Reden zum 100. Geburtstag von Einstein, Hahn, Meitner, von Laue, gehalten am 1. März 1979 in Berlin. Berlin 1979 (Dokumentationsreihe der Freien Universität Berlin)

Reichinstein, David: Albert Einstein – A Picture of His Life and His Conception of the World. Prague 1934

Reichinstein, David: Albert Einstein. Sein Lebensbild und seine Weltanschauung. Berlin 1932

Reiser, Anton: Albert Einstein – A Biographical Portrait. Thornton Butterworth, London 1931

Renn, Jürgen and Robert Schulmann: Albert Einstein/Mileva Marić. The Love Letters. Princeton 1992

Rolland, Romain and Stefan Zweig. Briefwechsel 1910–1940. Vol. 1: 1910–1923, vol. 2: 1924–1940. Berlin 1987

Rolland, Romain: Das Gewissen Europas. Tagebuch der Kriegsjahre 1914–1919. Aufzeichnungen und Dokumente zur Moralgeschichte Europas in jener Zeit. Vol. I, July 1914 to November 1915. Berlin 1963

Rolland, Romain: Das Gewissen Europas. Tagebuch der Kriegsjahre 1914–1919. Aufzeichnungen und Dokumente zur Moralgeschichte Europas in jener Zeit. Vol. II End November 1915 to 30 March 1917. Berlin 1965

Rolland, Romain: Das Gewissen Europas. Tagebuch der Kriegsjahre 1914–1919. Aufzeichnungen und Dokumente zur Moralgeschichte Europas in jener Zeit. Vol. III March 1917 to June 1919. Berlin 1974

Rolland, Romain: Le Bund Neues Vaterland (1914–1916). Lyon, Paris 1952

Rosenfeld, Günter: Sowjetunion und Deutschland 1922–1933. Cologne 1984

Rosenkranz, Ze'ev: Albert Einstein – Privat und ganz persönlich. Jüdische National- und Universitätsbibliothek Jerusalem 2004

Rosenthal-Schneider, Ilse: Begegnungen mit Einstein, von Laue und Planck – Realität und wissenschaftliche Wahrheit. Wiesbaden 1988

Ryan, Dennis P.: Einstein and the Humanities. New York 1987

Sayen, Jamie: Einstein in America. The Scientist's Conscience in the Age of Hitler and Hiroshima. New York 1985

Schilpp, Paul Arthur: Albert Einstein, Philosoph-Scientist. Salle, Indiana 1970

Schilpp, Paul Arthur (ed.): Albert Einstein als Philosoph und Naturforscher. Stuttgart 1955

Schlicker, Wolfgang: Albert Einstein: Physiker und Humanist. Berlin 1981

Schochow, Werner: Hugo Andres Krüß und die Preußische Staatsbibliothek. In: Bibliothek – Forschung und Praxis. Munich, New Providence, London, Paris no. 1, 1995, pp. 7–19

Schreiber, Georg: Deutsche Wissenschaftspolitik von Bismarck bis zum Atomwissenschaftler Otto Hahn. Cologne, Opladen 1952

Schreiber, Georg: Zwischen Demokratie und Diktatur. Persönliche Erinnerungen an die Politik und Kultur des Reiches von 1919–1944. Regensburg, Münster 1949

Schroeder-Gudehus, Brigitte: Internationale Wissenschaftsbeziehungen und auswärtige Kulturpolitik 1919–1933. Vom Boykott und Gegen-Boykott zu ihrer Wiederaufnahme. In: Rudolf Vierhaus and Bernhard vom Brocke (eds.): Forschung im Spannungsfeld von Politik und Gesellschaft. Aus Anlaß des 75jährigen Bestehens der Kaiser-Wilhelm-/ Max-Planck-Gesellschaft. Stuttgart 1990

Schulmann, Robert and Jürgen Renn (eds.): Einstein, Albert/Marić, Mileva: Am Sonntag küss' ich Dich mündlich. Die Liebesbriefe 1897–1903. Munich, Zurich 1994

Schulz, Friedrich and Erhard Schwarz: Einstein in Ahrenshoop. Entzückt von der herben Schönheit des Fischlandes. Edited by Heimatverein im Landkreis Nordvorpommern e.V. 1995

Schützeichel, Harald (ed.): Einstein, Albert. Zeiten des Staunens. Freiburg 1993

Schwarz, Richard: Albert and the War Department. ISIS, June 1989

Schwarzenbach, Alexis: Das verschmähte Genie. Albert Einstein und die Schweiz. Munich 2005

Seelig, Carl: Albert Einstein. Leben und Werk eines Genies unserer Zeit. Zurich 1960

Seelig, Carl (ed.): Helle Zeit – Dunkle Zeit, In memoriam A. Einstein. Zurich, Stuttgart, Vienna 1956

Seelig, Carl: Albert Einstein – A Documentary Biography. London 1956

Seelig, Carl: Albert Einstein. Eine dokumentarische Biographie. Zurich, Stuttgart, Vienna 1954

Seelig, Carl: Albert Einstein und die Schweiz. Zurich 1952

Solovine, Maurice: Albert Einstein – Lettres à Maurice Solovine. Paris 1956 (see also under Einstein)

Speziali, Pierre (ed.): Albert Einstein – Michele Besso, Correspondance 1903–1955. Paris 1972

Stachel, John: Einstein from "B" to "Z." Basel 2000

Stern, Fritz: Einstein's German World. Princeton 1999

Stern, Fritz: Der Traum vom Frieden und die Versuchung der Macht – Deutsche Geschichte im 20. Jahrhundert. Berlin 1999

Stern, Fritz: Freunde im Widerspruch. Haber und Einstein. In: Rudolf Vierhaus and Bernhard vom Brocke (eds.): Forschung im Spannungsfeld von Politik und Gesellschaft. Aus Anlaß des 75jährigen Bestehens der Kaiser-Wilhelm-/ Max-Planck-Gesellschaft. Stuttgart 1990, pp. 516–551

Stern, Fritz: Dreams and Delusions. New York 1987

Sternburg, Wilhelm von : "Es ist eine unheimliche Stimmung in Deutschland" – Carl von Ossietzky und seine Zeit. Berlin 1996

Stoltzenberg, Dietrich: Fritz Haber. Chemiker, Nobelpreisträger, Deutscher, Jude. New York, Basel, Cambridge, Tokyo 1994

Sugimoto, Kenji: Albert Einstein. A Photographic Biography, New York 1989

Sugimoto, Kenji: Albert Einstein. Die kommentierte Bilddokumentation. Mit 486 Fotos, Dokumenten, Zeichnungen und Grafiken sowie einer Zeittafel im Anhang. Gräfelfing near Munich 1987

Szöllösi-Janze, Margit: Von der Mehlmotte zum Holocaust. Fritz Haber und die chemische Schädlingsbekämpfung während und nach dem 1. Weltkrieg. In: J. Kocka, H. J. Puhle and K. Tenfelde (eds.): Von der Arbeiterbewegung zum modernen Sozialstaat. Festschrift für Gerhard A. Ritter zum 65. Geburtstag. Munich, London 1994, pp. 658–682

The Einstein family correspondence. including the Albert Einstein – Mileva Marić' love letters. The Property of the Einstein Family Correspondence Trust. CHRISTIE'S New York 1996

Trbuhović-Gjurić, Desanka: Im Schatten Albert Einsteins. Das tragische Leben der Mileva Einstein-Marić. Berlin 1993

Treder, Hans-Jürgen and Christa Kirsten (eds.): Albert Einstein in Berlin 1913–1933. Part I. Darstellung und Dokumente. Berlin 1979

Treder, Hans-Jürgen and Christa Kirsten (eds.): Albert Einstein in Berlin 1913–1933. Part II. Spezialinventar. Registen der Einstein-Dokumente in den Archiven der DDR; Registen von Sitzungsprotokollen der Berliner Akademie der Wissenschaften; Verzeichnis der Akademieschriften und der Berliner Patentschriften von A. Einstein. Berlin 1979

Treue, Wolfgang and Karlfried Gründer (eds.): Berliner Lebensbilder. Wissenschaftspolitik in Berlin. Berlin 1987

Über das persönliche und wissenschaftliche Wirken von Albert Einstein und Max von Laue. Berlin 1980 (Berliner wissenschaftshistorisches Kolloquium. Akademie der Wissenschaften der DDR, Institut für Theorie, Geschichte und Organisation der Wissenschaft)

Vallentin, Antonina: Das Drama Albert Einsteins. Stuttgart 1955

Vallentin, Antonina: The Drama of Albert Einstein. New York 1954

Vallentin, Antonina: Einstein – A Biography. London 1954

Wendel, Günter: Die Kaiser-Wilhelm-Gesellschaft 1911–1914. Berlin 1975

Werner, Petra: Ein Genie irrt seltener ... Otto Heinrich Warburg. 1991

White, Michael and John Gribbin: Einstein – A Life in Science. London 1993

Whitrow, G.J.: Einstein – The Man and His Achievement. New York 1967

Wickert, Johannes: Albert Einstein – mit Selbstzeugnissen und Bilddokumenten. Reinbeck bei Hamburg 1972

Wittwer, Wolfgang W.: Carl Friedrich Becker. In: Wolfgang Treue and Karlfried Gründer (eds.): Berliner Lebensbilder. Wissenschaftspolitik in Berlin. Berlin 1987

Zackheim, Michele: Einstein's Daughter. The Search for Lieserl. New York 1999

Zahn-Harnack, Agnes von: Adolf von Harnack. Berlin 1951

Zierold, Kurt: Forschungsförderung in drei Epochen. Deutsche Forschungsgemeinschaft. Geschichte – Arbeitsweise – Kommentar. Wiesbaden 1968

Zuelzer, Wolf: The Nicolai Case – A Biography. Detroit 1982

Zuelzer, Wolf: The Nicolai case. A Biography. Detroit 1982

# Notes

1 Current location: GStA: I. HA rep. 76, Vc sec. 1, part Vc, tit. XI, no. 55, re. Einsteins Relativitätstheorie.

2 Thüring, Bruno: A. Einsteins Umsturzversuch der Physik. 2nd ed., Dr. Georg Lüttke Verlag, Berlin, 1942.

3 The Weimar Republic denotes the period between 1919 and 1933 in the history of Germany. The National Assembly of the German Reich convened for the first time under a republican constitution, that is, as a parliamentary democracy, in 1919 in the Thuringian city of Weimar.

4 Published in: Grundmann/Griese/Steinberg: Relativitätstheorie und Weltanschauung. Zur philosophischen und wissenschaftspolitischen Wirkung Albert Einsteins. Berlin, 1967.

5 PA-AA: vol. 1 (R 64677) and vol. 2 (R 64678): Vorträge des Professors Einstein im Auslande.

6 PA-AA: Ausbürgerung, 2nd list A–G (R 99639).

7 A-Caputh: Bauakten Prof. Albert Einstein.

8 BLHA: rep. 2 A – Regierung Potsdam III F 11583 – re. Landverkauf an Professor Dr. Albert Einstein.

9 GStA: I. HA rep. 77, no. 6061.

10 BLHA: Pr. Br. rep. 2A – Regierung Potsdam I – no. 1165 – re. die politische Lage im Regierungsbezirk 1933–1934.

11 GStA: I. HA rep. 151, I A no. 8191 – re. Einbeziehung und Verwertung von Grundstücken im Regierungsbezirk Potsdam 1933–1938.

12 Herneck, Friedrich: Einstein privat. Herta W. erinnert sich an die Jahre 1927 bis 1933. Berlin 1978; Herneck, Friedrich: Einstein und sein Weltbild. Aufsätze und Vorträge von Friedrich Herneck. 2nd ed., Berlin, 1979.

13 The National Archives, Washington: Federal Bureau of Investigation/Bufile Number 61-7099.

14 Jerome, Fred: The Einstein File. J. Edgar Hoover's Secret War against the World's Most Famous Scientist. St. Martin's Press. New York, 2002.

15 Born, Max: Physik im Wandel meiner Zeit. Braunschweig, 1959, p. 243.

16 Born, Max: Physik im Wandel meiner Zeit. Braunschweig, 1959, p. 243.

17 GStA: I. HA rep. 76, Vc sec. 1, part Vc, tit. XI, no. 55 – re. Einsteins Relativitätstheorie.

18 Translator's note: the Oxford English Dictionary defines the word 'dossier' as: "A bundle of papers or documents referring to some matter; esp. a bundle of papers or information about a person."

19 I only refer here to the recent, very comprehensive biographies: Hermann, Armin: Einstein. Der Weltweise und sein Jahrhundert. Eine Biographie. Munich 1994; Fölsing, Albrecht: Albert Einstein. Eine Biographie. Frankfurt am Main, 1994.

20 According to a copy of the certificate in the Archive on the History of the Max Planck Society (MPG): V rep. 13, Einstein, no. 1.

21 The "Gründerzeit" was a period of rapid economic expansion, founded to some degree on the war reparations Germany garnered as victor of the Franco-German War of 1870/71.

22 See the short biographies of these persons in the name index.

23 MPG-A: V rep. 13, Fritz Haber no. 980.

24 Albert Einstein in Berlin 1913-1933. Part I. Darstellung und Dokumente. Edited by Christa Kirsten and Hans-Jürgen Treder. With an introduction by Hans-Jürgen Treder. Berlin, 1979.

25 Fölsing, Albrecht: Albert Einstein. Eine Biographie. Frankfurt am Main, 1994, pp. 43, 44. Hermann, Armin: Einstein. Der Weltweise und sein Jahrhundert. Eine Biographie. Munich: Piper, 1994, p. 90.

26 Adolf v. Harnack's memorandum in: 25 Jahre Kaiser Wilhelm-Gesellschaft zur Förderung der Wissenschaften. Vol. 1, Berlin, 1936, p. 40.

27 *Prussia:* A substate within the German Reich since the unification of the empire (1871–1945/47). The constitution of the Reich secured Prussian hegemony. It was bolstered by the circumstance that the offices of Prussian prime minister, Prussian foreign minister and Reich chancellor were held almost continuously by one and the same person. The staffs of the Prussian ministries and the corresponding offices of the Reich were tightly interlocked as well. The king of Prussia was, at the same time, the kaiser of Germany (abdication 1919). In 1871 the seat of the Prussian court, Berlin, became the capital city of the empire. Prussia was looked upon, both domestically and abroad, as the quintessence of ("Prusso-German") militarism and aggressive great-power politics. Law No. 46, issued by the Allied Control Council on 25 Feb. 1947, decreed the dissolution of the State of Prussia.

28 Bülow, Bernhard Fürst von: Denkwürdigkeiten. Vol. I, Berlin, 1930, p. 59.

29 Heinrich Mann: Der Untertan. 1914. First published 1919. (Translated by Ernest Boyd, adapted by Daniel Theisen: The Loyal Subject. New York, 1998.)

30 We can disregard private reasons – his love affair with his cousin and his shattered marriage (which *initially* did not preclude the possibility of his first wife staying in Switzerland). In career matters, Einstein hardly took the interests of his family members into consideration.

31 See G. Schreiber: Deutsche Wissenschaftspolitik von Bismarck bis zum Atomwissenschaftler Otto Hahn, Cologne-Opladen, 1952, p. 61.

32 For histories of the Kaiser Wilhelm Society, also describing the roles of von Rathenau and Harnack, see: 25 Jahre Kaiser Wilhelm Gesellschaft zur Förderung der Wissenschaften. Vols. 1 to 3. Berlin, 1936; Die Berliner Akademie der Wissenschaften in der Zeit des Imperialismus. Part I, 1900–1917. Berlin, 1975, pp. 200ff.; Brocke, Bernhard vom/Laitko, Hubert (eds.): Die Kaiser-Wilhelm-/Max-Planck-Gesellschaft und ihre Institute. Studien zu ihrer Geschichte: Das Harnack-Prinzip. Berlin, New York, 1996.

33 For an appreciation of Friedrich Althoff's science policy, see: Bernhard vom Brocke: Friedrich Althoff. In: Treue, Wolfgang/Gründer, Karlfried: Berliner Lebensbilder. Wissenschaftspolitik in Berlin. Berlin, 1987.

34 Owing to the extreme frequency of the surname *Schmidt,* the Prussian Ministry of Justice applied for a modification in 1920. His wife's maiden name *Ott* was appended. For the sake of clarity, Schmidt will always be referred to here as *Schmidt-Ott.*

35 Quote from Gutsche, Willibald: Ein Kaiser im Exil. Der letzte deutsche Kaiser Wilhelm II. in Holland. Eine kritische Biographie. Marburg, 1991, pp. 208, 209.
36 Reich Chancellor von Bethmann-Hollweg made this suggestion in his report dated 7 Apr. 1910 to the kaiser (in reply to the inquiry issued 18 Dec. 1909) (BA-B: R 1501, no. 108070 re. Die Kaiser Wilhelm-Gesellschaft, sheet 25).
37 Denkschrift Adolf v. Harnack. In: 25 Jahre Kaiser Wilhelm-Gesellschaft zur Förderung der Wissenschaften. Vol. 1, Berlin, 1936, p. 44.
38 BA-B: R 1501, no. 108070/1, re. Die Kaiser Wilhelm-Gesellschaft.
39 "Aufzeichnung, betreffend die Harnacksche Denkschrift wegen Gründung naturwissenschaftlicher Forschungsinstitute (Schreiben des Herrn Präsidenten des königlichen Staatsministeriums vom 19. Dezember 1909)" (BA-B: R 1501, no. 108970/1).
40 Treue, Wilhelm: Zur Frage der wirtschaftlichen Motive des deutschen Antisemitismus. In: Mosse, Werner E. (ed.): Deutsches Judentum in Krieg und Revolution 1916–1923. Tübingen, 1971, p. 389.
41 According to the letter by the head of the Imperial Cabinet of Civilian Affairs, Valentini, to the Minister of Culture (GStA: I. HA rep. 76, Vc sec. 2, tit. XXIII, lit. F, no. 1, vol. XI).
42 GStA: I. HA rep. 89, no. 21271.
43 The following description is based on the published minutes of the meeting in: Sitzungsberichte der Königlich Preußischen Akademie der Wissenschaften. 1912, pp. 35–55.
44 Sitzungsberichte der Königlich Preußischen Akademie der Wissenschaften. 1912, pp. 41–55.
45 Hartmann, Hans: Lexikon der Nobelpreisträger. Frankfurt am Main, Berlin, 1967, p. 293.
46 Hartmann, Hans: Lexikon der Nobelpreisträger. Frankfurt am Main, Berlin, 1967, p. 272.
47 Grüning, Michael: Ein Haus für Albert Einstein. Erinnerungen. Briefe. Dokumente. Berlin, 1990, p. 185.
48 Bartel, Hans-Georg: Walther Nernst. Leipzig, 1989, pp. 87ff. ("Nernst as an organizer of science").
49 Ostwald, Wilhelm: Lebenslinien - Eine Selbstbiographie, part II. Berlin, 1927, p. 308.
50 Ostwald, Wilhelm: Lebenslinien - Eine Selbstbiographie, part II. Berlin, 1927, p. 435.
51 Bartel, Hans-Georg: Walther Nernst. B. G. Leipzig, 1989, p. 85.
52 Stoltzenberg, Dietrich: Fritz Haber. Weinheim, New York, Basel, Cambridge, Tokyo, 1994, p. 200.
53 Stoltzenberg, Dietrich: Fritz Haber. Weinheim, New York, Basel, Cambridge, Tokyo, 1994, pp. 200, 202.
54 GStA: I. HA rep. 76, Vc tit. 8, sec. VIII, no. 13, vol. 1.
55 GStA: I. HA rep. 76, Vc tit. 8, sec. VIII, no. 13, vol. 1.
56 GStA: I. HA rep. 76, Vc tit. 8, sec. VIII, no. 13, vol. 1.
57 Kurz, Peter: Die berühmtesten Patentprüfer - drei biographische Skizzen. In: *Mitteilungen der deutschen Patentanwälte*. Vol. 85, 1994, p. 11.
58 Quoted from Hermann, Armin: Einstein. Der Weltweise und sein Jahrhundert. Munich, 1994, p. 117.
59 Hermann, Armin: Einstein. Der Weltweise und sein Jahrhundert. Munich, 1994, p. 136.
60 Albert Einstein. Briefe an Maurice Solovine. Berlin, 1960, p. 4.
61 Quoted from Hermann, Armin: Einstein. Der Weltweise und sein Jahrhundert. Eine Biographie. Munich, 1994, p. 138.
62 Quoted from Hermann, Armin: Einstein. Der Weltweise und sein Jahrhundert. Munich, 1994, p. 156.

63 Initially, Johannes Stark advocated the theory of relativity. Only later did he become a bitter opponent (see Hermann, Armin: Einstein. Der Weltweise und sein Jahrhundert. Munich, 1994, pp. 140ff., 160).
64 Quoted from Hermann, Armin: Einstein. Der Weltweise und sein Jahrhundert. Munich, 1994, p. 176.
65 Stoltzenberg, Dietrich: Fritz Haber. Weinheim, New York, Basel, Cambridge, Tokyo, 1994, p. 224.
66 Stoltzenberg, Dietrich: Fritz Haber. Weinheim, New York, Basel, Cambridge, Tokyo, 1994, p. 225.
67 Stoltzenberg, Dietrich: Fritz Haber. Weinheim, New York, Basel, Cambridge, Tokyo, 1994, p. 225.
68 Jacobus Henricus van 't Hoff (born on 30 Aug. 1852 in Rotterdam, deceased on 3 Mar. 1911 in Berlin), Nobel prize 1901 in chemistry "in recognition of his extraordinary achievement from his discovery of the laws of chemical dynamics and of osmotic pressure in solutions" (Hartmann, Hans: Lexikon der Nobelpreisträger. Frankfurt am Main, Berlin, 1967, p. 177).
69 GStA: I. HA rep. 76, Vc sec. 2, tit. XXIII, lit. F, no. 2, vol. 14.
70 In view of this personal network it is far from coincidental that all three of them – Schmidt-Ott, von Trott zu Solz and Valentini – were elected honorary members of the Berlin Academy of Sciences at the same time, in February 1914.
71 GStA: I. HA rep. 92, papers of Schmidt-Ott. B XXIIa, vol. 1.
72 Bartel, Hans-Georg: Walther Nernst. Leipzig, 1989, p. 90.
73 BBAdW: II-III-36.
74 BBAdW: II-III-41.
75 BBAdW: II-III-34.
76 BBAdW: II-V-132.
77 BBAdW: II-V-132.
78 BBAdW: II-V-102.
79 GStA: I. HA rep. 76, Vc sec. 2, tit. XXIII, lit. F, no. 2, vol. 14.
80 GStA: I. HA rep. 76, Vc sec. 2, tit. XXIII, lit. F, no. 2, vol. 14.
81 GStA: I. HA rep. 76, Vc sec. 2, tit. XXIII, lit. F, no. 2, vol. 14.
82 GStA: I. HA rep. 76, Vc sec. 2, tit. XXIII, lit. F, no. 2, vol. 14.
83 BBAdW: II-III-36.
84 BBAdW: II-III-36. English translation in: Collected Papers of Albert Einstein, vol. 5, Princeton, 1995, doc. 493.
85 GStA: I. HA rep. 76, Vc sec. 2, tit. XXIII, lit. F, no. 2, vol. 14.
86 GStA: I. HA rep. 76, Vc sec. 2, tit. XXIII, lit. F, no. 2, vol. 1.
87 Hermann, Armin: Einstein. Der Weltweise und sein Jahrhundert. Munich, 1994, p. 23.
88 The other coauthor was the historian Reinhold Koser.
89 Romain Rolland: Das Gewissen Europas. Tagebuch der Kriegsjahre 1914–1919, vol. I: July 1914 to November 1915. Berlin, 1963, p. 400 (English translation in: Collected Papers of Albert Einstein, vol. 8, doc. 65).
90 Hermann, Armin: Einstein. Der Weltweise und sein Jahrhundert. Munich, 1994, pp. 20, 21.
91 Einstein to Romain Rolland, 21 Aug. 1917 (Romain Rolland: Das Gewissen Europas. Tagebuch der Kriegsjahre, vol. III, Berlin, 1974, p. 220). English translation in: Collected Papers of Albert Einstein, vol. 8, Princeton, 1998, doc. 374.
92 For details on the drafting, evaluation and impact of this appeal, see Brocke, Bernhard vom: Wissenschaft und Militarismus. Der Aufruf der 93 'An die Kulturwelt' und der Zusammenbruch der internationalen Gelehrten Republik im Ersten Weltkrieg. In: Wilamowitz nach 50 Jahren. Darmstadt, 1985, pp. 649–719.
93 Flier (place of publication (among others): Deutsche Bücherei Leipzig) as well as in a slightly modified version in G. F. Nicolai: Die Biologie des Krieges. Zurich, 1919, pp. 7 ff.

94 Brocke, Bernhard vom: Wissenschaft und Militarismus. Der Aufruf der 93 ‚An die Kulturwelt' und der Zusammenbruch der internationalen Gelehrtenrepublik im Ersten Weltkrieg. In: Wilamowitz nach 50 Jahren. Darmstadt, 1985, p. 665.
95 Stoltzenberg, Dietrich: Fritz Haber. Weinheim, New York, Basel, Cambridge, Tokyo, 1994, p. 230, 231.
96 Stern, Fritz: Freunde im Widerspruch. Haber und Einstein. In: Forschung im Spannungsfeld von Politik und Gesellschaft. Geschichte und Struktur der Kaiser-Wilhelm-/Max-Planck-Gesellschaft, Stuttgart, 1990, p. 530.
97 Heilbron, J. L.: Max Planck. Ein Leben für die Wissenschaft 1858–1947. Stuttgart 1988, S. 253. See also the translation, Planck, Max. My Audience with Adolf Hitler. In: Hentschel, Klaus (ed.), Physics and National Socialism. An Anthology of Primary Sources. Birkhäuser Science Networks, vol. 18, Basel, Boston, Berlin, 1996, doc. 114.
98 Koppel's letter dated 13 Nov. 1916 regarding the founding of a KWI for War Technology. Koppel praised this reorientation of Haber's institute as "important and useful for the conduct of war." (GStA: I. HA rep. 92, Schmidt Ott C 84).
99 The Born–Einstein Letters. Correspondence between Albert Einstein and Max and Hedwig Born from 1916 to 1955 with commentaries by Max Born. Translated by Irene Born with a forward by Bertrand Russell. New York, 1971, p. 20.
100 Quoted after Stoltzenberg, Dietrich: Fritz Haber. Weinheim, New York, Basel, Cambridge, Tokyo, 1994, p. 310.
101 Aufruf an die Europäer, in: Nicolai, C.F.: Die Biologie des Krieges, 2nd ed., Zurich 1919, pp. 12–14. See also the English translation in: Collected Papers of Albert Einstein, vol. 6, Princeton, 1997, doc. 8.
102 Gülzow, Erwin: Der Bund „Neues Vaterland". Probleme der bürgerlich-pazifistischen Demokratie im ersten Weltkrieg. Dissertation. Humboldt University, Berlin, 1969, p. 51.
103 Hermann, Armin: Einstein. Der Weltweise und sein Jahrhundert. Munich, 1994, pp. 29, 30.
104 Goenner, Hubert/Castagnetti, Giuseppe: Albert Einstein as a Pacifist and Democrat during the First World War. Max Planck Institut for the History of Science, Berlin. Preprint 35 (1996).
105 For a comprehensive and detailed exposition, see Gülzow, Erwin: Der Bund 'Neues Vaterland.' Probleme der bürgerlich-pazifistischen Demokratie im ersten Weltkrieg. Dissertation. Humboldt University, Berlin, 1969.
106 Cited from excerpts in 'Das Werk des Untersuchungsausschusses der Verfassunggebenden Deutschen Nationalversammlung und des Deutschen Reichstages 1919–1928' in: LA-B: rep. 2000-21-01 no. 16.
107 Even its own publications give conflicting founding dates: In *Mitteilungen des Bundes Neues Vaterland*. New series, no. 1, revolution issue, November 1918, p. 4, refers to "November 1914," whereas the progress report of the "German League of Human Rights (formerly New Fatherland League)" for the year 1919 (BA-K: NL 199, no. 30) indicates "founding October 1914."
108 *Mitteilungen des Bundes Neues Vaterland*. New series, no. 1, revolution issue, November 1918, p. 4.
109 See the membership list of fall 1915 (BA-K: NL 199, no. 14).
110 BA-K: NL 199, no. 14.
111 LA-B: E rep. 200-21-01, no. 14.
112 LA-B: E rep. 200-21-01, no. 17.
113 LA-B: E rep. 200-21-01, no. 14.
114 BA-K: NL 199, no. 14.
115 BA-K: NL 199, no. 14.
116 Presumably the sociologist Rudolf Goldscheid.
117 BA-K: NL 199, no. 14.
118 BA-K: NL 1051 (W. Schücking), no. 65.

119 BA-K: NL 1051 (W. Schücking) no. 65. English translation in: Collected Papers of Albert Einstein, vol. 8, Princeton, 1998, doc. 131.

120 Rolland, Romain: Der freie Geist. Berlin, 1966, p. 35; Rolland, Romain: Das Gewissen Europas. Tagebuch der Kriegsjahre, vol. I. Berlin, 1963, p. 400. English translation in: Collected Papers of Albert Einstein, vol. 8, Princeton, 1998, doc. 65.

121 Rolland, Romain: Das Gewissen Europas. Tagebuch der Kriegsjahre, vol. I. Berlin, 1963, p. 594.

122 Rolland, Romain: Das Gewissen Europas. Tagebuch der Kriegsjahre, vol. I. Berlin, 1963, pp. 696–701.

123 BLHA: Pr. Br. rep. 30 Berlin C tit. 95, sec. 7, lit. F, no. 5 (15804).

124 See Becker, Werner: Die Rolle der liberalen Presse. In: Mosse, Werner E. (ed.): Deutsches Judentum in Krieg und Revolution 1916–1923. Tübingen, 1971, pp. 67ff.

125 *Mitteilungen des Bundes Neues Vaterland.* New series, no. 1, revolution issue, November 1918, p. 4.

126 BA-K: NL 199, no. 14.

127 LA-B: E rep. 200-21-01, no. 18.

128 BLHA: Pr. Br. rep. 30, Berlin C 1585.

129 Rolland, Romain: Das Gewissen Europas. Tagebuch der Kriegsjahre, vol. III. Berlin, 1974, p. 184.

130 Rolland, Romain: Das Gewissen Europas. Tagebuch der Kriegsjahre, vol. III. Berlin, 1974, p. 220. English translation in: Collected Papers of Albert Einstein, vol. 8, Princeton, 1998, doc. 374.

131 Rolland, Romain: Das Gewissen Europas. Tagebuch der Kriegsjahre, vol. III. Berlin, 1974, p. 221. English translation in: Collected Papers of Albert Einstein, vol. 8, Princeton, 1998, doc. 374.

132 See also the biographical profiles in the name index.

133 For more on this issue, see Fölsing, Albrecht: Albert Einstein. Frankfurt am Main, 1994, pp. 446–467.

134 MPG-A: V rep. 13 Fritz Haber, no. 977.

135 Romain Rolland: Das Gewissen Europas. Tagebuch der Kriegsjahre 1914–1919, vol. I, July 1914 through November 1915. Berlin, 1963, pp. 696–701.

136 Einstein to Romain Rolland, 21 Aug. 1917 (Romain Rolland: Das Gewissen Europas. Tagebuch der Kriegsjahre 1914–1919, vol. III. Berlin, 1974, p. 221). English translation (there dated as 22 Aug.) in: Collected Papers of Albert Einstein, vol. 8, Princeton, 1998, doc. 374.

137 Romain Rolland: Das Gewissen Europas. Tagebuch der Kriegsjahre 1914–1919, vol. III. Berlin, 1974, p. 243.

138 See, e.g., Romain Rolland. Das Gewissen Europas. Tagebuch der Kriegsjahre 1914–1919, vol. III. Berlin, 1974, pp. 629ff.

139 For more details, see: Brocke, Bernhard vom: Wissenschaft versus Militarismus: Nicolai, Einstein und die „Biologie des Krieges". In: Jahrbuch des deutsch-italienischen Instituts in Trient. X 1984. Bologna, 1985

140 Einstein to Lorentz, 2 Aug. 1915. In: Einstein, Albert: Über den Frieden. Weltordnung oder Weltuntergang? Edited by Otto Nathan and Heinz Norden. Foreword by Bertrand Russell. Berne, 1975, pp. 29, 30. English translation in: Collected Papers of Albert Einstein, vol. 8, Princeton, 1998, doc. 103.

141 Zuelzer, W.: The Nicolai Case. A Biography. Wayne State Univ. Press. Detroit 1982, chaps. 12 and 13; Rossiskii gosudarstwennyi woennyi archiw RGWA: Reichsgericht J 655/18 (signatures 567-3-4753 and 567-3-4754).

142 HUB-Archiv: UK N 54 vol. 3.

143 This may be the reason why Nicolai never published his letter to the minister of war in full and publically gave a misleading account of his escape route: that he had been "first with friends in

Munich, then in Grunewald near Berlin" – neither of which places could be associated with Einstein: his residence was in the Berlin suburb of *Schöneberg* not *Grunewald*. Nicolai's story was that he first attempted to cross the Swiss border before going north across the Danish border. (Nicolai: Warum ich aus Deutschland ging. Steen Hasselbalchs Forlag, Copenhagen (Institut für Zeitgeschichte, Munich: Nicolai papers RD 184/13, p. 29). The fact that Nicolai's biographer Zuelzer does not even mention the escape route Eilenburg–Munich is another indication that Munich was a red herring for the Ministry War. One of the people Zuelzer had interviewed was Margot Einstein – and she would have known what happened in May 1918. In his public justification, 'Why I left Germany,' Nicolai also vaguely alluded to a person who had advised him to flee: "Furthermore, the person in whose intuitive capacity to see the right way in darkness I trusted most firmly, and who had hitherto always advised me to relent, then told me: 'Now you may go.' " (Ibid., p. 28). In all likelihood this person had been Albert Einstein. One thing Nicolai could not prevent (and perhaps never knew about) was that a copy of his letter to the minister of war was soon sent to the University of Berlin and filed away in his personnel file ("Nicolai", HUB archive: UK N 54 vol. 19). Thus Einstein's political opponents learned that Albert Einstein was somehow involved in Professor Nicolai's escape. So their condemnation of the "traitor" Nicolai was also implicitly aimed at the "traitor" Einstein. I am indebted to the Berlin historian Ottokar Luban for important unpublished details about Einstein's role in Georg Nicolai's escape.

144 BA-B: R 1501 Geheime Registratur I.

145 MPG-A: V rep. 13 A. Einstein, no. 12. English translation in: Collected Papers of Albert Einstein, vol. 8, Princeton, 1998, doc. 489.

146 Werner, Petra: Ein Genie irrt seltener... Otto Heinrich Warburg. 1991, p. 119.

147 Quoted in Ardenne, Manfred von: Erinnerungen fortgeschrieben. Ein Forscherleben im Jahrhundert des Wandels der Wissenschaften und politischer Systeme. Düsseldorf 1997, p. 431. The original letter was the property of Manfred von Ardenne. "It had been Otto Warburg's expressed wish that this letter be given to me after his death." (Letter from Ardenne to me dated 21 Mar. 1997, S.G.) English translation in: Collected Papers of Albert Einstein, vol. 8, Princeton, doc. 491.

148 Werner, Petra: Ein Genie irrt seltener... Otto Heinrich Warburg. 1991, pp. 119–124.

149 GStA: I. HA rep. 76, Vc tit. XXIII, lit. F, no. 2, vol. XIV.

150 According to Fölsing, directorship of the yet to be founded KWI of Physics even formed a part of the offer made to Einstein in 1913. (Fölsing, Albrecht: Albert Einstein. Frankfurt am Main 1994, p. 461). It is possible that oral promises were made. There is no mention of it in the official correspondence on Einstein's appointment to the Berlin academy.

151 Nernst to Dr. Schmidt, 4 Feb. 1914 (GStA: I. HA rep. 76, Vc sec. 2, tit. XXIII, lit. A, no. 116).

152 GStA: I. HA rep. 76, Vc sec. 2, tit. XXIII, lit. A, no. 116.

153 GStA: I. HA rep. 76, Vc sec. 2, tit. XXIII, lit. A, no. 116.

154 GStA: I. HA rep. 76, Vc sec. 2, tit. XXIII, lit. A, no. 116.

155 Plesch, Janos: Janos erzählt von Berlin. Munich, 1958, p. 108.

156 GStA: I. HA rep. 76, Vc sec. 2, tit. XXIII, lit. A, no. 116.

157 Schmidt-Ott, department head at the Ministry of Culture.

158 GStA: I. HA rep. 76, Vc sec. 2, tit. XXIII, lit. A, no. 116.

159 GStA: I. HA rep. 76, Vc sec. 2, tit. XXIII, lit. A, no. 116.

160 GStA: I. HA rep. 76, Vc sec. 2, tit. XXIII, lit. A, no. 116. The institute's working approach served as a model for the subsequently founded national research foundation: 'Emergency Association of German Science' (Notgemeinschaft der Deutschen Wissenschaft). This is revealed in the wording of the "Petition: Establishment of an Institute for Theoretical Physics as an extention of the Kaiser Wilhelm Institute of Physics" dated 5 March 1929: "this form of activity, which is older than the Emergency Association of German Science." (GStA: I. HA rep. 76, Vc sec. 2, tit. XXIII, lit. A, no. 116).

161 Declaration of acceptance by Schmidt-Ott: letter to the president of the KWS, 13 Oct. 1917 (GStA: I. HA rep. 76, Vc sec. 2, tit. XXIII, lit. A, no. 116). He immediately appointed the assistant at the Ministry, Prof. Krüss as his proxy on the board of trustees.

162 After Siemens's death, Minister of State Schmidt-Ott assumed chairmanship of the trustees in 1920.

163 GStA: I. HA rep. 76, Vc sec. 2, tit. XXIII, lit. A, no. 116.

164 Because Einstein played no active role in the *further* fate of the institute, I dispense with discussing its later history.

165 Handbuch für das Deutsche Reich für das Rechnungsjahr 1914. Compiled by the Reich Office of the Interior. Berlin, 1914, p. 271.

166 BA-B: R 1501, no. 13148.

167 GStA: I. HA rep. 76, Vc sec. 2, tit. XXIII, lit. F, no. 2, vol. 14.

168 See chapter 2 for more on Einstein's citizenship.

169 BA-B: R 1501, no. 13148.

170 BA-B: R 1501, no. 13148.

171 BA-B: R 1501, no. 13149.

172 "The gentlemen from Berlin are speculating with me as if I were a prize laying hen. But I don't know whether I can still lay any eggs," Einstein once said to a friend of his. Quoted from Herneck, Friedrich: Albert Einstein. Ein Leben für Wahrheit, Menschlichkeit und Frieden, Berlin, 1963, p. 122.

173 Fölsing, Albrecht: Albert Einstein. Frankfurt am Main, 1994, pp. 439, 440.

174 Fölsing, Albrecht: Albert Einstein. Frankfurt am Main, 1994, pp. 446ff.

175 "I have been here in Berlin (Wilmersdorferstr.) since Sunday." (Einstein to his mother, 2 Apr. 1914. In: The Einstein Family Correspondence. Including the Albert Einstein – Mileva Marić Love Letters. The Property of the Einstein Family Correspondence Trust. Christie's, New York, 1996, p. 31.) On 1 April he visited Haber for the first time (so *initially,* Einstein did not occupy Haber's guest quarters, but was living with his uncle in the Wilmersdorf district). On Friday, 3 April 1914, he was welcomed as Koppel's guest for the first time.

176 Einstein to M. Besso, 15 Feb. 1915. Cited from Fölsing, Albrecht: Albert Einstein. Frankfurt am Main, 1994, p. 401. English translation in: Collected Papers of Albert Einstein, vol. 8, 1998, doc. 56.

177 The details are described in Highfield, Roger/Carter, Paul: The Private Lives of Albert Einstein. London, Boston, 1993, chaps. 7–8.

178 Fölsing, Albrecht: Albert Einstein. Frankfurt am Main 1994, p. 477.

179 SBPK: Acta PrSB Einsteinstiftung. English translation in: Collected Papers of Albert Einstein, vol. 8, 1998, doc. 502, dated to before 11 April 1918 – {crossed-out text}

180 SBPK: Acta PrSB Einsteinstiftung. English translation in: Collected Papers of Albert Einstein, vol. 8, 1998, doc. 508.

181 Fölsing, Albrecht: Albert Einstein. Frankfurt am Main, 1994, p. 480.

182 BBAdW: II-XIV-41.

183 Haber, Fritz: Fünf Vorträge aus den Jahren 1920–1923. Berlin, 1927, p. 24.

184 Haber, Fritz: Fünf Vorträge aus den Jahren 1920–1923. Berlin, 1927, p. 97.

185 Georg Schreiber's influence on cultural policy is discussed in: Morsey, Rudolf: Georg Schreiber. In: Treue, Wolfgang/Gründer, Karlfried: Berliner Lebensbilder. Wissenschaftspolitik in Berlin. Berlin, 1987.

186 Schreiber, Georg: Deutsche Wissenschaftspolitik von Bismarck bis zum Atomwissenschaftler Otto Hahn, Cologne-Opladen, 1952, p. 46.

187 Wolfgang Kapp (24 July 1858–12 June 1922), founder of the German Fatherland Party, on the radical right (1917); together with the general, W. Baron von Lüttwitz, he made an unsuccessful attempt to overthrow the Reich government (13 to 17 Mar. 1920).

188 The following discussion on the Emergency Association are based on the collection of essays: Hubert Laitko et al.: Wissenschaft in Berlin: Von den Anfängen bis zum Neubeginn nach 1945. Berlin, 1987, pp. 410–413.

189 The Prussian State Library on Unter den Linden in Berlin was inaugurated in 1914. A section of the building (the "academy wing") housed the Prussian Academy of Sciences. It was there that Einstein presented his general theory of relativity in 1916.

190 GStA: I. HA rep. 76, Vc sec. 1, tit. XI, part VI, no. 1, vol. XXI.

191 C. H. Becker: Kulturpolitische Aufgaben des Reiches, Leipzig, 1919, p. 2.

192 C. H. Becker: Kulturpolitische Aufgaben des Reiches, Leipzig, 1919, p. 5.

193 C. H. Becker: Kulturpolitische Aufgaben des Reiches, Leipzig, 1919, p. 49.

194 C. H. Becker: Kulturpolitische Aufgaben des Reiches, Leipzig, 1919, p. 18.

195 C. H. Becker: Kulturpolitische Aufgaben des Reiches, Leipzig, 1919, p. 16.

196 C. H. Becker: Kulturpolitische Aufgaben des Reiches, Leipzig, 1919, p. 15.

197 See also the readable survey on German science policy and organization after World War I: Treue, Wolfgang/Gründer, Karlfried (eds.): Berliner Lebensbilder. Wissenschaftspolitik in Berlin. Berlin, 1987.

198 GStA: I. HA rep. 92, papers of Schmidt-Ott, AL XXVII.

199 Alter, Peter: Die Kaiser-Wilhelm-Gesellschaft in den deutsch-britischen Wissenschaftsbeziehungen. In: Vierhaus, Rudolf/Brocke, Bernhard vom (eds.): Forschung im Spannungsfeld von Politik und Gesellschaft. Aus Anlaß des 75jährigen Bestehens der Kaiser-Wilhelm-/Max-Planck-Gesellschaft. Stuttgart, 1990, p. 726.

200 Der Völkerbund und die deutsche Wissenschaft. In: *Mitteilungen des Verbandes der deutschen Hochschulen*, no. 7, 1923.

201 Quoted from: Brocke, Bernhard vom: Wissenschaft und Militarismus. Der Aufruf der 93 'An die Kulturwelt' und der Zusammenbruch der internationalen Gelehrtenrepublik im Ersten Weltkrieg. In: Wilamowitz nach 50 Jahren. Darmstadt, 1985, p. 681.

202 See: Der Friedensvertrag zwischen Deutschland und der Entente. Deutsche Verlags Gesellschaft für Politik und Geschichte mbH Charlottenburg 1919, p. 147.

203 The relevant provisions of the treaty are article 282, point 20: "treaty of 20 May 1875, regarding the standardization and perfection of the metric system" and point 23: "treaty of 7 June 1905, regarding the creation of an International Institute of Agriculture in Rome."

204 At the convention held from 26 November to 1 December 1918, the Union Astronomique and the Association Internationale Géographique were founded as models of such unions.

205 The severe damage these actions caused the Entente countries themselves is indirectly confirmed by their inability at Versailles to disband the International Institute of Agriculture in

Rome and the International Committee of Weights and Measures.

206 Karo, Georg: Der geistige Krieg gegen Deutschland. 2nd ed., Halle, 1926, p. 5 (my percentage calculations, S.G.).

207 Cited from *Il Mondo*, 9 Sep. 1925: Die Boykottierung der deutschen Wissenschaft (in translation: GStA: I. HA rep. 76, Vc sec. 1, tit. XI, part VII).

208 See the reports by the German Embassy in Stockholm to the Foreign Office (files regarding the Nobel Prize, PA-AA: R 64994, vol. 3).

209 Anatole France received the Nobel prize in literature in 1921 even though there was already reason enough to award it in 1919.

210 On the differences of opinion on the details see: Hubert Laitko et al.: Wissenschaft in Berlin. Von den Anfängen bis zum Neubeginn nach 1945. Berlin, 1987, pp. 405, 414.

211 Karo, G. Der geistige Krieg gegen Deutschland. 2nd ed., Halle, 1926, p. 8.

212 BA-B: R 1501, no. 9004, no. 27.

213 *Deutsche Allgemeine Zeitung*, dated 29 Jul. 1925.

214 BA-B: R 1501, no. 9004, no. 27.

215 Karo, G.: Der geistige Krieg gegen Deutschland. 2nd ed., Halle, 1926, p. 19.

216 Roethe, G.: Vom Kriege gegen die deutsche Wissenschaft. In: *Deutsche Allgemeine Zeitung*, 27 Aug. 1925.

217 BBAdW: Reichszentrale für wissenschaftliche Berichterstattung no. 61.

218 GStA: I. HA rep. 92 Schmidt-Ott C 64 I (my emphasis, S.G.).

219 Schmidt-Ott: Erlebtes und Erstrebtes 1860–1950. Wiesbaden, 1952, p. 140.

220 Schmidt-Ott: Erlebtes und Erstrebtes 1860–1950. Wiesbaden, 1952, p. 153.

221 Schmidt-Ott: Erlebtes und Erstrebtes 1860–1950. Wiesbaden, 1952, p. 164.

222 Schmidt-Ott: Erlebtes und Erstrebtes 1860–1950. Wiesbaden, 1952, p. 173.

223 Schmidt-Ott: Erlebtes und Erstrebtes 1860–1950. Wiesbaden, 1952, p. 173.

224 Schmidt-Ott: Erlebtes und Erstrebtes 1860–1950. Wiesbaden, 1952, p. 189.

225 Schmidt-Ott: Erlebtes und Erstrebtes 1860–1950. Wiesbaden, 1952, p. 173.

226 On Schmidt-Ott's biography see also Buchardt, Lothar: Friedrich Schmidt-Ott. In: Treue, Wolfgang/Gründer, Karlfried: Berliner Lebensbilder. Wissenschaftspolitik in Berlin. Berlin, 1987.

227 GStA: I. HA rep. 92 Schmidt-Ott C 64 I – re.: Amerika-Institut.

228 My emphasis, S.G.

229 Kerkhof's letter dated 21 Jan. 1924 to the Reich Ministry of the Interior (BA-B: R 1501, no. 109004). Kerkhof quotes a "Dutch source" who wished to remain anonymous. But it presumably reflected Karl Kerkhof's own view exactly as well.

230 Minutes of the meeting at the Foreign Office on 6 February 1925 re.: Verhalten der deutschen Gelehrtenwelt gegenüber dem Auslande (PA-AA: R 64981).

231 The International Institute of Intellectual Cooperation in Paris subordinate to the Committee of the League of Nations. More details will be provided later. S.G.

232 Vallentin, Antonina: Das Drama Albert Einsteins. Stuttgart, 1955, p. 166.

233 SBPK: Acta VII 1m, Krüss.

234 Schochow, Werner: Hugo Andres Krüß und die Preußische Staatsbibliothek. In: *Bibliothek – Forschung und Praxis*. Munich, New Providence, London, Paris, no. 1/1995, p. 15.

235 BBAdW: Reichszentrale für wissenschaftliche Berichterstattung, no. 13.

236 MPG-A: V rep. 13 A. Einstein no. 21. See also English translations in: Collected Papers of Albert Einstein, vol. 7, Princeton, 2002, doc. 23; vol. 9, Princeton, 2004, doc. 149. Seelig, Carl: Albert Einstein. Eine dokumentarische Biographie, Zurich, Stuttgart, Vienna, 1954, p. 229.

237 Könneker, Carsten: Die andere Moderne. Roman und Nationalsozialismus im Zeichen der modernen

Physik. Eine Literatur- und Mentalitätsgeschichtliche Bestandsaufnahme. Cologne, 1999, p. 102.

238 Könneker, Carsten: Die andere Moderne. Roman und Nationalsozialismus im Zeichen der modernen Physik. Eine Literatur- und Mentalitätsgeschichtliche Bestandsaufnahme. Cologne, 1999.

239 Gehrcke, Ernst: Die Massensuggestion der Relativitätstheorie. Kulturhistorisch-psychologische Dokumente. Berlin, 1924.

240 Einsteins amerikanische Eindrücke. In: *Vossische Zeitung*, 10 Jul. 1921.

241 Quoted from Hermann, Armin: Einstein. Der Weltweise und sein Jahrhundert. Eine Biographie. Piper, Munich, 1994, p. 156.

242 To quote an appeal by the Central Association of German Citizens of the Jewish Faith in 1916.

243 GStA: I. HA rep. 76, Vc sec. 1, part Vc, tit. XI, no. 55.

244 Hentschel, Klaus: The Einstein Tower. An Intertexture of Dynamic Construction, Relativity Theory, and Astronomy. Stanford, 1997, p. 49.

245 Party memberships of the signatories of the proposal:

| | |
|---|---|
| Dr. Schlossmann: | GDP |
| Dr. Friedberg: | GDP |
| D. Rade: | GDP |
| Otto: | GDP |
| Dr. Thaer: | Populists |
| Dr. Fassbaender: | Center |
| Gottwald: | Center |
| Dr. Hoetzsch: | GNPP |
| Frau Dr. Wegscheider: | SDP |
| Hennig: | PSD |
| Lüdemann: | SDP |
| König (Frankfurt): | SDP |
| Dr. Weyl: | PSD |

246 Einstein to the minister of science, arts and culture dated 6 Dec. 1919 (GStA: I. HA rep. 76, Vc sec. 1, part Vc, tit. XI, no. 55). English translation in: Collected Papers of Albert Einstein, vol. 9, Princeton, 2004, doc. 194.

247 The architect Dr. Eisfelder was owner of the building at Haberlandstraße 5 until August 1920. (S.G.)

248 G. Müller to Krüss, 8 Jan. 1918 (SBPK: Acta Kaiser-Wilhelm-Institute: XXVI: Institut für Physik).

249 Einstein to Krüss, 10 Jan. 1918 (SBPK: Acta Kaiser-Wilhelm-Institute: XXVI: Institut für Physik). English translation in: Collected Papers of Albert Einstein, vol. 8, 1998, doc. 435.

250 GStA: I. HA rep. 76, Vc sec. 1, part Vc, tit. XI, no. 55.

251 GStA: I. HA rep. 76, Vc sec. 1, part Vc, tit. XI, no. 55.

252 Albert Einstein in Berlin 1913–1933. Part I, Berlin, 1979, p. 177.

253 GStA: I. HA rep. 76, Vc sec. 1, part Vc, tit. XI, no. 55.

254 GStA: I. HA rep. 76, Vc sec. 1, part Vc, tit. XI, no. 55.

255 GStA: I. HA rep. 76, Vc sec. 1, part Vc, tit. XI, no. 55.

256 E. Freundlich's arguments for an application for participation in an expedition to observe the solar eclipse on 20 Sep. 1922 (GStA: I. HA rep. 76, Vc sec. 2, tit. XI, part Vc no. 7, vol. III.)

257 For details see: Hentschel, Klaus: The Einstein Tower. An Intertexture of Dynamic Construction, Relativity Theory, and Astronomy. Stanford, 1997 (original ed., 1992). See also Jäger, Friedrich Wilhelm: Der Einsteinturm und die Relativitätstheorie. In: Der Einsteinturm in Potsdam. Architektur und Astrophysik. Published by: Astrophysikalisches Institut Potsdam. ARS NICOLAI and the authors, 1995, pp. 26 ff.

258 GStA: I. HA rep. 76, Vc sec. 1, tit. XI, part II, no. 6i, vol. I.

259 Hentschel, Klaus: Physik, Astronomie und Architektur. In: Der Einsteinturm in Potsdam. Architektur und Astrophysik. Published by: Astrophysikalisches Institut Potsdam. ARS NICOLAI and the authors, 1995, p. 49.

260 The architectural history of the tower is traced in:
- Barbara Eggers: Der Einsteinturm – die Geschichte eines "Monuments der Wissenschaft."
- Joachim Krause: Vom Einsteinturm zum Zeiss-Planetarium. Wissenschaftliches Weltbild und Architektur. Both in: Der Einsteinturm in Potsdam. Architektur und Astrophysik. Published by: Astrophysikalisches Institut Potsdam. ARS NICOLAI and the authors, 1995.

261 Quoted from Barbara Eggers: Der Einsteinturm – die Geschichte eines "Monuments der Wissenschaft" In: Der Einsteinturm in Potsdam. Architektur und Astrophysik. Published by: Astrophysikalisches Institut Potsdam. ARS NICOLAI and the authors, 1995, p. 78.

262 GStA: I. HA rep. 76, Vc sec. 1, tit. XI, no. 6i, vol. I.

263 A tally of all the funds would be meaningless here, owing to the drastic fluctuations in the exchange rate of the mark during the 1920s.

264 Hentschel, Klaus: Der Einstein-Turm. Heidelberg, Berlin, New York, 1992, p. 79. Compare the expanded English translation: The Einstein Tower. Stanford, 1997, chap. 7, pp. 63 ff.

265 Ministry of Science, Arts and Culture to the Revenue Office in Opladen, dated 27 Aug. 1921, regarding approval of the application by the board of trustees of the Einstein Donation Fund for exemption of the contribution by the Farben Factories, formerly Bayer & Co., from gift tax (GStA: I. HA rep. 76, Vc sec. 1, part Vc, tit. XI, no. 55).

266 GStA: I. HA rep. 76, Vc sec. 1, part Vc, tit. XI, no. 55.

267 GStA: I. HA rep. 76, Vc sec. 1, part Vc, tit. XI, no. 55.

268 BBAdW: Astrophysikalisches Observatorium, no. 147.

269 BBAdW: Astrophysikalisches Observatorium, no. 147.

270 BBAdW: Astrophysikalisches Observatorium, no. 147.

271 According to a letter by G. Müller and E. Finlay-Freundlich to the Ministry of Science, Arts and Culture, 7 Jan. 1921 (GStA: I. HA rep. 76, Vc sec. 1, tit. XI, no. 6i, vol. I).

272 BBAdW: Astrophysikalisches Observatorium, no. 147.

273 Heine, Jens Ulrich: Verstand & Schicksal. Die Männer der I.G. Farbenindustrie A.G. in 161 Kurzbiographien. Weinheim, New York, Basel, Cambridge, 1990. Section 1.1 on the founding firms.

274 Szöllösi-Janze, Margit: Berater, Agent, Interessent? Fritz Haber, die BASF und die staatliche Stickstoffpolitik im Ersten Weltkrieg. In: *Berichte zur Wissenschaftsgeschichte*, issue 2/3 – 1996, pp. 105–117.

275 Report on a meeting of the board of trustees of the Einstein Donation Fund on 5 Dec. 1924 at Potsdam in the foundation's tower telescope. In: BBAdW: Astrophysikalisches Observatorium, no. 147.

276 GStA: I. HA rep. 76, Vc sec. 1, tit. XI, no. 6i, vol. I.

277 GStA: I. HA rep. 76, Vc sec. 1, tit. XI, no. 6i, vol. I.

278 Ludendorff forwarded a transcription of this letter to the official expert at the Ministry of Culture, Leist, on 16 Aug. 1928 (GStA: I. HA rep. 76, Vc sec. 1, tit. XI, no. 6i, vol. I). English translation in: Hentschel, Klaus: The Einstein Tower. Stanford, 1997, p. 102.

279 See the article by Eric G. Forbes on E. Freundlich in: Dictionary of Scientific Biography. Vol. V. New York, 1972, pp. 181–184.

280 Hans Ludendorff was the brother of General Erich Ludendorff and shared with him views of the political right.

281 Letter by the minister to A. Einstein, 20 Feb. 1923 with a note in the file (undated) and notification by the minister of culture to the Foreign Office, from 15

Mar. 1923 (GStA: I. HA rep. 76, Vc sec. 2, tit. XI, part Vc, no. 7, vol. III).

282 GStA: I. HA rep. 76, Vc sec. 2, tit. XI, part Vc, no. 7, vol. III.

283 GStA: I. HA rep. 76, Vc sec. 1, tit. XI, part II, no. 6i, vol. I.

284 Einstein to Arnold Berliner, 30 Jan. 1929, quoted from Hentschel, Klaus: The Einstein Tower. Stanford, 1997, p. 102.

285 GStA: I. HA rep. 76, Vc sec. 1, tit. XI, part Vc, no. 7, vol. III.

286 Carl Zeiss Jena GmbH company archive holding: SU-Rück.-Akte, no. 15 "Potsdam."

287 GStA: I. HA rep. 76, Vc sec. 1, tit. XI, no. 6i, vol. I.

288 Hentschel, Klaus: Physik, Astronomie und Architektur. In: Der Einsteinturm in Potsdam. Architektur und Astrophysik. Published by: Astrophysikalisches Institut Potsdam. ARS NICOLAI and the authors, 1995, p. 51.

289 GStA: I. HA rep. 76, Vc sec. 1, tit. XI, part II, no. 6i, vol. II. English translation in: Hentschel, Klaus: The Einstein Tower. Stanford, 1997, p. 132.

290 This is meant in a figurative sense. There is no documentary evidence that his works were among the ones demonstratively burned on Opernplatz in Berlin on 10 May 1933.

291 The Born–Einstein Letters. Correspondence between Albert Einstein and Max and Hedwig Born from 1916 to 1955 with commentaries by Max Born. Translated by Irene Born with a forward by Bertrand Russell. New York, 1971, p. 35.

292 Haber to Einstein (MPG-A: V rep. 13 Fritz Haber, no. 978).

293 Einstein to Paul and Maja Winteler, 11 Nov. 1918. Quoted from Fölsing, Albrecht: Albert Einstein. Eine Biographie. Frankfurt am Main, 1994, p. 475. English translation in: Collected Papers of Albert Einstein, vol. 8, Princeton, 1998, doc. 652.

294 Einstein to Pauline Einstein, 11 Nov. 1918. Quoted from Fölsing, Albrecht: Albert Einstein. Eine Biographie. Frankfurt am Main, 1994, p. 475. English translation in: Collected Papers of Albert Einstein, vol. 8, Princeton, 1998, doc. 651.

295 Einstein to Michele Besso, 4 Dec. 1918. Quoted from Fölsing, Albrecht: Albert Einstein. Eine Biographie. Frankfurt am Main, 1994, p. 475. English translation in: Collected Papers of Albert Einstein, vol. 8, Princeton, 1998, doc. 663.

296 Moszkowski, Alexander: Albert Einstein. Einblick in seine Gedankenwelt. Berlin, 1922, p. 235.

297 Einstein, Albert: Über den Frieden. Weltordnung oder Weltuntergang. Berne, 1975, p. 58.

298 Seelig, Carl: Albert Einstein. Eine dokumentarische Biographie. Zurich, Stuttgart, Vienna, 1954, p. 115.

299 The Born–Einstein Letters. Correspondence between Albert Einstein and Max and Hedwig Born from 1916 to 1955 with commentaries by Max Born. Translated by Irene Born with a forward by Bertrand Russell. New York, 1971, pp. 21 f.

300 Grüning, Michael: Ein Haus für Albert Einstein. Erinnerungen. Briefe. Dokumente. Berlin, 1990, p. 148.

301 Rolland, Romain: Das Gewissen Europas. Tagebuch der Kriegsjahre 1914–1919. Aufzeichnungen und Dokumente zur Moralgeschichte Europas in jener Zeit. Vol. III. Berlin, 1974, p. 836.

302 Seelig, Carl: Albert Einstein. Zurich, Stuttgart, Vienna, 1954, p. 114.

303 Liebknecht was neither Jewish nor partly Jewish. Stefan Zweig was by no means the only one to think he was.

304 Romain Rolland, Stefan Zweig. Briefwechsel 1910–1940. Vol. 1 1910–1923. Berlin, 1987, p. 508.

305 Cited from Friedländer, Saul: Die politischen Veränderungen der Kriegszeit. In: Mosse, Werner E. (ed.): Deutsches

Judentum in Krieg und Revolution 1916–1923. Tübingen, 1971, p. 49.

306 Einstein, Albert: Mein Weltbild. Edited by Carl Seelig. Frankfurt/Main, 1955, p. 94 (From the essay 'Antisemitismus und akademische Jugend.')

307 Elbogen, Ismar/Sterlin, Eleonore: Die Geschichte der Juden in Deutschland, p. 280.

308 A letter by Einstein and L. Landau dated 19 Feb. 1920 to the Prussian minister of culture (GStA: I. HA rep. 76, Va sec. 1, tit. VII, no. 78, vol. 7). English translation in: Collected Papers of Albert Einstein, vol. 9, Princeton, 2004, doc. 317.

309 GStA: I. HA rep. 76, Va sec. 1, tit. VII, no. 78, vol. 7.

310 This refers to *Georg Friedrich Nicolai* – son of the baptized Jew, Gustav Lewinstein.

311 In: *Berliner Tageblatt.* Friday, 27 Aug. 1920.

312 *Eclair,* Paris, 14 Apr. 1922.

313 PA-AA: R 64677.

314 Roderich-Stoltheim, F.: Einsteins Truglehre. Allgemeinverständlich dargelegt und widerlegt. Leipzig, 1921, p. 22.

315 Veröffentlichungen der Deutschen Gesellschaft für Weltätherforschung und anschauliche Physik (undated), issue 6.

316 Roderich-Stoltheim, F.: Einsteins Truglehre. Leipzig, 1921, p. 3.

317 Roderich-Stoltheim, F.: Einsteins Truglehre. Leipzig, 1921, p. 10.

318 *Deutsche Zeitung,* 26 Sep. 1920.

319 Quoted from Könneker, Carsten: Die andere Moderne. Roman und Nationalsozialismus im Zeichen der modernen Physik. Eine Literatur- und Mentalitätsgeschichtliche Bestandsaufnahme. Cologne, 1999, p. 199.

320 According to: Dokumentation über die Verfolgung jüdischer Bürger von Ulm/Danube, 1961, p. 7.

321 Roderich-Stoltheim, F.: Einsteins Truglehre. Leipzig, 1921, p. 9.

322 Roderich-Stoltheim, F.: Einsteins Truglehre. Leipzig, 1921, p. 19.

323 Hentschel: Interpretationen und Fehlinterpretationen der speziellen und der allgemeinen Relativitätstheorie durch Zeitgenossen Albert Einsteins. Basel, Boston, Berlin, 1990.

324 Lehmann-Rußbüldt, Otto: Der Kampf der Deutschen Liga für Menschenrechte, vormals Bund Neues Vaterland, für den Weltfrieden, Berlin, 1927, p. 100.

325 Hartmann, Hans: Lexikon der Nobelpreisträger. Frankfurt/Main, Berlin, 1967, p. 227.

326 Poliakov, Léon/Wulf, Joseph: Das Dritte Reich und seine Denker. Dokumente und Berichte. Wiesbaden 1989, p. 293. For an English translation of the article by Lenard and Stark, see also Hentschel, Klaus (ed.): Physics and National Socialism. An Anthology of Primary Sources. Birkhäuser Science Networks vol. 18, Basel, Boston, Berlin, 1996, doc. 3.

327 BA-B: formerly Berlin Document Center: Philipp Lenard.

328 *Deutsch-Völkische Monatshefte,* 1921, issue no. 1, p. 32.

329 *Deutsch-Völkische Monatshefte,* 1921, issue no. 1, p. 32.

330 'Die Relativitätstheorie eine Massensuggestion.' In: Gehrcke: Kritik der Relativitätstheorie. Gesammelte Schriften über absolute und relative Bewegung. Berlin, 1924.

331 It was not intended for publication. Stefan Zweig wrote on 10 Sep. 1920 to Romain Rolland: "I signed the telegram [...] on the condition that it not become publically known." (Romain Rolland, Stefan Zweig. Briefwechsel 1910–1940. Vol. 1, 1910–1923. Berlin, 1987, p. 576).

332 See the first "official" English translation (partly illegible carbon copy dated 2 Dec. 1953) of the article among the FBI files: Albert Einstein, part 8 of 9, Bufile no. 61-7099 (attached to the

blacked-out Weyland report of 4. Sep. 1953).

333 In the article Einstein misspells his name as "Gehrke."

334 Reprinted in Gehrcke, E.: Kritik der Relativitätstheorie. Gesammelte Schriften über absolute und relative Bewegung. Berlin, 1924.

335 Gehrcke, Ernst: Die Massensuggestion der Relativitätstheorie. Kulturhistorisch-psychologische Dokumente. Berlin, 1924, p. 12.

336 HUB: personnel file, 33 vol. 1: Gehrcke, Ernst.

337 Hedwig Born to Albert Einstein, 8 Sep. 1920 (The Born–Einstein Letters. Correspondence between Albert Einstein and Max and Hedwig Born from 1916 to 1955 with commentaries by Max Born. Translated by Irene Born with a forward by Bertrand Russell. New York, 1971, p. 34).

338 The Born–Einstein Letters. Correspondence between Albert Einstein and Max and Hedwig Born from 1916 to 1955 with commentaries by Max Born. Translated by Irene Born with a forward by Bertrand Russell. New York, 1971, p. 35.

339 See Fölsing, Albrecht: Albert Einstein. Eine Biographie. Frankfurt am Main 1994, p. 525.

340 SBPK: Acta Kaiser-Wilhelm-Institute: XXVI: Institut für Physik.

341 GStA: I. HA rep. 76, Vc sec. 1, part Vc, tit. XI, no. 55.

342 GStA: I. HA rep. 76, Vc sec. 1, part Vc, tit. XI, no. 55.

343 GStA: I. HA rep. 76, Vc sec. 1, part Vc, tit. XI, no. 55.

344 *Wolffs Telegraphisches Büro,* 7 Sep. 1920.

345 GStA: I. HA rep. 76, Vc sec. 1, part Vc, tit. XI, no. 55.

346 BBAdW: II-III-38.

347 BBAdW: II-III-38.

348 Planck to Roethe, 14 Sep. 1920 (BBAdW: II-III-38).

349 GStA: I. HA rep. 76, Vc sec. 1, part Vc, tit. XI, no. 55.

350 *Berliner Tageblatt,* 24 Sep. 1920.

351 Kleinert, Andreas: Paul Weyland, der Berliner Einstein-Töter. In: Naturwissenschaft und Technik in der Geschichte. 25 Jahre Lehrstuhl für Geschichte der Naturwissenschaft und Technik. Edited by Helmuth Albrecht, Stuttgart, 1993, pp. 199–232.

352 Christiania is the former name of *Oslo* (Norway).

353 Hans Ludendorff to the Interior Ministry, 2 Jun. 1922. Quoted from Kleinert, Andreas: Paul Weyland, der Berliner Einstein-Töter. In: Naturwissenschaft und Technik in der Geschichte. 25 Jahre Lehrstuhl für Geschichte der Naturwissenschaft und Technik. Edited by Helmuth Albrecht, Stuttgart, 1993, p. 208.

354 GStA: I. HA rep. 76, Vc sec. 1, part Vc, tit. XI, no. 55.

355 BA-B: ZA II 14504.

356 Einstein to Wilhelm Orthmann on 20 Sep. 1922 (HUB: Bestand ASTA, no. 129.).

357 Seelig, Carl: Albert Einstein. Zurich, Stuttgart, Vienna, 1954, p. 213.

358 Einstein, Albert: Briefe an Maurice Solovine. Berlin, 1960, p. 42.

359 English translation in: Collected Papers of Albert Einstein, vol. 8, Princeton, 1998, doc. 278.

360 Eckart, D.: Der Bolschewismus von Moses bis Lenin, Zwiegespräche zwischen Adolf Hitler und mir, Munich, 1924, pp. 12 f.

361 Eckart, D.: Der Bolschewismus von Moses bis Lenin. Munich, 1924, p. 12.

362 See 'Albert Einstein – 'Weiser von Zion'.' In Könneker, Carsten: Die andere Moderne. Roman und Nationalsozialismus im Zeichen der modernen Physik. Eine Literatur- und Mentalitätsgeschichtliche Bestandsaufnahme. Cologne, 1999, pp. 194 ff.

363 Born, Max: Physik im Wandel meiner Zeit. Braunschweig, 1959, p. 243.

364 Archiv der Leopoldina Halle: Matrikel, no. 3879.
365 GStA: I. HA rep. 76e, sec. 1, tit. XI, part VC, no. 55.
366 Regarding Manchester see Einstein's letter to Haber, 9 Mar. 1921 (MPG-A: V rep. 13 F. Haber no. 978).
367 Report by the German Embassy in Tokyo dated 3 Jan. 1923 (GStA: I. HA rep. 76e, sec. 1, tit. XI, part VC, no. 55, f 157, 158).
368 Report by the German Embassy in Buenos Aires dated 30 April 1925 (PA-AA: R 64678).
369 Einstein's letter to his sons, 17 Dec. 1922 (The Einstein Family Correspondence. Including the Albert Einstein - Mileva Marić love letters. The Property of the Einstein Family Correspondence Trust. Christie's, New York, 1996, p. 57).
370 Letter to Solovine, 8 Mar. 1921. In Einstein, Albert: Briefe an Solovine. Berlin 1960, p. 26.
371 Haber himself pointed out that he was "internationally encumbered by the poison-gas war" within the context of his integration into the efforts to break the boycott of German science (Haber to Donnevert, von Harnack, Heilbron, Planck, Richter and Schmidt-Ott, 19 Dec. 1925. SBPK: Acta PrSB. Völkerbund I, vol. 1).
372 Schroeder-Gudehus, Brigitte: Internationale Wissenschaftsbeziehungen und auswärtige Kulturpolitik 1919–1933. Vom Boykott und Gegen-Boykott zu ihrer Wiederaufnahme. In: Vierhaus, Rudolf/Brocke, Bernhard vom (eds.): Forschung im Spannungsfeld von Politik und Gesellschaft. Aus Anlaß des 75jährigen Bestehens der Kaiser-Wilhelm-/Max-Planck-Gesellschaft. Stuttgart, 1990, pp. 869, 966.
373 GStA: I. HA rep. 76, Vc sec. 1, tit. XI, part VI, no. 1, vol. XXI.
374 Gumbel, E.J.: Laßt Köpfe rollen. Faschistische Morde 1924–1931. Flugschrift im Auftrage der Deutschen Liga für Menschenrechte. Verlag der Deutschen Liga für Menschenrechte; Gumbel, E.J.: Verschwörer. Beiträge zur Geschichte und Soziologie der deutschen nationalistischen Geheimbünde seit 1918. Malik Verlag, 1924; Gumbel, E.J.: Vier Jahre politischer Mord. Berlin, 1922.
375 MPG-A: V rep. 13 A. Einstein, no. 26.
376 "Eing. 7/7 22. Planck": A handwritten note by Planck.
377 HUB: Bestand ASTA 129.
378 HUB: Bestand ASTA 129.
379 GStA: I. HA rep. 76e, sec. 1, tit. XI, part VC, no. 55.
380 A mispelling of *Advertiser.*
381 Einstein to Legation Councillor Dr. Soehring at the Foreign Office, 24 Apr. 1924 (PA-AA: R 64677).
382 Einstein to Ambassador Dr. Solf, quoted in the report by the German embassy in Tokyo, dated 3 Jan. 1923 (PA-AA: R 64677).
383 For more details on the travel itineraries see: Fölsing, Albrecht: Albert Einstein. Eine Biographie. Frankfurt am Main, 1994, pp. 601 ff.
384 GStA: I. HA rep. 76, Vc sec. 1, part Vc, tit. XI, no. 55 - re.: Einsteins Relativitätstheorie.
385 A misspelling of H.A. Lorentz.
386 GStA: I. HA rep. 76, Vc sec. 1, part Vc, tit. XI, no. 55.
387 Correction: Albert.
388 Transcription. GStA: I. HA rep. 76e, sec. 1, tit. XI, part VC, no. 55.
389 A misspelling of "gravitation."
390 Neurath Reichsprotektor von Böhmen und Mähren. In: *Völkischer Beobachter,* 19 Mar. 1939.
391 Weizmann, Chaim: Memoiren. Das Werden des Staates Israel. Zurich, 1953, p. 427.
392 Seelig, Carl (ed.): Helle Zeit - Dunkle Zeit. Zurich, Stuttgart, Vienna, 1956, pp. 70 ff.
393 *Jüdische Rundschau,* 9 Apr. 1929.
394 For details see: Blumenfeld, Kurt: Erlebte Judenfrage. Ein Vierteljahrhun-

dert deutscher Zionismus. Stuttgart, 1962, pp. 126 ff.

395 Seelig, Carl (ed.): Helle Zeit - Dunkle Zeit. Zurich, Stuttgart, Vienna, 1956, p. 78.

396 Seelig, Carl (ed.): Helle Zeit - Dunkle Zeit. Zurich, Stuttgart, Vienna, 1956, p. 79.

397 Einstein, Albert: Briefe an Maurice Solovine. Berlin, 1960, p. 26.

398 Weizmann, Chaim: Memoiren. Das Werden des Staates Israel. Zurich, 1953, p. 515.

399 MPG-A: V rep. 13, Fritz Haber, no. 978.

400 MPG-A: V rep. 13 Fritz Haber, no. 978.

401 Mispelling of "Weyland."

402 Hermann, Armin: Einstein. Der Weltweise und sein Jahrhundert. Munich, Zurich, 1995, p. 269.

403 Quoted from Hermann, Armin: Einstein. Der Weltweise und sein Jahrhundert. Munich, Zurich, 1995, p. 269.

404 Grüning, Michael: Ein Haus für Albert Einstein. Berlin, 1990, p. 35.

405 Grüning, Michael: Ein Haus für Albert Einstein. Berlin, 1990, p. 39.

406 Grüning, Michael: Ein Haus für Albert Einstein. Berlin, 1990, p. 41.

407 Grüning, Michael: Ein Haus für Albert Einstein. Berlin, 1990, p. 159.

408 "Charles Proteus Steinmetz (1865–1923). Born in Breslau, Steinmetz had left Germany in 1888 as the politically undesirable editor of a socialist journal. He became director of the General Electric company in Schenectady, where he died in 1923, with over 200 patents to his name." (Sugimoto, Kenji: Albert Einstein. New York, 1989, p. 75).

409 *Presse der Sowjetunion,* 1957, no. 133, p. 2897.

410 Even so, the total financial gain was lower than originally planned, "for instead of the anticipated four to five million dollars, by the end of the year only three quarters of a million had been definitively received" (Fölsing, Albrecht: Albert Einstein. Eine Biographie. Frankfurt am Main, 1994, p. 581).

411 Seelig, Carl: Albert Einstein. Zurich, Stuttgart, Vienna, 1954, p. 205.

412 Living dates under "Ussuschkin": BA-B: R 1501, no. 25673/28.

412 Living dates under "Ussuschkin": BA-B: R 1501, no. 25673/28.

413 Kusnezow, B. G.: Einstein, Moscow, 1962, p. 253 (in Russian).

414 Hermann, Armin: Einstein. Der Weltweise und sein Jahrhundert. Munich, Zurich, 1995, p. 271.

415 Quoted after Gehrcke, Ernst: Die Massensuggestion der Relativitätstheorie. Kulturhistorisch-psychologische Dokumente. Berlin, 1924, p. 39.

416 Quoted from: BA-B: 61 Re1 (Reichslandbund Personal Eig-Eit 108).

417 BBAdW: minutes of the plenary meeting of 14 Jul. 1921.

418 Fölsing, Albrecht: Albert Einstein. Eine Biographie. Frankfurt am Main, 1994, p. 584.

419 BBAdW: II-III-39.

420 Misspelling in the original for "Ligue pour les droits de l'homme."

421 Translation: "The interest of science dictates that relations be reestablished between German scholars and ourselves. You could help better than anyone, and you will do a very great service to your colleagues in Germany and in France and above all to our common ideal by accepting."

422 Kessler, Harry Graf: Walther Rathenau - Sein Leben und sein Werk. Frankfurt am Main 1988, chapter X: Die neue Außenpolitik. Der Kampf um den Frieden.

423 Kessler, Harry Graf: Walther Rathenau. Frankfurt am Main, 1988 (Berlin, 1928), p. 225.

424 Kessler, Harry Graf: Walther Rathenau. Frankfurt am Main, 1988 (Berlin, 1928), p. 225.

425 Kurt Blumenfeld wrote: "I had asked Einstein to go to Rathenau with me to influence him into giving up his office as foreign minister. Einstein shared my view." (Blumenfeld, Kurt:

Erlebte Judenfrage. Ein Vierteljahrhundert deutscher Zionismus. Stuttgart, 1962, p. 142).

426 Einstein, Albert: In Memoriam Walther Rathenau. In: *Neue Rundschau* 1922, vol. 33, issue 8, August 1922, pp. 815/816.

427 Weizmann, Chaim: Memoiren. Das Werden des Staates Israel. Zurich, 1953, p. 426.

428 MPG-A: V rep. 13 Fritz Haber, no. 978.

429 Kessler, Harry Graf: Tagebücher 1918–1937. Edited by Wolfgang Pfeiffer-Belli. Frankfurt am Main, 1961, p. 241.

430 Kessler, Harry Graf: Tagebücher 1918–1937. Frankfurt am Main, 1961, p. 243.

431 PA-AA: R 64677.

432 Gehrcke, Ernst: Die Massensuggestion der Relativitätstheorie. Kulturhistorisch-psychologische Dokumente. Berlin, 1924, p. 73.

433 Kessler, Harry Graf: Tagebücher 1918–1937. Frankfurt am Main, 1961, p. 278.

434 Kessler, Harry Graf: Tagebücher 1918–1937. Frankfurt am Main, 1961, pp. 276, 277.

435 Painlevé was later also intermittently minister of war – from 28 Nov. 1925 to 22 Oct. 1929 – and minister of aviation from 13 Dec. 1930 to 24 Jan. 1931 as well as from 4 Jun. 1932 to 29 Jan. 1933.

436 Kessler, Harry Graf: Tagebücher 1918–1937. Frankfurt am Main, 1961, p. 278.

437 PA-AA: R 64677.

438 Report by the German embassy in Paris dated 1 Apr. 1922 to the Foreign Office (PA-AA: R 64677).

439 Einstein, Langevin and Nordmann (Langevin and Nordmann had picked up Einstein at the Belgian border) feared that the crowd at the train station was a mob of protesters. But they were mistaken. It was a completely friendly welcoming party. Even this misjudgment shows how explosive the situation in France was at the time of Einstein's visit.

440 PA-AA: R 64677.

441 Quoted from *Neue Zürcher Zeitung*, 27 Mar. 1922: Kleine Chronik. Kulturdokumente.

442 Quoted from the translation by Gehrcke, Ernst: Die Massensuggestion der Relativitätstheorie. Kulturhistorisch-psychologische Dokumente. Berlin, 1924, p. 77.

443 PA-AA: R 64677 (my emphasis, S.G.). Copy of the report in: BA-B: R 1501, no. 109003, vol. 1, sheet 113 Rs., 114.

444 Friese, Eberhard: Kontinuität und Wandel. Deutsch-japanische Kultur- und Wissenschaftsbeziehungen nach dem Ersten Weltkrieg. In: Vierhaus, Rudolf/Brocke, Bernhard von (eds.): Forschung im Spannungsfeld von Politik und Gesellschaft. Aus Anlaß des 75jährigen Bestehens der Kaiser-Wilhelm-/Max-Planck-Gesellschaft. Stuttgart, 1990, p. 810.

445 Heilbron to Solf, 27 Sep. 1922 (PA-AA: R 64677).

446 Friese, Eberhard: Kontinuität und Wandel. Deutsch-japanische Kultur- und Wissenschaftsbeziehungen nach dem Ersten Weltkrieg. In: Vierhaus, Rudolf/Brocke, Bernhard von (eds.): Forschung im Spannungsfeld von Politik und Gesellschaft. Aus Anlaß des 75jährigen Bestehens der Kaiser-Wilhelm-/Max-Planck-Gesellschaft. Stuttgart, 1990, p. 805.

447 BA-K: N 1053, no. 111.

448 SBPK. Harnack papers: W. Solf.

449 Solf to von Jagow, 20 Oct. 1922 (BA-B: 90 So/1 FC, no. 1205/3474).

450 BA-B: 90 So/1 FC, no. 1205/3474.

451 BA-K: N 1053 Nr, 124 (correspondence with Walther Rathenau).

452 Solf to Prince Hatzfeld, 20 Oct. 1922 (BA-B: 90 So/1 FC, no. 1205/3474)

453 BA-K: N 1053, no. 111.

454 Soehring to Einstein, 24 Jul. 1922 (PA-AA: R 64677).

455 PA-AA: R 64677.

456 *RM* = Reich Ministry, S.G.

457 BA-K: N 1053, no. 101.

458 BA-K: N 1053, no. 101.

459 Heilbron to Solf, 28 Sep. 1922 (PA-AA: R 64677).
460 BBAdW: II-III-39.
461 Note by the German consul general for China dated 13 Nov. 1922 (BA-B: Deutsche Botschaft China, no. 3508).
462 Report by the German consulate general for China, 6 Jan. 1922 to the Foreign Office (PA-AA: R 64677); copy of the report in: BA-B: Deutsche Botschaft China, no. 3508.
463 BA-B: 90 So 1 FC (film) – papers of Dr. Solf.
464 Einstein's letter to his sons, 17 Dec. 1922 from Kyoto (The Einstein family correspondence. Including the Albert Einstein – Mileva Marić love letters. The Property of the Einstein Family Correspondence Trust. Christie's, New York, 1996, p. 57). By that point Einstein had already delivered 13 lectures.
465 Sugimoto, Kenji: Albert Einstein. New York, 1989, p. 79.
466 PA-AA: R 64677; copy = BA-K: N 1053, no. 101. Original: "in the thousands."
467 See section 2.5.1 where Solf reports on Maximilian Harden's statements, p. 115.
468 BA-K: N 1053, no. 101.
469 BA-K: N 1053, no. 101.
470 BA-K: N 1053, no. 101.
471 BA-K: N 1053, no. 101.
472 BA-K: N 1053/93.
473 My emphasis, S.G.
474 "Eivstein," as printed on the telegram.
475 BA-B: Deutsche Botschaft China no. 3508.
476 From the report by the German consulate general for China, 6 Jan. 1922 (BA-B: Deutsche Botschaft China, no. 3508).
477 Dr. Pfister, 8 Oct. 1922 to the consul general (BA-B: Deutsche Botschaft China, no. 3508).
478 From the report by the German consulate general for China, 6 Jan. 1922 (PA-AA: R 64677); copy of the report in: BA-B: Deutsche Botschaft China, no. 3508.
479 German consulate general for China, Shanghai, 13 November 1922. Note. (BA-B: Deutsche Botschaft China, no. 3508).
480 Dr. Solf's report from Tokyo had stated: "Einstein is traveling from here to Dutch India and then on to Palestine!"
481 German consulate general in Batavia, 29 Jan. 1923 to the German embassy in Tokyo (PA-AA: R 64677).
482 PA-AA: R 64677.
483 PA-AA: R 64677; copy = GStA: I. HA rep. 76, Vc sec. 1, part Vc, tit. XI, no. 55 f 150. English translation in: Glick, Thomas F.: Einstein in Spain. Relativity and the Recovery of Science. Princeton, 1988, p. 327.
484 Elsa Einstein to Hermann Struck in Haifa/Palestine (MPG-A: V rep. 13 A. Einstein, no. 31).
485 Fraenkel, Heinrich/Manvell, Roger: Hermann Göring. Hannover, 1964.
486 Hassel, Ulrich von: Die Hassel-Tagebücher 1938–1944. Aufzeichnungen vom Andern Deutschland. Edited by Friedrich Freiherr Hiller von Gaertringen. Berlin, 1988, e.g., pp. 211, 265, 277, 281, 299, 339, 365.
487 Neue Deutsche Biographie. Vol. 8. Berlin, 1969, pp. 44–45. See also: Hassel, Ulrich von: Die Hassel-Tagebücher 1938–1944. Aufzeichnungen vom Andern Deutschland. Edited by Friedrich Freiherr Hiller von Gaertringen. Berlin, 1988.
488 PA-AA: R 64677. English translation in: Glick, Thomas F.: Einstein in Spain. Relativity and the Recovery of Science. Princeton, 1988, pp. 327–329.
489 PA-AA: R 64677 (my emphasis, S.G.).
490 BBAdW: II-III-40.
491 PA-AA: Nadolny papers – vol. 2.
492 Planck to Einstein, 10 Nov. 1923 (Fölsing, F.: Albert Einstein. Frankfurt am Main, 1993, p. 620).
493 Fölsing, F.: Albert Einstein. Frankfurt am Main, 1993, p. 621.
494 Dr. Pauli, Buenos Aires to the Foreign Office, 22 Sep. 1922 (PA-AA: R 64677).

495 My emphasis, S.G.
496 Dr. Pauli, Buenos Aires to the Foreign Office, 22 Sep. 1922 (PA-AA: R 64677).
497 Dr. Pauli, Buenos Aires to the Foreign Office, 22 Sep. 1922 (PA-AA: R 64677).
498 Report by the German legation in Buenos Aires to the Foreign Office. Buenos Aires, 26 Sep. 1922 (PA-AA: R 64677).
499 Report by the German legation in Buenos Aires to the Foreign Office, Buenos Aires, 26 Sep. 1922 (PA-AA: R 64677).
500 German legation Buenos Aires to the Foreign Office, 14 May 1924 (PA-AA: R 64677).
501 PA-AA: R 64677.
502 Note by Legation Councillor Soehring (?) from 6 Jan. 1925 (PA-AA: R 64677).
503 Record from 21 Jan. 1925 (PA-AA: R 64678).
504 Foreign Office to Albert Einstein, 5 Feb. 1925 (PA-AA: Abt. VI/Kunst und Wissenschaft no. 518: Vorträge des Professors Einstein im Auslande. Vol. 2. R 64678).
505 PA-AA: R 64678.
506 Galloni, Ernesto: Alberto Einstein, su visita a la Argentina. In: *Anales de la Academia National de Ciencias Exaktas, Físicas a Naturales*. Buenos Aires. Vol. 32. Buenos Aires, 1980, pp. 263 ff.
507 German legation in Buenos Aires, 30 Apr. 1929 to the Foreign Office, Berlin (PA-AA: R 64678).
508 Report by the German legation in Montevideo, 4 Jun. 1925 to the Foreign Office (PA-AA: R 64678).
509 Report by the German legation in Rio de Janeiro, 20 May 1925 (PA-AA: R 64678).
510 His travels as a member of the Committee on Intellectual Cooperation of the League of Nations will be treated further below.
511 GStA: I. HA rep. 76, Vc sec. 1, part Vc, tit. XI, no. 55.
512 Neue Deutsche Biographie. Vol. 9, Berlin, 1972, p. 368.
513 Krüger, Peter/Hahn, Erich J.C.: Der Loyalitätskonflikt des Staatssekretärs Bernhard Wilhelm von Bülow im Frühjahr 1933. In: *Vierteljahreshefte für Zeitgeschichte*. Vol. 2, 1972, pp. 395, 396.
514 Dr. Friedrich von Prittwitz und Gaffron: Zwischen Petersburg und Washington. Ein Diplomatenleben. Munich, 1952, p. 191.
515 GStA: I. HA rep. 76, Vc sec. 1, part Vc, tit. XI, no. 55.
516 GStA: I. HA rep. 76, Vc sec. 1, part Vc, tit. XI, no. 55 (my emphasis, S.G.). The *identity* of the author of this report unfortunately could not be determined.
517 Akten zur Deutschen Auswärtigen Politik 1918–1945 im Archiv des Auswärtigen Amts: Serie B: 1925–1933. Vol. XVII. Göttingen, 1982, pp. 86, 87.
518 Bracher, Dietrich (ed.): Lebensbilder aus dem deutschen Widerstand 1933–1945. Mainz, 1984, pp. 119–121.
519 GStA: I. HA rep. 76, Vc sec. 1, part Vc, tit. XI, no. 55.
520 GStA: I. HA rep. 76, Vc sec. 2, tit. XXIII, lit. F, no. 2, vol. 15.
521 Württemberg was one of the former states in southwestern Germany composing the Reich since 1871. It was united with the *Land* of Baden in 1952 to form one of the federal provinces, Baden-Württemberg.
522 For a facsimile of the release, see Sugimoto, Kenji: Albert Einstein. A Photographic Biography, Including over 400 Photographs, Documents, Drawings, and Graphics and a Chronological Appendix. New York, 1989, p. 24. English translation in: Collected Papers of Albert Einstein, vol. 1, Princeton, 1987, doc. 20.
523 Sugimoto, Kenji: Albert Einstein. New York, 1989, p. 25. English translation in: Collected Papers of Albert Einstein, vol. 1, Princeton, 1987, doc. 60. A subsequent note that he had acquired Swiss citizenship by naturalization (see the note in the file by Senior Civil Servant

von Rottenburg at the Ministry of Culture, dated 19 Jun. 1923) is consequently erroneous. Einstein officially applied for and was granted citizenship in due form.

524 Einstein to Academy Secretary Lüders, 24 Mar. 1923 (BBAdW: II-III-40).

525 See, e.g., Romain Rolland – hence a Frenchman – on 11 Sep. 1917: "Foerster is, although German (Prussian) by birth, a naturalized Austrian, because he held lectures in Vienna for a year. (The same legal provisions are valid in Germany, where whoever is appointed to a professorship receives German citizenship from that mere fact. But he loses it again if he returns to his native country after resigning his offices.)" (Romain Rolland: Das Gewissen Europas. Tagebuch der Kriegsjahre, vol. III, Berlin, 1974, p. 243).

526 Einstein to Academy Secretary Lüders, 24 Mar. 1923 (BBAdW: II-III-40).

527 GStA: I. HA rep. 76, Vc sec. 2, tit. XXIII, lit. F, no. 2, vol. 15.

528 GStA: I. HA rep. 76, Vc sec. 2, tit. XXIII, lit. F, no. 2, vol. 15.

529 President of the PTR to the secretary of state at the Reich Office of the Interior, 24 Oct. 1916 (BA-B: Reichsamt des Innern, no. 13148 – re.: Das Kuratorium der physikalisch-technischen Reichsanstalt. Jan. 1903–Dec. 1916).

530 BA-B: Reichsamt des Innern, no. 13148.

531 BA-B: Reichsamt des Innern, no. 13149 – re.: Das Kuratorium der physikalisch-technischen Reichsanstalt.

532 Telegram from the German embassy in Paris to the Foreign Office, 27 Mar. 1922 (PA-AA: R 64677).

533 Telegram from the Foreign Office to the German embassy in Paris, 27 Mar. 1922 (PA-AA: R 64678).

534 Report by the German embassy in Paris to the Foreign Office, 29 Apr. 1922 (PA-AA: R 64678).

535 Letter by Ministerial Head of Department Heilbron to Ambassador Solf, 27 Sep. 1922 (PA-AA: R 64678).

536 The Prussian minister of science, arts and culture to the Foreign Office, 6 Dec. 1922 (PA-AA: R 64677 or R 64994, vol. 3).

537 BBAdW: II-III-39.

538 Report by the German envoy, Nadolny dated 12 Dec. 1922 to the Foreign Office (GStA: I. HA rep. 76, Vc sec. 2, tit. XXIII, lit. F, no. 2, vol. 15).

539 PA-AA: R 64994, vol. 3.

540 GStA: I. HA rep. 76 Vc sec. 2, tit. 23, lit. F, no. 2, vol. 15, sheet 83 (copy in PA-AA: Ausbürgerung 83-76, Fall: Einstein Professor).

541 According to the copy from 1933 (PA-AA: Ausbürgerung 83-76, Fall: Einstein Professor).

542 The Foreign Office to the Prussian minister of science, arts and culture, 31 Mar. 1923 (PA-AA: R 64677).

543 Verbal note by the Foreign Office, 9 Apr. 1923 (PA-AA: R 64677).

544 BBAdW: II-III-40. Einstein uses the not quite correct address "Sekretär" (instead of "Sekretar").

545 Hermann, Armin: Einstein. Der Weltweise und sein Jahrhundert. Munich, 1994, p. 299.

546 Note in the file by the senior civil servant at the Ministry of Culture, von Rottenburg, 19 Jun. 1923 (GStA: I. HA rep. 76, Vc sec. 2, tit. XXIII, lit. F, no. 2, vol. 15).

547 Einstein's affidavit for the academy's files, 7 Feb. 1924 (BBAdW: II-III-40).

548 Rendition by Soehring from 6 Jan. 1925 (PA-AA: R 64678).

549 PA-AA: R 64678.

550 The stamped out-going date was 22 Jan. 1925.

551 PA-AA: R 64678.

552 GStA: I. HA rep. 76, Vc sec. 2, tit. XXIII, lit. F, no. 2, vol. 15.

553 GStA: I. HA rep. 76, Vc sec. 2, tit. XXIII, lit. F, no. 2, vol. 15.

554 GStA: I. HA rep. 76, Vc sec. 2, tit. XXIII, lit. F, no. 2, vol. 15.

555 GStA: I. HA rep. 76, Vc sec. 2, tit. XXIII, lit. F, no. 2, vol. 15.

556 GStA: I. HA rep. 76, Vc sec. 2, tit. XXIII, lit. F, no. 2, vol. 15.

557 The undersecretary general of the League of Nations, Dufour-Feronce renders this institution throughout in capital letters: "Kommission für Geistige Zusammenarbeit"; likewise, "Internationales *Institut* für Geistige Zusammenarbeit."

558 Einstein, Albert: Über den Frieden. Weltordnung oder Weltuntergang? Edited by Otto Nathan and Heinz Norden. Foreword by Bertrand Russell. Berne, 1975, p. 129. The quote originates from Einstein's letter to Carl Seelig written in the early 1950s (Seelig, Carl: Albert Einstein. Eine dokumentarische Biographie. Zurich, Stuttgart, Vienna, 1954, p. 209).

559 Elsa Einstein to Alfred Kerr (AdK-A: Alfred-Kerr-Archiv).

560 On 26 May 1923 von Bülow was already counting on Germany soon becoming a member of the League of Nations (PA-AA: R 65511).

561 "Observations on the International Committee on Intellectuel Cooperation" (personal comments by the undersecretary general of the League of Nations, P. Nitobe, dated 18 Aug. 1922). SdN-Archives: R 1031/13/22452/14297.

562 "Observations [...]" (SdN-Archives: R 1031/13/22452/14297).

563 Société des Nations: L'Organisation du Travail intellectuel. Rapport présenté par M. Léon Bourgeois, représentant de la France, adopté par le conseil le 2 Septembre 1921 (SdN-Archives: R 1029/13c/20801/14297).

564 "Observations [...]" (SdN-Archives: R 1031/13/"22452/14297).

565 Dr. Margarete Rothbarth: Internationale geistige Zusammenarbeit. Preprint from: Wörterbuch des Völkerrechts und der Diplomatie. Berlin and Leipzig, 1928, p. 2.

566 Report by the German consulate from 19 May 1922 to the Foreign Office (PA-AA: R 64677).

567 Report by the German consulate from 19 May 1922 to the Foreign Office (PA-AA: R 64677).

568 Report by the German consulate from 19 May 1922 to the Foreign Office (PA-AA: R 64677).

569 Dr. Margarete Rothbarth: Internationale geistige Zusammenarbeit. Preprint from: Wörterbuch des Völkerrechts und der Diplomatie. Berlin and Leipzig, 1928, p. 2.

570 "Observations [...]" (SdN-Archives: R 1031/13/22452/14297).

571 For biographical details see in the name index at the back of this volume.

572 "Observations [...]" (SdN-Archives: R 1031/13/22452/14297).

573 "Observations [...]" (SdN-Archives: R 1031/13/22452/14297).

574 "Observations [...]" (SdN-Archives: R 1031/13/22452/14297).

575 PA-AA: R 64677.

576 Report by the German consulate from 19 May 1922 to the Foreign Office (PA-AA: R 64677): A remark by Einstein in an article on his Parisian impressions ('Einstein über seine Pariser Eindrücke') in the *Vossische Zeitung* dated 18 Apr. 1922, corroborates this. He said: "During my short stay I only met with scientists; I also made the acquaintance of a few representatives of the League of Nations, but they were not true politicians."

577 "Observations [...]" (SdN-Archives: R 1031/13/22452/14297).

578 SdN-Archives: R 1029/13/20823/14297.

579 The signature has been removed from the original in the Archive of the League of Nations. The following is based on a copy of the letter.

580 Report by the German consulate from 19 May 1922 to the Foreign Office (PA-AA: R 64677).

581 PA-AA: R 64677.

582 "office" meaning membership in the *committee.*
583 PA-AA: R 64677.
584 SdN-Archives: R 1029/13c/20823/14297.
585 Clark, Ronald W.: Albert Einstein. Ein Leben zwischen Tragik und Genialität. Munich, 1995, p. 256.
586 Translation from the original French (SdN-Archives: R 1029/13/20823/14297). See also Clark, Ronald W.: Albert Einstein. Leben und Werk. Eine Biographie. Esslingen, 1974, pp. 258–259.
587 "notre ami retire démission." SdN-Archives: R 1029/13/20823/14297.
588 PA-AA: R 64677.
589 On 1 Aug. 1922, the first day of the committee meeting, Einstein took part in an antiwar demonstration in the Berliner Lustgarten.
590 "Mr. A. Einstein, professor of physics at the University of Berlin, member of the Royal Academy, Amsterdam, the Royal Academy, London, and the Academy of Sciences, Berlin."
591 Comert to Einstein, 3 Aug. 1923 (SdN-Archives: R 1029/13c/20823/14297).
592 SdN-Archives: S 408 No 5 VIII.
593 German consulate in Geneva, 3 Sep. 1922 to the Foreign Office (PA-AA: R 65510).
594 German consulate in Geneva, 3 Sep. 1922 to the Foreign Office (PA-AA: R 65510).
595 German consulate in Geneva, 3 Sep. 1922 to the Foreign Office (PA-AA: R 65510).
596 German consulate in Geneva, 27 Nov. 1922 to the Foreign Office (PA-AA: R 65510).
597 Copy = SdN-Archives: R 1029/13/20825/14297 and S 408, no. 5 VIII.
598 Einstein, Albert: Briefe an Maurice Solovine. Berlin, 1960, p. 44.
599 Staatsbibliothek Unter den Linden, manuscripts department. Autograph collection: Einstein.
600 Einstein to Pierre Comert, 4 Jul. 1922, quoted from Fölsing, Albrecht: Albert Einstein. Eine Biographie. Frankfurt am Main, 1994, p. 596.
601 Comert's letter to Einstein (SdN-Archives: R 1023/13/20823/14297).
602 Einstein, Albert: Über den Frieden. Berne, 1975, p. 80.
603 SdN-Archives: R 1032/13/27987/14297.
604 SdN-Archives: R 1032/13/27987/14297.
605 Einstein, Albert: Über den Frieden. Berne, 1975, p. 81.
606 SdN-Archives: R 1029/13c/20823/14297.
607 SdN-Archives: R 1029/13c/20823/14297.
608 SdN-Archives: S 408 No 5 VIII. (The files of the League of Nations apparently only have an English translation of the letter.)
609 Appointments to the Committee on Intellectual Cooperation, 16 June 1924 (SdN-Archives: 13/36655/14397). Emphasis in the original.
610 SdN-Archives: R 1029/13/20823/14297.
611 SdN-Archives: R 1029/13/20823/14297.
612 PA-AA: R 65511.
613 *Frankfurter Zeitung*. Evening edition, 29 June 1924.
614 German consulate on 12 Aug. 1924 to the Foreign Office (PA-AA: R 65511).
615 Dr. Margarete Rothbarth: Internationale geistige Zusammenarbeit. Preprint from: Wörterbuch des Völkerrechts und der Diplomatie. Berlin and Leipzig, 1928, p. 2.
616 Dr. Margarete Rothbarth: Internationale geistige Zusammenarbeit. Preprint from: Wörterbuch des Völkerrechts und der Diplomatie. Berlin and Leipzig, 1928, p. 3.
617 Einstein, Albert / Freud, Sigmund: Warum Krieg? International Institute of Intellectual Cooperation, League of Nations, 1933. Numbered print run of only 2,000 copies.
618 SBPK: Völkerbund I, vol. 1.
619 Einstein, Albert: Über den Frieden. Berne. 1975, pp. 95, 96 (my emphasis, S.G.).
620 Einstein, Albert: Über den Frieden. Berne, 1975, p. 88.

621 Record: "Deutschlands Stellung zur Völkerbundskommission für geistige Zusammenarbeit," 26 Feb. 1926 (PA-AA: R 65516). My emphasis, S.G.
622 Rocco is named as "successor to a Mussolini opponent" in: Einstein, Albert: Über den Frieden. Berne, 1975, p. 97. This can only have referred to Professor *Ruffini* from Turin.
623 Consulate in Geneva to the Foreign Office, 31 Jul. 1925 (PA-AA: R 65514).
624 Krüss to Einstein, 2 May 1929 (SBPK: Acta PrSB, Völkerbund II, folder 2, vol. 4).
625 SdN-Archives: S 408, no. 5 VIII.
626 German consulate for the Cantons of Geneva, Neuchâtel, Vaud and Valais to the Foreign Office, 4 Jan. 1923 (PA-AA: R 64677).
627 Einstein, Albert: Über den Frieden. Berne, 1975, p. 92.
628 SdN-Archives: R 1072/13c/49461/37637.
629 SdN-Archives: R 1072/13c/49461/37637.
630 SdN-Archives: R 1072/13c/50712/37638; copy = R 1072/13c/50712/37637.
631 SdN-Archives: R 1072/13c/51225/37637.
632 Einstein, Albert: Über den Frieden. Berne, 1975, p. 91.
633 GStA: I. HA rep. 76, Vc sec. 1, tit. XI, part VII, no. 10, supplementary issue.
634 SBPK: Völkerbund I, vol. 1.
635 SdN-Archives: R1072/13c/50712/37637.
636 German consulate in Geneva to the Foreign Office, 31 Jul. 1925 (PA-AA: R 65514).
637 Krüss to Soehring, 7 Jun. 1924 (PA-AA: R. 65511).
638 Foreign Office to the German consulate in Geneva, 10 Jun. 1924 (PA-AA: R 65511).
639 German consulate in Geneva to the Foreign Office, 25 Jun. 1924 (PA-AA: R 65511).
640 German consulate in Geneva to the Foreign Office, 25 Jun. 1924 (PA-AA: R 65511).
641 Krüss to Oprescu, 22 Oct. 1924 (SdN-Archives: R 1076/13c/40176/25762).
642 SdN-Archives: R 1076/13c/40176/25762.
643 Minutes of the meeting at the Foreign Office of 6 Feb. 1925 (PA-AA: R64981). My emphasis, S.G.
644 Letter from the German consulate to Oprescu, 10 Aug. 1925 (SdN-Archives: R 1076/13c/45446/41587).
645 SdN-Archives: R 1076/13c/45446/41587.
646 SdN-Archives: R 1076/13c/45446/41587.
647 Undated letter by Krüss to Oprescu (SdN-Archives: R 1076/13c/45446/41587).
648 SdN-Archives: R 1076/13c/45446/41587.
649 SdN-Archives: R 1035/13c/4062/14297.
650 PA-AA: R 65516. The German consulate's verdict on Schulze-Gaevernitz from 31 Jul. 1925 was: "Based on the composition of the staff at the newly founded International Institute, it seems quite certain that the importance of the German university professor, Schulze-Gaevernitz, exceeds the academic standing of his future Parisian colleagues significantly" (PA-AA: R 65514). Soehring's repudiative stance (at the Foreign Office) was therefore clearly politically motivated.
651 PA-AA: R 65516.
652 PA-AA: R 65516.
653 Dr. Margarete Rothbarth: Internationale geistige Zusammenarbeit. Preprint from: Wörterbuch des Völkerrechts und der Diplomatie. Berlin and Leipzig, 1928, p. 5.
654 Foreign Office to the German embassy in Paris, 30 Jul. 1927 (PA-AA: Deutsche Botschaft Paris, no. 654c III2 adhib. 3).
655 Dufour-Feronce to Ambassador Arco von Malzan in Washington, 28 Dec. 1926 (PA-AA: Dufour-Feronce papers).
656 SdN-Archives: R 2224/5B/2423/2423 (my emphasis, S.G.).
657 SBPK: Acta PrSB. Völkerbund. Institut de Coop. Allgemeines.
658 Einstein added the postscript by *hand*.
659 SBPK: Völkerbund II, folder 3.
660 SdN-Archives: R 1074/13c/57907/37637.
661 SdN-Archives: R 1074/13c/57907/37637.

662 Undated letter by Krüss to Oprescu (probably from 1925). SdN-Archives: R 1074/13c/ 57907/37637.

663 German consulate in Geneva to the Foreign Office, 19 Feb. 1926 (PA-AA: R65516).

664 SBPK: Acta PrSB. Völkerbund I, vol. 2.

665 SdN-Archives: R 2219/5B/6953/1397.

666 SdN-Archives: R 2219/5B/6953/1397.

667 SdN-Archives: R 2219/5B/6953/1397 and S 408, no. 5 VIII.

668 SBPK: Acta Pr SB. Comité de direction.

668 SBPK: Acta Pr SB. Comité de direction.

669 SBPK: Völkerbund I, vol. 2.

670 PA-AA: Deutsche Botschaft Paris, no. 654c III 2 adhib. 3.

671 Grüning, Michael: Ein Haus für Albert Einstein. Erinnerungen Briefe Dokumente. Verlag der Nation, Berlin, 1990, p. 350.

672 SdN-Archives: R 2219/5B/6953/6397.

673 We discover here in passing that Haberlandstrasse 5 in Berlin was not Albert Einstein's main domicile but *Waldstrasse 7 in Caputh, as his last officially registered place of residence*. Judging from the sender's addresses on Einstein's correspondence of this time, the transfer of his registration at the local police from Berlin to Caputh took place in the first half of the year 1932. This is one more indicator of his attachment to his beloved summer villa! The exact dates of registration at the neighborhood police in Schöneberg, a suburb of Berlin, unfortunately cannot be verified from the existing municipal records, however. "As a consequence of war losses, the few reminants of the registration records predating 8 May 1945, the registration records from Berlin (East), Berlin (West), and the recently compiled registration data have no information on Albert Einstein * 14 Mar. 1879." (Reply by the Landeseinwohneramtes Berlin to the author's inquiry from 6 Mar. 2003).

674 SBPK: Acta PrSB, Völkerbund I, vol. 4.

675 Grüning, Michael: Ein Haus für Albert Einstein. Berlin, 1990, p. 351.

676 Krüss to Haber, 22 Jun. 1931 (SBPK: Völkerbund I, vol. 4).

677 Einstein, Albert: Über den Frieden. Berne, 1975, pp. 127, 128. No specific date for this letter is indicated. I think it would have been sent out at the *end of 1930, at the earliest*.

678 SdN-Archives: R 2219/5B/6953/1397.

679 SdN-Archives: R 2219/5B/6953/1397.

680 SdN-Archives: R 2219/5B/6953/1397.

681 SdN-Archives: R 2219/5B/6953/1397.

682 SdN-Archives: S 408, no. 5 VIII/8.

683 Einstein, Albert: Über den Frieden. Berne, 1975, p. 91.

684 SdN-Archives: S 408, no. 5 VIII/8.

685 SdN-Archives: S 408, no. 4.

686 SdN-Archives: S 408, no. 4.

687 SdN-Archives: S 408, no. 4.

688 Dufour-Feronce to Ambassador von Hoesch, 31 Jul. 1928 (PA-AA: Deutsche Botschaft Paris, no. 654c III2 adhib. 3).

689 SBPK: Acta PrSB, Völkerbund I, vol. 5.

690 SdN-Archives: S 408, no. 4.

691 Mispelled as Könen in the original.

692 Prof. Konen was a noteworthy opponent of Einstein and his theory of relativity. He later served in this capacity as advisor to the ministry under Goebbels. But this only temporarily. In 1933 already, Konen was forced into retirement.

693 Krüss to Dufour-Ference, 4 Jul. 1932 (SBPK: Völkerbund I, vol. 5).

694 SdN-Archives: R 2251/5B/39386/21266.

695 Oath of office by Dr. H.A. Krüss (SBPK: Acta PrSB. Krüss Papers, vol. 2a, no. I 3).

696 Compare Einstein's letter to Minister Rocco in Rome. In: Einstein, Albert: Mein Weltbild. Edited by Carl Seelig. Frankfurt/Main 1955, pp. 20, 21.

697 SBPK: Acta, Krüss papers, I 3 [3].

698 SBPK: Acta, Krüss papers, I 3 [3].

699 Note in the file by the secretary of the Academy of Sciences, E. Heymann, dated 11 Apr. 1933 (BBAdW: II-III-57).

700 SBPK: Acta Pr StB, Krüss papers, I 3 [3].

701 SBPK: Acta Pr StB, Krüss papers, I 3 [3].
702 SBPK: Acta PrSB, Völkerbund I, vol. 5.
703 *New Yorker Staatszeitung* from 27 Oct. 1933 described the Prussian State Library as a haven for precious books and reported on Dr. Krüss's visit to the library conference.
704 The terms used in the German translation are: "*Lehrfreiheit* und *Lernfreiheit*."
705 SBPK: Acta PrSB, Völkerbund I, vol. 5.
706 Quoted from the letter by the Foreign Office dated 14 Jun. 1933 to the Reich Ministry of the Interior and the Prussian Ministry of Science, Arts and Culture (GSTA: I. HA rep. 76 Va sec. 1, tit. 4, no. 1, vol. 13).
707 SBPK: Acta PrSB, Völkerbund I, vol. 5.
708 SBPK: Acta PrSB, Völkerbund II, vol. 6.
709 German consulate in Geneva to the Foreign Office, 19 Feb. 1926 (PA-AA: R 65516).
710 Note in the file by the secretary of the Academy of Sciences, E. Heymann, dated 11 Apr. 1933 (BBAdW: II-III-57).
711 Schochow, Werner: Hugo Andres Krüß und die Preußische Staatsbibliothek. In: *Bibliothek – Forschung und Praxis*. Munich, New Providence, London, Paris, no. 1/1995, p. 15.
712 Krüss to the Foreign Office, 9 Nov. 1933 (PA-AA: Deutsche Botschaft Paris, no. 654c III2 adhib. 3).
713 SBPK: Völkerbund II, folder 2, vol. 6.
714 SBPK: Völkerbund II, folder 2, vol. 6.
715 Quoted from Clark, Ronald W.: Albert Einstein. Ein Leben zwischen Tragik und Genialität. Munich, 1995, pp. 261, 262.
716 Quoted from Clark, Ronald W.: Albert Einstein. Ein Leben zwischen Tragik und Genialität. Munich, 1995, p. 262.
717 SdN-Archives: S 408, no. 5 VIII/8.
718 Einstein, Albert/Freud, Sigmund: Warum Krieg? Internationales Institut für geistige Zusammenarbeit, Völkerbund 1933. Limited edition of only 2,000 copies. Printed by Imprimerie Darantière, Dijon (France) March 1933.
719 The Dawes plan was an international treaty concerning German reparations payments after World War I. It was signed on 16 Aug. 1924 in London and came into force on 1 Sep. 1924. It fixed the annual installments Germany was to pay until 1928/29 at 2.5 billion reichsmarks. An international loan of (800 million goldmarks), primarily advanced by the USA to stabilize the German currency, was an integral part of the plan.
720 See Kracauer, Siegfried: Die Angestellten. Kulturkritischer Essay. New edition, Leipzig/Weimar, 1981.
721 LA-B: rep. 211 Acc. 1674, no. 488.
722 LA-B: rep. 211 Acc. 1674, no. 488. My emphasis, S.G.
723 LA-B: rep. 211 Acc. 1674, no. 488.
724 BA-B: R 43 I/1923, vol. 5.
725 BA-B: R 43 I/1923, vol. 5.
726 Fölsing, Albrecht: Albert Einstein. Eine Biographie. Frankfurt am Main, 1994, p. 691.
727 The letter is apparently lost. At least it was not locatable in LA-B.
728 Magistrat von Berlin Kämmerei/Hauptfinanzverwaltung (LA-B: A rep. 005-03-01).
729 Herneck, Friedrich: Albert Einstein und das politische Schicksal seines Sommerhauses in Caputh bei Potsdam (in Herneck, Friedrich: Einstein und sein Weltbild. Buchverlag der Morgen, Berlin, 1979, pp. 256–273). Herneck could only rely on oral information because "archival documentation on it are missing." The present author (S.G.) could not find any official documents either. Herneck's rendition certainly fits within the context of other procedures.
730 BLHA: Pr. Br. rep. 2 A Regierung Potsdam III F, no. 11583.
731 BLHA: Pr. Br. rep. 2 A Regierung Potsdam III F, no. 11583. The purchase was a protracted procedure and the actual "conveyance" of the property purchased by Margot Marianoff and Ilse

Kayser only took place on 22 May 1933. The cause of this delay was private: In practical matters the young ladies apparently had more in common with their stepfather Einstein than with their natural mother. First they could not find the purchase and sale agreement among Einstein's papers (BLHA: Pr. Br. rep. 2 A Regierung Potsdam III F, no. 11583) while Einstein was away in America for the winter 1930/31. Then the authorities inadvertently sent them the wrong necessary paperwork, etc. The purchase of the complete property (including the additional purchases) in Caputh was only finally closed when political preparations for its *confiscation* were already in place.

732 BLHA: Pr. Br. rep. 2 A Regierung Potsdam III F, no. 11583.

733 BLHA: Pr. Br. rep. 2 A Regierung Potsdam III F, no. 11583.

734 BLHA: Pr. Br. rep. 2 A Regierung Potsdam III F, no. 11583.

735 BLHA: Pr. Br. rep. 2 A Regierung Potsdam III F, no. 11583.

736 BLHA: Pr. Br. rep. 2 A Regierung Potsdam III F, no. 11583.

737 The meaninglessness of indicating current values here is revealed by a simple comparison of the property prices. A square meter of building land in Caputh cost on 1 Jan. 2003 between 90 and 140 euros. Comparing the prices of the land and Einstein's summer house against Einstein's income is more instructive. The Einsteins' tax bracket for 1936 was assessed at 16,000 reichsmarks (estimated value). Albert Einstein's taxable annual income in 1931 came to 21,428.- reichsmarks. This would mean that the house and property together were worth about nine months' salary.

738 BLHA: Pr. Br. rep. 2 A Regierung Potsdam III F, no. 11583.

739 BLHA: Pr. Br. rep. 2 A Regierung Potsdam III F, no. 11583.

740 I only add here: After the land had been bought (with Albert Einstein's money but in the name of his daughters) the construction could begin. A note in the file from 25 Oct. 1929 points out that the land had meanwhile been built up (BLHA: Pr. Br. rep. 2 A Regierung Potsdam III F, no. 11583). Michael Grüning has reported in detail about the villa designed by Konrad Wachsmann and the building process. (Grüning, Michael: Der Wachsmann-Report. Auskünfte eines Architekten. Berlin, 1985. Grüning, Michael: Ein Haus für Albert Einstein. Erinnerungen. Briefe. Dokumente. Berlin, 1990).

741 Quoted from Highfield, Roger/Carter, Paul: The Private Lives of Albert Einstein. London, Boston, 1993, p. 162.

742 MPG-A: V, rep. 13 - Fritz Haber, no. 980.

742 MPG-A: V, rep. 13 - Fritz Haber, no. 980.

743 *Jüdische Rundschau*, 19 Mar. 1929.

744 Einstein to Ussishkin, 19 Mar. 1929. In: *Jüdische Rundschau*, 9 Apr. 1929.

745 Einstein to Haber dated 9 Mar. 1921 (MPG-A: V rep. 13 Fritz Haber no. 978).

746 Letter to Solovine, 8 Mar. 1921. In Einstein, Albert: Briefe an Solovine. Berlin, 1960, p. 26.

747 Fölsing, Albrecht: Albert Einstein. Eine Biographie. Frankfurt am Main, 1994, p. 577.

748 Fölsing, Albrecht: Albert Einstein. Eine Biographie. Frankfurt am Main, 1994, p. 681. His reply was: "Unfortunately, I do not see myself - either by my sexual or my musical abilities - in a position to follow your kind invitation."

749 Kessler, Harry Graf: Tagebücher 1918–1937. Frankfurt am Main, 1961, p. 240, 241.

750 Kessler, Harry Graf: Tagebücher 1918–1937. Frankfurt am Main, 1961, p. 278.

751 Armin Hermann contends it was the banker Erich Mendelssohn (Hermann, Armin: Einstein. Der Weltweise und sein Jahrhundert. Munich, 1994, p. 273).

752 Kessler, Harry Graf: Tagebücher 1918–1937. Frankfurt am Main, 1961, p. 520.
753 BLHA: Pr. Br. rep. 2 A Regierung Potsdam III F, no. 11583.
753 BLHA: Pr. Br. rep. 2 A Regierung Potsdam III F, no. 11583.
754 Fölsing, Albrecht: Albert Einstein. Eine Biographie. Frankfurt am Main, 1994, p. 697.
755 Address at the grave of H.A. Lorentz. In: Einstein, Albert: Mein Weltbild. Edited by Carl Seelig. Frankfurt/Main, 1955, p. 32.
756 For details see: Fölsing, Albrecht: Albert Einstein. Eine Biographie. Frankfurt am Main, 1994, chapter VI on unified field theory during a time of inner strife (pp. 611 ff.).
757 Einstein, Albert: Über den Frieden. Berne, 1975, p. 172.
758 GStA: I. HA rep. 76, Vc sec. 1, part Vc, tit. XI, no. 55.
759 Secretary of State Lammers was replaced by Wilhelm Stuckart at the beginning of July 1933.
760 GStA: I. HA rep. 76, Vc sec. 1, part Vc, tit. XI, no. 55.
761 Misspelling of Planck.
762 Elsa Einstein to Antonina Vallentin, 6 Jun. 1932 (MPG-A: V rep. 13 A. Einstein, no. 105).
763 Elsa Einstein to Antonina Vallentin, 11 Apr. 1933 (MPG-A: V rep. 13 A. Einstein, no. 105).
764 *Mitteilungen des Bundes Neues Vaterland.* New series, no. 1, revolution issue, November 1918, p. 3.
765 Press reports include: *Leipziger Tageblatt* from 13 Nov. 1918 and *Tägliche Rundschau* from 11 Nov. 1918.
766 *Mitteilungen des Bundes Neues Vaterland.* New series, no. 1, revolution issue, November 1918, p. 10.
767 *Mitteilungen des Bundes Neues Vaterland.* New series, no. 1, revolution issue, November 1918, pp. 12–14.
768 The Born–Einstein Letters. Correspondence between Albert Einstein and Max and Hedwig Born from 1916 to 1955 with commentaries by Max Born. Translated by Irene Born with a forward by Bertrand Russell. New York, 1971, pp. 149 f.
769 Angress, Werner T.: Juden im politischen Leben der Revolutionszeit. In: Mosse, Werner E. (ed.): Deutsches Judentum in Krieg und Revolution 1916–1923. Tübingen, 1971, p. 297.
770 Progress report by the German League of Human Rights (BA-K: NL 199, no. 30).
771 The progress report by the German League of Human Rights suggests the year 1920. Kessler's diary clearly indicates, however, that the trip took place in 1921.
772 BA-K: NL 199, no. 30.
773 Kessler, Harry Graf: Tagebücher 1918–1937. Frankfurt am Main, 1961, pp. 241–244.
774 *Vorwärts. Berliner Volksblatt.* The central organ of the Social Democratic Party of Germany, 21 February 1922. The appeal was also disseminated in the form of a flyer and solicited more signatures (BA-K: NL 199, no. 30).
775 Minutes dated 5 Jan. 1923 (BA-K: NL 199, no. 30).
776 Minutes dated 5 Jan. 1923 (BA-K: NL 199, no. 30).
777 Report by the headquarters of the German League of Human Rights, July 1923 (BA-K: NL 199, no. 30).
778 BA-K: NL 199, no. 30. According to a note by the Reich commissioner for the surveillance of public order dated 14 June 1927, the club's offices were located in the building at Wilhelmstr. 48. This building was, the note continues, "without a doubt [...] the center for communist welfare organizations and communist intellectual propaganda." These "radical pacifist, partly communist organizations," located at the same address, included among others, the League of Human Rights, the World League against Imperialism and Colonial Oppression, and the Association

of Worker Photographers headed by Willi Münzenberg (BA-B: R 1507/1050d, sheet 58).

779 Dr. Rudolf Kuczynski.

780 Mühsam, Erich; born on 6 Apr. 1878 in Berlin, deceased (murdered) on 10 or 11 Jul. 1934 in the concentration camp at Oranienburg, German writer. 1919 member of the Central Council of the Bavarian Soviet Republic; following its collapse, sentenced to fifteen years fortress confinement, six years of which he served; 1933 arrested again; this radical anarchist authored satirical ballads, plays and essays (Meyers Lexikonverlag).

781 BA-K: NL 199, no. 30.

782 BA-K: N 1057, no. 22.

783 Gesellschaft der Freunde des neuen Rußland. Aufruf (GStA: I. HA rep. 76, Vc sec. 2, tit. XXIII, lit. A, no. 134).

784 Letter by the Society for Eastern European Studies to Prof. Richter, head of department at the Prussian Ministry of Science, Arts and Culture dated 7 Mar. 1931: "The institute has been maintained from the very beginning by funds of the Reich (Foreign Office)." The Society for Eastern European Studies was responsible for the "welfare of the institute." (GStA: I. HA rep. 76, Vc sec. 2, tit. XXIII, lit. A, no. 134).

785 Bericht über das Russische Wissenschaftliche Institut in Berlin, 20 Jun. 1933 (GStA: I. HA rep. 76, Vc sec. 2, tit. XXIII, lit. A, no. 134).

786 GStA: I. HA rep. 76, Vc sec. 2, part 23, lit. A, no. 134.

787 GStA: I. HA rep. 76, Vc sec. 2, tit. XXIII, lit. A, no. 134.

788 BLHA: Pr. Br. rep. 30, Berlin tit. 95, sec. 9, no. 43.

789 BA-B: R 1507/ 1050d, sheet 89.

790 BA-B: R 1507/1050d and BA-B: R 1507/alt 134/37.

791 Gelegentliches von Albert Einstein. Zum 50. Geburtstag 14.3.1929. Dargelegt von der Soncino Gesellschaft der Freunde des jüdischen Buches zu Berlin (Berlin, 1929), pp. 20, 21.

792 SAPMO: NY 4126/13/40.

793 Lenin, W.I.: Werke. Berlin, 1961, vol. 32, p. 541.

794 Lenin: Werke. Berlin, 1959, vol. 24, p. 10.

795 For Lenin's commentary: W.I. Lenin: Werke. Berlin, 1961, vol. 32, pp. 540–544.

796 BA-K: N 1053, no. 101.

797 Grüning, Michael: Ein Haus für Albert Einstein. Erinnerungen. Briefe. Dokumente. Berlin, 1990, pp. 357, 370–371, 374.

798 *Das Neue Rußland*, 8 ser., issue 8/9, p. 40, Berlin 1931.

799 Einstein, Albert: On Peace. New York, 1960.

800 Clark, Ronald W.: Albert Einstein. The Life and Times. London, 1973, esp. the chapter on the 'call for peace.'

801 Gumbel, E.J.: Vier Jahre politischer Mord. Berlin, 1922; Gumbel, E.J.: Verschwörer. Beiträge zur Geschichte und Soziologie der deutschen nationalistischen Geheimbünde seit 1918. Malik Verlag, 1924.

802 *Die Menschenrechte. Organ der Deutschen Liga für Menschenrechte*, 15 July 1931. Gumbel's book cited below is: Gumbel, Emil Julius: Verräter verfallen der Feme. Opfer, Mörder, Richter 1919–1929. Berlin, Malik-Verlag, 1929.

803 *Umfrage.* Soll Deutschland Kolonialpolitik treiben? In: *Europäische Gespräche*, no. 12/ 1927, p. 611.

804 *Umfrage.* Soll Deutschland Kolonialpolitik treiben? In: *Europäische Gespräche*, no. 12/ 1927, p. 626.

805 Einstein, Albert: Über den Frieden. Berne, 1975, p. 112.

806 Einstein, Albert: Über den Frieden. Berne, 1975, p. 129.

807 Einstein, Albert: Über den Frieden. Berne, 1975, pp. 130–131.

808 Einstein, Albert: Über den Frieden. Berne, 1975, p. 113.

809 Einstein, Albert: Über den Frieden. Berne, 1975, p. 142.

810 BA-B: R 1501, no. 13208 betr. Internationale Arbeiterhilfe.

811 Grüning, Michael: Ein Haus für Albert Einstein. Berlin, 1990, p. 367.

812 Grüning, Michael: Ein Haus für Albert Einstein. Berlin, 1990, p. 395.

813 Grüning, Michael: Ein Haus für Albert Einstein. Berlin, 1990, p. 398.

814 Grüning, Michael: Ein Haus für Albert Einstein. Berlin, 1990, p. 418.

815 BA-B: R 1507/alt 134/72.

816 BA-B: R 1501, no. 25988.

817 *Deutsche Zeitung*, 3 Sep. 1930.

818 BA-B: ZB I 429 vol. 1.

819 BA-B: ZB I 429 vol. 1.

820 GStA: I. HA rep. 76, Vc sec. 1, part Vc, tit. XI, no. 55, f 359, 360, 361.

821 Clark, Ronald W.: Albert Einstein. Ein Leben zwischen Tragik und Genialität. Munich, 1995, p. 267.

822 Ein Prozeß gegen die IAH. In: *Die Rote Fahne*, 12 Jun. 1932.

823 The source of this appeal is a film completed on 22 Jun. 1959 by the BDC, at that time still under American authority (BA-B: 62 FC. NS 26 – NSDAP-Hauptarchiv, no. 4104/13974 P). The BDC original is not locatable among the BA-B holdings.

824 Grüning, Michael: Ein Haus für Albert Einstein. Berlin, 1990, p. 198.

825 AdK-A: papers of Heinrich Mann SB 301.

826 Grüning, Michael: Ein Haus für Albert Einstein. Berlin, 1990, p. 198.

827 BA-B: R 58/4182. The files do not indicate whether the text continued. The file R 58/4182 contains material that the police had found during a search through the offices of the League of Human Rights, Berlin, Monbijouplatz 10, on 4 Mar. 1933 (R 58/4182). The state of the file suggests that GLHR documents predating 4 Mar. 1933 were destroyed.

828 *Vorwärts*, evening edition, no. 292/B 146 from 23 Jun. 1932.

829 *Vorwärts*, evening edition, no. 296/B 145 from 25 Jun. 1932.

830 Der Weg zur Einheitsfront. Eine Erklärung des Bundesvorstandes des ADGB. In: *Vorwärts*, morning edition, no. 289/A 146 from 22 Jun. 1932.

831 A poster with the invitation and issues was similar, if not quite as direct and threatening. It read:

"*Albert Einstein, Heinrich Mann, Ernst Toller, Arnold Zweig, Kaethe Kollwitz, Bar[on] von Schoenaich* et al. have issued an urgent appeal:
*CPG and SDP should join together as a united bloc on the basis of common candidate lists.*
*Intellectual workers!* Writers, teachers, doctors, engineers, technicians, students. *What is your position on this proposal? On the anti-fascist campaign?* Do you want to fall in with the anti-fascist united front? And how should this united front be formed? Employment and the future are at stake! The sinister men of the reactionaries are threatening to steal our say!
*So – make known your position on these issues at the major public meeting* of all intellectual workers and the liberal professions. Monday, the 18th of July 1932, 8 PM in the Spichern Halls, Spichernstr. 2. Subway Nürnberger Platz
Leftist Cartel of Intellectual Workers and the Liberal Professions"

(poster, Märkisches Museum, Berlin). The announced speakers were: Johannes R. Becher, Maria Hodann, Walter Hammer, Otto Lehmann-Russbüldt, Dr. Joh. König, Karl Olbrisch, Theodor Plivier, Dr. Fritz Schiff and Kurt Klaber.

832 BA-B: R 58/4182.

833 BA-B: R 58/4182.

834 AdK-A: I/ 396.

834 AdK-A: I/ 396.

835 That was why the Reich Central Security Office (Reichssicherheitshauptamt) later took over "that part of the card catalog of the Reich Commissioner for

the Surveillance of Public Order of interest to political security" (BA-B: R 58/254).

836 Bund neues Vaterland (jetzt) Liga für Menschenrechte, BA-B: R 1507, no. 485.

837 My emphasis, S.G.

838 BA-B: R 1507, no. 485.

839 BA-B: R 1507, no. 485.

840 BA-B: R 1507, no. 485.

841 BA-B: R 1507, no. 485.

842 Membership list of the board of trustees of the Children's Homes of Red Aid (Kuratorium der Kinderheime der Roten Hilfe) among the files of the Reichskommissariat für die Überwachung der öffentlichen Ordnung: BA-B: R 1507/67159, no. 262.

843 BA-B: R 1507/67159, no. 262.

844 Instruction of 17 Dec. 1936 signed by Dr. Best on "Übernahme des abwehrpolitisch interessierenden Teils der Kartei des Reichskommissariats zur Überwachung der öffentlichen Ordnung" (BA-B: R 58/254).

845 BLHA: Pr. Br. rep. 30 Berlin C, tit. 95, sec. 9, no. 43 – re.: Polizeipräsidium Berlin Abt. I A: Die Gesellschaft der Freunde des neuen Rußland und der Bund der Freunde der Sowjetunion 1923–1930.

846 BA-B: R 1507/1050d, sheet 89.

847 BA-B: R 1507, no. 485.

848 GStA: I. HA rep. 77, tit. 4043, no. 427.

849 GStA: I. HA rep. 77, tit. 4043, no. 427.

850 GStA: I. HA rep. 77, tit. 4043, no. 228.

851 Einstein, Albert: Über den Frieden. Weltordnung oder Weltuntergang? Edited by Otto Nathan and Heinz Norden. Foreword by Bertrand Russell. Berne, 1975, p. 245.

852 Einstein, Albert: Über den Frieden. Berne, 1975, p. 113.

853 Romain Rolland, Stefan Zweig: Briefwechsel 1910–1940. Vol. 2, 1924–1940. Berlin, 1987, pp. 535, 536.

854 Johann von Leers: since 1933 head of the Foreign Policy and Foreign Studies Department at the Deutschen Hochschule für Politik in Berlin; Reich head of Nazi indoctrination; a major (Sturmbannführer) in the SS.

854 Johann von Leers: since 1933 head of the Foreign Policy and Foreign Studies Department at the Deutschen Hochschule für Politik in Berlin; Reich head of Nazi indoctrination; a major (Sturmbannführer) in the SS.

855 GStA: I. HA rep. 76, Vc sec. 1, part Vc, tit. XI, no. 55.

856 GStA: I. HA rep. 76, Vc sec. 1, part Vc, tit. XI, no. 55.

857 GStA: I. HA rep. 76, Vc sec. 1, part Vc, tit. XI, no. 55.

858 GStA: I. HA rep. 76, Vc sec. 1, part Vc, tit. XI, no. 55.

859 GStA: I. HA rep. 76, Vc sec. 1, part Vc, tit. XI, no. 55.

860 Die Astronomie lehrt jetzt: Die Welt wird immer größer. Ein Thema zum Schwindligwerden. GStA: I. HA rep. 76, Vc sec. 1, part Vc, tit. XI, no. 55.

861 GStA: I. HA rep. 76, Vc sec. 1, part Vc, tit. XI, no. 55.

862 Denkschrift betreffend die Personal-Erneuerung an den Deutschen Hochschulen in den Naturwissenschaftlich-Mathematischen Fächern. GSTA: I. HA rep. 76, Va sec. 1, tit. IV, no. 1, vol. 13.

863 GSTA: I. HA rep. 76, Va sec. 1, tit. IV, no. 1, vol. 13.

864 GStA: I. HA rep. 76, Vc sec. 1, part Vc, tit. XI, no. 55.

865 BA-B: R 1501, no. 25720/1.

866 GStA: I. HA rep. 77, no. 6061. The date indicated on the letter is 30 April 1933. The German legation in Brussels already had a transcription of it in hand on 1 April 1933, consequently the letter was already available before that date, presumably on 30 March 1933. The legation had received the letter "with the request to forward."

867 Information from the Senior Mayor of Ulm, Ivo Gönner, dated 17 Feb. 2003 to the author.

868 The date on the original letter was apparently corrected from the 12th to the 11th (April 1933).

869 Elsa Einstein to Antonina Luchaire, 11 Apr. 1933. MPG-A: V rep. 13 A. Einstein, no. 105/17.
870 Grüning, Michael: Ein Haus für Albert Einstein. Erinnerungen. Briefe. Dokumente. Berlin, 1990, p. 151.
871 Heilbron, J.L.: The Dilemmas of an Upright Man. Max Planck as Spokesman for German Science. Berkeley, Los Angeles, London, 1986, p. 155.
872 *Berliner Börsen-Zeitung*, no. 333, dated 20 Jul. 1933 (my emphasis, S.G.).
873 BA-B: R 43 II 600 (my emphasis, S.G.).
874 BA-B: R 43 II 600 (my emphasis, S.G.).
875 BBAdW: II-III-57. English translation in: Hentschel, Klaus (ed.): Physics and National Socialism. An Anthology of Primary Sources. Birkhäuser Science Networks, vol. 18, Basel, Boston, Berlin, 1996, doc. 6, p. 19.
876 Ficker to Planck, 29 Mar. 1933; Planck to Ficker, 31 Mar. 1933 (BBAdW: II-III-57, f 3).
877 Minutes of the plenary session of the Academy of Sciences on 30 March 1933 (BBAdW: II-V-102).
878 BBAdW: II-III-57. English translation in: Hentschel, Klaus (ed.): Physics and National Socialism. An Anthology of Primary Sources. Birkhäuser Science Networks, vol. 18, Basel, Boston, Berlin, 1996, doc. 6, pp. 19–20.
879 BBAdW: II-III-57.
880 BBAdW: II-III-57.
881 Minutes of the special plenary meeting of the Academy of Sciences on 6 April 1933 (BBAdW: II-III-57).
882 BBAdW: II-III-57.
883 BBAdW: II-III-57.
884 Max Planck to Ficker, 13 Apr. 1933 (BBAdW: II-III-57).
885 Max Planck to Ficker, 13 Apr. 1933 (BBAdW: II-III-57).
886 BBAdW: II-III-57.
887 Albert Einstein in Berlin 1913–1933. Part I. Akademie-Verlag Berlin, 1979, p. 256.
888 Minutes of the plenary meeting of the Academy of Sciences on 11 May 1933 (BBAdW: II-V-102). It was general knowledge how much Planck esteemed Einstein. So these words should also be interpreted as a sign of his mental anguish. With Hitler's accession to power with its consequence of Einstein's departure, the hardest days of his life had *also begun for Max Planck.*
889 Plenary meeting of the Academy of Sciences on 11 May 1933 (BBAdW: II-V-102).
890 BBAdW: II-III-57.
891 Seelig, Carl: Albert Einstein. Eine dokumentarische Biographie. Zurich, Stuttgart, Vienna, 1954, p. 77.
892 Planck, Max: My Audience with Adolf Hitler. In: Hentschel, Klaus (ed.): Physics and National Socialism. An Anthology of Primary Sources. Birkhäuser Science Networks, vol. 18, Basel, Boston, Berlin, 1996, doc. 114.
893 The biographical details on Max Planck are based on Heilbron, J.L.: The Dilemmas of an Upright Man. Max Planck as Spokesman for German Science. Berkeley, Los Angeles, London, 1986.
894 Albert Einstein in Berlin 1913–1933. Part I. Berlin, 1979, pp. 251, 248.
895 MPG-A: V rep. 13, F. Haber, no. 983.
896 Albert Einstein in Berlin 1913–1933. Part I. Berlin, 1979, p. 270.
897 MPG-A: V rep. 13, F. Haber, no. 983.
898 Weizmann, Chaim: Memoiren. Das Werden des Staates Israel. Zurich, 1953, p. 515.
899 Invitation = BA-K: N1053, no. 107.
900 BA-K: N1053, no. 107.
901 Albert Einstein in Berlin 1913–1933. Part I. Berlin, 1979, p. 269.
902 Hilberg, Raul: Die Vernichtung der europäischen Juden. Vol. 1. Frankfurt am Main, 1990, p. 90.
903 Planck, Max: My Audience with Adolf Hitler. In: Hentschel, Klaus (ed.): Physics and National Socialism. An Anthology of Primary Sources. Birkhäuser Science Networks, vol. 18, Basel, Boston, Berlin, 1996, doc. 114, p. 360.
904 MPG-A: V rep. 13, A. Einstein, no. 65.

905 PA-AA: R 45490 as well as GStA: I. HA rep. 77, no. 6061. Einstein addressed his letter to the "Deutsche Generalkonsulat"; it would have been correctly addressed to the German legation (Gesandtschaft). The letter is undated but the legation's letter to the Foreign Office in this matter provides the date 28 March.

906 PA-AA: R 45490.

907 PA-AA: R 45490.

908 The Municipality of Caputh, where the Einstein's summer villa was located, belonged at that time within the administrative district of Belzig.

909 PA-AA: R 45490.

910 PA-AA: R 45490.

911 GStA: I. HA rep. 77, no. 6061.

912 GStA: I. HA rep. 77, no. 6061. The Prussian minister of the interior sent a copy on 24 Jul. 1933 to the Foreign Office (PA-AA: R 99639).

913 GStA: I. HA rep. 77, no. 6061. Why there were "no more objections" on 14 Nov. 1933 is somewhat questionable. Just recently, 3 Nov. 1933 the Fiscal Court at the Brandenburg Revenue Office rejected the appeal submitted by Elsa and Albert Einstein against the decision by the Beelitz Revenue Office from 27 Jun. 1933 and imposed the Reich evasion penalty tax. Thus legal confirmation was provided that Einstein had (allegedly) *not* satisfied his tax obligations.

914 GStA: I. HA rep. 77, no. 6061.

915 R 99639.

916 Reich minister of the interior to Prussian minister of the Interior, 22 Jul. 1933 (PA-AA: Inland II A/B 83–76, Ausbürgerung, 2nd list A–G, R 99639).

917 BA-B: R 1501, no. 25708.

918 Handwritten note on the express letter.

919 Secretary of State von Bülow (Foreign Office) to Secretary of State Pfundtner (Reich Ministry of the Interior), 17 Aug. 1933. (PA-AA: Inland II A/B 83–76, Ausbürgerung, 2nd list A–G, R 99639).

920 Neue Deutsche Biographie. Duncker & Humblot, Berlin, vol. 2, 1972, pp. 731, 732.

921 Krüger, Peter/Hahn, Erich J.C.: Der Loyalitätskonflikt des Staatssekretärs Bernhard Wilhelm von Bülow im Frühjahr 1933. In: *Vierteljahreshefte für Zeitgeschichte*. Vol. 20, 1972, pp. 393, 394.

922 Krüger, Peter/Hahn, Erich J.C.: Der Loyalitätskonflikt des Staatssekretärs Bernhard Wilhelm von Bülow im Frühjahr 1933. In: *Vierteljahreshefte für Zeitgeschichte*. Vol. 20, 1972, p. 402.

923 Von Bülow's note in the files, 22 Sep. 1933 (PA-AA: Inland II A/B 83–76, Ausbürgerung, 2nd list A–G, R 99639).

924 Note in the files dated 26 Sep. 1933.

925 The Prussian minister of the interior to the Foreign Office, 30 Oct. 1933 (PA-AA: Inland II A/B 83–76, Ausbürgerung, 2nd list A–G, R 99639).

926 The Prussian minister of the interior to the Foreign Office, 30 Oct. 1933 (PA-AA: Inland II A/B 83–76, Ausbürgerung, 2nd list A–G, R 99639).

927 To the Reich minister of the interior, 18 Nov. 1933 (PA-AA: Inland II A/B 83–76, Ausbürgerung, 2nd list A–G, R 99639).

928 BA-B: R 1501, no. 25953.

929 President of the Office of Revenue of the District of Berlin to the Reich minister of finance, 21 Sep. 1933 (BA-B: R 1501, no. 25626/1).

930 BA-B: R 1501, no. 25953.

931 BA-B: ZR 795 A 2.

932 GStA: I. HA rep. 77, no. 6061.

933 BA-B: ZR 795 A 2.

934 Seelig, Carl: Albert Einstein. Eine dokumentarische Biographie. Zurich, Stuttgart, Vienna, 1954, p. 254.

935 PA-AA: Inland II A/B 83–76, Ausbürgerung, 2nd list A–G, R 99639.

936 PA-AA: Inland II A/B 83–76, Ausbürgerung, 2nd list A–G, R 99639.

937 PA-AA: R 99639.

938 PA-AA: Inland II A/B 83–76, Ausbürgerung, 2nd list A–G, R 99639.

939 von Bülow to Mr. von Kotze (PA-AA: Inland II A/B 83–76, Ausbürgerung, 2nd list A–G, R 99639).

940 In the calendar year of 1930 Einstein earned an annual salary at the academy of 16,264.08 reichsmarks. The amount of 1,112.40 reichsmarks went to taxes and 133.40 reichsmarks to national welfare (Reichshilfe) (BBAdW: II-III-39).

941 Geheimes Staatspolizeiamt. Nachweisung über beschlagnahmtes Vermögen staatsfeindlicher Organisationen und Einzelpersonen (GStA: I. HA rep. 151, IA no. 8191).

942 GStA: I. HA rep. 151, IA no. 8191.

943 GStA: I. HA rep. 151, IA no. 8191. This communication by the Gestapo was the basis of Einstein's suit, filed in 1952 (on 30 Jul. 1952 by proxy, R.A. Held – New York) at the Claims Office in Berlin, followed by an appeal and final recognition of his restitution claims (decision by the Reparations Division of the District Court of Berlin dated 18 Jul. 1956). It was a stroke of luck that he had received and kept the seizure order of 10 May 1933: The Dresdner Bank informed the restitution offices of Berlin on 23 Jun. 1955 "that no record can be found among our extant files about an account or security deposit account held by the above-named either here or at one of our branch offices in Berlin." The Berliner Handelsgesellschaft sent a similar notice on 22 Jun. 1955. Because the District Court of Berlin issued its final verdict only on 18 Jul. 1956, *after* Einstein's death, Einstein never saw the money confiscated from him in 1933 again. The Reparations Division of the Berlin District Court rejected another claim (for securities valued at $ 3,000, $ 45.67 and $ 46.17) from deposits at the Dresdner Bank on 15 Oct. 1958, because no documentation could be found on its seizure by the Deutsche Reich and what had remained in the bank safe had been confiscated by the occupying Russian forces at the end of the war.

944 GS = Gesetzessammlung (statute books); RGBl = Reichsgesetzblatt (Reich law gazette).

945 All the documents from the Office of the State Police cited here refer exclusively to *Else* Einstein, not *Elsa* Einstein.

946 Office of the Secret State Police to the Prussian minister of the interior, 24 Nov. 1933 (GStA: I. HA rep. 151, IA no. 7976).

947 According to information provided to the author based on "the best documentation," on 16 Oct. 1995 by a reputable German bank preferring to remain anonymous.

948 As of 1 Jan. 2003 the average official value for 1 sq. meter of building land in Caputh was 110 euros, ranging from a maximum of 140 euros to a minimum of 90 euros. A piece of property of 2,759 sq. meters would therefore be worth between 386,260 euros and 248,310 euros (thus before the conversion into euros, roughly between 755,000 and 485,000 deutschmarks). This does not mean that 1 reichsmark averaged about 72 deutschmarks at the 2003 conversion rate; the point is that 1,300 reichsmarks for a sailboat or 8,277 reichsmarks for a piece of building land were not trivial sums. The buyer of the sailboat was certainly not a "profiteer."

949 GStA: I. HA rep. 151, I A no. 8191 (my emphasis, S.G.).

950 BA-B: R 1501 no. 25953.

951 BA-B: ZR 795 A2 (my emphasis, S.G.).

952 Hilberg, Raul: Die Vernichtung der europäischen Juden. Vol. 1. Frankfurt am Main, 1990, p. 140.

953 Hilberg, Raul: Die Vernichtung der europäischen Juden. Vol. 1. Frankfurt am Main, 1990, p. 141.

954 Based on the decision by the Fiscal Court at the Revenue Office of the Province of Brandenburg dated 3 Nov. 1933 (PA-AA: Inland II A/B 83–76, Ausbürgerung, 2nd list A–G, R 99639).

955 Based on the decision by the Fiscal Court at the Revenue Office of the Province of Brandenburg dated 3 Nov. 1933 (PA-AA: Inland II A/B 83–76, Ausbürgerung, 2nd list A–G, R 99639).
956 PA-AA: Inland II A/B 83–76, Ausbürgerung, 2nd list A–G, R 99639.
957 President of the Revenue Office of the Province of Brandenburg to the Reich minister of finance, 11 Jan. 1934 (PA-AA: Inland II A/B 83–76, Ausbürgerung, 2nd list A–G, R 99639).
958 GStA: I. HA rep. 151, IA no. 8191 – re.: Einziehung und Verwertung von Grundstücken im Regierungsbezirk Potsdam 1933–1938.
959 BLHA: Pr. Br. rep. 2 A, Regierung Potsdam III F, no. 11583.
960 As already mentioned in the foreword, my intense search for the Gestapo file "Einstein" perused by Friedrich Herneck in 1961 was in vain. Mr. Uwe Lobeck, head of the Friedrich Herneck Archive, Dresden, made available to me an account of the confiscation procedure of the summer villa based on Friedrich Herneck's notes. This was the basis of subsequent descriptions preceding the interrogation of Herta Schiefelbein. See also Herneck, Friedrich: Albert Einstein und das politische Schicksal seines Sommerhauses in Caputh bei Potsdam (In: Herneck, Friedrich: Einstein und sein Weltbild. Aufsätze und Vorträge von Friedrich Herneck. Berlin, 1979, pp. 256–273).
961 BA-B: formerly BDC – file: SA Graf Helldorf.
962 A-Caputh: no. 808.
963 Herneck, Friedrich: Einstein privat. Herta W. erinnert sich an die Jahre 1927 bis 1933. Berlin, 1978, p. 161.
964 GStA: I. HA rep. 151, IA no. 8191.
965 According to information by Uwe Lobeck, head of the Friedrich Herneck Archive, Dresden.
966 GStA: I. HA rep. 151, IA no. 8191.
967 Prussian minister of finance to the president of the District of Potsdam, 2 May 1935 (GStA: I. HA rep. 151, IA no. 8191).
968 GStA: I. HA rep. 151, IA, no. 8191.
969 A-Caputh: no. 808.
970 See A-Caputh, folder 811, sheet 64: "Grundstücksverhältnisse Einstein." Total sales:
(1) 1508 sq.m. from A. Stern, purchaser: Margot Marianoff (née Einstein) and Ilse Kayser (née Einstein), conveyance on 18 Jul. 1930.
(2) 683 sq.m. on 22 May 1933 from the State Forestry Administration of the Kunersdorf Chief Forestry Department. Purchaser: Margot Marianoff (née Einstein) and Ilse Kayser (née Einstein), conveyance on 22 May 1933.
(3) 568 sq.m. from Robert Wolff. Purchaser: Ilse Kayser (née Einstein). Purchase and sale agreement of 9 Nov. 1932, conveyance on 30 Jan. 1933.
971 A-Caputh: no. 808.
972 A-Caputh: no. 808.
973 GStA: I. HA rep. 151, IA no. 8191.
974 GStA: I. HA rep. 151, IA no. 8191.
975 GStA: I. HA rep. 151, IA no. 8191.
976 GStA: I. HA rep. 151, IA no. 8191.
977 A-Caputh: no. 808.
978 A-Caputh: no. 808. The "Wirtschaftspartei" was a right-wing party attracting disenchanted middle-class businessmen and property owners. Like Mr. Wolff, most of its membership soon joined the governing Nazi party (NSGWP).
979 A-Caputh: no. 808.
980 President of the District of Potsdam to the Prussian Minister of Finance, 18 July 1935 (GStA: I. HA rep. 151, IA no. 8191).
981 GStA: I. HA rep. 151, IA no. 8191.
982 Kreisamtleiter des Gaus Kurmark der NSDAP to Landrat des Kreises Zauch-Belzig, dated 11 Mar. 1935 (GStA: I. HA rep. 151, IA no. 8191).
983 Letter by die Deutsche Studentenschaft – Gruppe Deutsche Hochschule für Politik to the minister of finance,

4 Apr. 1935 (GStA: I. HA rep. 151, IA no. 8191).

984 GStA: I. HA rep. 151, IA no. 8191.

985 Purchase and sale agreement of 27 Aug. 1936 (GStA: I. HA rep. 151, IA no. 8191).

986 See the letter by the administrator of the District of Zauch-Belzig dated 19 May 1936 to the mayor of Caputh (A-Caputh: no. 808).

987 Statistical survey of the inhabitants of the Children's Country Home Caputh, 20 Apr. 1934 (A-Caputh: no. 808).

988 "Ein verlorenes Paradies." Das jüdische Kinder- und Landschulheim Caputh. Dokumente einer anderen pädagogischen Praxis (no place of pub., undated), section on 'Chronik – Jahre in Caputh.'

989 BLHA: Pr. Br. rep. 2 A Regierung Potsdam. I pol. no. 1919 – re.: Einzelaktionen gegen Juden.

990 A-Caputh: no. 808.

991 BLHA: rep. 41, Caputh no. 14.

992 BLHA: rep. 41, Caputh no. 14.

993 BLHA: rep. 41, Caputh no. 14.

994 Quoted from the letter by the administrator of the District of Zauch-Belzig, dated 27 Aug. 1935 (BLHA: rep. 41, Caputh no. 14).

995 "Ein verlorenes Paradies." Das jüdische Kinder- und Landschulheim Caputh. Dokumente einer anderen pädagogischen Praxis (no place of publ., undated).

996 LA-B: Magistrat von Groß-Berlin – Abteilung Finanzen und Grundstücksfragen, no. 21/65.

997 LA-B: A rep. 092, no. 23765 Littmann, Hildegart.

998 LA-B: A rep. 092, no. 8836.

999 "Ein verlorenes Paradies. Das jüdische Kinder- und Landschulheim Caputh. Dokumente einer anderen pädagogischen Praxis (no place of pub., undated).

1000 Grüning, Michael: Ein Haus für Albert Einstein. Erinnerungen. Briefe. Dokumente. Berlin, 1990, p. 73. For details on the design of the sailboat see ibid., pp. 210 ff.

1001 According to an official note dated 16 Apr. 1934. BLHA: Pr. Br. rep. 2 A, Regierung Potsdam. I pol. no. 1165.

1002 Thus the contention that Einstein did nothing about his financial affairs is refuted (Hermann, Armin: Einstein. Der Weltweise und sein Jahrhundert. Eine Biographie. Munich, Zurich, 1995, p. 410). It may well be true that Einstein occasionally left that impression ("Let the Germans gobble up my little bit of money.") His actual conduct was different.

1003 BLHA: Pr. Br. rep. 2 A, Regierung Potsdam. I pol. no. 1165.

1004 It was not possible to ascertain from the files whether the source of the details in the press report had been the conversation between Kayser and Schuhmann, and if so, how they had been relayed.

1005 BLHA: Pr. Br. rep. 2 A, Regierung Potsdam. I pol. no. 1165.

1006 BLHA: Pr. Br. rep. 2 A, Regierung Potsdam. I pol. no. 1165.

1007 A-Caputh: no. 808.

1008 BLHA: Pr. Br. rep. 2 A, Regierung Potsdam. I pol. no. 1165.

1009 BLHA: Pr. Br. rep. 2 A, Regierung Potsdam. I pol. no. 116.

1010 Nor did the advertisement indicate – quite appropriately – the name of the sailboat's former owner, *Einstein*. The correspondence reveals that it was not a secret among the bidders, even so. In a letter to the Ministry of Finance dated 25 Apr. 1934, for instance, the dentist Dr. Fiebig from Nowawes specifically mentioned that "the sailboat formerly owned by Professor Einstein was being offered" in the *Potsdamer Tageszeitung* of 28 Feb. 1934.

1011 BLHA: Pr. Br. rep. 2 A, Regierung Potsdam. I pol. no. 1165.

1012 BLHA: Pr. Br. rep. 2 A, Regierung Potsdam. I pol. no. 1165.

1013 BLHA: Pr. Br. rep. 2 A, Regierung Potsdam. I pol. no. 1165.

1014 BLHA: Pr. Br. rep. 2 A, Regierung Potsdam. I pol. no. 1165.

1015 BLHA: Pr. Br. rep. 2 A, Regierung Potsdam. I pol. no. 1165.

1016 BLHA: Pr. Br. rep. 2 A, Regierung Potsdam. I pol. no. 1165.

1017 BLHA: Pr. Br. rep. 2 A, Regierung Potsdam. I pol. no. 1165.

1018 BLHA: Pr. Br. rep. 2 A, Regierung Potsdam. I pol. no. 1165.

1019 BLHA: Pr. Br. rep. 2 A, Regierung Potsdam. I pol. no. 1165.

1020 According to the letter by the minister of finance dated 2 May 1934 to the district president of Potstam (BLHA: Pr. Br. rep. 2 A, Regierung Potsdam. I pol. no. 1165).

1021 BLHA: Pr. Br. rep. 2 A, Regierung Potsdam. I pol. no. 1165.

1022 Deductions from the sales price were: the outstanding amount of 197.50 reichsmarks "to the shipbuilder Schümann [*sic*] for storage of the boat" and 3 reichsmarks for the sales advertisement in the *Potsdamer Tageszeitung* (BLHA: Pr. Br. rep. 2 A, Regierung Potsdam. I pol. no. 1165).

1023 Herneck, Friedrich: Einstein und sein Weltbild. Aufsätze und Vorträge von Friedrich Herneck. Berlin, 1979, p. 272.

1024 Schwarzenbach, Alexis: Das verschmähte Genie. Albert Einstein und die Schweiz. Deutsche Verlags-Anstalt, Munich. 2005, pp. 136, 137. The mailing date is indicated here as 9 March 1933 from East Coq-sur-Mer, near Ostende. Because Einstein only arrived in Belgium – from the USA – toward the end of March 1933, the sending date was probably 29 March 1933.

1025 Schwarzenbach, Alexis: Das verschmähte Genie. Munich, 2005, pp. 138, 139.

1026 Schwarzenbach, Alexis: Das verschmähte Genie. Munich, 2005, p. 140.

1027 Schwarzenbach, Alexis: Das verschmähte Genie. Munich, 2005, pp. 141ff.

1028 Schwarzenbach, Alexis: Das verschmähte Genie. Munich, 2005, p. 153.

1029 PA-AA: R 45490.

1030 Koch, Martin: Aufstieg und Fall einer Pseudowissenschaft. Briefwechsel zwischen Philipp Lenard und Johannes Stark wirft neues Licht auf die 'Deutsche Physik.' In: *Neues Deutschland,* 11/12 Nov. 2000.

1031 Johannes Stark: 'Weiße Juden' in der Wissenschaft. In: *Das Schwarze Korps,* 15 Jul. 1937, p. 6 (quoted from: Poliakov, Léon/Wulf, Joseph: Das Dritte Reich und seine Denker. Dokumente und Berichte. Wiesbaden. 1989, pp. 299, 300). English translation: 'White Jews' in science: 'Science' is politically bankrupt. In: Hentschel, Klaus (ed.): Physics and National Socialism. An Anthology of Primary Sources. Birkhäuser Science Networks, vol. 18, Basel, Boston, Berlin, 1996, doc. 56.

1032 Prussian minister of science, arts and culture to University Professor von Laue, Berlin Zehlendorf, 10 Nov. 1934 (BBAdW: papers of Laue U VIII 1 1943, no. 1).

1033 Prussian minister of science, arts and culture to University Professor von Laue, Berlin Zehlendorf, 10 Nov. 1934 (BBAdW: papers of Laue U VIII 1 1934, no. 5).

1034 Albert Einstein in Berlin 1913–1933. Part I. Berlin, 1979, pp. 268 and 269.

1035 Koch, Martin: Aufstieg und Fall einer Pseudowissenschaft. Briefwechsel zwischen Philipp Lenard und Johannes Stark wirft neues Licht auf die »Deutsche Physik«. In: *Neues Deutschland,* 11/12 November 2000.

1036 The National Archives, Washington: Federal Bureau of Investigation/Bufile Number 61-7099.

1037 The Counter Intelligence Corps (CIC) was operative between 1945–65 as a defense organization of the US Armed

Forces. Afterwards it was split up into smaller agencies. G-2 = Department G-2, Documents Section. G-2 Docs. Sect./SHAEF (Supreme Headquarters Allied Expeditionary Forces) was transferred to Frankfurt am Main in May 1945 and established its headquarters in Fechenheim. After the SHAEF was dissolved in 1945, some of the G-2 Docs. Sect. staff was ordered to Berlin and moved into the 6889 Berlin Document Center on Wasserkäfigsteig no. 1 in Zehlendorf, Berlin in August 1945. It was attached to the newly formed American military government OMGUS (Office of Military Government of the United States). 1 Nov. 1946 the department was renamed 7771 Document Center and when it took over the archive from the US State Department in 1953, its name was reduced to Berlin Document Center. (See: 47 Jahre Berlin Document Center. In: *Der Archivar,* 1992, col. 34). According to investigations by the East German Ministry of State Security (in search of recruits [gesellschaftliche Mitarbeiter]), the Document Center was "only indirectly related to the American secret service.") ("Closing remark" dated 26 Jul. 1967: BStU ZA AS 2490/67 f. 18 ff.). "In general, the staff of the Info. Center was negatively disposed (less personally than politically) to the CIC people." (BStU: ZA AS 2490/67).

1038 Einstein 1b.pdf: 14–16.

1039 Another report dated 25 Jan. 1951 renders this name as "Schauerhammer."

1040 Misspelling of Bobek. In the following misspellings are adopted *where I quote from FBI reports.*

1041 Misspelling of Houtermans.

1042 Security Apparatus of the Administration of the State of Berlin-Brandenburg.

1043 Klara was the alias of the fourth department of the Red Army's General Staff (Soviet military intelligence).

1044 Einstein 1b.pdf: 39–46.

1045 Per original. The OMS was the international communications department of the Communist International.

1046 "Front groups" can only refer to the CPG's "Revolutionary Working Groups" (Aufbruch-Arbeitskreise).

1047 Original spelling for Dünow.

1048 Einstein 9a.pdf: 34.

1049 Einstein 9a.pdf: 34.

1050 Einstein 9a.pdf: 41.

1051 It was permissible to write cables "in plain text or in code," either "solely in one language or in a mixture of plain text and code" (Reichs-Telegramm-Adressbuch nach amtlichen Quellen bearbeitet. 1929, p. 6). At the end of the 1920s the transmission of images was also allowed. "Permissible image telegrams include anything that can be transmitted by telegraph as an image, e.g., any type of picture incl. photographs [...], drawings, maps, written and printed material." (Reichs-Telegramm-Adressbuch nach amtlichen Quellen bearbeitet. 1929, p. 10).

1052 Landesarchiv Berlin: rep. 211, acc. 1674, no. 488.

1053 My emphasis, S.G.

1054 'Besuch bei Einstein.' My emphasis, S.G.

1055 The floor plan of Einstein's apartment, first published here, agrees largely with the drawing that Friedrich Herneck made on the basis of conversations with the former housemaid Herta Schiefelbein (Herneck, Friedrich: Einstein privat. Herta W. erinnert sich an die Jahre 1927 bis 1933. Buchverlag der Morgen, Berlin, 1978, p. 29). Aside from a number of details, there is one main discrepancy: The "salon" in Herneck's drawing is much larger than the library – a difference of individual perspective: the housemaid experienced the living-room as the center of private and social life much more intensively than the library.

1056 Grüning, Michael: Ein Haus für Albert Einstein. Erinnerungen. Briefe. Dokumente. Berlin, 1990, p. 10.

1057 Hoffmann, Banesh: Albert Einstein. Schöpfer und Rebell. In collaboration with Helen Dukas. Zurich, 1976, p. 179.

1058 Herneck, Friedrich: Einstein privat. Herta W. erinnert sich an die Jahre 1927 bis 1933. Berlin, 1978, p. 16.

1059 Herneck drew in the street in his drawing. The windows of the daughters' room, livingroom, library and Albert Einstein's bedroom opened onto Haberlandstrasse, the windowless walls of the daughters' and Elsa's bedrooms followed Aschaffenburger Strasse.

1060 Herneck, Friedrich: Einstein privat. Herta W. erinnert sich an die Jahre 1927 bis 1933. Berlin, 1978, p. 24.

1061 – Unless Herneck was already *familiar* with the apartment layout and just put into Herta Schiefelbein's mouth what he already knew. Herneck's published drawing agrees so remarkably well with the stove fitter, Georg Schwingel's sketch that this cannot be excluded.

1062 There was a balcony in front of the "salon" (Grüning, Michael: Ein Haus für Albert Einstein. Erinnerungen. Briefe. Dokumente. Berlin, 1990, p. 37).

1063 The building blueprint from 1907 indicates its original purpose as a livingroom or "salon" (Landesarchiv Berlin: rep. 211, acc. 1674, no. 488).

1064 Grüning, Michael: Ein Haus für Albert Einstein. Erinnerungen. Briefe. Dokumente. Berlin, 1990, p. 139.

1065 "His study of Spartan simplicity, reminiscent of a student's lodging, next to it a small, secluded room for taking naps" ('Einstein-Milieu.' In: *Umschau*, 9 Mar. 1929, p. 194).

1066 Landesarchiv Berlin: rep. 211, acc. 1674, no. 488.

1067 Herneck, Friedrich: Einstein privat. Herta W. erinnert sich an die Jahre 1927 bis 1933. Berlin, 1978, pp. 21, 22.

1068 Grüning, Michael: Ein Haus für Albert Einstein. Erinnerungen. Briefe. Dokumente. Berlin, 1990, p. 154.

1069 Herneck, Friedrich: Einstein privat. Herta W. erinnert sich an die Jahre 1927 bis 1933. Berlin, 1978, pp. 150, 151.

1070 Grüning, Michael: Ein Haus für Albert Einstein. Erinnerungen. Briefe. Dokumente. Berlin, 1990, p. 146.

1071 BA-B: R 1501/alt 10/65, vol. 1.

1072 BA-B: R 1501/alt 10/65, vol. 1.

1073 The "Karl-Liebknecht-Haus" (former seat of the CPG's Central Committee).

1074 SAPMO: Dy30/IV2/11/v. 990.

1075 "Lala" = Walter Caro.

1076 The League of the Intellectual Professions (*Bund geistiger Berufe*) was not identical to the CIW but a similar communist-instilled association with an "apolitical" front.

1077 Cadre file Kromrey = SAPMO: Dy30/IV2/11/v. 990.

1078 BA-B: ZC 13817, vol. 5.

1079 BA-B: NJ 2844, vol. 1.

1080 See footnote 1073 above.

1081 BA-B: NJ 2844, vol. 3.

1082 BA-B: ZC 12528 vol. 3.

1083 BA-B: ZC 12528 vol. 3. The name Kattner does not appear in this document.

1084 Letter dated 8 Feb. 1934 to the then chief of the Gestapo, Diels (BA-B: ZC 20050, vol. 1).

1085 BA-B: NJ 5891, vol. 1.

1086 BA-B: NJ 2844, vol. 1.

1087 At the SED's Central Committee.

1088 In BStU: SV 279/87, BStU: KS 296/64, BStU: AP 1310/98, and elsewhere.

1089 BA-B: NJ 2844, vol. 1.

1090 Clark, Ronald W.: Albert Einstein. Ein Leben zwischen Tragik und Genialität. Munich, 1995, pp. 252, 253.

1091 Hermann, Armin: Einstein. Der Weltweise und sein Jahrhundert. Eine Biographie. Munich, 1996, p. 322.

1092 MPG: V 13 Einstein.

1093 Herneck, Friedrich: Einstein privat. Herta W. erinnert sich an die Jahre 1927 bis 1933. Buchverlag der Morgen, Berlin, 1978, p. 11.

1094 Herneck, Friedrich: Einstein privat. Herta W. erinnert sich an die Jahre 1927 bis 1933. Buchverlag der Morgen, Berlin, 1978, p. 12.
1095 Herneck, Friedrich: Einstein privat. Herta W. erinnert sich an die Jahre 1927 bis 1933. Buchverlag der Morgen, Berlin, 1978, p. 27.
1096 Grüning, Michael: Ein Haus für Albert Einstein. Erinnerungen. Briefe. Dokumente. Berlin, 1990, p. 534.
1097 Grüning, Michael: Ein Haus für Albert Einstein. Erinnerungen. Briefe. Dokumente. Berlin, 1990, p. 534.
1098 Grüning, Michael: Ein Haus für Albert Einstein. Erinnerungen. Briefe. Dokumente. Berlin, 1990, p. 536.
1099 Grüning, Michael: Ein Haus für Albert Einstein. Erinnerungen. Briefe. Dokumente. Berlin, 1990, p. 535.
1100 Kraushaar, Luise: Bericht über meine konspirative Arbeit zwischen April 1931 bis Ende 1937 (Vertraulich). 11 Feb. 1986 (BStU: SV 1/81 vol. 262).
1101 BB = "Betriebsberichterstattung" (actually, industrial espionage). S.G.
1102 Volksgerichthof verdict dated 15 Apr. 1937 (BA-B: ZC 6083, vol. 1. The decree by the Reich Central Security Office (RSHA) to keep all the files related to legal proceedings concerning this "BB-Ressort" under lock and key indicates the high importance attached to it. On 26 Sep. 1938, therefore, *before* the war had begun and long *before* the bombing raids, it was determined that the ninety-five ring binders with the proceedings on undercover agents ("V-Leute"), industrial espionage ("BB-Sachakten"), "Klara and Grete files," "BB archive," "T files," "M School" (Military Political School in Moscow), "Lenin School," "Omsk" and "WEB" be "stored in a steel file cabinet in an underground bomb shelter in the event of an air-raid alarm." (BA-B: ZR 592/A3).
1103 Kraushaar, Luise: Bericht über meine konspirative Arbeit zwischen April 1931 bis Ende 1937 (Vertraulich). 11 Feb. 1986 (BStU: SV 1/81 vol. 262).
1104 Helga von Hammerstein (born 1913). S.G.
1105 BStU: SV 1/81, vol. 243.
1106 The forgery workshop at Kaiser Allee no. 48a was raided on 28 Nov. 1932 (BA-B: NJ 5891, vol. 1). Consequently, the apartment of Albert Einstein's secretary must have been used until the end of 1932 – up to the departure of Helen Dukas (together with Elsa and Albert Einstein) on 10 Dec. 1932.
1107 According to notes by the mentioned officer of the East German Ministry of State Security.
1108 BStU: SV 1/81 vol. 262.
1109 According to Helen Dukas's declaration on 15 Jan. 1936 before the immigration authorities of the USA, she was at that time thirty-nine years old, 5 foot 5 inches tall, and weighed 101 pounds (1.65 m and 45.8 kg). (Einstein's FBI file: Einstein 1b.pds: 23).
1110 The current address is: Am Volkspark no. 51.
1111 According to the building file at the City Hall, Wilmersdorf, Berlin, perused by me on 8 Feb. 2000.
1112 The move to Hindenburgstrasse seems to have been prompted solely because the building at Rosenheimer Strasse no. 29, dating to 1908, was in a bad condition (leaking chimneys, sagging floors, etc.) and the owners were unwilling to renovate (LA-B: rep. 21, acc. 1674 Bauakte Rosenheimer Str. 29). Political and other reasons are not likely.
1113 BStU: SV 1/81, vol. 151. See also Dünow's statements during his Gestapo interrogation on 29 Dec. 1933 (BA-B: ZC 5709, vol. 5).
1114 BA-B: NJ 1, vol. 5.
1115 Einstein 1a.pdf: 61.
1116 D. Marianoff and P. Wayne: Einstein. An Intimate Study of a Great Man. New York. French edition: Einstein. Dans

l'intimité. Édition Jeheber. Genève, Paris, 1951.

1117 According to information provided by the Civil Registry in Tempelhof, Berlin, to the author dated 31 Jul. 2001. Likewise per FBI information (Einstein 1a.pdf: 61).

1118 OGPU = United State Political Administration (Soviet Secret Service 1923–1934). GStA: I. HA St 18/176.

1119 GStA: I. HA St 18/176.

1120 Quoted from Hermann, Armin: Einstein. Der Weltweise und sein Jahrhundert. Eine Biographie. Munich, 1996, p. 334.

1121 Herneck, Friedrich: Einstein privat. Herta W. erinnert sich an die Jahre 1927 bis 1933. Berlin, 1978, p. 151.

1122 Quoted according to Clark, Ronald W.: Albert Einstein. Ein Leben zwischen Tragik und Genialität. Munich, 1995, p. 409.

1123 Grüning: Ein Haus für Albert Einstein. Erinnerungen. Briefe. Dokumente. Berlin, 1990, p. 215.

1124 Kippenberger's coworker and girlfriend Änne Kerf (later Anna Christina *Kjossewa*) recalled in 1983 that a meeting had taken place on 27 Feb. 1933 in an "apartment located in Schöneberg" (in agreement with Franz Feuchtwanger, whose other details about this meeting she contests, however). BStU: SV 1/81, vol. 261.

1125 LA-B: rep. 211, acc. 1674, no. 488.

1125 LA-B: rep. 211, acc. 1674, no. 488.

1126 LA-B: rep. 211, acc. 1674, no. 488.

1126 LA-B: rep. 211, acc. 1674, no. 488.

1127 Lehmann-Russbüldt: Meine Erinnerungen an Ernst Reuter. In: Landesarchiv Berlin: rep. 200-21-01, no. 16.

1128 SAPMO: NY 4126/13/40.

1129 Lenin, W.I.: Werke. Berlin, 1961, vol. 32, p. 541.

1130 Lenin: Werke. Berlin, 1959, vol. 24, p. 10.

1131 For Lenin's commentary: W.I. Lenin: Werke. Berlin, 1961, vol. 32, pp. 540–544.

1132 In *Berliner Tageblatt*. Friday, 27 Aug. 1920.

1133 BA-K: NL 199, no. 30.

1134 BA-B: R 1507 no. 1050d.

1135 BA-K: NL 199, no. 30.

1136 BA-K: N 1057, no. 22.

1137 BA-K: NL 199, no. 30.

1138 Membership list of the board of trustees of the Children's Homes of Red Aid (Kuratorium der Kinderheime der Roten Hilfe) among the files of the Reichskommissariat für die Überwachung der öffentlichen Ordnung: BA-B: R 1507/67159, no. 262.

1139 BA-B: R 1507/67159, no. 262.

1140 Instruction of 17 Dec. 1936 signed by Dr. Best on "Übernahme des abwehrpolitisch interessierenden Teils der Kartei des Reichskommissariats zur Überwachung der öffentlichen Ordnung" (BA-B: R 58/254).

1141 *Die Menschenrechte.* Organ der Deutschen Liga für Menschenrechte. Berlin, 1 Dec. 1926, p. 10.

1142 The fact that Health Official Dr. Magnus Hirschfeld, Käthe Kollwitz, Heinrich Mann, Thomas Mann, Prof. Max Reinhardt, the banker Hugo Simon and Prof. Heinrich Zille also undersigned this appeal shows how successful Red Aid was even among the middle class.

1143 BA-B: R 1507/1050d.

1144 Gelegentliches von Albert Einstein. Zum 50. Geburtstag 14.3.1929. Dargelegt von der Soncino Gesellschaft der Freunde des jüdischen Buches zu Berlin (Berlin, 1929), pp. 20, 21.

1145 Ein Prozeß gegen die IAH. In: *Die Rote Fahne,* 12 Jun. 1932.

1146 GStA: I. HA rep. 77, tit. 4043, no. 206.

1147 GStA: I. HA rep. 77, tit. 4043, no. 206.

1148 GStA: I. HA rep. 77, tit. 4043, no. 206.

1149 For instance: "In view of the political activities of the Jew Einstein (e.g., Red Aid) the property served, with the owner's knowledge, subversive purposes to the extent that Einstein worked there." (President of the District of

Potsdam to the Prussian minister of finance, 18 Jul. 1935; GStA: I. HA rep. 151, IA no. 8191).

1150 BA-B: R 1507/alt 134/72.

1151 BStU: SV 1/81 vol. 158; same as: BA-B: R 58/3218.

1152 BA-B: R 3017 ORA VGH no. 3.

1153 BStU: AP 1034/60, identical with the document in SV 1/81, vol. 306.

1154 Landrock, Konrad: Friedrich Georg Houtermans (1903–1966) – ein bedeutender Physiker des 20. Jahrhunderts. In: *Naturwissenschaftliche Rundschau.* issue 4/2003, pp. 187–199.

1155 Klaus Hentschel. Letter to the editor. In: *ISIS,* vol. 81, no. 307 (1990), pp. 279, 280.

1156 Kleinert, Andreas: Paul Weyland, der Berliner Einstein-Töter. In: Naturwissenschaft und Technik in der Geschichte. 25 Jahre Lehrstuhl für Geschichte der Naturwissenschaft und Technik. Ed. by Helmuth Albrecht, Stuttgart, 1993.

1157 "Tshistka" = "purge", S.G.

1158 BStU: AP 1034/60.

1159 BA-B: R 1501/alt 10/65, vol. 1.

1160 SAPMO: Dy 30/IV2/11/v. 990.

1161 BStU: ZA AU 42/56, vol. I.

1162 BStU: AOP 22/67, vol. II.

1163 "Student" – alias for Wilhelm Fricke.

1164 BStU: Ministerium für Staatssicherheit, AOP 22/67, vol. 10.

1165 BStU: SV 3/85, vol. 1.

1166 BStU: SV 3/85, vol. 1.

1167 BStU: AOP 22/67, vol. III.

1168 BStU: AOP 22/67, vol. 13 (my emphasis, S.G.).

1169 Pleticha, Heinrich: Weltgeschichte. Vol. 12. Bertelsmann Lexikon Verlag GmbH, Gütersloh, 1996, p. 59.

1170 Pleticha, Heinrich: Weltgeschichte. Vol. 12. Bertelsmann Lexikon Verlag GmbH, Gütersloh, 1996, p. 60.

1171 Herneck, Friedrich: Einstein und sein Weltbild. Aufsätze und Vorträge von Friedrich Herneck. Berlin, 1979, pp. 275, 276.

1172 Fölsing, Albrecht: Albert Einstein. Eine Biographie. Frankfurt am Main, 1995, p. 809.

# Bibliographical Name Index

Persons mentioned in the main narrative and in the appendix are listed below in alphabetical order, with brief biographical information. *Italics* signal additional biographical details in the main text.

Printing: Krips bv, Meppel
Binding: Stürtz, Würzburg

Northern Michigan University
3 1854 008 150 313